T0396611

Materials Horizons: From Nature to Nanomaterials

Series Editor

Vijay Kumar Thakur, School of Aerospace, Transport and Manufacturing, Cranfield University, Cranfield, UK

Materials are an indispensable part of human civilization since the inception of life on earth. With the passage of time, innumerable new materials have been explored as well as developed and the search for new innovative materials continues briskly. Keeping in mind the immense perspectives of various classes of materials, this series aims at providing a comprehensive collection of works across the breadth of materials research at cutting-edge interface of materials science with physics, chemistry, biology and engineering.

This series covers a galaxy of materials ranging from natural materials to nanomaterials. Some of the topics include but not limited to: biological materials, biomimetic materials, ceramics, composites, coatings, functional materials, glasses, inorganic materials, inorganic-organic hybrids, metals, membranes, magnetic materials, manufacturing of materials, nanomaterials, organic materials and pigments to name a few. The series provides most timely and comprehensive information on advanced synthesis, processing, characterization, manufacturing and applications in a broad range of interdisciplinary fields in science, engineering and technology.

This series accepts both authored and edited works, including textbooks, monographs, reference works, and professional books. The books in this series will provide a deep insight into the state-of-art of Materials Horizons and serve students, academic, government and industrial scientists involved in all aspects of materials research.

Review Process

The proposal for each volume is reviewed by the following:

1. Responsible (in-house) editor
2. One external subject expert
3. One of the editorial board members.

The chapters in each volume are individually reviewed single blind by expert reviewers and the volume editor.

Akarsh Verma · Sushanta K. Sethi ·
Shigenobu Ogata
Editors

Coating Materials

Computational Aspects, Applications and Challenges

Editors
Akarsh Verma
Department of Mechanical Engineering
University of Petroleum and Energy Studies
Dehradun, India

Department of Mechanical Science
and Bioengineering
Osaka University
Osaka, Japan

Shigenobu Ogata
Department of Mechanical Science
and Bioengineering
Osaka University
Osaka, Japan

Sushanta K. Sethi
Department of Metallurgical Engineering
and Materials Science
Indian Institute of Technology Bombay
Mumbai, India

ISSN 2524-5384 ISSN 2524-5392 (electronic)
Materials Horizons: From Nature to Nanomaterials
ISBN 978-981-99-3548-2 ISBN 978-981-99-3549-9 (eBook)
https://doi.org/10.1007/978-981-99-3549-9

This Springer imprint is published by the registered company Springer Nature Singapore Pte Ltd.
The registered company address is: 152 Beach Road, #21-01/04 Gateway East, Singapore 189721,
Singapore

Contents

Chapter 1
Introduction to Coatings: Types and Their Synthesis

Jovale Vincent Tongco, Sushant K. Sethi, Anil Kumar, Akarsh Verma, and Uday Shankar

1 Introduction

Polymeric coatings on solid substrates play a significant role in physical, chemical, and biochemical sciences [1]. Ideally, the coatings are substrate and application dependent, regardless the physical characteristics and chemical compositions. Before designing a coating, one should know the interactions of the proposed coating material with the substrate. For this, researcher used to predict the interaction energy with aluminum, as it exhibits the lowest surface energy than that of other commonly used substrates [2]. They assume that if the proposed coating demonstrates excellent adhesion ability with Al, then it may surely show good adhesion with other substrates. Several times, the chemical functionalization sometimes is required to enhance the performance property of the proposed coating material. For example, researchers use fluorine functionalization of several fillers to enhance the phobic nature of the coating

J. V. Tongco
Department of Forest, Rangeland and Fire Sciences, University of Idaho, 875 Perimeter Drive, Moscow, ID 83844, USA

S. K. Sethi
Department of Metallurgical Engineering and Materials Science, Indian Institute of Technology Bombay, Mumbai 400076, India

A. Kumar
Department of Food and Nutrition, Kunsan National University, Gunsan, South Korea

A. Verma
Department of Mechanical Engineering, University of Petroleum and Energy Studies, Dehradun 248007, India

Department of Mechanical Science and Bioengineering, Osaka University, Osaka 560-8531, Japan

U. Shankar (✉)
Department of Applied Mechanics, Motilal Nehru National Institute of Technology Allahabad, Teliarganj, Prayagraj, Uttar Pradesh 211004, India
e-mail: udayshankar@mnnit.ac.in

© The Author(s), under exclusive license to Springer Nature Singapore Pte Ltd. 2023
A. Verma et al. (eds.), *Coating Materials*, Materials Horizons: From Nature to Nanomaterials, https://doi.org/10.1007/978-981-99-3549-9_1

Table 1 List of polymers used for various types of coatings and the reason behind this

Polymers	Applications	Reason
Polydimethyl siloxane (PDMS)	Self-clean, fire-retardant	Low surface energy, superhydrophobic, and thermal stability
Polyvinylidene fluoride (PVDF)	Self-clean, anti-corrosion	Low surface energy, superhydrophobic
Polytetra fluoroethylene (PTFE)	Self-clean	
Polyethyleneimine	Anti-viral	Excellent antiviral activity
Poly-L-lysine		
Diethylaminoethyl-dextran		
Poly (amidoamine)		
Polyphenol ether		
Epoxy	Anti-corrosion	Prohibit water, oxygen, and other chemicals from making contact and reacting with the substrates
Urethane		
Silicone		
Polyvinyl chloride (PVC)		
Acrylic		
Nitrocellulose		
Phenolic resins		

materials [3, 4]. A list of commonly used polymers for various coating applications is summarized in Table 1.

A few researchers use various computational techniques for the optimization of best-suited coating formulation and then fabricate the optimized one and make a comparative analysis with the experimental outcomes [2, 5, 6]. Researchers use computational tools to predict blend compatibility, polymer-filler interactions, and several other properties estimations [7–14]. However, the various computational techniques employed in different sectors of coatings have been discussed in the subsequent chapters. In this chapter, we discuss various types of coatings such as self-clean, anti-corrosion, anti-viral, and fire-retardant coatings, their synthetization, and performance behaviors.

2 Self-cleaning Coatings

A great deal of research has been done on self-cleaning coatings due to their cost-effective applications and minimal maintenance requirements. Self-cleaning coatings are mostly hydrophobic due to their application in outdoor environments where rain and water-based dirt and particulates collect. The most important applications are outside of the house and building windows and roofing, automobile windshields, and water-proof surfaces. Hydrophobic coatings also gained interest in their ability to impede ice crystal growth due to freezing because of the reduced contact time the water droplet has on the surface coated [15]. The same study showed a developed material capable of having 97% transparency (perfect for windows and car windshields), good adhesion with glass (durability and longevity), and a water contact angle of 135 ± 2 (high hydrophobicity leading to its self-cleaning ability). The material was developed by copolymerization of non-polar compounds diluted and co-polymerized in Xylene, polymerized by thermo-electrically controlled unit (80–90 °C) with Di-tertiary butyl peroxide as initiator, and lastly application of activated TiO_2 and application on to glass substrates for characterization (FT-IR, SEM, contact angle, and optical properties) and thermal analysis (TGA, DSC).

Self-cleaning coatings are also special due to their ability to protect and maintain the shiny appearance and luster of building walls [5, 16–25]. Walls lose these properties mainly due to dust and carbonaceous particle accumulation carried by rain and moisture. Polymer-based paints only protect to a certain extent as they are also susceptible to wear and tear in the long run. Development of robust superhydrophobic coatings on building walls with minimal interactions with pigments and paints and capable of self-cleaning through natural rainwater or mechanical water sprinklers are taking ground in recent years [26]. Dip and spray coating of hydrophobic silicon nanoparticles makes this building wall coating possible.

Another interesting application of self-healing coatings is sports footwear and equipment. Sports shoes and equipment easily gets dirty due to normal use in dusty and muddy places. Regular washing of sports shoes and equipment leads to a shorter lifespan and quality degradation. Superhydrophobic materials (Fluoro-containing silica nanoparticles, polydimethylsiloxane) were dip, spin, or spray coated on these materials to increase their ability to repel muddy water [26].

Various applications aside, fabrication and design of self-cleaning coatings should be simple and cost-effective as it will defeat the purpose of self-maintenance of the coated surface if the process of synthesizing the coatings is complicated, expensive, and full of unwanted side products. Such design of simple processes involves the fabrication of self-cleaning omniphobic coatings. Omniphobic coatings are recently developed as a versatile approach to the self-cleaning mechanism of surface protection as it repels both polar and non-polar liquids. In the past decades, fluorinated copolymers are utilized and designed with rough surfaces, high static contact angles, and ultralow sliding angles for both polar and non-polar liquids for optimal self-cleaning ability. Fabrication of coatings and materials with rough surfaces involves a complicated process and often results in opaque products that do not satisfy a given use case (windows and car windshields). A recent study focused on the development of smooth surfaces coupled with low surface energy materials producing dynamic dewetting properties for polar and non-polar liquids making them slide off the surface more easily. The study employed the development of materials called "slippery liquid-infused porous surfaces (SLIPS)". Further modifications are required as the SLIPS incorporated material is still visually translucent (refer to Fig. 1). To improve transparency, the polyethyleneimine component of poly (ethylene imine)-graft-PDMS enhances the compatibility of PDMS and epoxy matrix (polar) which

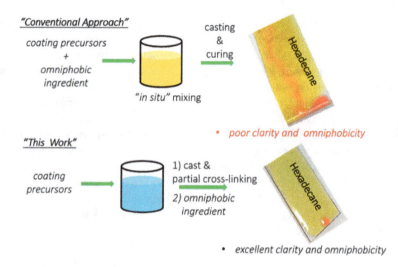

Fig. 1 Comparing the conventional approach and the top-layer approach of the study on the omniphobic epoxy surface fabrication (Adapted from reference [27])

suppresses the phase separation of the PDMS in the matrix producing clear and durable transparent films [27, 28].

3 Anti-corrosion Coatings

Corrosion is evident in metallic and steel structures and research on preventing and mitigating this natural and spontaneous process is prevalent in the modern world. Prolonging the life and structural integrity of steel and metallic materials in building and construction is vital in terms of safety and economics, not to mention the environmental impact of these materials and their potential waste and by-products. Problems originating from corrosion all over the world cost the economy around 2.5 trillion dollars in the US alone. Globally, around 10% of the annual production of metal products used in industry is lost because of corrosion. Preserving metals and steels through the development of anti-corrosion coatings is starting to gain traction in recent years [29].

The global railway system is an indispensable way for people and goods to move around. One good example of applying anti-corrosion coating is on the parts of railways exposed to atmospheric environments. Protection of these parts of railways can help save governments and private companies considerable amounts of money through minimal maintenance and replacements. The simplest way to protect the railways is by employing alloying techniques as some metals (such as Cr and P) naturally resistant to environmental weathering and atmospheric conditions. Most of the time this works, but coatings are still necessary for prolonging the life of railways and other metals (bridges, tunnel entrances, signages, guardrails, etc.) used by the railway system. In this regard, using different types of organic, inorganic, and metallic coatings depending on the usage enhances the anti-corrosion capabilities of the metallic portions of the railway system. Organic coatings, such as epoxy, urethane, acrylics, polysiloxanes, and fluoropolymers are better applied on bridge structures like girder and stringer elements, beams, guardrails, and parapets. The mechanism of organic coatings against corrosion in this specific application protects the underlying metallic element through barrier effects and shielding concerning water and oxygen that produces metallic oxides. Inorganic coatings such as silicates and ceramics also work well on bridge structures. The main track elements (rails and fasteners) are best coated with metallic coatings like zinc, aluminum, and zinc-aluminum alloys. It is also possible to utilize hybrid (combined) coatings such as metallic-organic and metallic-metallic coating systems on more complicated parts like supports, brackets, and cross beams.

Another interesting take on anti-corrosion coatings is the use of graphene and other graphene-based materials. Graphene can be used to develop coatings for protecting metals with the additional utility of being functionalized, enhancing strength, improving anti-corrosion performance, and prolonging the life of the coated metal surfaces. Graphene and its derivatives are relatively non-reactive, so it does not react with the underlying metal surface and the corrosive agents applied on the surface of the coating. They also impart additional flexibility and mechanical strength to the metal. Graphene-based materials such as graphene oxide and reduced graphene oxide can be easily incorporated into composite anti-corrosive coatings because of their capability to disperse easily in polymeric matrices [30]. Graphene products with different properties and morphologies are produced and used in the past decade as a material for designing and fabricating anti-corrosion coatings.

Anti-corrosion coatings are also helpful if they are applied on hard-to-reach surfaces and places where frequent maintenance is difficult. A recent study demonstrated this notion by applying anti-corrosion coating on a glass and metal surface underwater. The design of the coating used for this specific application is challenging because of the difficulties encountered in the fluidity and stability of the coating underwater. The coating the group designed is stable underwater and exhibits photo-thermal healing and self-repair, prolonging the life of the submerged material and improving its anti-corrosion properties [31]. The study employed a hydrophobic material such as paraffin wax with a polypropylene network as starting matrix for the thermally induced phase separation (TIPS) method to fabricate the novel composite coatings. The paraffin wax in the matrix undergoes a solid-to-liquid transition at higher temperatures caused by the photo-thermal effect. The semi-melted paraffin wax then flows through the polypropylene network onto the damaged portion of the coating. The proposed mechanism for coating self-repair and healing is shown in Fig. 2.

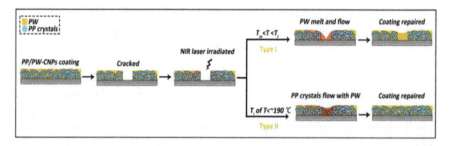

Fig. 2 Proposed mechanism for the self-repair of paraffin wax-polypropylene network (composite coating) under NIR laser irradiation (Adapted from reference [31])

4   Anti-viral Coatings

The recent pandemic demonstrated that new strains of viruses, especially those easily transmitted, could potentially affect the global healthcare system and the economy. It is imperative to develop the first line of defense against known and unknown viruses, and surface coatings are one of the best ways because they can prevent fomites as a transmission method for viruses to proliferate. One thing to consider in the development of anti-viral coatings is its general mode of action against wide-ranging viruses in the environment. The goal is not to completely eradicate, but in a way minimize infection that could ease the burden of the next line of defenses against the virus. In this regard, copper is widely utilized and considered a potent anti-viral agent due to its documented detrimental effects on viruses and their reproduction. Certain limitations exist for the usage of copper directly, such as its application costs and its incompatibility with certain surfaces and conditions. A recent study developed a novel method of organic coating fabrication employing intelligent release of Cu^{2+} pigment based on cation exchange resin dispersed in polyvinyl co butyral-co-vinyl alcohol-co-vinyl acetate (PVB) binder applied as paint to a stainless-steel surface. The material was tested on SARS-CoV-2, inactivating the viruses within 4 h of incubation. The organic coating showed positive results in terms of inactivating SARS-CoV-2 compared to pure copper coating and unpigmented PVB surfaces [32]. The schematic representation of the coating and viral interface is shown in Fig. 3.

The approach used in the study demonstrating the utilization of a simple paint system, based on a polymeric binder with embedded "smart release" pigment holding Cu^{2+} as an anti-viral agent liberated by ion exchange through direct contact, holds great promise in the field of anti-viral coatings as a cost-effective and rapid release system leading to efficient virus inactivation on high touch surfaces.

Polymers exhibiting anti-viral properties range from natural polymers to engineered thermoplastics [33]. Polysaccharides such as glutamine and fluctosan exhibit anti-viral effects due to the presence of the sulfonic functional group, which works by diminishing viral adhesion to the coated surface. Synthetic polymers such as polyethyleneimine, poly-L-lysine, diethylaminoethyl-dextran, poly(amidoamine), and polyphenol ether can be incorporated into biopolymeric matrices to produce

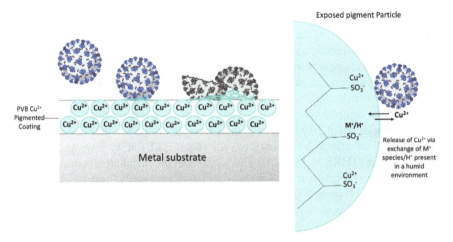

Fig. 3 Schematic representation of Cu^{2+}-CEP/PVB composite coating interface and the associated release mechanism of cation-exchange anti-viral (Adapted from reference [32])

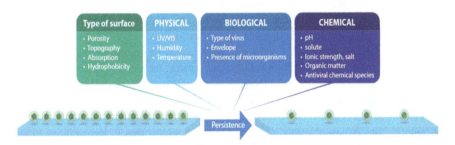

Fig. 4 The types, properties, and characteristics of surfaces influencing the persistence of viruses (Adapted from reference [34])

coatings capable of reducing viral proliferation on surfaces through ionic effects [34]. The persistence of viruses is affected by the physical and chemical properties of the surfaces, as described in Fig. 4.

5   Fire-Retardant Coatings

Wood has been used for centuries as building and furniture material. The inherent flammability of wood is due to its organic nature and poses a real fire hazard to people and the environment. Most of the research on fire-retardant coatings found in literature involves protecting wood and other cellulosic materials. Other methods exist for wood protection, but the coating process is the most cost-efficient as protecting the surface of materials prevents the spread of fire into the bulk matter. Some coating materials also exhibit intumescence, a property in which the coating expands in contact with fire, sealing and protecting the material by sealing gaps underneath.

A recent study demonstrated the use of melamine phosphate as an intumescent fire retardant, graphite nanoplates as an enhancer for flame retardancy and conductive filler, and acrylic resin as an agent for foam formation. The materials are combined and prepared as a simple blend to fabricate the coating exhibiting excellent heat resistance, electromagnetic interference shielding properties, and ultimately flame retardancy. The addition of 20 wt% of both melamine phosphate and graphite nanoplates on the blend enhanced the limiting oxygen index (LOI) up to 30% and the heat resistant index (THRI) to 189.1 °C. This shows that the blending procedure and the materials with synergistic effects in terms of flame retardancy, heat resistance, and even additional properties like EMI shielding can be attained and used on materials requiring these properties such as building and construction policies, production of safe furniture and decorations, and even novel protective coatings for electronics [35].

Novel phosphorus and boron-containing flame retardants (PPEA) were considered by another research highlighting the synthesis route through the esterification of cyclic phosphate esters (PEA) and polyethyleneglycolborate (PEG-BA) with varying mass ratios. The synthesized materials were characterized by Fourier transform infrared (FT-IR) and 1H nuclear magnetic resonance (1H NMR) spectroscopy. The synthesized PEA and PPEAs were combined with an amino resin blend for the fabrication of transparent flame-retardant coatings on wood surfaces. Results of transparency and fire resistance analysis showed that the coating containing PEG-BA exhibited a higher degree of transparency and better flame retardancy than coatings with less or no PEG-BA. The results also indicated that the flame-retardant property of adding PEG-BA diminishes at a certain value, which was found to be at PEA/PEG-BA 80:20. Higher amounts of PEG-BA led to a decline of the synergistic effects of PEG-BA addition resulting to decrease in heat resistance and flame retardancy [36]. Figure 5 shows the synthesis route for producing PEG-BA as described in the study.

Inorganic materials can also be incorporated into intumescent fire-retardant coatings such as layered double hydroxides. The addition of $MgAlCO_3$ and $CaAlCO_3$ layered double hydroxides (LDH) synergistically enhanced the intumescent fire-retardant performance of the coatings fabricated. The optimized amounts added to yield the best thermal insulation and heat resistance according to fire performance tests for the $MgAlCO_3$-LDH and $CaAlCO_3$-LDH are 2.4 wt% and 6.7 wt%, respectively. Further characterizations through cone calorimetry demonstrated that

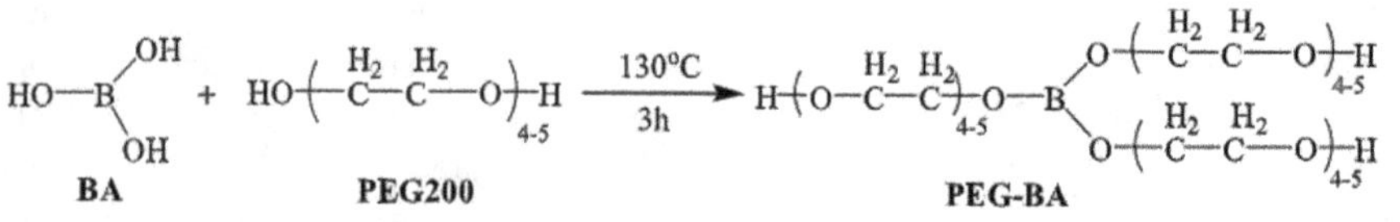

Fig. 5 Synthesis of polyethylene glycol borate (Adapted from reference [36])

MgAlCO$_3$-LDH is better at decreasing the specific extinction of the area of the fire-retardant coatings. On the other hand, CaAlCO$_3$-LDH is more advantageous in terms of other inherent properties such as effective heat combustion, heat release rate, and total heat released. The protective mechanism is due to the formation of amorphous structures after fire exposure, as observed using X-ray diffraction (XRD) analysis. Furthermore, the LDH fillers are capable of suppressing the oxidation of the coatings and could prevent the degradation of and improve the thermal insulation of the coating [37]. Lastly, the authors have a vast experience in the field of atomistic-scale modeling, graphene-based materials and characterizing the polymer composites [38–112].

6 Summary and Conclusion

This chapter discusses the major coating types, fundamentals, their synthetization, characterizations, and applicable techniques from a practical point of view. Coatings such as self-clean, anti-viral, fire-retardant, and anti-corrosion are the prime sectors of the coating industries. For self-clean polymers with low surface energy such as PDMS, PVDF, and PTFE are widely been used to display excellent water repellent nature. For anti-corrosion coatings epoxy, urethane, silicone, acrylic, and polyvinyl chloride are commonly used polymers. Similarly for anti-viral coatings researchers use glutamine, fluctosan polyethyleneimine, poly-L-lysine, diethylaminoethyl-dextran, poly (amidoamine), and polyphenol ether are widely used since they demonstrate anti-viral activity. For fire-retardant coatings, polymers with high thermal stability such as PDMS are commonly used.

Acknowledgements Author "Akarsh Verma" is grateful to the monetary support provided by the University of Petroleum and Energy Studies (UPES)-SEED Grant program.

Conflict of Interest "There are no conflicts of interest to declare by the authors."

References

1. Wei Q, Haag R (2015) Universal polymer coatings and their representative biomedical applications. Mater Horiz 2:567–577. https://doi.org/10.1039/C5MH00089K
2. Sethi SK, Soni L, Shankar U, Chauhan RP, Manik G (2020) A molecular dynamics simulation study to investigate poly(vinyl acetate)-poly(dimethyl siloxane) based easy-clean coating: an insight into the surface behavior and substrate interaction. J Mol Struct 1202:127342. https://doi.org/10.1016/j.molstruc.2019.127342
3. Howarter JA, Genson KL, Youngblood JP (2011) Wetting behavior of oleophobic polymer coatings synthesized from fluorosurfactant-macromers. ACS Appl Mater Interfaces 3:2022–2030. https://doi.org/10.1021/am200255v

4. Sedai BR, Khatiwada BK, Mortazavian H, Blum FD (2016) Development of superhydrophobicity in fluorosilane-treated diatomaceous earth polymer coatings. Appl Surf Sci 386:178–186. https://doi.org/10.1016/j.apsusc.2016.06.009

5. Sethi SK, Shankar U, Manik G (2019) Fabrication and characterization of non-fluoro based transparent easy-clean coating formulations optimized from molecular dynamics simulation. Prog Org Coat 136:105306. https://doi.org/10.1016/j.porgcoat.2019.105306

6. S.K. Sethi, S. Kadian, Anubhav, Goel, R.P. Chauhan, G. Manik, Fabrication and Analysis of ZnO Quantum Dots Based Easy Clean Coating: A Combined Theoretical and Experimental Investigation, ChemistrySelect. 5 (2020) 8942–8950. https://doi.org/10.1002/slct.202001092.

7. Gogoi R, Sethi SK, Manik G (2021) Surface functionalization and CNT coating induced improved interfacial interactions of carbon fiber with polypropylene matrix: a molecular dynamics study. Appl Surf Sci 539. https://doi.org/10.1016/j.apsusc.2020.148162

8. Sethi S, Soni L, Manik G (2018) 5th annual international conference on materials science, Metal & Manufacturing – M3 2018 - blend compatibility studies using atomistic and mesoscale molecular dynamics simulations. https://doi.org/10.5176/2251-1857_M318.35

9. Verma A, Jain N, Sethi SK (2022) Modeling and simulation of graphene-based composites. Innovat Graphene-Based Polymer Composit 167–198. https://doi.org/10.1016/B978-0-12-823789-2.00001-7

10. Sethi SK, Gogoi R, Verma A, Manik G (2022) How can the geometry of a rough surface affect its wettability?-a coarse-grained simulation analysis. Prog Org Coat 172:107062. https://doi.org/10.1016/J.PORGCOAT.2022.107062

11. Maurya AK, Gogoi R, Sethi SK, Manik G (2021) A combined theoretical and experimental investigation of the valorization of mechanical and thermal properties of the fly ash-reinforced polypropylene hybrid composites. J Mater Sci 56:16976–16998

12. Verma A, Parashar A, Packirisamy M (2019) Effect of grain boundaries on the interfacial behaviour of graphene-polyethylene nanocomposite. Appl Surf Sci 470:1085–1092

13. Dhiman A, Gupta A, Sethi SK, Manik G, Agrawal G (2022) Encapsulation of wax in complete silica microcapsules. J Mater Res 2022:1–14. https://doi.org/10.1557/S43578-022-00865-Y

14. Shankar U, Sethi SK, Singh BP, Kumar A, Manik G, Bandyopadhyay A (2021) Optically transparent and lightweight nanocomposite substrate of poly(methyl methacrylate-co-acrylonitrile)/MWCNT for optoelectronic applications: an experimental and theoretical insight. J Mater Sci 56:17040–17061

15. Shahzadi P, Gilani SR, Rana BB, Ghaffar A, Munir A (2021) Transparent, self-cleaning, scratch resistance and environment friendly coatings for glass substrate and their potential applications in outdoor and automobile industry. Sci Rep11:1. 11:1–14. https://doi.org/10.1038/s41598-021-00230-9

16. Sethi S, Dhinojwala A (2009) Superhydrophobic conductive carbon nanotube coatings for steel. Langmuir 25:4311–4313. https://doi.org/10.1021/la9001187

17. Sethi SK, Manik G (2021) A combined theoretical and experimental investigation on the wettability of MWCNT filled PVAc-g-PDMS easy-clean coating. Prog Org Coat 151:106092. https://doi.org/10.1016/j.porgcoat.2020.106092

18. Sethi SK, Manik G (2018) Recent progress in super hydrophobic/hydrophilic self-cleaning surfaces for various industrial applications: a review. Polym Plast Technol Eng 57:1932–1952. https://doi.org/10.1080/03602559.2018.1447128

19. Kadian S, Sethi SK, Manik G (2021) Recent advancements in synthesis and property control of graphene quantum dots for biomedical and optoelectronic applications. Mater Chem Front 5. https://doi.org/10.1039/d0qm00550a

20. Sethi SK, Gogoi R, Manik G (2021) Plastics in self-cleaning applications, reference module in materials science and materials. Engineering. https://doi.org/10.1016/B978-0-12-820352-1.00113-9

21. Sethi SK, Kadian S, Manik G (2022) A review of recent progress in molecular dynamics and coarse-grain simulations assisted understanding of wettability. Archiv Comput Methods Eng 2021(1):1–27. https://doi.org/10.1007/S11831-021-09689-1

22. Sethi SK, Singh M, Manik G (2020) A multi-scale modeling and simulation study to investigate the effect of roughness of a surface on its self-cleaning performance. Mol Syst Des Eng. 5:1277–1289. https://doi.org/10.1039/d0me00068j
23. Verma A, Baurai K, Sanjay MR, Siengchin S (2020) Mechanical, microstructural, and thermal characterization insights of pyrolyzed carbon black from waste tires reinforced epoxy nanocomposites for coating application. Polym Compos 41(1):338–349
24. Agrawal G, Samal SK, Sethi SK, Manik G, Agrawal R (2019) Microgel/silica hybrid colloids: bioinspired synthesis and controlled release application. Polymer (Guildf) 178:121599. https://doi.org/10.1016/j.polymer.2019.121599
25. Kataria A, Verma A, Sethi SK, Ogata S (2022) Introduction to interatomic potentials/forcefields. In Forcefields for atomistic-scale simulations: materials and applications. Springer Nature Singapore, Singapore, pp 21–49
26. Latthe SS, Sutar RS, Kodag VS, Bhosale AK, Kumar AM, Kumar Sadasivuni K, Xing R, Liu S (2019) Self–cleaning superhydrophobic coatings: potential industrial applications. Prog Org Coat 128:52–58. https://doi.org/10.1016/J.PORGCOAT.2018.12.008
27. Khan F, Rabnawaz M, Li Z, Khan A, Naveed M, Tuhin MO, Rahimb F (2019) Simple design for durable and clear self-cleaning coatings. ACS Appl Polym Mater 1:2659–2667. https://doi.org/10.1021/acsapm.9b00596
28. Verma A, Budiyal L, Sanjay MR, Siengchin S (2019) Processing and characterization analysis of pyrolyzed oil rubber (from waste tires)-epoxy polymer blend composite for lightweight structures and coatings applications. Polym Eng Sci 59(10):2041–2051
29. Lazorenko G, Kasprzhitskii A, Nazdracheva T (2021) Anti-corrosion coatings for protection of steel railway structures exposed to atmospheric environments: a review. Constr Build Mater 288:123115. https://doi.org/10.1016/J.CONBUILDMAT.2021.123115
30. Kulyk B, Freitas MA, Santos NF, Mohseni F, Carvalho AF, Yasakau K, Fernandes AJ, Bernardes A, Figueiredo B, Silva R, Tedim J (2022) A critical review on the production and application of graphene and graphene-based materials in anti-corrosion coatings. Crit Rev Solid State Mater Sci 47(3):309–355. https://doi.org/10.1080/10408436.2021.1886046
31. Shen T, Liang ZH, Yang HC, Li W (2021) Anti-corrosion coating within a polymer network: enabling photothermal repairing underwater. Chem Eng J 412:128640. https://doi.org/10.1016/J.CEJ.2021.128640
32. Saud Z, Richards CAJ, Williams G, Stanton RJ (2022) Anti-viral organic coatings for high touch surfaces based on smart-release, Cu2+ containing pigments. Prog Org Coat 172:107135. https://doi.org/10.1016/J.PORGCOAT.2022.107135
33. Arpitha GR, Verma A, MR S, Gorbatyuk S, Khan A, Sobahi TR, Asiri AM, Siengchin S (2022) Bio-composite film from corn starch based vetiver cellulose. J Nat Fibers 19(16):14634–14644
34. Rakowska PD, Tiddia M, Faruqui N, Bankier C, Pei Y, Pollard AJ, Zhang J, Gilmore IS (2021) Antiviral surfaces and coatings and their mechanisms of action. Commun Mater 2:1. 2(2021) 1–19. https://doi.org/10.1038/s43246-021-00153-y
35. Liang C, Du Y, Wang Y, Ma A, Huang S, Ma Z (2021) Intumescent fire-retardant coatings for ancient wooden architectures with ideal electromagnetic interference shielding. Adv Compos Hybrid Mater. 4:979–988
36. Yan L, Xu Z, Deng N (2019) Effects of polyethylene glycol borate on the flame retardancy and smoke suppression properties of transparent fire-retardant coatings applied on wood substrates. Prog Org Coat 135:123–134. https://doi.org/10.1016/J.PORGCOAT.2019.05.043
37. Hu X, Zhu X, Sun Z (2020) Fireproof performance of the intumescent fire retardant coatings with layered double hydroxides additives. Constr Build Mater 256:119445. https://doi.org/10.1016/J.CONBUILDMAT.2020.119445
38. Deji R, Verma A, Kaur N, Choudhary BC, Sharma RK (2022) Density functional theory study of carbon monoxide adsorption on transition metal doped armchair graphene nanoribbon. Mater Today Proceed 54:771–776
39. Deji R, Verma A, Choudhary BC, Sharma RK (2022) New insights into NO adsorption on alkali metal and transition metal doped graphene nanoribbon surface: A DFT approach. J Mol Graph Model 111:108109

40. Verma A, Jain N, Parashar A, Singh VK, Sanjay MR, Siengchin S (2020) Lightweight graphene composite materials. In Lightweight polymer composite structures. CRC Press, pp 1–20
41. Verma A, Parashar A (2020) Characterization of 2D nanomaterials for energy storage. In Recent advances in theoretical, applied, computational and experimental mechanics. Springer, Singapore, pp 221–226
42. Deji R, Jyoti R, Verma A, Choudhary BC, Sharma RK (2022) A theoretical study of HCN adsorption and width effect on co-doped armchair graphene nanoribbon. Comput Theor Chem 1209:113592
43. Verma A, Kumar R, Parashar A (2019) Enhanced thermal transport across a bi-crystalline graphene–polymer interface: an atomistic approach. Phys Chem Chem Phys 21(11):6229–6237
44. Deji R, Verma A, Kaur N, Choudhary BC, Sharma RK (2022) Adsorption chemistry of co-doped graphene nanoribbon and its derivatives towards carbon based gases for gas sensing applications: quantum DFT investigation. Mater Sci Semicond Process 146:106670
45. Verma A, Parashar A (2018) Reactive force field based atomistic simulations to study fracture toughness of bicrystalline graphene functionalised with oxide groups. Diam Relat Mater 88:193–203
46. Verma A, Parashar A, van Duin AC (2022) Graphene-reinforced polymeric membranes for water desalination and gas separation/barrier applications. In Innovations in graphene-based polymer composites. Woodhead Publishing, pp 133–165
47. Verma A, Parashar A (2017) The effect of STW defects on the mechanical properties and fracture toughness of pristine and hydrogenated graphene. Phys Chem Chem Phys 19(24):16023–16037
48. Verma A, Parashar A (2018) Molecular dynamics based simulations to study failure morphology of hydroxyl and epoxide functionalised graphene. Comput Mater Sci 143:15–26
49. Verma A, Parashar A (2018) Molecular dynamics based simulations to study the fracture strength of monolayer graphene oxide. Nanotechnology 29(11):115706
50. Verma A, Parashar A (2018) Structural and chemical insights into thermal transport for strained functionalised graphene: a molecular dynamics study. Mater Res Express 5(11):115605
51. Verma A, Gaur A, Singh VK (2017) Mechanical properties and microstructure of starch and sisal fiber biocomposite modified with epoxy resin mechanical properties and microstructure modified with epoxy resin 6(1), pp 500–520
52. Kataria A, Verma A, Sanjay MR, Siengchin S (2022) Molecular modeling of 2D graphene grain boundaries: mechanical and fracture aspects. Mater Today Proc 52:2404–2408
53. Verma A, Ogata S (2023) Magnesium based alloys for reinforcing biopolymer composites and coatings: a critical overview on biomedical materials. Adv Indust Eng Polymer Res. https://doi.org/10.1016/j.aiepr.2023.01.002
54. Arpitha GR, Jain N, Verma A, Madhusudhan M (2022) Corncob bio-waste and boron nitride particles reinforced epoxy-based composites for lightweight applications: fabrication and characterization. Biomass Convers Biorefinery 1–8. https://doi.org/10.1007/s13399-022-037 17-1
55. Kumar G, Mishra RR, Verma A (2022) Introduction to molecular dynamics simulations. In Forcefields for atomistic-scale simulations: materials and applications. Springer, Singapore, pp 1–19. https://doi.org/10.1007/978-981-19-3092-8_1
56. Bharath KN, Madhu P, Gowda TY, Verma A, Sanjay MR, Siengchin S (2021) Mechanical and chemical properties evaluation of sheep wool fiber–reinforced vinylester and polyester composites. Mater Perform Charact 10(1):99–109
57. Verma A, Parashar A, Jain N, Singh VK, Rangappa SM, Siengchin S (2020) Surface modification techniques for the preparation of different novel biofibers for composites. In Biofibers and biopolymers for biocomposites. Springer, Cham, pp 1–34
58. Verma A, Singh VK (2016) Experimental investigations on thermal properties of coconut shell particles in DAP solution for use in green composite applications. J Mater Sci Eng 5(3):1000242

59. Singh K, Jain N, Verma A, Singh VK, Chauhan S (2020) Functionalized graphite–reinforced cross-linked poly (vinyl alcohol) nanocomposites for vibration isolator application: morphology, mechanical, and thermal assessment. Mater Perform Charact 9(1):215–230
60. Verma A, Singh VK, Arif M (2016) Study of flame retardant and mechanical properties of coconut shell particles filled composite. Res Rev J Mater Sci 4(3):1–5
61. Verma A, Jain N, Rastogi S, Dogra V, Sanjay SM, Siengchin S, Mansour R (2020) Mechanism, anti-corrosion protection and components of anti-corrosion polymer coatings. In Polymer coatings. CRC Press, pp 53–66
62. Chaudhary A, Sharma S, Verma A (2022) WEDM machining of heat treated ASSAB'88 tool steel: a comprehensive experimental analysis. Mater Today Proceed 50:946–951
63. Bisht N, Verma A, Chauhan S, Singh VK (2021) Effect of functionalized silicon carbide nanoparticles as additive in cross-linked PVA based composites for vibration damping application. J Vinyl Add Tech 27(4):920–932
64. Verma A, Jain N, Parashar A, Singh VK, Sanjay MR, Siengchin S (2020) Design and modeling of lightweight polymer composite structures. In Lightweight polymer composite structures. CRC Press, pp 193–224
65. Dogra V, Kishore C, Verma A, Rana AK, Gaur A (2021) Fabrication and experimental testing of hybrid composite material having biodegradable bagasse fiber in a modified epoxy resin: evaluation of mechanical and morphological behavior. Appl Sci Eng Progress 14(4):661–667
66. Verma A, Parashar A, Packirisamy M (2018) Tailoring the failure morphology of 2D bicrystalline graphene oxide. J Appl Phys 124(1):015102
67. Singla V, Verma A, Parashar A (2018) A molecular dynamics based study to estimate the point defects formation energies in graphene containing STW defects. Mater Res Express 6(1):015606
68. Verma A, Negi P, Singh VK (2019) Experimental analysis on carbon residuum transformed epoxy resin: chicken feather fiber hybrid composite. Polym Compos 40(7):2690–2699
69. Verma A, Singh VK (2018) Mechanical, microstructural and thermal characterization of epoxy-based human hair–reinforced composites. J Test Eval 47(2):1193–1215
70. Jain N, Verma A, Ogata S, Sanjay MR, Siengchin S (2022) Application of machine learning in determining the mechanical properties of materials. In: Machine learning applied to composite materials. Springer, Singapore, pp 99–113
71. Verma A, Singh VK, Verma SK, Sharma A (2016) Human hair: a biodegradable composite fiber–a review. Int J Waste Resour 6(206):1000206
72. Kataria A, Verma A, Sanjay MR, Siengchin S, Jawaid M (2022) Physical, morphological, structural, thermal, and tensile properties of coir fibers. In: Coir fiber and its composites. Woodhead Publishing, pp 79–107
73. Jain N, Verma A, Singh VK (2019) Dynamic mechanical analysis and creep-recovery behaviour of polyvinyl alcohol based cross-linked biocomposite reinforced with basalt fiber. Mater Res Express 6(10):105373
74. Chaurasia A, Verma A, Parashar A, Mulik RS (2019) Experimental and computational studies to analyze the effect of h-BN nanosheets on mechanical behavior of h-BN/polyethylene nanocomposites. J Phys Chem C 123(32):20059–20070
75. Verma A, Negi P, Singh VK (2018) Physical and thermal characterization of chicken feather fiber and crumb rubber reformed epoxy resin hybrid composite. Adv Civil Eng Mater 7(1):538–557
76. Bharath KN, Madhu P, Gowda TG, Verma A, Sanjay MR, Siengchin S (2020) A novel approach for development of printed circuit board from biofiber based composites. Polym Compos 41(11):4550–4558
77. Verma A, Joshi K, Gaur A, Singh VK (2018) Starch-jute fiber hybrid biocomposite modified with an epoxy resin coating: fabrication and experimental characterization. J Mech Behav Mater 27(5–6):20182006
78. Verma A, Singh C, Singh VK, Jain N (2019) Fabrication and characterization of chitosan-coated sisal fiber–Phytagel modified soy protein-based green composite. J Compos Mater 53(18):2481–2504

79. Rastogi S, Verma A, Singh VK (2020) Experimental response of nonwoven waste cellulose fabric–reinforced epoxy composites for high toughness and coating applications. Mater Perform Charact 9(1):151–172
80. Verma A, Zhang W, Van Duin AC (2021) ReaxFF reactive molecular dynamics simulations to study the interfacial dynamics between defective h-BN nanosheets and water nanodroplets. Phys Chem Chem Phys 23(18):10822–10834
81. Verma A, Jain N, Parashar A, Gaur A, Sanjay MR, Siengchin S (2021) Lifecycle assessment of thermoplastic and thermosetting bamboo composites. In: Bamboo fiber composites. Springer, Singapore, pp 235–246
82. Arpitha GR, Verma A, Sanjay MR, Siengchin S (2021) Preparation and experimental investigation on mechanical and tribological performance of hemp-glass fiber reinforced laminated composites for lightweight applications. Adv Civil Eng Mater 10(1):427–439
83. Verma A, Samant SS (2016) Inspection of hydrodynamic lubrication in infinitely long journal bearing with oscillating journal velocity. J Appl Mech Eng 5(3):1–7
84. Verma A, Jain N, Rangappa SM, Siengchin S, Jawaid M (2021) Natural fibers based biophenolic composites. In: Phenolic polymers based composite materials. Springer, Singapore, pp 153–168
85. Verma A, Parashar A, Singh SK, Jain N, Sanjay SM, Siengchin S (2020) Modeling and simulation in polymer coatings. In: Polymer coatings. CRC Press, pp 309–324
86. Verma A, Singh VK Experimental characterization of modified epoxy resin assorted with almond shell particles. ESSENCE-Int J Environ Rehabil Conserv 7(1):36–44
87. Verma A (2022) A perspective on the potential material candidate for railway sector applications: PVA based functionalized graphene reinforced composite. Appl Sci Eng Progress 15(2):5727–5727
88. Verma A, Jain N, Singh K, Singh VK, Rangappa SM, Siengchin S (2022) PVA-based blends and composites. In: Biodegradable polymers, blends and composites. Woodhead Publishing, pp 309–326
89. Raja S, Verma A, Rangappa SM, Siengchin S (2022) Development and experimental analysis of polymer based composite bipolar plate using Aquila Taguchi optimization: design of experiments. Polym Compos 43(8):5522–5533
90. Prabhakaran S, Sharma S, Verma A, Rangappa SM, Siengchin S (2022) Mechanical, thermal, and acoustical studies on natural alternative material for partition walls: a novel experimental investigation. Polym Compos 43(7):4711–4720
91. Verma A, Parashar A, Packirisamy M (2019) Role of chemical adatoms in fracture mechanics of graphene nanolayer. Mater Today Proceed 11:920–924
92. Chaudhary A, Sharma S, Verma A (2022) Optimization of WEDM process parameters for machining of heat treated ASSAB'88 tool steel using Response surface methodology (RSM). Mater Today Proceed 50:917–922
93. Verma A, Jain N, Sanjay MR, Siengchin S Viscoelastic properties of completely biodegradable polymer-based composites. In: Vibration and damping behavior of biocomposites. CRC Press, pp 173–188
94. Verma A, Jain N, Mishra RR (2022) Applications and drawbacks of epoxy/natural fiber composites. In: Handbook of epoxy/fiber composites. Springer Singapore, Singapore, pp 1–15
95. Lila MK, Verma A, Bhurat SS (2022) Impact behaviors of epoxy/synthetic fiber composites. In: Handbook of epoxy/fiber composites. Springer Singapore, Singapore, pp 1–18
96. Chaturvedi S, Verma A, Singh SK, Ogata S (2022) EAM inter-atomic potential—its implication on nickel, copper, and aluminum (and their alloys). In: Forcefields for atomistic-scale simulations: materials and applications. Springer, Singapore, pp 133–156
97. Verma A, Sharma S (2022) Atomistic simulations to study thermal effects and strain rate on mechanical and fracture properties of graphene like BC3. In: Forcefields for atomistic-scale simulations: materials and applications. Springer, Singapore, pp 237–252
98. Chaturvedi S, Verma A, Sethi SK, Ogata S (2022) Defect energy calculations of nickel, copper and aluminium (and their alloys): molecular dynamics approach. In: Forcefields for atomistic-scale simulations: materials and applications. Springer, Singapore, pp 157–186

99. Shankar U, Gogoi R, Sethi SK, Verma A (2022) Introduction to materials studio software for the atomistic-scale simulations. In: Forcefields for atomistic-scale simulations: materials and applications. Springer, Singapore, pp 299–313

100. Shankar U, Sethi SK, Verma A (2022) Forcefields and modeling of polymer coatings and nanocomposites. In: Forcefields for atomistic-scale simulations: materials and applications. Springer, Singapore, pp 81–98

101. Verma A, Ogata S (2022) Computational modelling of deformation and failure of bone at molecular scale. In: Forcefields for atomistic-scale simulations: materials and applications. Springer, Singapore, pp 253–268

102. Homer ER, Verma A, Britton D, Johnson OK, Thompson GB (2022) Simulated migration behavior of metastable $\Sigma3$ (11 8 5) incoherent twin grain boundaries. In: IOP conference series: materials science and engineering, vol 1249, no 1, p 012019

103. Chaturvedi S, Verma A, Sethi SK, Rangappa SM, Siengchin S (2022) Stalk fibers (rice, wheat, barley, etc.) composites and applications. In: Plant fibers, their composites, and applications. Woodhead Publishing, pp 347–362

104. Verma A, Parashar A, Packirisamy M (2018) Atomistic modeling of graphene/hexagonal boron nitride polymer nanocomposites: a review. Wiley Interdiscip Rev Comput Mol Sci 8(3):e1346

105. Thimmaiah SH, Narayanappa K, Thyavihalli Girijappa Y, Gulihonenahali Rajakumara A, Hemath M, Thiagamani SMK, Verma A (2022) An artificial neural network and Taguchi prediction on wear characteristics of Kenaf–Kevlar fabric reinforced hybrid polyester composites. Polymer Compos 44(1):261–273

106. Kataria A, Chaturvedi S, Chaudhary V, Verma A, Jain N, Sanjay MR, Siengchin S (2023) Cellulose fiber-reinforced composites—history of evolution, chemistry, and structure. In: Cellulose fibre reinforced composites. Woodhead Publishing, pp 1–22. https://doi.org/10.1016/B978-0-323-90125-3.00012-4

107. Chaturvedi S, Kataria A, Chaudhary V, Verma A, Jain N, Sanjay MR, Siengchin S (2023). Bionanocomposites reinforced with cellulose fibers and agro-industrial wastes. In: Cellulose fibre reinforced composites. Woodhead Publishing, pp 1–22. https://doi.org/10.1016/B978-0-323-90125-3.00017-3

108. Shankar U, Bandyopadhyay A (2022) Plastics in high chemical resistant applications. Encycl Mater Plast Polym 4:187–199. https://doi.org/10.1016/B978-0-12-820352-1.00135-8

109. Shankar U, Oberoi D, Bandyopadhyay A (2022) A review on the alternative of indium tin oxide coated glass substrate in flexible and bendable organic optoelectronic device. Polym Adv Technol 33(10):3078–3111

110. Shankar U, Gupta CR, Oberoi D, Singh BP, Kumar A, Bandyopadhyay A (2020) A facile way to synthesize an intrinsically ultraviolet-C resistant tough semiconducting polymeric glass for organic optoelectronic device application. Carbon 168:485–498

111. Shankar U, Bandyopadhyay A One-dimensional polymeric nanocomposites for soft electronics, one-dimensional polymeric nanocomposites, Taylor & Francis. Chapter 27. https://www.routledge.com/One-Dimensional-Polymeric-Nanocomposites-Synthesis-to-Emerging-Applications/Gupta-Nguyen/p/book/9781032116211

112. Arpitha GR, Mohit H, Madhu P, Akarsh Verma* (2023) Effect of sugarcane bagasse and alumina reinforcements on physical, mechanical, and thermal characteristics of epoxy composites using artificial neural networks and response surface methodology. Biomass Conversion and Biorefineryhttps://doi.org/10.1007/s13399-023-03886-7

Chapter 2
Coatings: Types and Synthesis Techniques

Vaishally Dogra, Chandra Kishore, Abhilasha Mishra, Akarsh Verma, and Amit Gaur

1 Introduction

A coating is a covering done on a material, usually, a substrate, generally used as a protective layer against thermal, electrical or mechanical wear and tear. The coating material plays a decisive role in the mechanical properties of the material being coated over. Coatings have a wide scope in the manufacturing industry; selective coatings are used to manufacture high-quality corrosion-resistant surfaces, lab-on-a-chips, microfluidics, etc. [1].

V. Dogra (✉) · C. Kishore
Department of Mechanical Engineering, Graphic Era (Deemed to be University), Dehradun, India
e-mail: vaishali.dogra@gmail.com

V. Dogra
Department of Mechanical Engineering, Graphic Era Hill University, Dehradun, India

A. Mishra
Department of Chemistry, Graphic Era (Deemed to be University), Dehradun, India

A. Verma
Department of Mechanical Engineering, University of Petroleum and Energy Studies, Dehradun 248007, India

Department of Mechanical Science and Bioengineering, Osaka University, Osaka 560-8531, Japan

A. Gaur
G. B. Pant, University of Agriculture and Technology, Pantnagar, India

© The Author(s), under exclusive license to Springer Nature Singapore Pte Ltd. 2023
A. Verma et al. (eds.), *Coating Materials*, Materials Horizons: From Nature to Nanomaterials, https://doi.org/10.1007/978-981-99-3549-9_2

1.1 Types of Coatings

1. **Graphene-based polymer coatings**

   Graphene is a promising filler material for being added to coating due to its remarkable and best properties like recyclable and barrier properties, better mechanical stability, and best electrical insulation. Due to its vast properties, graphene is used in numerous applications like batteries, capacitors, coatings, etc. graphene derivatives are used for antiviral and antimicrobial applications in the form of surface disinfectant coatings [2].

   Graphene-based nanomaterial coatings also find its use in the form of corrosion-resistant and self-healing coatings. Fig. 1 depicts how graphene and graphene derivatives are produced from raw graphite. Research has revealed that graphene-based nanocomposite can be used as organic anti-corrosion coatings, fire retardant coatings and also surface protection lubricating coatings [3]. Graphene oxide-based nanomaterial developed by grafting enhanced the passive corrosion protection. It can also provide active protection from corrosion by adding inhibitors and it can also achieve self-healing in areas where damage occurs [4–23].

2. **Nanomaterial-based coatings**

   The enhanced properties of nanomaterial-based coatings include antifogging, long durability, antiskid, better mechanical properties, good adhesion, scratch resistant and many more to count. Because of the eco-friendly nature of green nanomaterials, they find a vast and variety of applications in coating industries. Green nanomaterials like TiO_2 and ZnO have high refractive index for which

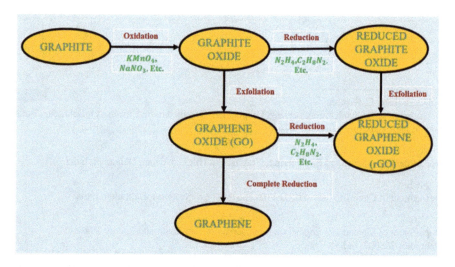

Fig. 1 Depiction of production of graphene and its derivatives from raw graphite [3] [Reproduced with permission from Elsevier]

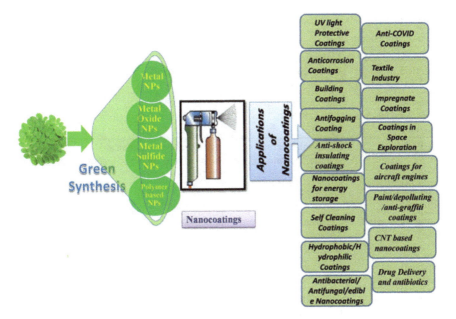

Fig. 2 Numerous applications of green nanomaterials for coatings [24] [Reproduced with permission from Elsevier]

they are used as light protective coatings against ultraviolet rays acting as physical sunscreens. Fig. 2 shows vast applications of green nanomaterials in terms of coatings. Green nanomaterials like Au, Fe_2O_3, TiO_2, ZnO can also be used as sensor coatings. Water breeding, water sheeting, anticorrosion, antifouling and antifogging, energy storage and conversion, self-cleaning, hydrophobic and hydrophilic coatings, waterborne coatings are also some major applications of nanomaterial-based coatings [24].

Nanomaterials are also used as icephobic coatings, which reduces ice and snow accumulation in high-altitude areas. The currently available icephobic coatings reduce the snow losses with approximately 45–65% efficiency according to the altitudes, which helps to increase the PV power output [25]. SnO_2-based coating microfiber interferometer provides a simple and quick method for the detection of ammonia gas concentration having a sensitivity of 0.58 pm/ppm. This application helps in reducing environmental pollution and human health problems related to ammonia gas like liver failure, respiratory irritation when inhaled and more [26, 27].

3. **Superhydrophobic coatings**

The coating surfaces having water contact angle more than 150° are called as superhydrophobic coatings, and these coatings consist of lotus effect. Good water-repellent and self-cleaning property of these coatings makes them find applications in large industrial applications in different fields. Table 1 illustrates some of the common general application fields of these coatings.

Table 1 Some common applications of superhydrophobic coatings

Field of application	Use	Characteristic	References
Architectural and historical stone surfaces	Limestone and marble conservation	Bioremediation treatments, salt inhibitors, protective oxalate layers	[28]
Aerospace engineering-related activities	Interfacial slip and drag reduction	Creates drag reduction by exhibiting apparent slip velocity	[29]
Medical sector	Tissue engineering and biomedicine	Killing of bacteria on clinical devices surfaces, tuning of surface chemistry and pattern	[30]
Effectual oil removal	Oil–water separation	Repels water with high contact angle and absorbs any kind of oil	[31]
Varnish and paint industries	Corrosion inhibition, surface wetting behavior	Limiting interactions with corrosive species	[32]
Hospitals and medical sectors	Antibacterial coatings	Resist bacterial adhesion	[33]
Optics, optoelectronic devices, aerospace	Antireflective polymer coatings	Improves efficiency of solar cells	[34]

4. **Self-healing coatings**

Self-healing coating plays a vital role in corrosion protection and oxidation under high-temperature oxidation environments. These coatings automatically recover cracks and damages and have been widely used in industrial and biomedical applications. Self-healing coatings include two mechanisms: oxidation-induced healing wherein micro-sized cracks are healed and precipitation-induced healing where the crack planes are bridged with precipitated phases. Al-modified SiC coatings were investigated under different heat treatment conditions [35], and crack healing was achieved under high-temperature oxidation environment.

Bio-based self-healing coating material prepared [36] from renewable castor oil and multifunctional alamine have improved crosslinking density and glass transition temperature and exhibits good self-healing performance. Bio-based epoxy coatings also help in corrosion prevention [37].

5. **Smart anticorrosion coatings**

For metallic structures, corrosion results in deterioration, which can lead to various damages, and also leads to economic and environmental damages. Anti-corrosion coatings with some coating inhibitors on metallic surfaces can be a solution to all these problems. Self-healing anti-corrosion smart coating was developed [38] that can release organic inhibitor, which forms a layer during the process if corrosion occurs. The coating fabricated depicted a remarkable healing efficiency of approximately 98%. Nanotubes, lignin, polyuria, chitosan, mesoporous silica, hydroxides, polydopamine, etc. are some of the micro carriers that inhibit corrosion and act as micro carriers in anti-corrosion coatings [39].

Yitrium doped ZnO nanoparticles epoxy coating developed for mild steel by co-precipitation method resulted out to be anti-corrosive filler in coating [40]. Graphene is also gaining interest in anti-corrosion coatings for metallic surfaces.

6. **Waterproof coatings**

   Waterproof coatings find application in a number of fields like waterproof breathable clothes, environmental protective building material and an alternative to plastic-based coatings. Waterborne polyurethane with polyethylene glycol macromere provides a breathable coating with minimal water smell and water resistance reduction [41]. Water penetration in concretes can lead to various aggressive conditions in building construction which can be corrected or deflated by extending the service life of concrete structures with the help of waterproofing barriers. The different types of waterproofing barriers including polymeric barriers, crystalline cementitious barriers, polymeric-modified cementitious coating and hydrophobic cementitious coating are depicted in Fig. 3 showing their levels of penetration.

   These different WP coatings include different materials like blends of fine sand, cement, fibers and some dry powder active ingredients, different polymers like synthetic latex, natural rubber, polyvinyl acetate, epoxy resins, coal-tar, etc., thermoplastics, PVC, butyl rubber sheets and more [42]. Many waterproof coatings are researched and developed for various applications of industry [43, 44].

7. **Self-cleaning and anti-reflective coatings**

   Antireflective coatings (ARC) play a very important role in optical industries. Solar cell is the major application of ARC, which helps in increasing the cell life and optical performance leading to a sustainable production of energy. Sarkın et al. [45] and Cherupurakal et al. [46] have reviewed the advances in superhydrophobic polymers for anti-reflective self-cleaning coatings on photovoltaic

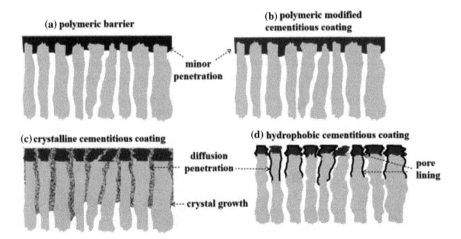

Fig. 3 Water penetration levels of waterproofing (WP) barriers [42] [Reproduced with permission from Elsevier]

panels. Silica, Titania, Zinc, alumina, etc. are some of the common materials used for these types of coatings prepared by different syntheses like sol-gel method, etc.

Metal oxides and their coatings have a major role in ARC coatings on an industrial scale due to their high thermal efficiency. Commonly used metal oxides and their composites involved in these coatings are SiO_2, Al_2O_3, SiC, SnO_2, MgF_2 and ZrO_2. These coatings help in improving the performance of optical components in terms of productivity [47].

1.2 Different Coating Syntheses and Deposition Methods

1. **Thin film vapor deposition (Chemical and Physical)**
   Vapor deposition method is used for creating coatings and thin films having vast application from microelectronics to metallurgical industries. The unique characteristic of this method is to manipulate structure of films during their growth allowing for creation of new material with advance properties like high efficiency, thermal conductivity, magneto resistance and optical emission. In this process, the reaction of chemical constituents takes place near vapor phase forming a solid deposit. Fig. 4 depicts the events in sequence that takes place during a chemical vapor deposition (CVD) method [48]. The precursor activation in this process can be by different methods like thermal activation, photon activation or plasma activation.

   CVD method was used as an effective and scalable approach for film fabrication of ZnO at low atmospheric pressure and temperature providing a dense and high quality of ZnO film on glass substrate, and also this method allowed high saving of energy [49].

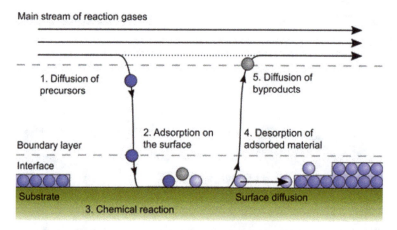

Fig. 4 Sequenced steps involved during a chemical vapor deposition process [48] [Reproduced with permission from Elsevier]

2. **Sputtering**

   Surface finishing, processing of semiconductor, jewelry making are the common industries finding diverse application of thin-film fabrication techniques, sputtering being the widely used one. The ejected atoms after the acceleration of ionized atoms on a surface are condensed onto a sample to form a thin film, and this process is known as sputter deposition [50]. Thin film deposition of metals, oxides and nitrides, metal alloys, nonmetals like polytetrafluoroethylene (PTFE) requires sputtering deposition method providing advantages like long lifetime, uniform thickness and thickness control. Reactive, magnetron, bias, ion-beam and diode sputtering are the different types of sputtering [48].

3. **Sol-gel coatings**

   A low-cost effective and simple method of coating preparation of oxide materials is the sol-gel method including four main steps; hydrolysis, condensation, particle growth and gel formation. After sol-gel preparation, the coating of the developed material is done by dip coating, spin coating or spray coating as the commonly coating methods. This method of coating synthesis finds applications mainly in super hydrophilic and superhydrophobic surfaces, smart solar-responsive windows, self-healing coatings, nanocontainers and antireflective surfaces [48]. Fig. 5 illustrates the process of coating preparation by sol-gel method [51]. Recent progresses in sol-gel-based ceramic coatings have been reviewed in reference [52].

   Sol-gel-based fire protective coating has been reported in reference [53], which acts as a thermal insulator improving the flame-retardant properties of the treated substrate. The coating developed for polyurethane (PU) foam was a mixture of tetraethoxysilane, methyl triethoxysilane, 3-aminopropyl triethoxysilane and diethyl phosphite and the fabricated coating protected the underlying PU foam from burning.

4. **Ion deposition**

   In ion-deposition method, an ion source directs the ions at the substrate during the deposition. Sputter arrangement or an evaporation source can be the film source material. The main advantage of this method is that it can be integrated into an existing vacuum deposition system [48]. Ion-beam assisted deposition (IBAD) technique (refer to Fig. 6.) implements the combination of ion irradiation and metal evaporation. This combined method includes a main feature of not definitive formation of film structure at nucleation stage and also during the film growth process, change in component composition and structure of crystal takes place as compared to that formed during its initial stages due to which this the vastly used deposition method for thin film production [54].

5. **Thermal spraying**

   This technique uses heat (thermal energy) and also kinetic energy to accelerate feedstock and deposit it onto the surface of the material to be coated known as the substrate. The use of such a process to obtain coating on the substrate can improve certain properties of the substrate such as wear resistance, corrosion resistance, etc. The technique of thermal spraying also resolves problems

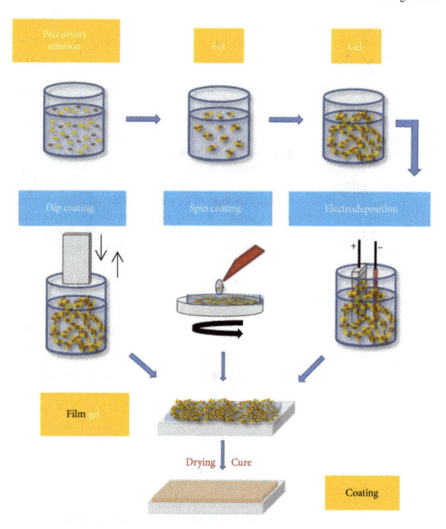

Fig. 5 Coatings manufacturing by sol-gel method [51] [Reproduced with permission from Elsevier]

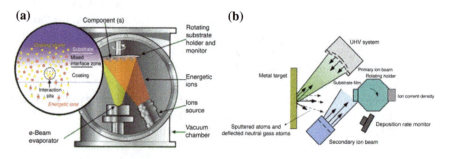

Fig. 6 IBAD method of ion deposition [54] [Reproduced with permission from Elsevier]

involving processing and binding by directly depositing powder on the said substrate, simultaneously controlling the coating layer thickness [55].

There are many more types of coatings and methods of deposition and synthesis like dip coating, spin coating, spray coating, induction cladding, plasma electrolytic oxidation, etc. Above discussed are some of the common and generally used methods used in industries related to coating manufacturing and applications. Lastly, the authors have a vast experience in the field of molecular dynamics and polymer composites [56–113].

Conflict of Interest "There are no conflicts of interest to declare by the authors."

References

1. Speidel A et al (2020) Towards selective compositionally graded coatings by electrochemical jet processing. Procedia CIRP 95:833–837
2. Ayub M et al (2021) Graphene-based nanomaterials as antimicrobial surface coatings: a parallel approach to restrain the expansion of COVID-19. Surf Interfaces 27(September):101460. https://doi.org/10.1016/j.surfin.2021.101460
3. Sharma S et al (2021) New perspectives on graphene / graphene oxide based polymer nanocomposites for corrosion applications : the relevance of the graphene / polymer barrier coatings. Prog Organic Coat 154(November 2020):106215. https://doi.org/10.1016/j.por gcoat.2021.106215
4. Yan D et al (2021) Dual-functional graphene oxide-based nanomaterial for enhancing the passive and active corrosion protection of epoxy coating. Compos B 222(May):109075. https://doi.org/10.1016/j.compositesb.2021.109075
5. Deji R, Verma A, Kaur N, Choudhary BC, Sharma RK (2022) Density functional theory study of carbon monoxide adsorption on transition metal doped armchair graphene nanoribbon. Mater Today Proceed 54:771–776
6. Deji R, Verma A, Choudhary BC, Sharma RK (2022) New insights into NO adsorption on alkali metal and transition metal doped graphene nanoribbon surface: a DFT approach. J Mol Graph Model 111:108109
7. Verma A, Jain N, Parashar A, Singh VK, Sanjay MR, Siengchin S (2020) Lightweight graphene composite materials. In: Lightweight polymer composite structures. CRC Press, pp 1–20
8. Verma A, Parashar A (2020) Characterization of 2D nanomaterials for energy storage. In: Recent advances in theoretical, applied, computational and experimental mechanics. Springer, Singapore, pp 221–226
9. Deji R, Jyoti R, Verma A, Choudhary BC, Sharma RK (2022) A theoretical study of HCN adsorption and width effect on co-doped armchair graphene nanoribbon. Comput Theor Chem 1209:113592
10. Verma A, Parashar A, Packirisamy M (2019) Effect of grain boundaries on the interfacial behaviour of graphene-polyethylene nanocomposite. Appl Surf Sci 470:1085–1092
11. Deji R, Verma A, Kaur N, Choudhary BC, Sharma RK (2022) Adsorption chemistry of co-doped graphene nanoribbon and its derivatives towards carbon based gases for gas sensing applications: Quantum DFT investigation. Mater Sci Semicond Process 146:106670
12. Verma A, Jain N, Sethi SK (2022)\ Modeling and simulation of graphene-based composites. In: Innovations in graphene-based polymer composites. Woodhead Publishing, pp 167–198

13. Verma A, Parashar A, van Duin AC (2022) Graphene-reinforced polymeric membranes for water desalination and gas separation/barrier applications. In: Innovations in graphene-based polymer composites. Woodhead Publishing, pp 133–165
14. Verma A, Parashar A (2017) The effect of STW defects on the mechanical properties and fracture toughness of pristine and hydrogenated graphene. Phys Chem Chem Phys 19(24):16023–16037
15. Verma A, Parashar A (2018) Molecular dynamics based simulations to study failure morphology of hydroxyl and epoxide functionalised graphene. Comput Mater Sci 143:15–26
16. Verma A, Parashar A (2018) Molecular dynamics based simulations to study the fracture strength of monolayer graphene oxide. Nanotechnology 29(11):115706
17. Verma A, Parashar A (2018) Structural and chemical insights into thermal transport for strained functionalised graphene: a molecular dynamics study. Mater Res Express 5(11):115605
18. Verma A, Parashar A, Packirisamy M (2018) Tailoring the failure morphology of 2D bicrystalline graphene oxide. J Appl Phys 124(1):015102
19. Verma A, Parashar A (2018) Reactive force field based atomistic simulations to study fracture toughness of bicrystalline graphene functionalised with oxide groups. Diam Relat Mater 88:193–203
20. Singla V, Verma A, Parashar A (2018) A molecular dynamics based study to estimate the point defects formation energies in graphene containing STW defects. Mater Res Express 6(1):015606
21. Verma A, Parashar A, Packirisamy M (2019) Role of chemical adatoms in fracture mechanics of graphene nanolayer. Mater Today Proceed 11:920–924
22. Verma A, Parashar A, Packirisamy M (2018) Atomistic modeling of graphene/hexagonal boron nitride polymer nanocomposites: a review. Wiley Interdiscip Rev Comput Mol Sci 8(3):e1346
23. Verma A, Kumar R, Parashar A (2019) Enhanced thermal transport across a bi-crystalline graphene–polymer interface: an atomistic approach. Phys Chem Chem Phys 21(11):6229–6237
24. Gautam YK, Sharma K, Tyagi S, Kumar A (2022) Green nanomaterials for industrial applications *Applications of green nanomaterials in coatings*. Elsevier Inc. https://doi.org/10.1016/B978-0-12-823296-5.00014-9
25. Manni M, Chiara M, Nocente A, Lobaccaro G (2022) The influence of icephobic nanomaterial coatings on solar cell panels at high latitudes. 248(November): 76–87
26. Fu H, You Y, Wang S, Chang H (2022) SnO 2 nanomaterial coating micro-fiber interferometer for ammonia concentration measurement. Opt fiber Technol 68(December 2021):102819. https://doi.org/10.1016/j.yofte.2022.102819
27. Verma A, Gaur A, Singh VK (2017) Mechanical properties and microstructure of starch and sisal fiber biocomposite modified with epoxy resin mechanical properties and microstructure modified with epoxy resin. 6(1):500–520
28. Cappelletti G, Fermo P, Golgi V (2016) Smart composite coatings and membranes *15*. Elsevier Ltd., Hydrophobic and superhydrophobic coatings for limestone and marble conservation. https://doi.org/10.1016/B978-1-78242-283-9.00015-4
29. Rose JBR (2019) Superhydrophobic polymer coatings *Interfacial slip-and-drag reduction by superhydrophobic polymer coating*. Elsevier Inc. https://doi.org/10.1016/B978-0-12-816671-0.00014-X
30. Yahyaei H, Makki H, Mohseni M (2019) Superhydrophobic polymer coatings *Superhydrophobic coatings for medical applications*. Elsevier Inc. https://doi.org/10.1016/B978-0-12-816671-0.00015-1
31. Latthe SS et al (2019) Superhydrophobic polymer coatings. *Superhydrophobic surfaces for oil-water separation*. Elsevier Inc. https://doi.org/10.1016/B978-0-12-816671-0.00016-3
32. Superhydrophobic Film Coatings for Corrosion Inhibition (2019) 23
33. Eduok, Ubong, Jerzy Szpunar, and Eno Ebenso. 2019. Superhydrophobic polymer coatings *Superhydrophobic antibacterial polymer coatings*. Elsevier Inc. https://doi.org/10.1016/B978-0-12-816671-0.00012-6

34. Sahoo S, Pradhan S, Das S (2019) Superhydrophobic polymer coatings *Superhydrophobic Antireflective polymer coatings with improved solar cell efficiency*. Elsevier Inc. https://doi.org/10.1016/B978-0-12-816671-0.00013-8
35. Huang J, Guo L, Zhong L (2022) Synergistic healing mechanism of self-healing ceramics coating. Ceram Int 48(5):6520–6527. https://doi.org/10.1016/j.ceramint.2021.11.198
36. Wei X et al (2021) Bio-based self-healing coating material derived from renewable castor oil and multifunctional alamine. Eur Polymer J 160(May):110804. https://doi.org/10.1016/j.europolymj.2021.110804
37. Pulikkalparambil H, Parameswaranpillai J, Siengchin S (2021) UV light triggered self-healing of green epoxy coatings. Constr Build Mater 305(September):124725. https://doi.org/10.1016/j.conbuildmat.2021.124725
38. Auepattana-aumrung K, Crespy D (2023) Self-healing and anticorrosion coatings based on responsive polymers with metal coordination bonds. Chem Eng J 452(P1):139055. https://doi.org/10.1016/j.cej.2022.139055
39. Cao Y et al (2022) Novel long-acting smart anticorrosion coating based on PH-controlled release polyaniline hollow microspheres encapsulating inhibitor. J Mol Liq 359:119341. https://doi.org/10.1016/j.molliq.2022.119341
40. Jiheng D, Zhao H, Yu H (2021) Epidermis microstructure inspired mica-based coatings for smart corrosion protection. Progress in organic coatings 152(January):106126. https://doi.org/10.1016/j.porgcoat.2020.106126
41. Jeong JH et al (2016) Waterborne polyurethane modified with poly(ethylene glycol) macromer for waterproof breathable coating. Progress in organic coatings: 1–7.https://doi.org/10.1016/j.porgcoat.2016.10.004
42. Al-jabari M (2022) Integral waterproofing of concrete structures *12. Waterproofing coatings and membranes*. LTD. https://doi.org/10.1016/B978-0-12-824354-1.00012-X
43. Wang S et al (2021) Experimental research on waterproof enhanced biomass-based building insulation materials. Energy Build 252:111392. https://doi.org/10.1016/j.enbuild.2021.111392
44. Hong J, Wook S (2012) A metal sputtered waterproof coating that enhances hot water stability. J Ind Eng Chem 18(4):1496–1498. https://doi.org/10.1016/j.jiec.2012.02.008
45. Sarkın AS, Ekren N, Sağlam S (2020) A review of anti-reflection and self-cleaning coatings on photovoltaic panels. Solar Energy 199(June 2019):63–73. https://doi.org/10.1016/j.solener.2020.01.084
46. Cherupurakal N, Mozumder MS, Mourad AHI, Lalwani S (2021) Recent advances in super-hydrophobic polymers for antireflective self-cleaning solar panels. Renew Sustain Energy Rev 151(November 2020):111538. https://doi.org/10.1016/j.rser.2021.111538
47. Dogra V, Verma D, Kishore C (2022) Inorganic anticorrosive materials (IAMs) *A prospective utilization of metal oxides for self-cleaning and antireflective coatings*. Inc. https://doi.org/10.1016/B978-0-323-90410-0.00008-8
48. Abu-thabit NY, Makhlouf ASH (2020) Advances in smart coatings and thin films for future industrial and biomedical engineering applications *Fundamental of smart coatings and thin films : synthesis, deposition methods, and industrial applications*. Elsevier Inc. https://doi.org/10.1016/B978-0-12-849870-5.00001-X
49. Vega NC, Straube B, Marin-ramirez O, Comedi D (2023) Low temperature chemical vapor deposition as a sustainable method to obtain c -oriented and highly UV luminescent ZnO thin films. Mater Lett 333:133684. https://doi.org/10.1016/j.matlet.2022.133684
50. Simon AH (2018) Handbook of thin film deposition *Chapter 7*. Elsevier Inc., Sputter processing. https://doi.org/10.1016/B978-0-12-812311-9.00007-4
51. Zno T (2022) Nanomaterial-Based Coatings
52. Zanurin A et al (2022) Research progress of sol-gel ceramic coating: a review. Mater Today Proceed 48:1849–1854. https://doi.org/10.1016/j.matpr.2021.09.203
53. Gossiaux A, Duquesne S, Dewailly B, Bachelet P (2022) Transparent fire protective sol-gel coating for wood panels. 110(February)

54. Guglya A, Lyubchenko E (2018) Emerging applications of nanoparticles and architecture nanostructures *Ion-beam-assisted deposition of thin films*. Elsevier Inc. https://doi.org/10.1016/B978-0-323-51254-1/00004-X
55. Ham G-s et al (2020) Fabrication, microstructure and wear properties of novel Fe-Mo-Cr-C-B metallic glass coating layers manufactured by various thermal spray processes. Mater Des 195:109043. https://doi.org/10.1016/j.matdes.2020.109043
56. Bharath KN, Madhu P, Gowda TY, Verma A, Sanjay MR, Siengchin S (2021) Mechanical and chemical properties evaluation of sheep wool fiber–reinforced vinylester and polyester composites. Mater Perform Charact 10(1):99–109
57. Verma A, Parashar A, Jain N, Singh VK, Rangappa SM, Siengchin S (2020) Surface modification techniques for the preparation of different novel biofibers for composites. In: Biofibers and biopolymers for biocomposites. Springer, Cham, pp 1–34
58. Verma A, Singh VK (2016) Experimental investigations on thermal properties of coconut shell particles in DAP solution for use in green composite applications. J Mater Sci Eng 5(3):1000242
59. Singh K, Jain N, Verma A, Singh VK, Chauhan S (2020) Functionalized graphite–reinforced cross-linked poly (vinyl alcohol) nanocomposites for vibration isolator application: morphology, mechanical, and thermal assessment. Mater Perform Charact 9(1):215–230
60. Verma A, Singh VK, Arif M (2016) Study of flame retardant and mechanical properties of coconut shell particles filled composite. Res Rev J Mater Sci 4(3):1–5
61. Verma A, Jain N, Rastogi S, Dogra V, Sanjay SM, Siengchin S, Mansour R (2020) Mechanism, anti-corrosion protection and components of anti-corrosion polymer coatings. In: Polymer coatings. CRC Press, pp 53–66
62. Chaudhary A, Sharma S, Verma A (2022) WEDM machining of heat treated ASSAB'88 tool steel: a comprehensive experimental analysis. Mater Today Proceed 50:946–951
63. Bisht N, Verma A, Chauhan S, Singh VK (2021) Effect of functionalized silicon carbide nanoparticles as additive in cross-linked PVA based composites for vibration damping application. J Vinyl Add Tech 27(4):920–932
64. Verma A, Jain N, Parashar A, Singh VK, Sanjay MR, Siengchin S (2020) Design and modeling of lightweight polymer composite structures. In: Lightweight polymer composite structures. CRC Press, pp 193–224
65. Dogra V, Kishore C, Verma A, Rana AK, Gaur A (2021) Fabrication and experimental testing of hybrid composite material having biodegradable bagasse fiber in a modified epoxy resin: evaluation of mechanical and morphological behavior. Appl Sci Eng Progress 14(4):661–667
66. Verma A, Baurai K, Sanjay MR, Siengchin S (2020) Mechanical, microstructural, and thermal characterization insights of pyrolyzed carbon black from waste tires reinforced epoxy nanocomposites for coating application. Polym Compos 41(1):338–349
67. Verma A, Budiyal L, Sanjay MR, Siengchin S (2019) Processing and characterization analysis of pyrolyzed oil rubber (from waste tires)-epoxy polymer blend composite for lightweight structures and coatings applications. Polym Eng Sci 59(10):2041–2051
68. Verma A, Negi P, Singh VK (2019) Experimental analysis on carbon residuum transformed epoxy resin: chicken feather fiber hybrid composite. Polym Compos 40(7):2690–2699
69. Verma A, Singh VK (2018) Mechanical, microstructural and thermal characterization of epoxy-based human hair–reinforced composites. J Test Eval 47(2):1193–1215
70. Jain N, Verma A, Ogata S, Sanjay MR, Siengchin S (2022) Application of machine learning in determining the mechanical properties of materials. In: Machine learning applied to composite materials. Springer, Singapore, pp 99–113
71. Verma A, Singh VK, Verma SK, Sharma A (2016) Human hair: a biodegradable composite fiber–a review. Int J Waste Resour 6(206):1000206
72. Kataria A, Verma A, Sanjay MR, Siengchin S, Jawaid M (2022) Physical, morphological, structural, thermal, and tensile properties of coir fibers. In: Coir fiber and its composites. Woodhead Publishing, pp 79–107
73. Jain N, Verma A, Singh VK (2019) Dynamic mechanical analysis and creep-recovery behaviour of polyvinyl alcohol based cross-linked biocomposite reinforced with basalt fiber. Mater Res Express 6(10):105373

74. Chaurasia A, Verma A, Parashar A, Mulik RS (2019) Experimental and computational studies to analyze the effect of h-BN nanosheets on mechanical behavior of h-BN/polyethylene nanocomposites. J Phys Chem C 123(32):20059–20070

75. Verma A, Negi P, Singh VK (2018) Physical and thermal characterization of chicken feather fiber and crumb rubber reformed epoxy resin hybrid composite. Adv Civil Eng Mater 7(1):538–557

76. Bharath KN, Madhu P, Gowda TG, Verma A, Sanjay MR, Siengchin S (2020) A novel approach for development of printed circuit board from biofiber based composites. Polym Compos 41(11):4550–4558

77. Verma A, Joshi K, Gaur A, Singh VK (2018) Starch-jute fiber hybrid biocomposite modified with an epoxy resin coating: fabrication and experimental characterization. J Mech Behav Mater 27(5–6):20182006

78. Verma A, Singh C, Singh VK, Jain N (2019) Fabrication and characterization of chitosan-coated sisal fiber–Phytagel modified soy protein-based green composite. J Compos Mater 53(18):2481–2504

79. Rastogi S, Verma A, Singh VK (2020) Experimental response of nonwoven waste cellulose fabric–reinforced epoxy composites for high toughness and coating applications. Mater Perfor Charact 9(1):151–172

80. Verma A, Zhang W, Van Duin AC (2021) ReaxFF reactive molecular dynamics simulations to study the interfacial dynamics between defective h-BN nanosheets and water nanodroplets. Phys Chem Chem Phys 23(18):10822–10834

81. Verma A, Jain N, Parashar A, Gaur A, Sanjay MR, Siengchin S (2021) Lifecycle assessment of thermoplastic and thermosetting bamboo composites. In: Bamboo fiber composites. Springer, Singapore, pp 235–246

82. Arpitha GR, Verma A, Sanjay MR, Siengchin S (2021) Preparation and experimental investigation on mechanical and tribological performance of hemp-glass fiber reinforced laminated composites for lightweight applications. Adv Civil Eng Mater 10(1):427–439

83. Verma A, Samant SS (2016) Inspection of hydrodynamic lubrication in infinitely long journal bearing with oscillating journal velocity. J Appl Mech Eng 5(3):1–7

84. Verma A, Jain N, Rangappa SM, Siengchin S, Jawaid M (2021) Natural fibers based bio-phenolic composites. In: Phenolic polymers based composite materials. Springer, Singapore, pp 153–168

85. Verma A, Parashar A, Singh SK, Jain N, Sanjay SM, Siengchin S (2020) Modeling and simulation in polymer coatings. In: Polymer coatings. CRC Press, pp 309–324

86. Verma A, Singh VK Experimental characterization of modified epoxy resin assorted with almond shell particles. ESSENCE-Int J Environ Rehabil Conserv 7(1):36–44

87. Verma A (2022) A perspective on the potential material candidate for railway sector applications: PVA based functionalized graphene reinforced composite. Appl Sci Eng Progress 15(2):5727–5727

88. Verma A, Jain N, Singh K, Singh VK, Rangappa SM, Siengchin S (2022) PVA-based blends and composites. In: Biodegradable polymers, blends and composites. Woodhead Publishing, pp 309–326

89. Raja S, Verma A, Rangappa SM, Siengchin S (2022) Development and experimental analysis of polymer based composite bipolar plate using Aquila Taguchi optimization: design of experiments. Polym Compos 43(8):5522–5533

90. Prabhakaran S, Sharma S, Verma A, Rangappa SM, Siengchin S (2022) Mechanical, thermal, and acoustical studies on natural alternative material for partition walls: a novel experimental investigation. Polym Compos 43(7):4711–4720

91. Sethi SK, Gogoi R, Verma A, Manik G (2022) How can the geometry of a rough surface affect its wettability?-A coarse-grained simulation analysis. Prog Org Coat 172:107062

92. Chaudhary A, Sharma S, Verma A (2022) Optimization of WEDM process parameters for machining of heat treated ASSAB'88 tool steel using response surface methodology (RSM). Mater Today Proceed 50:917–922

93. Verma A, Jain N, Sanjay MR, Siengchin S Viscoelastic properties of completely biodegradable polymer-based composites. In: Vibration and damping behavior of biocomposites. CRC Press, pp 173–188

94. Verma A, Jain N, Mishra RR (2022) Applications and drawbacks of epoxy/natural fiber composites. In: Handbook of epoxy/fiber composites. Springer Singapore, Singapore, pp 1–15

95. Lila MK, Verma A, Bhurat SS (2022) Impact behaviors of epoxy/synthetic fiber composites. In: Handbook of epoxy/fiber composites. Springer Singapore, Singapore, pp 1–18

96. Chaturvedi S, Verma A, Singh SK, Ogata S (2022) EAM inter-atomic potential—its implication on nickel, copper, and aluminum (and their alloys). In: Forcefields for atomistic-scale simulations: materials and applications. Springer, Singapore, pp 133–156

97. Verma A, Sharma S (2022) Atomistic simulations to study thermal effects and strain rate on mechanical and fracture properties of graphene like BC3. In: Forcefields for atomistic-scale simulations: materials and applications. Springer, Singapore, pp 237–252

98. Chaturvedi S, Verma A, Sethi SK, Ogata S (2022) Defect energy calculations of nickel, copper and aluminium (and their alloys): molecular dynamics approach. In: Forcefields for atomistic-scale simulations: materials and applications. Springer, Singapore, pp 157–186

99. Shankar U, Gogoi R, Sethi SK, Verma A (2022) Introduction to materials studio software for the atomistic-scale simulations. In: Forcefields for atomistic-scale simulations: materials and applications. Springer, Singapore, pp 299–313

100. Shankar U, Sethi SK, Verma A (2022) Forcefields and modeling of polymer coatings and nanocomposites. In: Forcefields for atomistic-scale simulations: materials and applications. Springer, Singapore, pp 81–98

101. Verma A, Ogata S (2022) Computational modelling of deformation and failure of bone at molecular scale. In: Forcefields for atomistic-scale simulations: materials and applications. Springer, Singapore, pp 253–268

102. Homer ER, Verma A, Britton D, Johnson OK, Thompson GB (2022) Simulated migration behavior of metastable $\Sigma3$ (11 8 5) incoherent twin grain boundaries. In: IOP conference series: materials science and engineering (vol 1249, no 1, p 012019)

103. Chaturvedi S, Verma A, Sethi SK, Rangappa SM, Siengchin S (2022) Stalk fibers (rice, wheat, barley, etc.) composites and applications. In: Plant fibers, their composites, and applications. Woodhead Publishing, pp 347–362

104. Arpitha GR, Verma A, MR S, Gorbatyuk S, Khan A, Sobahi TR, Asiri AM, Siengchin S (2022) Bio-composite film from corn starch based vetiver cellulose. J Nat Fibers 19(16):14634–14644

105. Thimmaiah SH, Narayanappa K, ThyavihalliGirijappa Y, Gulihonenahali Rajakumara A, Hemath M, Thiagamani SMK, Verma A (2022) An artificial neural network and Taguchi prediction on wear characteristics of Kenaf–Kevlar fabric reinforced hybrid polyester composites. Polym Compos 44(1):261–273

106. Kataria A, Verma A, Sethi SK, Ogata S (2022) Introduction to interatomic potentials/forcefields. In Forcefields for atomistic-scale simulations: materials and applications. Springer, Singapore, pp 21–49

107. Kataria A, Verma A, Sanjay MR, Siengchin S (2022) Molecular modeling of 2D graphene grain boundaries: mechanical and fracture aspects. Mater Today Proc 52:2404–2408

108. Verma A, Ogata S (2023) Magnesium based alloys for reinforcing biopolymer composites and coatings: a critical overview on biomedical materials. Adv Indust Eng Polym Res. https://doi.org/10.1016/j.aiepr.2023.01.002

109. Arpitha GR, Jain N, Verma A, Madhusudhan M (2022) Corncob bio-waste and boron nitride particles reinforced epoxy-based composites for lightweight applications: fabrication and characterization. Biomass conversion and biorefinery, pp 1–8. https://doi.org/10.1007/s13399-022-03717-1

110. Kumar G, Mishra RR, Verma A (2022) Introduction to molecular dynamics simulations. In: Forcefields for atomistic-scale simulations: materials and applications. Springer, Singapore, pp 1–19. https://doi.org/10.1007/978-981-19-3092-8_1

111. Kataria A, Chaturvedi S, Chaudhary V, Verma A, Sanjay NMKR, Siengchin S (2023) Cellulose fiber-reinforced composites—history of evolution, chemistry, and structure. In: Cellulose fibre reinforced composites. Woodhead Publishing, pp 1–22. https://doi.org/10.1016/B978-0-323-90125-3.00012-4

112. Chaturvedi S, Kataria A, Chaudhary V, Verma A, Sanjay NJMR, Siengchin S (2023) Bionanocomposites reinforced with cellulose fibers and agro-industrial wastes. In: Cellulose fibre reinforced composites. Woodhead Publishing, pp 1–22. https://doi.org/10.1016/B978-0-323-90125-3.00017-3

113. Arpitha GR, Mohit H, Madhu P, Verma* A (2023) Effect of sugarcane bagasse and alumina reinforcements on physical, mechanical, and thermal characteristics of epoxy composites using artificial neural networks and response surface methodology. Biomass Conversion and Biorefinery.https://doi.org/10.1007/s13399-023-03886-7

Chapter 3
Modern Coating Processes and Technologies

Ankit Kumar, Jyoti Jaiswal, Kazuyoshi Tsuchiya, and Rahul S. Mulik

1 Introduction

Coating technology is a contemporary field that is constantly evolving, in part because of the creation of new materials and, more specifically, because of recent advances in nanotechnology and nanoscience [1–3]. Smart coatings can react to environmental elements like radiation, temperature gradients, biomarkers, and internal and external stress states [4, 5]. New application possibilities are made possible by modern coatings, including waterproof and self-cleaning/self-healing systems [6–8]. For biological and medicinal purposes, new coatings are created with a focus on antibacterial properties [8, 9]. Manufacturing of coatings and related technologies are also impacted by the development. Modern coating production technologies are complemented by new techniques for the investigation and characterization of materials [10,

A. Kumar · R. S. Mulik (✉)
Department of Mechanical and Industrial Engineering, IIT Roorkee, Roorkee 247667, India
e-mail: rsm@iitr.ac.in

A. Kumar
e-mail: akumar29@me.iitr.ac.in

J. Jaiswal (✉) · K. Tsuchiya (✉)
Micro/Nano Technology Center, Tokai University, 4-1-1 Kitakaname, Hiratsuka 259-1292, Kanagawa, Japan
e-mail: jyoti.jaiswal@rgu.ac.in

K. Tsuchiya
e-mail: tsuchiya@tokai-u.jp

J. Jaiswal
Centre for Advanced Research, Department of Physics, Rajiv Gandhi University, Arunachal Pradesh 791112, India

K. Tsuchiya
Department of Mechanical Engineering, Tokai University, 4-1-1 Kitakaname, Hiratsuka 259-1292, Kanagawa, Japan

© The Author(s), under exclusive license to Springer Nature Singapore Pte Ltd. 2023
A. Verma et al. (eds.), *Coating Materials*, Materials Horizons: From Nature to Nanomaterials, https://doi.org/10.1007/978-981-99-3549-9_3

11]. Therefore, improve our understanding of their micro- and nanostructure as well as numerous properties at various time and length scales [12]. It is important to note that corrosion/wear resistance, and by extension coating results, is a system attribute rather than a feature inherent to materials, and therefore the reaction to extrinsic influences should be thoroughly understood [13, 14].

Demand for coatings is constantly rising as new applications necessitate strict regulations (e.g., in microelectronics industry). Self-shielding as well as self-healing coatings are also necessary for smartphones, super capacitors and other electronic devices [15, 16]. Additionally, investments are being made in R&D for products including wear- and self-cleaning coatings, corrosion-resistant coatings, and biomedical coatings [13, 17]. The sectors with the greatest market influence will be those related to personal health, aircraft, home applications, food packaging, information technology and automotive [17–20].

The market for paints and coatings are anticipated to rise from $211.9 billion in 2021 to $230.22 billion in 2022 at a multiple annual growth rate (CAGR) of 8.6%. At a CAGR of 7.8%, the market is anticipated to reach $311.47 billion in 2026 [21, 22]. It is anticipated to have significant growth over the projection period beginning in 2020 because of a number of developments led by a robust recovery in global manufacturing and building construction. Particularly in western Europe, Japan and North America [23]. Additionally, advanced coating techniques with increased efficiency and clean procedures are being introduced daily [19, 24]. So, with the help of renowned researchers from academic and industrial centers, we developed this collection [25].

A coating is a covering that is put on an object's surface, also recognized as the substrate. The coating may be used for practical, ornamental, or a combination of both purposes. It is possible to apply coatings as liquids, gases, or solids, such as powder coatings [26]. The term "coating" in the context of corrosion refers to the application or coating of thin layers of a covering material over the surface of any object, typically to improve important features and create a barrier against surface deterioration caused by interactions with the environment [27]. According to the discipline of surface engineering, a coating is a layer of substance that is applied to a substrate to enhance its surface characteristics for wear and corrosion [28]. There are many variables that might affect the choice of coating, including the service environment, substrate material compatibility, component shape and size, life expectancy, and cost [29].

The word "coating" typically refers to

- A thin material(s) layer [30],
- Inorganic lining, such as that found in glass and porcelain,
- Organic coating such as paint, varnish, or polymeric substance, and
- Optical film covering, as well as printed or etched covering [31].

The coating could be used to offer

- Scratch resistance as well as abrasion resistance,
- The ability to withstand heat and fire,

- Protection from the flow of electric charge,
- Wettability and sealing capacity,
- Aesthetics and physical appearance are enhanced,
- Corrosion resistance,
- Protection from ordinary abrasion, erosion, pitting, and cavitation,
- Aesthetics and physical appearance are enhanced,
- Release or nonstick characteristics [10, 17, 19, 31–35].

2 Types of Coatings Process

2.1 Physical Vapor Deposition (PVD) Coating

The thermal evaporation of aluminium onto thin polymeric webs, such as polypropylene (PP) and polyester (PET), has created large quantities of decorative films, capacitor films, ornamental films, and some window films [36]. Extensive web conduct expertise has been paired with deposition procedures including magnetron sputtering, chemical vapor deposition and plasma-enhanced electron beam evaporation to produce a variety of brand-new, intriguing coating materials, with oxides as well as nitrides of the majority of elements. More specifically, combining these coating layers obsessed by a complex coating pile produced novel goods like electrochromic devices, architectural glazing films with low emissivity, solar heat reflecting properties, and high-performance optical reflectors. Inimitable coating properties, such as, flexible glossy barriers aimed at moisture, transparent electrodes, gases, and amorphous lenient magnetic coating materials for safety devices, can be achieved with the help of these methods [37].

The PVD method has a reputation for providing materials exposed to corrosive fluids with wear and corrosive resistance as well as thin film as a protective layer on their surfaces [38, 39]. Its applications range from aesthetic items to industrial equipment. This approach has the advantage of allowing the coating layers' mechanical, corrosion, and cosmetic qualities to be changed as needed. Generally speaking, PVD coatings are deposited in a very high vacuum condition as well as involves the transition of liquid/solid coating materials to a vaporization point, trailed through a metal vapor concentration that results in the formation of a solid and dense film. The sputtering and evaporation are the two kinds of PVD that are most known. Since PVD coating layers tend to be thin, multilayered coatings are always necessary, and considerable consideration should be given to the choice of materials. In addition to their ornamental functions, several PVD-coated fragments serve as components with high rates of wear, which causes abrasion on the surface and wears away the fabricated coating film. This process lessens the parts' ability to resist corrosion and increases their susceptibility to corrosive fluids. A typical schematic perspective of several types of electron beam PVD devices is shown in Fig. 1 [40]. A physical evaporation mechanism dominates this method's coating growth. The thermal energy necessary for evaporation might be provided by a variety of supply units,

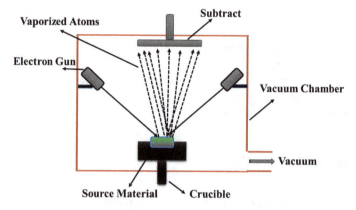

Fig. 1 An electron beam is used as the heat source in this graphical illustration of a PVD (Physical Vapor Deposition) equipment [42]

including electron beams, heating wires, laser beams, molecular beams, etc. [41]. The source material's atoms, which might be solid or liquid, are heated to the point of evaporation by this thermal energy. Through the vacuum, the vaporized atoms go a considerable distance before depositing on the substrate.

The kinetic energy (KE) of the incoming flux during growth via thermal evaporation is of the directive of 0.1 eV and depends on the temperature of the evaporate source. The incident flux's kinetic energy can be raised to several hundred eV when using plasma or ion beam deposition procedures. Over trapping, preferred sputtering, improved atom diffusion, dynamic collisional mixing, and low energy (often <100 eV) ion irradiation during vapor phase film formation has been found to be beneficial in controlling modifying the physical properties of as-fabricated layers [36]. The deposition of thin films is essential for many different technologies, including, coatings, sensors, electronic devices, displays, and optical apparatus. There is still a great deal of interest in alternate approaches that might be more affordable, more dependable, or capable of making pictures with innovative or enhanced properties, even when well-established procedures for the fabrication of better-quality coating films already exist. This PVD chapter focuses on certain areas where new technologies are upending conventional methods for producing thin films [43, 44].

2.1.1 Magnetron Sputtering Coating

A plasma-based deposition method called sputtering accelerates energetic ions in the direction of a target. Atoms are expelled (or sputtered) off the surface when the ions hit the target. These atoms move in the direction of the substrate and combine with the developing film. In the deposition process known as magnetron sputtering, a gaseous plasma is produced and contained in a space covering the material to be deposited, or the "target." The target's surface is worn away via high-power particles

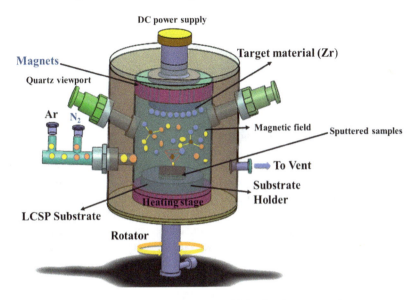

Fig. 2 Illustration of Magnetron sputtering set up to fabrication the nano coating

in the plasma also the released atoms then move over the vacuum atmosphere as well as deposit on substrate to fabrication a thin coating. Figure 2 depicts the samples utilized in this work and the production of the sputtering coating [45, 46].

A chamber is first evacuated to a high vacuum in a typical sputtering deposition procedure to reduce the partial pressures of all background gases and any impurities. After the base pressure is established, the plasma-containing sputtering gas is injected into the chamber, and the total pressure is then normally controlled in the milli-Torr range by a pressure control system. High voltage is given to the anode, which is typically linked to the chamber as electrical ground, and the cathode, which is typically placed behind the sputtering target, to formation of the plasma formation method. The sputtering gas's electrons, which are present there, are propelled away from the cathode, resulting in collisions with surrounding sputtering gas atoms [47, 48].

Ionization results from the electrostatic repulsion that these collisions produce, which "knocks off" electrons from the gas atoms that are sputtering. Now that the negatively charged cathode is being approached by the positively sputtering gas atoms, high energy collisions with the targets surface result. Atoms at the surface of the target may be ejected into the vacuum environment as a result of each of these impacts with adequate kinetic energy to influence the surface of the substrate. High molecular weight gases like argon or xenon are often chosen as the sputtering gas in order to enable as many high energy collisions as possible, which will boost the rate of deposition. Additionally, during film formation, gases like oxygen or nitrogen can be added to the chamber if a reactive sputtering process is needed [48–51].

Sputtering deposition technique is frequently used to deposit magnetic deposited films, hard coatings (tools, engine parts), solid lubricants as well as attractive coatings. Additionally, it is frequently used to deposit architectural glass coatings, reflective coatings, and thin film metallization on semiconductor materials [49, 52, 53].

2.1.2 PLD (Pulsed Laser Deposition)

PLD, a form of physical vapour deposition, is a process in which a target for the material to be deposited is in contact with a high-power pulsed laser beam that has been focussed inside of a vacuum chamber. This substance is vaporised from the target (in the form of plasma plume), where it is then applied to a substrate as a thin film (silicon wafer facing the target). This procedure can take place in an extremely low vacuum or in the existence of a background gas for example oxygen, which is frequently employed to thoroughly oxygenate the formed films when depositing oxides [54, 55].

In comparison to many other deposition procedures, the basic setup is straightforward, but the physical processes governing film formation and laser-target interaction are highly intricate. The energy from the laser pulse is transformed into electronic excitation, thermal, chemical, and mechanical energy when it is absorbed by the target, leading to exfoliation, evaporation, and plasma production. The expelled species grow into a plume in the surrounding vacuum before depositing on the typically high temperature substrate. The plume contains a variety of energetic species, molecules, including atoms, electrons, ions, particulates, clusters, and molten globules. PLD's intricate workings include the laser-induced ablation of the target material, the creation of a plasma plume including highly energetic ions, electrons, and neutrals, as well as the crystallisation of the film at high temperature substrate. In general, the pulsed laser deposition method can be broken down into four stages: laser absorption on the target surface, laser ablation of the target material, plasma generation. Plasma dynamics, ablation material deposition on the substrate, film growth as well as nucleation on the substrate surface [56].

Individually of these stages is essential for the final film's crystallinity, homogeneity, and stoichiometry. Additionally shown in Fig. 3 is a schematic illustration of the PLD setup used to generate the nanocoating on the Si substrate. The PLD approach is a versatile, optimistic, straightforward, and often used process for producing metal oxide and composite metal oxide. Herein, the target material is evaporated either in an ultrahigh vacuum or in the existence of gases like oxygen using a powerful pulsed laser [57]. Reactive pulsed deposition involves depositing material as it is being ablated. The ablated material interacts with the gas molecules and is then deposited on the substrate. Because the ionised ions in this method have a high kinetic energy, crystallisation is accomplished at lower deposition temperatures than in other PVD processes.

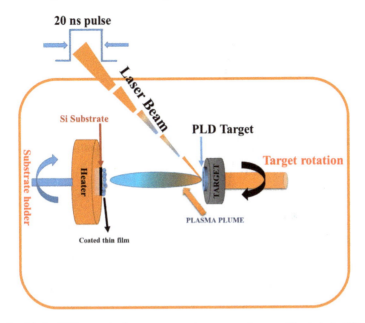

Fig. 3 Graphical, a PLD setup is shown to create the nano coating on a Si substrate [58]

2.1.3 Thermal Evaporation

Polymeric coatings are also created by thermal evaporation. The low thermal conductivity of polymers restricts the amount of thermal evaporation that can occur when they are heated resistively. On the other hand, the breakdown of the molecules restricts the amount of organic material that may be evaporated using an electron beam. However, there are other applications, such as PTFE and nylon in metal-polymer nanocomposite films [59, 60], where polymeric films have been successfully generated by heat evaporation. The primary issue is the molecular weight drop during thermal heating. Vacuum thermal evaporation has been used to successfully deposit small polymers up to several thousand g mol^{-1}, which are equivalent to tens of monomeric units [61]. PTHs are crucial molecules in the creation of polymeric semiconducting materials. Thermal evaporation has been used to successfully encapsulate PTH molecules at 300 °C. One nanometre per minute deposition rate has been attained. Without further annealing, the films were quite crystalline. These evaporated PTH coatings have been used to demonstrate polymer solar cell systems [62, 63].

A further eminent technique for coating thin layers is thermal evaporation, in which the source material evaporates in a vacuum as a result of high temperature heating, enabling the vapor particles to move more freely and directly contact a substrate where they transform back into a solid form. Using a charge holding boat or protected coil that comes in the shape of a powder or solid bar is how this approach works. In order to obtain the high melting temperatures necessary for metals, a

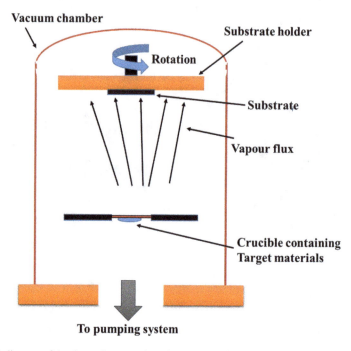

Fig. 4 A diagram of the thermal evaporation phenomena [59]

significant amount of direct current (DC) is applied to the resistive boat/coil, where the high vacuum (below 104 Pa) supports the metal's evaporation and further delivers it to the substrate. Low melting point materials benefit the most from this approach [60, 64]. A graphic illustration of the thermal evaporation system as shown in Fig. 4 [59].

Thermal evaporation technique is achieved at high temperature an evaporated substance that has been placed in an evaporation source to a high temperature (typically resistive). It is done at pressure 10^{-5} torr in a vacuum. The gentlest PVD approach is thermal evaporation, with evaporated particle energies at 1500 K. (0.12 eV). This technique uses the least amount of power and is the simplest PVD technique. Metal boats and heaters built of refractory wires (such as Wo, Mo), which can be shielded by a passive coating material, are the main categories of simple evaporation sources (2). For precise evaporation procedures like molecular beam epitaxy, more complex effusion cells with an external heater and a crucible made of passive coating material (such boron nitride) are utilized [64, 65].

2.1.4 Electron Beam Evaporation

With electron beam evaporation, a concentrated beam of high energy electrons can evaporate the source material. The thermionic emission of electrons that results from

Fig. 5 Diagram illustrating the electron beam evaporation system's phenomenon [59]

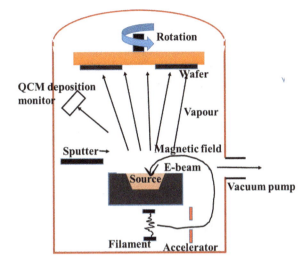

a hot filament can, after acceleration, supply enough energy for evaporating any substance. 10 kW is provided upon impact in a distinctive scenario where 1 A of emission is accelerated over a 10 kV voltage drop. As shown in Fig. 5, the filament is placed out of the way of the evaporant to prevent melting, and an electron beam is drawn to the surface via magnetic field (*B*). The fact in this below Fig. 5 indicates the direction of the magnetic field. The Lorentz force, *F*, is the result of the interaction of the electric (*E*) and magnetic fields on an electron (1) [59, 66]:

$$F = F_E + F_B = q_e + q_e E + q_e (v \times B) \tag{1}$$

According to Fig. 5, the cross-product vector, F_B, is oriented perpendicular to both *v* and *B*. Accelerated electrons gone from the filament or cathode by the first force term in Eq. (1). The second force term states that the electrons' newly gained speed causes them to be sideways rebounded as they pass the lines of the magnetic field. The centrifugal force of electrons with a radius of curvature r balances the second force [67].

Web coating offers more potential markets now that dependable, high-power electron beam guns are more widely available. It is possible to program the electron beam guns to scan different places with different dwell periods inside a sizable crucible containing the material to be deposited. Since the maximum scan width is only approximately 1 m, multiple parallel electron beams typically two are required for wide web coating. With web speeds of 12 m/s and above, electron beam technology is most likely the fastest deposition source currently accessible. In theory, it is also the most cost-effective approach [68, 69]. Even so, there aren't many of these machines in use, despite the fact that they may displace resistance and induction heating in the aluminizing industry. It has been deemed to be a step too far due to the significant expenditure required compared to a standard coater, the high level of technological complexity, and some conservatism on the part of the metallizes. The majority of pure metals, together with those has greater melting points scale can be coated using this source. If the vapour pressures are close enough together, alloys like Cr and Ni as well as a variety of oxides and nitrides can be evaporated for the occasional disintegration of the compounds. Transparent barrier coatings, Magnetic data storage and ultrahigh-rate aluminization are examples of current applications [69].

2.1.5 Reactive Sputter Deposition

Reactive sputtering involves sputtering thin coatings of compounds onto substrates from metallic (as opposed to non-metallic) targets in the existence of a reactive gas that is typically combined with an inert working gas (invariably Ar). Here is a brief list of the most popular reactively sputtered compounds and the reactive gases used:

a. Al_2O_3, In_2O_3, SnO_2, SiO_2, Ta_2O_5—Oxides coatings (Oxygen)
b. TaN, TiN, AlN, Si_3N_4, CN_x—Nitrides coatings (nitrogen, ammonia)
c. TiC, WC, SiC—Carbides coatings (methane, acetylene, propane)
d. H_2S, CdS, CuS, ZnS—Sulfides coatings.

Compound coatings are frequently made via reactive sputtering [70, 71]. A pure target is used to sputter metal, and the process is then given enough reactive gas to create the required chemical at the substrate (Fig. 6). When there is an air leak or a high-water upbringing pressure, reactive sputtering frequently occurs as an unwanted byproduct of sputtering. In each instance, the insertion of gas species causes the film to deviate from the intended purity. The addition of a reactive gas causes the film-depositing atoms to interact with the gas atoms to generate compound films with various stoichiometry [71, 72]. Despite the injection of additional gas, nearby is currently no increase in chamber pressure due to the gas atoms have been absorbed by the coating. As the rise of reactive gas flow, the films react more intensely until they ultimately reach their "final" state of reaction at an appropriately highly sensitive gas drift. Usually, this is a chemical that is stable or "terminal." Beyond this limit, the depositing layer is unable to absorb any more reactive gas atoms. Now, a reacted, compound layer is formed on the cathode surface with every new flow of reactive gas. The rate at which metal atoms spatter from the cathode is slowed down by the fact that this compound nearly always has a minor sputtering yield than the pure metal cathode. The film can absorb the reactive gas at a slower rate when the metal deposition rate is slower, which raises the reactive species background level even more. This further slows down the metal sputtering rate by triggering further reaction at the cathode surface. The deposition process significantly slows down as the cathode transitions from a metallic to a compound state [73, 74].

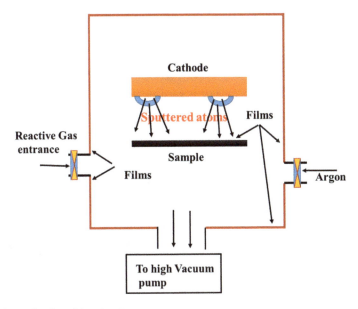

Fig. 6 A reactive deposition chamber

Hysteresis Effects

Figure 7a displays the flow, and Fig. 7b shows the chamber pressure. By speeding up the system's pumps so that the amount of gas withdrawn by the pumps is significantly more than the amount of gas chemically consumed, the harshness of the hysteresis consequence can be lessened [67, 75]. By doing this, when the deposition target switches from the metallic to the composite manner, the destabilizing pressure swings at the target are significantly decreased. Hysteresis can be eliminated by swelling pumping speed, although doing so is expensive due to the cost of an additional pump's capacity and an increase in gas consumption. Through dilution, the additional gas flow has the benefit of lowering pollution after vessel outgassing and leaks.

Despite the hysteresis effect, it is possible to synthesis any material composi-tions by controlling the responsive gas partially. The reactive gas, partial pressure is retained constant while the power to the target of sputtering is also maintained continual, a balance between the feeding as well as availability of the reactive gas is preserved. The gas atoms flux at each surface is determined by the fractional pressure. The quantity of that gas available will be controlled if the partial pressure is handled. A controlled partial-pressure will temporarily restrict flow to maintain constant partial pressure in the event of a method interruption, such as an arc on the target. The flow will be increased once more to maintain the correct controlled pressure once the plasma is recovered (quenched arc) and the metal is sputtered at its fastest pace. It is anticipated that partial-pressure control will be stable. The material is removed from the target in a mostly uniform fashion, with the exception of disruptions like arcs, and at the surface of the substrate, the target and gas atoms

Fig. 7 Hysteresis curve for reactive sputtering used for the deposition rate (**a**) and sputtering chamber pressure (**b**) [67, 75]

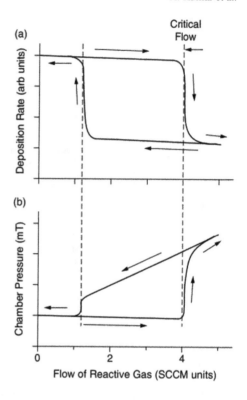

arrive in the correct ratios to create the stoichiometric compound. A species-selective method for continuously monitoring the gases in the process chamber is needed for partial-pressure control [71]. The quadrupole mass analyzer, which can distinguish gases based on their mass ratios and frequently produces a distinct signal for each gas present, is the instrument that is utilized the most commonly. When the analyzer achieves an acceptable signal-to-noise ratio within a time window that is brief enough to allow for the flow to be changed before the underlying process instabilities drive the partial pressure too far from equilibrium, good partial-pressure management is obtained. The necessary reactive gas atoms must arrive at the substrate at a rate that corresponds to the metal atom arrival rate in order to create the stoichiometric product. The resultant film will not have the desired composition if these arrival rates are not balanced. Therefore, feasible to sustain the necessary inside gas flow atom rate such that, when they combine with the incoming metal atoms and the appropriate material phase is created by managing the reactive gas, partial pressure around the substratum surface [76].

It is important to properly regulate the responsive gas partial pressure to prevent pressure swings of a large magnitude. Even brief low-pressure intervals because gas shortages that result in numerous monolayers of metal-rich configuration, whereas a brief excess causes deposition rates to slow down and may even result in the creation of an undesirable phase. These areas of non-stoichiometric configuration will impair coating performance and frequently lead to early coating failure and maybe even component failure. It's not always simple to keep the metal arrival rate constant. If the energy delivered to the magnetron sputtering target stays constant, a properly prepared metal target in a pure argon environment will have a stable metal removal rate. The constant metal removal rate at the cathode leads to a constant flow of accessible metal and a consistent transport rate to the substrate. The energy of atoms arriving can be increased in a variety of ways. Biasing the sample to a negative voltage while it is being deposited is the easiest. By accelerating plasma ions toward the material due to the bias, more energy is deposited near the surface. With increasing deposition rate, increasing ion bombardment becomes necessary. The necessary bias current density for depositions rate is high (up to around a m/ min) approaches 2 mA/cm^2. Unfortunately, it is challenging to obtain high amounts of bias current with the standard deposition techniques mentioned above. Although it is advantageous for the cathode to be sputtered at a high rate when the plasma is constrained close to it, this makes it difficult to pull ions to the sample region, which is located many centimeters distant [71, 76].

2.1.6 Cathodic Arc Deposition (CAD)

A different type of vapor deposition is CVD. The semiconductor industry frequently uses this technique, which uses a high vacuum to build a solid, high resistance and high-quality thin film coating layer on any surface [3]. Mechanical parts that are in regular touch and require protection from corrosion and wear can employ CVD. This process involves exposing a wafer-shaped substrate to a mixture of non-toxic starting materials, where a chemical reaction causes a layer to be deposited on the material's surface. However, some of the derivatives of these chemical processes, which are eliminated by the vacuum pump's continuous inflow, may still be present in the chamber. Since then, numerous groups from all over the world have looked into the method since it holds out the possibility of an effective source of highly charged material for dense, good adhesive coatings with a wide variety of compositions. These techniques have greatly enhanced the coatings' quality, and information are now emerging on the use of the cathodic arc approach in the more difficult technological fields of electronics and optics [77, 78].

The high deposition energy of the abbreviating atoms, which is necessary for coating layer growth to support adhesion and to obstruct columnar growth, is what draws people to the CAD method. These energies depend on the substance and for carbon and iron, respectively, vary from 25 to 55 eV. Ion-surface interaction theory-based calculations of the energies needed for high-quality film formation further suggest that energies of the order of 25–100 eV are also preferable for activating surface atom movements while retaining better crystallinity of coated layers. The creation of macro particles (MPs), with sizes ranging from 0.1 to 10 μm in growing films, is a drawback of the CAD process. Because MPs produce rapid changes in microstructure and texture, surface roughness, and coating micro holes, they are generally thought to be harmful in applications. A novel sort of compact,

Fig. 8 Reactor for
cathodic-arc deposition [77]

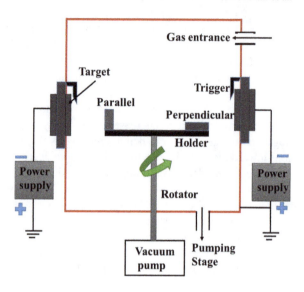

MP-inhibiting pulsed cathodic arc deposition system that is appropriate for use in
conjunction with an ultra-high vacuum (UHV) system is presented in this work.
Additionally, by co-depositing from two arc sources, nanoscale single and multilayer
coatings with alternative layers of super-lattice structure were created. In addition,
by describing the fundamental ideas behind pulsed plasma creation, a technique for
regulating nano-scale layer thickness with pulsed discharge is described [61, 79].

In a vacuum (~1 Pa), the cathodic arc deposition occurs between two metallic
electrodes. A low-voltage, high-current plasma discharge is known as a cathodic
arc. By adding a reactive gas to the argon plasma, such as nitrogen, the process
can be rendered reactive. Cathode spots, which range in size from 1 to 10 μm, are
isolated locations. A cathode spot may carry currents of 1–10 A. A concentrated
plasma of cathode material is produced by a typical arc discharge current of a few
hundred amps (A). From a few millimeter cathodes toward typically big plants that
are a meter in size with sufficient power sources, the equipment is available (Fig. 8).
The extracted material may take the form of atoms, molecules, or quickly solidifying
metallic droplets with a diameter of 0.1–10 μm. One of the most commonly employed
industrial processes, this one is appropriate for coatings like TiN [80].

2.2 Chemical Vapor Deposition (CVD)

The CVD process is a potent one for creating solid thin films and coatings of
the highest caliber. Although prevalent in contemporary sectors, it is always being
improved as new materials are suited to it. With the exact fabrication of both inorganic thin films of 2D materials and high-purity polymeric thin films that can be

conformably applied to diverse substrates, CVD synthesis is currently reaching new heights. This introduction provides a broad overview of CVD technology, discussing instrument design, process management, material characterization, and repeatability challenges. The greatest practices for research, including substrate preparation, high-temperature growth, also post-growth activities, are provided utilizing graphene, 2D transition metal dichalcogenides (TMDs), and polymeric thin films as exemplary examples. Current trends and scaling-up problems are also discussed. By evaluating current restrictions and optimizations, we also offer insight into potential future directions for the coating technique, such as reactor design for low-temperature and high-throughput manufacturing of thin films [46, 81, 82]. Typically, PVD does not involve chemical reactions and instead uses evaporation and sputtering processes. The creation of organic and inorganic films on metals, semiconductors, and other materials uses CVD extensively in industry. A variety of processes, including, atmospheric-pressure chemical vapour deposition (APCVD), plasma-assisted chemical vapour deposition (PACVD), low-pressure chemical vapour deposition (LPCVD), plasma-enhanced chemical vapour deposition (PECVD) and laser-enhanced chemical vapour deposition, are included in CVD (LECVD). Additionally, hybrid techniques that combine aspects of both physical and CVD have also been developed. This CVD chapter focuses on several areas where new technologies are upending traditional methods for producing thin films [83, 84]. Basic schematic diagram of CVD with gas discontinuous flow shown in Fig. 9.

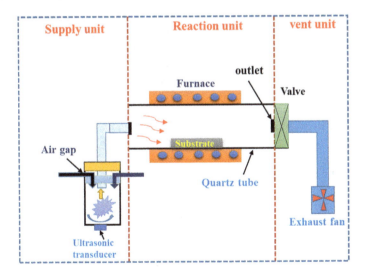

Fig. 9 Basic diagram of CVD using Cu substrate and gas discontinuous flow [85]

2.2.1 APCVD (Atmospheric-Pressure Chemical Vapor Deposition)

A synthesis process in which one or more volatile precursors are sprayed onto the substrate at atmospheric pressure and react or break down on the surface to form a deposit. Due to its extremely low diffusivity coefficient, air pressure chemical vapour deposition (APCVD) is one of the best syntheses and a crucial stage in the production of graphene-based devices. Many other types of technology, including solid-state electronic devices, are beginning to appreciate the benefits of high-temperature APCVD procedures (Fig. 10) for the manufacture of thin films. The application of graphene-based goods in optoelectronics, flexible electronics, and energy harvesting is particularly promising, especially for metal oxide semiconductor (MOS) transistors [86, 87].

Aimed at the large-scale manufacture of good quality graphene on a variety of metallic substrates, including Pt, Ir, Ni, and Cu, APCVD has been regarded as the most promising and reasonably priced method [88, 89]. Due to its low carbon solubility, controllable surface, and affordability for manufacturing monolayer graphene, Cu is particularly regarded as the ideal option [89, 90]. There have been numerous attempts to produce large-scale single crystal graphene with the fewest grain boundaries imaginable. Here, two typical approaches to realization: the first technique focuses on the expansion of a single domain with potential. In spite of the cm-scale provinces that have been attained, this process is not the best one for large-scale growth in practice because it is difficult to control the quantity of nucleation seeds and because the self-limiting growth factors are unclear, which makes the situation worse and necessitates a very long growth time (over 24 h) [91]. In order to produce uniform single-crystalline graphene, the second method entails aligning the graphene domain orientations on any substrate and then combining them atomically [92]. This seems to be ideal for cultivation on an industrial basis.

The creation of single-layer graphene (SLG) or floating-layer graphene (FLG) on any substrate by the use of a chemical synthesis technique called APCVD is possible. Precursors of graphene are deposited chemically under atmospheric pressure under precise reaction environments [93]. Because of its adaptability, APCVD can be used in complexly mixed homogeneous gas-phase and heterogeneous surface processes. Interacting chemicals increased and the resulting homogeneous nucleation

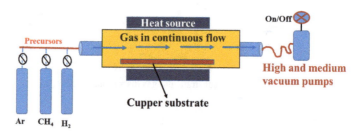

Fig. 10 Graphene fabrication by APCVD utilizing a Cu substrate and discontinuous gas flow is shown schematically [89]

grew more prominent when the partial pressure and/or temperature of the homo-geneous gas-phase processes rose. This homogenous nucleation must be reduced to generate high-quality graphene. An eight-step generic process for growing a uniform, extremely crystalline layer of graphene on the surface of catalytic metal substrates using APCVD consists of the following eight steps: (1) Mass transfer of the reactant, (2) film precursor reaction, (3) gas molecule diffusion, (4) elimination of byproduct (5) substrate diffusion, (6) surface reaction, (7) desorption of product, and (8) precursor adsorption [90].

2.2.2 Low-Pressure Chemical Vapor Deposition (LPCVD)

System pressures used for LPCVD deposition typically vary from 0.1 to 10 Torr. The reader will likely remember that this application is categorized as a medium vacuum one. Reactor configurations that have been used for LPCVD thin film processes include vertical flow batch reactors, single-wafer reactors, and resistance heated tubular hot-wall reactors [94]. A schematic of a horizontal hot-wall tube reactor used for LPCVD processing can be found in Fig. 11 [95].

These reactors were particularly effective at depositing silicon nitride, LPCVD silicon dioxide, and polysilicon thin films and could treat 100 or more wafers at once. The silicon dioxide, low temperature oxide, LTO, to silicon nitride LPCVD reactions require temperatures between 425 and 740 °C, with some processes occa-sionally functioning at temperatures greater than 800 °C (high temperature oxide, silicon dioxide, HTO). Electrical resistive heaters are used to reach these process temperatures [96, 97]. In this set up, the reaction chamber is fed the precursor gas either from an entrance at the tube's front or through a series of lengthy injector tubes that run the whole length of the wafer boat. Both of these gas inlet arrangements, nevertheless, cause issues for the user. If the precursor is only injected into the tube at the top, it will react with the surfaces of the wafer and other pieces of equipment as it travels down the tube, lowering the concentration of the precursor in the gas phase. The wafers near the front of the tube have thicker deposited films as a result of greater precursor concentrations. As a result, the film thickness profile for an

Fig. 11 A graphical representation of horizontal hot-wall LPCVD reactor [95]

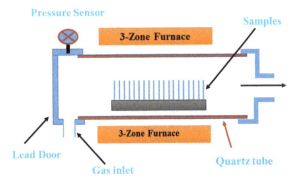

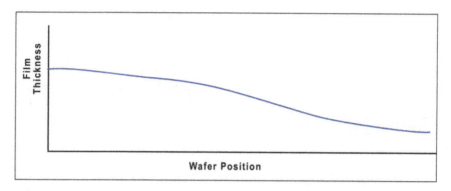

Fig. 12 Wafer load film thickness profile during LPCVD at constant temperature without precursor injectors [99]

LPCVD process in which a constant temperature is maintained over the wafer load and the precursor is input at the door resembles that depicted in Fig. 12. The depletion problem might also be resolved by using temperature ramps over the wafer load (increasing temperatures), or by using gas injectors in lower temperature operations [98, 99].

These solutions, however, failed as device designs shrank in the 1990s and the early 2000s because they caused unacceptably wide variations in the properties of the films (between films deposited at low and temperature) or, in the case of injector use, problems with reproducibility and expensive maintenance [98]. Additionally, when processing devices at lesser design instructions, the batch hot-wall techniques' relatively high time-at-temperature load could no longer be sustained. Last but not least, several hot-wall LPCVD techniques demanded "caged boat" wafer carriers (In situ doped polysilicon and doped TEOS oxide,), which formed particle levels which were unacceptably high for processing sophisticated devices. The majority of fab environments no longer use horizontal hot wall systems due to these and other factors [100].

A number of alternative equipment designs for LPCVD processes have been investigated as a result of the shortcomings of standard LPCVD equipment in advanced device manufacturing. Vertical flow small batch systems, Cold-wall and cold-wall, inductively linked systems have attained some acceptability and are still in use today. Modern fabs have primarily switched to the use of single wafer cluster instruments for CVD and other processing requirements due to their demonstrated advantages in effective particle, process control wafer handling and process integration [101].

2.2.3 Plasma-Enhanced Chemical Vapour Deposition (PECVD)

Chemical vapor deposition techniques like plasma-enhanced chemical vapor deposition (PECVD) are used to produce thin coatings on a substrate that transition from a gaseous to a solid state. Following the formation of a plasma of the interacting gases, chemical reactions are involved in the process. Radio frequency (RF) direct current (DC) or alternating current (AC) discharge among two electrodes, the gap filled with the reactive gases, is typically how the plasma is produced [102].

A plasma is any gas in which a significant percentage of the atoms or molecules are ionised. With typical capacitor discharges, fractional ionisation in plasma used for coating deposition and interrelated materials processing. While inductive plasmas and arc discharges can start at atmospheric pressure, processing plasmas are normally operated at pressures of a few millitorr to a few torr [103]. Because of how inefficiently electrons and neutral gas exchange energy due to the electrons' relative lightness in comparison to atoms and molecules, plasmas with low fractional ionisation are very interesting for the processing of diverse materials. As a result, while the neutral atoms remain at room temperature, the electrons can be kept at very high equivalent temperatures of tens of thousands of kelvins, which is equivalent to several electron-volts of average energy. These energising electrons can trigger a variety of reactions, such as the detachment of initial molecules and the production of significant amounts of permitted radicals that would otherwise be highly implausible at low temperatures [104]. The graphical representation of typical PECVD reactor revealed in Fig. 13.

Because electrons are more moveable than ions, deposition within a discharge has a second advantage. The plasma is therefore often more positive than anything it comes into touch with since, in the absence of this, a significant number of electrons flow from plasma to object. A tiny sheath region normally contains the voltage differential between the plasma and the objects. Ionized atoms/molecules that wordy to the sheath region's edge are driven in the direction of the surrounding surface by an electrostatic force. Thus, energetic ion bombardment occurs on any surfaces that are in interaction with the plasma [105, 106].

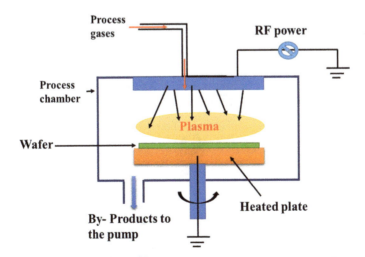

Fig. 13 Graphical view of a typical PECVD reactor [105]

The floating potential is normally only 10–20 V, although modifications to reactor geometry and configuration can result in substantially larger sheath potentials. Thus, during the deposition process, films may be exposed to intense ion bombardment. By removing impurities and increasing the density of the film, this bombardment can enhance the mechanical and electrical properties of the film. The ion density in a high-density plasma can be so high that the deposited film sputters significantly, which can be exploited to assist planarize the film and fill trenches or holes. Films are frequently deposited conformably (covering sidewalls) and onto wafers containing metal layers or other temperature-sensitive features using plasma deposition in the semiconductor manufacturing process. In comparison to deposition with sputtering and thermal/electron-beam evaporation, PECVD also fabricated some of the quickest deposition speeds while preserving film quality (such as roughness, defects/voids), frequently at the sacrifice of homogeneity [106].

2.2.4 Laser-Enhanced Chemical Vapor Deposition (LECVD or LCVD)

A modified CVD procedure for the production of thin films is laser chemical vapour deposition (LCVD). An LCVD device consists of a chamber with reagent gas inlets. Reagent gases can thus break down when heated by a concentrated laser beam to create metallic or ceramic films. An LCVD device can enable patterning and writing by the movement of the focussed laser beam relative to the substrate if it is equipped with a local laser heating system. A thin solid film and a by-product are produced at the end of the process, which primarily involves the introduction of two reagent gases into the chamber. The schematic view of an LECVD chamber is shown in Fig. 14 [107, 108].

Fig. 14 The schematic view of a LECVD chamber [108]

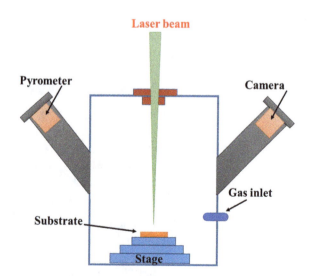

The limited quantity of material that can be transported to the area of interest by using gaseous metal–organic precursors limits the capacity of LCVD to form patterns over an standing structure. As a result, in order to prevent contamination and dilution of the source precursor [108]. It must also be performed in a vacuum. In LCVD, the precursor gas can decompose by thermal or photoactive (photolytic) processes (pyrolytic LCVD). It is possible to make sub-micrometer 3D structures in any case utilizing a variety of substrates and materials (mostly metals, semiconductors, and some oxides). Although commercial uses for additive repairing of device arrays for flat panel displays have been developed, LCVD's use in other micro-AM applications has been, at best, minimal [109].

An LCVD process is often divided into two subcategories:

Photolytic LECVD: Here, the reagent gases absorb the concentrated laser beam's energy, which causes the gas molecules to break down and produce a thin solid coating on the substrate. Furthermore, given that gas molecules absorb energy during the photolytic LCVD process, laser wavelengths should be purposefully chosen to rely on the materials. UV lasers like KrF, Ar^+, and ArF are frequently used in the procedure.

Pyrolytic LECVD: The laser beam is focused on the intended deposit areas throughout this operation. As a result, a thin solid film is coated on the substrate as the temperature locally rises on the substrate until it meets the necessary threshold. Nd:YAG and CO_2 continuous-wave infrared lasers are frequently used in the procedure.

In the photolytic LCVD process, the laser absorption may expand along the laser beam rather than remaining confined to the beam focal point, resulting in a loss of resolution and an increase in thickness and width. However, because the heating in the pyrolytic LCVD process is more confined, it allows for better resolution (up to 5 μm) [110, 111].

2.3 Confined-Plume Chemical Deposition Technique

The preparation of known highly hard, very hard, or ultra-incompressible ceramic compositions as microcrystalline coatings in one step on either inorganic/organic provisions is stated. By using a mid-infrared pulsed laser to irradiate pre-ceramic chemical precursors sandwiched between IR-transmissive hard/soft supports under temporal and spatial confinement at a laser wavelength resonant with a precursor vibrational band, it is possible to deposit crystalline ceramic coatings in one step without significantly damaging the support material's thermal properties. The final crystalline ceramic product is formed by confined-plume, chemical deposition (CPCD) observed in Fig. 15, which is initiated by reaction plume development at the laser/precursor/beam crossing point. By moving the laser beam in a raster pattern across a sample specimen, continuous ceramic coatings are created [112, 113].

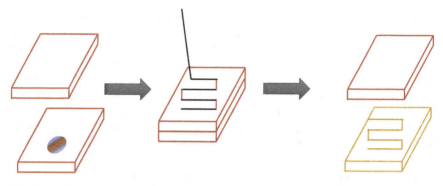

Fig. 15 An illustration of the confined-plume chemical deposition method [112]

2.4 Dip Coating

Due of its simplicity and convenience, dip coating has been widely used for research objectives. However, the coatings made in this manner are of variable quality; as a result, this method is unsuitable for industrial procedures [114, 115]. Figure 16 displays the crucial variables that can impact the coatings created by dip coating. Typically, the five steps of the dip coating procedure are as below [115]:

1. *Immersion*: The substrate materials are dipped into the coating solution at a consistent speed. Prior to this stage, a pre-treatment process would be carried out depending on the type of substrate.
2. *Start-up*: After a predetermined amount of time, the substrate begins to be taken out of the solution.
3. *Deposition*: The coating layer has a slower draw and is thinner. Whereas it is being removed, the thin film coating starts to be placed on the substrate. How quickly the substrate is being dragged out has a direct impact on the coating's thickness.

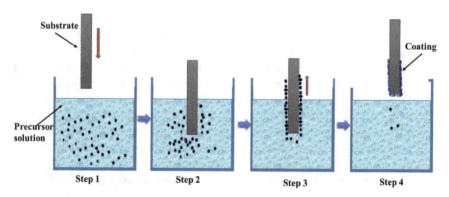

Fig. 16 Process flow diagram for continuous dip coating [116]

4. *Drainage*: This process involves draining extra liquid from the substrate's surface.
5. *Evaporation*: A thin film of solvent is forming as it begins to evaporate from the substrate's surface. This process could occur in step 3 if the solvent is volatile.

This method is more-easier and more convenient than the alternatives because it doesn't require complicated equipment. Due to the ease of the process, the coating layer that is created could not be of high quality. The thickness distribution may be a problem for coatings made of composite materials based on graphene. The created coating layer for pure graphene-based material coatings might not be sufficiently dense to exhibit better capabilities, and the applied substrate cannot be incredibly large and complex. Although dip coating is better suitable for usage in the lab, it can nevertheless be used on a wide scale to make items that meet low standard criteria at a reasonable price. This process is more suitable to create this kind of coating because graphene-based composite coatings have substantially greater viscosities than pure graphene-based coatings, resulting in stronger interfacial adhesion in the direction of a substrate as well as an added even coating layer. It might be necessary to do another treatment step to create a firm coating layer [116].

2.5 Decorative and Barrier Coatings

2.5.1 Decorative Coatings

Modest metallization of polymer coating films and sheets led to the first decorative coatings for packaging, which gave the web a shiny metallic appearance. This shiny metallic web made it possible to use a material that stood out from the competition when packaging, boxing, or labelling products. As with any products that are in competition, as soon as one manufacturer created a brighter box or label, others hurried to imitate it. Variants were created in an effort to distinguish products as more companies began to use metallic effects in their packaging designs [117]. A dyed substrate material was used to create colored metallic coatings instead of the bright metallic silver reflecting substance, or the metallic coating was covered with a transparent colored lacquer to provide colour. The most important development has been the development of embossed holographic substrate materials, which can also be produced in silver or colourful metallic versions [118].

The most well-known colour of vapor-deposited hard coatings, such as ZrN, TiN, and a big amount of ternary or multicomponent composites of the transition, is the golden hue. Such coatings combine their attractive qualities with their great wear resistance and better corrosion resistance when used for decorative reasons. In this study, the corrosion behaviour of coating/substrate structures and their testing procedures are covered after discussing colours and how they depend on composition and process parameters. In the event that less noble substrate materials need to be coated, examples of decorative hard coating applications and particular coating sequences [118, 119].

Initially, optical coatings were simply applied by evaporating the substance using an electron beam evaporation source or a resistant-heated source. The coatings created in this way are almost always nonstoichiometric. The reactive evaporation method was then applied to enhance the coatings' stoichiometry and, consequently, their refractive index. The development of plasma-enhanced evaporation methods was the next step in improving stoichiometry control. The activated reactive evaporation process, which was previously addressed, is one such method that has been utilized to coating a number of optical films. TiO_2, ZrO_2, and HfO_2 high-threshold optical films have been created [118, 120].

2.5.2 Barrier Coatings

The converting industry is interested in a wide range of products, together with transparent glass-barrier coatings, solar cells, antiabrasion/antireflective coatings for windows, and packaging film and piece of sheets, thanks to developments in source and materials technology for vacuum, high-speed, and web-coating operations. Plasma polymerization and plasma deposition processes can be used to produce plastic films with glass-barrier coatings, creating new business potential for the coatings sector. The physical properties of antiabrasion/antireflection coatings on plastic film for glazing applications should be as follows: a refractive index between 1.3 and 1.4 (the lower the better); If the materials is multilayered or has a graded refractive index, at least the top layer must have a thickness of 0.25 visible light wavelength, a hard surface with a low coefficient of friction, and weather and pollution resistance. The required material, coating grade and material structure are taken into consideration when designing the source and deposition environment. Three plasmas are used to create high-speed barrier and anti-abrasion/antireflection coatings from oxides: one plasma to ionize a sizable section of the deposition material, two plasmas are used: one to anneal the coating to produce tighter packing and one to clean the substrate next to the source and provide the polar surface needed for good nucleation and adhesion [121, 122].

High vacuum is typically used throughout the coating process to prevent background gases from interacting with the substance being coated. High vacuum aids in preventing photon and electron heating of the substrate. Well-nucleated, low-lateral stress oxide barriers can be processed by web-handling machine rollers because they have a 5% elongation. Even though the coatings are amorphous, poor nucleation results in non-barrier, columnar formations. The pretreatment of plasma with helium or methane, the barrier characteristics of plasma-activated CVD of SiO_x on polyester can be increased by an order of magnitude. Although corona treatment at atmospheric pressure is still necessary, the pretreatment is best carried out in-line [123, 124].

Tetraalkoxy-titanium compounds are the best choice for applications needful a high refractive index since they are nonvolatile, nontoxic, and can be used in place of siloxanes to create titanium-dioxide micro/nano coating used for the initial material to provide intermediate refractive indices. These procedures have three advantages over coating of electron beam: they can use a larger range of starting materials and reaction conditions; they are conformed as they are produced at relatively high pressures; and they are less expensive. As a result, a wider range of chemical-bond structures can be created in the coating; alternatively, the coating can be adjusted for the polymer's flexibility or the oxides' hardness; thirdly, because the plastic being coated is under much less heat stress, cooling occurs more quickly; and finally, these processes put less heat stress on the plastic being coated. Adatoms generate liquid particles in the gas phase that are small and intense enough to prevent "snow" formation, energetic enough to allow adatom movement for annealing, and large enough to significantly decrease the heat of strengthening load on the coating surface [125, 126].

2.6 Cathodic Arc Plasma Deposition (CAPD)

Sputter ion plating, evaporative ion plating, and cathodic arc plasma deposition (CAPD) are all members of the same family of ion plating procedures as the cathodic arc plasma deposition (CAPD) deposition technique of thin film [127]. The cathodic arc plasma deposition method uses deposition species that are intensely ionized and have higher ion energies than conventional ion plating techniques [128]. The ion plating techniques have been industrialized to take benefit of the special features of the process development and to satisfy the requirements for specialized coatings, such as wear resistance, excellent adhesion, decorative capabilities and corrosion resistance. After demonstrating its extraordinary performance in cutting tool applications, cathodic arc technology is now finding much broader applications in the coating of erosion resistance, ornamental coatings, architectural, corrosion resistance and solar coatings [129]. Schematic diagram of CAPD apparatus revealed in Fig. 17.

The source chamber, one or more vacuum arcs, cathode, arc power supplies, anode arc, ignitor, and substrate bias power supplies all contribute to the material

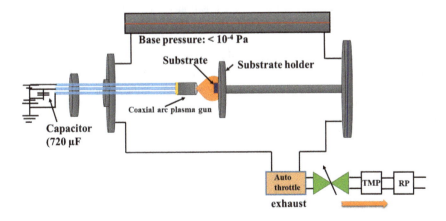

Fig. 17 Schematic diagram of CAPD apparatus [130]

evaporation during the CAPD process. Depending on the source material, arcs are maintained at voltages between 15 and 50 V with typical arc currents between 30 and 400 A. The number of arc spots that form on the cathode surface when high currents are employed depends on the cathode material. This is demonstrated by the random movement of spots on the cathode's surface, which frequently occur at rates of tens of meters per second or faster. External factors including electrostatic fields, magnetic fields and coating gas pressures can also affect the arc spot's mobility and speed. By way of the arc spot moves over the cathode surface, a series of quick flash evaporation events remove materials from the source. With the proper boundary shields and/or magnetic fields, arc spots that are continual due to the material plasma produced by the arc itself can be managed [130, 131].

2.7 Spin Coating Method

The method of spin coating is frequently used to deposit thin films to substrates. High-speed spinning of a material and solvent solution results in an even coating because of centripetal force and the liquid's surface tension. Spin coating produces a thin film with a thickness of a few nanometers to a few microns after any remaining solvent has evaporated. Spin coating is employed in a vast array of technological fields and industrial sectors. Spin coating's main advantage over other techniques is its capacity to swiftly and easily manufacture extremely uniform films. Spin coating is widely utilized in organic electronics and nanotechnology, building on many of the methods employed in other semiconductor industries. However, significant variations in approach are necessary because to the relatively thin layers, high uniformity needed for optimal device preparation, as well as the need for self-assembly and organization to take place during the casting process. The spin-coating technique is used to create thin, homogeneous films with thicknesses between micro- and nanometers. The sample is rotated while the substrate is positioned on a chuck, and the centrifugal force propels the liquid sharply outward. The key factors contributing to the flat deposition on the surface are viscous force and surface tension. Lastly, evaporation results in the development of the thin film. Fluid dispenses, spin up, steady fluid outflow, spin off, and evaporation are some of the stages that make up spin coating [132].

Spin coating has the benefit of producing very fine, thin, and uniform coatings; nevertheless, it has the drawback of making large area samples problematic [133]. The desired film thickness can be obtained using the spin-coating technique. The following Eq. (2) illustrates how these parameters determine thickness and how they alter the thickness of the layer [134, 135]:

$$h = \left(1 - \frac{\rho_A}{\rho_A}\right) \cdot \left(\frac{3\eta m}{2\rho_{A0}\omega^2}\right)^2 \tag{2}$$

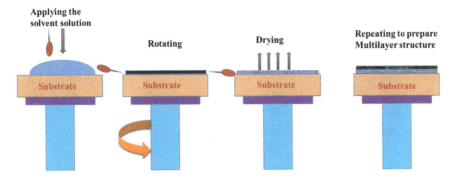

Fig. 18 Graphical diagram of spin-coating method [137]

where, h is thickness, ρ_A is volatile liquid density, is solution viscosity, m is evaporation rate, and ω is angular speed. An easier Eq. (3) has been proposed as follows because the evaporation rate is determined experimentally [136]:

$$h = A\omega^{-B} \tag{3}$$

B is a constant that was determined empirically. Most of the time, B is in the range between 0.4 and 0.7. This equation makes it obvious that the film will be thinner the faster the substrate rotates [137]. Figure 18 displays a schematic diagram of the spin-coating technique.

2.8 Plasma Electrolytic Oxidation (PEO) Process

Specifically for light metals, the plasma electrolytic oxidation (PEO) Process creates thick, dense metal oxide coatings with the goal of enhancing corrosion and wear resistance. The coating produced by the PEO technique is rather superior to that produced by standard anodic oxidation. Mechanical, petrochemical, and biological sectors, to name a few, are only a few that use it extensively. Numerous studies have been conducted to examine the coating performance created by the PEO method [138].

PEO is a high voltage electrochemical technique that includes creating a plasma discharge at the metal-electrolyte interface. This hardens and dandifies the substrate surface into a ceramic oxide layer, protecting it from thermal expansion damage. When the voltage applied is greater than the "breakdown value," plasma discharges happen (typically several hundred volts). A rapidly expanding area of surface engineering nowadays is plasma electrolytic oxidation, mainly for the treatment of Ti, Al and Mg alloys. Figure 19 depicts the schematic for the PEO processing cell [139, 140].

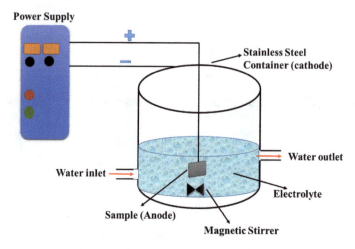

Fig. 19 The graphical representation of the PEO processing cell [140, 141]

In a number of ways, the PEO procedure is drastically dissimilar from the conventional anodising procedure. The following are the process's primary attributes:

- The method uses alkaline electrolytes, whose composition is crucial for both the speed of coating and the characteristics of the resulting anode layer. These electrolytes are suitable for the environment.
- High voltage alternating current is used in the process. Because of the high voltage, a micro-plasma forms around the electrodes and diffuses into the anodic layer of the aluminium substrate, where it reacts and forms more anodic layers.
- As the voltages are greater than the breakdown voltage of the layer created, open channels are not required to maintain the process, allowing for the formation of thick, dense, non-porous layers.
- The technique has a very high productivity since it uses controlled high voltage power.
- Maintaining the bath temperature precisely is not required. In reality, temperature changes of up to 10–20 °C can still provide excellent coatings, substantially simplifying the procedure.
- The content of the electrolyte varies greatly liable on the kind of coating. It is possible to alter the pace of growth of the oxide as well as contaminate it with substances that give the oxide extra qualities depending on the reagents used (Biocidal, self-lubrication properties etc.) [141].

Due to the good density of the coating, there are almost no fluctuations in the anodized item's dimensions, and a fully completed part can be coated without the need for any significant post-finishing procedures. About 15% of the exterior layer created by the PEO procedure is a soft outer coating that may be removed by polishing or grinding, leaving an exceptionally hard ceramic layer behind (2000HV) [142].

2.9 Micro-arc Oxidation (MAO) Coating

Concerning the makeup of coating layers, the MAO technique is regarded as a flexible coating method. Figure 20 depicts a schematic of the procedure. In general, MAO creates micro-arcs as plasma channels by using a large voltage differential between the anode and cathode. Contingent on the strength of the micro-arcs, a piece of the surface melts when these arcs strike the substrate. When coating materials are deposited in the working electrolyte on the substrate surface, plasma channels release some of their pressure at the same time. Oxides are produced and deposited on the surface of the substrate materials as a result of an oxidation reaction started by the oxygen that already exists inside the electrolyte. This technique is diverse due to the adaptability of incorporating anticipated substances as a solute in the working electrolyte. Al, Mg, Ti, and their alloys are currently the materials that are most frequently coated with MAO [143, 144]. The most significant quality of an MAO-treated layer is high corrosion resistance. Additionally, because it has a porous nature, the coating layer on biomedical implants and fixations promotes high levels of bone ingrowth [145, 146].

The coated surface of MAO has great hardness and adhesion qualities, and it has varying levels of porosity throughout its structure. The coating itself is where this kind of multi-structural nature originates. A surface that has been treated with MAO at various frequencies produces porous structures with various porosities. The substrate is covered with a barrier inner layer, a solid layer of metallic oxides, during the initial coating processes. The subsequent coating stages result in the creation of the top layer

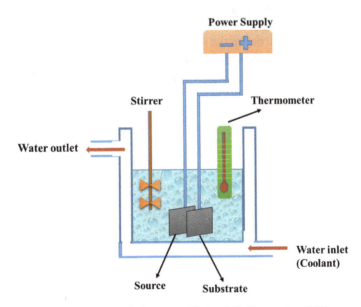

Fig. 20 Schematic representation of micro-arc oxidation (MAO) procedure [143]

structure is porous, which has a claimed thickness of up to 100 μm [147]. In bio applications, surface adherence is improved because of its porous structure. Current density, voltage, process duration, electrolyte type, pulse current as well as AC/DC types of current, are the factors determining coating quality [146, 148]. However, a number of researchers used a variety of treatment parameter ranges, and it has been asserted that in every study, coated samples' corrosion characteristics increased while metallic ion release was noticeably reduced [77]. The MAO process's major drawback would be its restriction to substrate coating materials, which are primarily valve metals like Al, Ti, Ta, Nb, Zr, and Mg [149, 150].

2.9.1 Electrodeposition Coating Techniques

By delivering a current through an electrodeposition, electrochemical cell is a conventional electrochemical process that uses a redox reaction to create a thin, homogeneous substrate coating electrode [151]. High-valence/Low-valence (metallic) materials are deposited by reduction or oxidation at the substrate, respectively. Different types of electrode materials, such as transition metal oxides, transition metal sulphides, metals/alloys and conducting polymers, have been successfully created using electrodeposition processes. Thin films for thermoelectric purposes have frequently been created via electrodeposition [100]. Nanowires can also be made by electrodeposition (graphical representation Shown in Fig. 21) using the template [101, 152]. Direct current (DC) and pulse electrodeposition are the two main categories of electrodeposition. The thin layer is probably going to develop a dendritic structure from DC electrodeposition, which will result in an uneven surface. This problem can be effectively solved by pulse electrodeposition. There is ample time for the ions to distribute on the surface of the electrode and for crystallization process at the alloy surface during the off (resting) phase.

Materials are thought to be protected through electrodeposition by depositing metallic ions on a substrate. Ions transfer in the unit cell during this phase as a result of the difference in potential between the anode and cathode poles. The submerged

Fig. 21 Simple electrodeposition system [157]

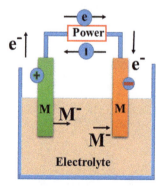

sample eventually develops a coating layer by pleasing in ions from the other electrode over time. Popular electrodeposition materials have been the subject of extensive research [153]. Co/Pt, Ni–P–W, Ag/Pd, Cu/Ag, Ni–P/Sn, Cu/Ni, Co/Ag and Ni–P are among the common metals that have undergone extensive study [25]. These tests demonstrate that the electrodeposited layers pointedly improve the corrosive resistance properties of the metal surface. It has also been shown that this technique works well for making superhydrophobic polymeric coatings like polythiophene [154]. Transition metal oxides, such as NiO, CuO, Co_3O_4, and $NiCo_2O_4$, are obtained through annealing in cathodic electrodeposition by electrochemical reduction of NO_3 [155]. Anodic electrodeposition is the process of precipitating high-valence oxides or oxyhydroxides (such as $FeOOH$, $NiOOH$, etc.) by oxidising transition metal ions as of a solvable lower oxidation state (like Nb_2O_5, Fe_2O_3, etc.) [156].

Below is a description of two commonly used electrolytic deposition (ELD), electrodeposition techniques and electrophoretic deposition (EPD).

2.9.2 Electrolytic Deposition (ELD) Coating

ELD is an electrochemical procedure used to create a uniformly layered, dense metallic coating on conductive substrates. When installed inside an electrochemical unit cell, substrate and deposition materials are chosen for the cathode as well as anode. Figure 22 provides a broad overview of the procedure. Applying a potential difference between the anode and cathode poles causes metallic ions to move in the direction of the working electrolyte and then in the direction of the substrate. The super-saturation of the electrolyte, which is required for the deposition phase, is brought on by the charging current in the circuit. With this method, the number of metallic ions in the electrolyte stays constant while the coating is being applied [158]. There have been rumors about the development of alternative applications for this technique, solid-oxide fuel cells, high-temperature, electronics, biomedical, and including optical, even though it is mostly utilized for ornamental and low-corrosion/wear applications. The MAO technique can be more closely compared to the deposition of ceramic materials on metallic surfaces by further raising the potential difference in electrolytic unit cells. According to Tian et al. [159], who coated steel pipes with Ni–Co–Al_2O_3, the corrosion of substrate exposed to oil sand slurry was noticeably accelerated. Yang et al. [160] found considerable corrosion and corrosion resistance that was strengthened by erosion after depositing Ni–Co–SiC on carbon steel pipes that were exposed to oil sand slurry. On a mild steel substrate that was coated in Zn–Ni–Al_2O_3, Fayomi et al. [161] observed the same outcomes. Additionally, Redondo et al. [162] created a corrosion-resistant polypyrrole (PPy) layer on a copper substrate using a dihydrogen phosphate solution.

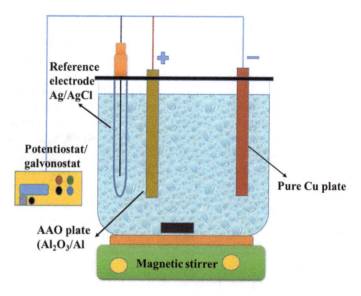

Fig. 22 An electrodeposition arrangement for copper metal particles on aluminium oxide is shown schematically [25]

2.9.3 Electrophoretic Deposition (EPD) Coating

EPD is a different sort of electrodeposition that creates thicker layers of colloidal coating. By coagulating colloidal particles in a unit cell with an electric field similar to an ELD, thin films are produced on surfaces shown in Fig. 23. EPD is a multi-phase method in which electrolyte particles suspended in solution are propelled toward a single electrode by an external electric field. A bigger coagulated particle is formed when the moving particles congregate in one electrode. The larger particles settle on the electrode's surface, which is a substrate that needs to be coated.

Finally, a thick coating layer with a powder-like structure will be built up on the substrate. A schematic of the EPD process's operation is shown in Fig. 5. Densification techniques (such as furnace curing, light curing, sintering, etc.) are suggested to raise the protective layer's quality. There have already been several EPD applications introduced, including coating, selective deposition, gradated material deposition, porous structure deposition, and biological applications [163]. Typically, borides, carbides, oxides, phosphates, and metals are employed as materials in EPD. Sol–gel and electrophoretic deposition (EPD) were used to create corrosion-resistant coatings on stainless steel AISI 304, and Castro et al. [164] observed two and four times increases in corrosion resistance for each of these techniques, respectively. An AISI 316 L stainless steel was coated with chitosan for biomedical purposes in another investigation by Gebhart et al. [165] They noted beneficial impacts of this coating on the substrate's corrosion behaviour. Additionally, they claimed that the main element influencing coating characteristics including hydrophobicity, thickness, and structure is the applied electric field in EPD. Chen et al. [166] treated TC$_4$ Ti-alloy

Fig. 23 Electrophoretic deposition process illustration [25, 169]

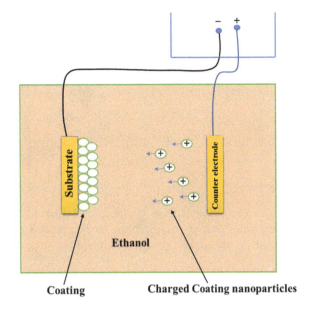

orthopedic implants with graphene. They asserted that the synthetic joint implants coated in graphene have considerably longer lives. They found that any corrosion that occurred on substrates was caused by microcracks in coating surfaces. SiC particles were successfully deposited on paper-based friction materials as a consequence of Fei et al. [167] work into the wear resistance of EPD coatings, considerably enhancing the material's resistance to wear. Table 1 lists the characteristics and parts of the ELD and EPD procedures.

Charged particles travel toward the opposing electrode during electrophoretic deposition (EPD), deposit, and procedure a uniform coating film. It offers the ensuing advantages: it is a quick process, necessitates basic equipment, allows for substrate selection flexibility, and emits little to no environmental impact. Additionally, it can be changed to meet specific requirements, such as a special application technique for complex geometries with good control over the film thickness and shape. According to the electrodes and energetic particles that are deposited, there are two different types of EPD coatings [170], as depicted in Fig. 24. In the cathodic EPD, positive

Table 1 Techniques for electrodeposition characteristics [168]

Property	ELD	EPD
Electrolytic conductivity	High	Low
Surface charge	Medium	High
Approximate rate of deposition	0.1 μm/min	1000 μm/min
Ionic electrolytic strength	High	Low
Coating elements	Ions	Solid particles
Preferred electrolyte	Water	Organic

Fig. 24 Schematic of the electrophoretic deposition process [172]

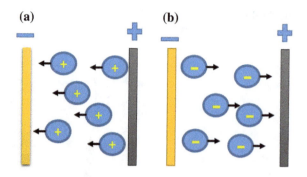

particles deposition happens on the cathode (negative electrode), whereas negative nanoparticles coating occurs on the anode (positive electrode). The driving forces behind EPD are particle charges and mobilities in aqueous medium with an electric field [171].

The uses of EPD have grown over time. It has been used in new composites, fuel cells, microelectronic devices, antioxidant ceramic coatings, bioactive coatings and purposeful films in addition to typical applications, such as automotive primers. The EPD process has the advantage of producing highly anticorrosion coatings in a cost-effective and environmentally responsible way. In the auto industry, it has a crucial role in corrosion protection. Several examples of the numerous applications for this technique include the thick silica film, superconducting films, hydroxyapatite coating, luminescent materials, gas diffusion electrodes and sensors, nanosize zeolite membrane, carbon nanotube film, ceramic particles onto fabrics, and microgels [173].

2.9.4 Thermal Spray Coating Technique

The term "thermal spray coating" refers to a group of procedures where a set of designed materials are melted using a plasma, chemical combustion and electric heat source the melted materials are then sprayed onto the surface to create a protective layer. These sorts of corrosive and wear-resistant coatings are dependable ones. Using this method, a high-speed jet is used to spray the components onto the substrate after heating them to a molten or semi-solid phase. Typically, plasma discharge or chemical combustion is employed as the heat source for this. The thickness reached with thermal spray coating methods can be anywhere between 20 m to several millimeters, which is a significant improvement above the thickness provided by PVD, electroplating processes or CVD [174]. Additionally, a relatively large surface area of a substrate can be easily covered by a thermal spray coating using a variety of materials, including refractory metals and metallic alloys, polymers, ceramics and composites [174, 175]. A material's performance in industrial applications is influenced by its surface characteristics. It is feasible to affordably replace a poor base material with a coating that has higher surface qualities and performance since surface modification technologies are so readily available. High-entropy alloys (HEAs) are a brand-new class of materials with unique traits and properties that are showing promise as thermal spray coatings for extreme conditions. Surface modification is very advantageous in conditions involving HEAs. Due to reports of their remarkable qualities in both bulk and coating forms, HEAs are increasingly being used as feedstock for coating processes. Recent findings that thermal sprayed HEA coatings outperformed traditional materials have sped up research in this area [176]. Based on their properties and process requirements, thermal spray coatings are divided into various categories. The most common categories include plasma, warm/cold, wire arc spraying detonation, high-velocity oxyfuel (HVOF), high-velocity air fuel (HVAF), and flame [175].

2.9.5 HVOF (High-Velocity Oxy-Fuel Coating)

An HVOF coating procedure is shown schematically in Fig. 25. In a specially built combustion chamber, a mixture of fuel such as propane, hydrogen, acetylene, methane, or oxygen, and natural gas in gas or phase of liquid burn continuously to produce a high-pressure vapor of hot gas. The ignition products are discharged from the combustion chamber through a nozzle to produce a spray that travels at a rate of more than 1000 m/s. Following combustion, powdered coating ingredients are fed into this hot jet stream to speed their partial melting as they exit the nozzle tip. The semisolid particles are forced against the substrate by the hot jet, which forms a coating layer with various thicknesses of up to several millimetres. The benefit of this method is that corrosion- and wear-resistant layers can be deposited using coating materials like hydroxyapatite (HA), Zr, W, Al, Cr, and their carbides and oxides, or polymeric materials like nylon 11/silica nanocomposites [177, 178]. Additionally, the coating layer is highly dense and adheres effectively to the substrate. HVOF offers a layered coating, seen in Fig. 25. The coating layer could be applied to non-conductive materials like polymers and ceramics that can withstand the high jet stream and particle velocity and temperature. Numerous studies have examined HVOF coatings' wear and corrosive resistance in diverse applications. These investigations showed that the coating layers created using this approach worked effectively and enhanced the corrosion-wear characteristics of substrates [179, 180]. Table 2 offers a summary of these studies.

2.9.6 Plasma Spray Coating

A graphic representation of a plasma spray coating apparatus is observed in Fig. 26. This procedure can be carried out in either an atmosphere of vacuum. A plasma gun is used in this method to produce a high-temperature DC/induction plasma (up to 10,000 K), which is capable of melting refractory metals, ceramics, and polymers with ease. Gas, water, or a combination of these two, known as hybrid plasma, are the components utilized to stabilize plasma. In this hot plasma stream, which is fed with the materials to be deposited, the high temperature melts the feedstock. The melted

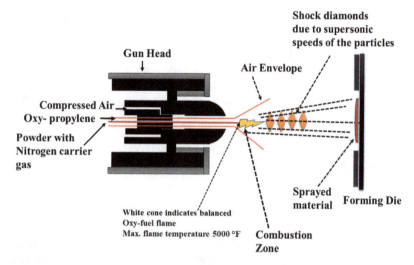

Fig. 25 Representation of setup high-velocity oxy-fuel (HVOF) coating system [181]

Table 2 HVOF provides electrochemical corrosion measurements for various coating compositions [182]

Coating configuration	Corrosion amount (mm/year)		
	0.1 M NaOH	0.1 M H$_2$SO$_4$	Sea water
WC–Co–Cr	–	–	0.32
Cr$_3$C$_2$–NiCr	0.17	0.077	–
WC–Co	–	–	0.76
Cr$_2$O$_3$–Al$_2$O$_3$–TiO$_2$	3.2×10^{-5}	3.6×10^{-5}	
WC–Cr$_3$C$_2$–Ni	0.38	0.15	
Cr$_2$O$_3$	7.6×10^{-3}	$\times 10^{-3}$	–

droplets are quickly fabricated on the substrate in opposition to the coating setup due to the rapid speed of plasma at the tip of a converging nozzle. This technique' adaptability enables the use of a variety of feedstock types, including powder, slurry, suspensions, and liquids [183]. Due to high temperature and surface tension, the resultant coating layer has a strong resistance to corrosion and wear and can stick to the substrate. Numerous investigations on various materials, including chromium oxide and NiCr alloys, indicated a considerable improvement in wear and corrosion resistance. For applications including electrical barriers coatings, antistick films for papers and rollers, wear-resistant surface coating for plastic mouldings, and corrosion protection for metal substrates, plasma spraying of polymers, particularly PEEK, has been employed (nylon, PVDF). Vacuum plasma spraying, then again, is a low-temperature procedure that is typically employed to change a substrate's surface using compounds that cannot react under atmospheric pressure [184]. Surface modification of engineering rubbers, fibers, metals, polymers and plastics are the most common

Fig. 26 Illustration of the plasma spray coating system and its components [187]

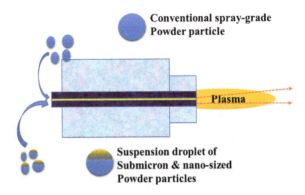

use of vacuum plasma spraying. A material may experience cross-linking, a reduction in friction, an increase in adhesion, etc. during this process [185, 186].

2.9.7 Cold Spray Coating

The method of cold spray coating depends on particle impact and solid mechanics. This procedure does not use a heat source to coat surfaces, in contrast to HVOF and plasma spray coating processes. Particle size, target temperature, coating material characteristics, and a critical velocity all play a role in the overall workings of cold spray coating. To provide the appropriate kinetic energy, powder materials are delivered into a stream of helium and nitrogen-filled high-velocity media. This energy deforms the particles during particle–substrate impacts and bonds them to the substrate. The penetration of the particles inside the substrate may be another mechanism for this process. The surface is coated with the preferred materials using a high flow rate of accelerated particles. Numerous types of polymers, metals, ceramics, composite materials, and metallic alloys are among the most used powder materials [188, 189]. Additionally, mild metals like Al and Cu are among the most researched substrate materials, yet reports of the coating of some hard materials like Ti and W can be found in literature. In several investigations, the accelerating medium's temperature has been raised to improve process effectiveness. Although this procedure is less expensive and simpler than other thermal spray procedures, it has a relatively small operational window. Figure 27 depicts a graphic diagram of the cold spray coating process [190].

2.9.8 Warm Spray Coating

A graphic illustration of the heated spray coating technique is revealed in Fig. 28. Low working temperatures reduce the effectiveness and dependability of thermal spray coating techniques, as was mentioned in relation to cold spray coating. High temperatures, however, melt the feedstock and bring about new chemical processes,

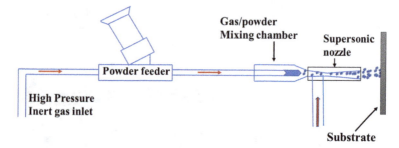

Fig. 27 An illustration of the cold spray coating method is shown in the above figure [191]

Fig. 28 An illustration of
the warm spray coating
process setup [25]

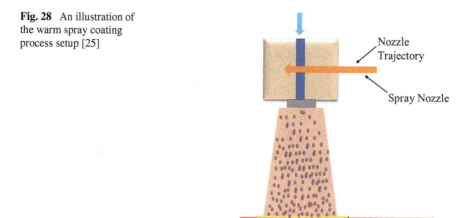

which may result in oxidation or a change in the substrates' or particles' physical characteristics. A novel method known as a warm spray coating was devised to address this issue. By adding nitrogen to the fluid mixture, this HVOF coating modification benefits from a lower combustion chamber temperature. As a result, this technique offers a high coating process efficiency and falls in between cold spray coating and HVOF coating [192]. However, due to the low temperature and the presence of oxygen in the accelerating stream, and as described in the literature, the created coating layer contains more contaminants than the coating layers produced by the other two procedures. These porosities and oxide phases are depicted in Fig. 28. Adopting this method has the advantage of covering materials such as bio-metal-glasses, Ti and its alloys, plastic products, and polymers like PEEK that easily oxidize at extreme temperature or materials that cannot survive high working temperatures [192, 193]. Numerous studies have examined the corrosion properties of cold/warm spray coatings on numerous materials, for instance Ti, bio-metal-glasses, WC–Co tungsten carbide, etc., under various corrosion conditions, and they have discovered an improvement in the substrates' corrosion resistance despite the fact that they are not frequently used in enormously harsh environments.

2.9.9   Arc Wire Spray Coating

Arc wire spray coating is a different kind of thermal spray coating method (Fig. 29). From this procedure, two metallic consumable wires that are charged with a DC source create an arc between them, melting the feeder wires in the process. The results of this melting process are then pumped with the given pressure of compressed gas out of a convergent nozzle tip and toward the target. Although this procedure is flexible enough to use a variety of metallic alloys as coating layers, it is only compatible with conductive wires and materials [194]. This problem was addressed by the introduction of an adapted kind of arc wire plasma with one consumable wire that creates an arc with a non-consumable metallic cathode. This process continues with the same remaining phases as the first iteration. This technique is well recognized for covering internal surfaces like engine blocks and other structures. While the internal surfaces are covered with a wear- and corrosion-resistant metallic alloy, it offers a lighter metal as the complete block. This adaptability can considerably lower production costs. Almost any conductor can be utilized in this process as the feedstock, including Ni, Mo, Al, Zn, and other metallic alloys such Ti and Ni alloys [195]. The literature has also recorded the use of cored wire. Figure 29 depicts the microstructure of a substrate that has been sprayed with arc wire. Up until this point, some of the well-known thermal spray coating techniques have been discussed. It is without a doubt possible to compare different techniques in terms of their underlying principles, the effectiveness and quality of the coating, the processing time, and the ease of application [196].

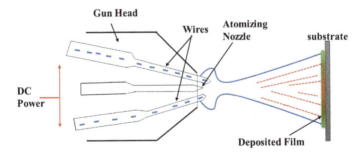

Fig. 29 Arc wire spray coating setup and working mechanism diagram [196]

3 Summary and Conclusions

An effective coating deposition on a substrate depends on a number of factors, including the feedstock form (rods, wire, precursors, powder etc.), the substrate material, the deposition process, and the deposition materials. The most crucial processes, however, involve deposition because they deal with the chemistry of materials and the alloying of compositional elements in the coating layer. Additionally, one can choose the optimal choice for deposition based on the properties of various feedstock and substrate materials. Physical/chemical vapour deposition (PVD/CVD), electrophoretic deposition (EPD), electrodeposition (including electrolytic deposition (ELD) and, micro-arc oxidation (MAO), dip coating, the Confined-Plume Chemical Deposition technique, and various thermal spraying processes are the most effective and extensively researched deposition methods (i.e., plasma, warm, cold, HVOF and arc wire spraying). Advance material selection is crucial to achieving the best coating efficiency because the procedures indicated use various mechanisms to deposit particular types of materials on surfaces. Some of the processes convert feedstock to liquids and semisolids in the form of particles, droplets, and clusters using thermal sources. Others deposit materials without a chemical change of state by using the difference in electrochemical charges between the poles. The thickness, microstructure, and usefulness of the coating layers vary depending on the feedstock, method, and substrate materials of deposition. Additionally, while the majority of techniques may deposit metallic alloys, polymers, and ceramics independent of their physical properties, some are specialized to metallic feedstocks, which are electrically conductive.

References

1. Streitberger H-J, Goldschmidt A (2018) BASF handbook basics of coating technology. European Coatings
2. Sozer N, Kokini JL (2012) The applications of nanotechnology. Chemical analysis of food: techniques and applications, pp 145–170
3. Fernando RH (2009) Nanocomposite and nanostructured coatings: recent advancements. ACS Publications
4. Sánchez-Romate XXF, Suárez AJ, Prolongo SG (2020) Smart coatings with carbon nanoparticles. In: 21st century surface science-a handbook. IntechOpen
5. Snihirova D, Lamaka SV, Montemor MF (2016) Smart composite coatings for corrosion protection of aluminium alloys in aerospace applications. Smart Compos Coat Membr 85–121
6. Nundy S, Ghosh A, Mallick TK (2020) Hydrophilic and superhydrophilic self-cleaning coatings by morphologically varying ZnO microstructures for photovoltaic and glazing applications. ACS Omega 5(2):1033–1039
7. Kumar A, Malik G, Goyal K, Sardana N, Chandra R, Mulik RS (2021) Controllable synthesis of tunable aspect ratios novel h-BN nanorods with an enhanced wetting performance for water repellent applications. Vacuum 184:109927. https://doi.org/10.1016/j.vacuum.2020.109927
8. Gite VV, Sohn D, Tatiya P, Marathe RJ (2021) Insights of technologies for self-healing organic coatings. In: Handbook of modern coating technologies. Elsevier, pp 37–65

9. Martínez-Pérez M, Esteban J, Peremarch CP-J (2021) The role of antibacterial coatings in the development of biomaterials. In: Handbook of modern coating technologies, pp 1–36

10. Makhlouf ASH, Tiginyanu I (2011) Nanocoatings and ultra-thin films: technologies and applications. Elsevier

11. Nguyen-Tri P, Nguyen TA, Carriere P, Ngo Xuan C (2018) Nanocomposite coatings: preparation, characterization, properties, and applications. Int J Corr 2018

12. Saha B, Toh WQ, Liu E, Tor SB, Hardt DE, Lee J (2015) A review on the importance of surface coating of micro/nano-mold in micro/nano-molding processes. J Micromech Microeng 26(1):013002

13. Mishra T, Mahato M, Tiwari SK (2021) Oxide-based self-cleaning and corrosion protective coatings. In: Handbook of modern coating technologies. Elsevier, pp 135–173

14. Koutsomichalis A, Vaxevanidis N, Petropoulos G, Xatzaki E, Mourlas A, Antoniou S (2009) Tribological coatings for aerospace applications and the case of WC–Co plasma spray coatings. Tribol Ind 31(1&2):37

15. Li L, Bai Y, Li L, Wang S, Zhang T (2017) A superhydrophobic smart coating for flexible and wearable sensing electronics. Adv Mater 29(43):1702517

16. Sharma M, Adalati R, Kumar A, Mehta M, Chandra R (2022) Composite assembling of oxide-based optically transparent electrodes for high-performance asymmetric supercapacitors. ACS Appl Mater Interfaces

17. Makhlouf ASH, Abu-Thabit NY (2019) Advances in smart coatings and thin films for future industrial and biomedical engineering applications. Elsevier

18. Sharafudeen R (2020) Smart superhydrophobic anticorrosive coatings. In: Advances in smart coatings and thin films for future industrial and biomedical engineering applications. Elsevier, pp 515–534

19. Makhlouf ASH, Gajarla Y (2020) Advances in smart coatings for magnesium alloys and their applications in industry. In: Advances in smart coatings and thin films for future industrial and biomedical engineering applications. Elsevier, pp 245–261

20. Basheer AA (2020) Advances in the smart materials applications in the aerospace industries. Aircr Eng Aerosp Technol 92(7):1027–1035

21. Khan A, Ahmed N, Rabnawaz M (2020) Covalent adaptable network and self-healing materials: current trends and future prospects in sustainability. Polymers 12(9):2027

22. Tarasevich T (2022) Photopolymers: environmentally benign technology for a variety of industries. State University of New York at Albany

23. Goldschmidt A, Streitberger H-J (2003) BASF handbook on basics of coating technology. William Andrew

24. Makhlouf A (2011) Current and advanced coating technologies for industrial applications. In: Nanocoatings and ultra-thin films. Elsevier, pp 3–23

25. Fotovvati B, Namdari N, Dehghanghadikolaei A (2019) On coating techniques for surface protection: a review. J Manuf Mater Process 3(1):28

26. Howarth G, Manock H (1997) Water-borne polyurethane dispersions and their use in functional coatings. Surf Coat Int 80(7):324–328

27. Sharafudeen R (2020) Smart hybrid coatings for corrosion protection applications. In: Advances in smart coatings and thin films for future industrial and biomedical engineering applications. Elsevier, pp 289–306

28. Mousavi R, Bahrololoom M, Deflorian F (2016) Preparation, corrosion, and wear resistance of Ni–Mo/Al composite coating reinforced with Al particles. Mater Des 110:456–465

29. Ong G, Kasi R, Subramaniam R (2021) A review on plant extracts as natural additives in coating applications. Prog Org Coat 151:106091

30. Funke W (1996) Thin-layer technology in organic coatings. Prog Org Coat 28(1):3–7. https://doi.org/10.1016/0300-9440(95)00583-8

31. Wang S, Ji R, Lu W (2021) Applications of optical coatings on spectral selective structures. In: Handbook of modern coating technologies. Elsevier, pp 269–319

32. Tsvetkova I, Krasil'Nikova L, Khoroshavina Y, Galushko A, Frantsuzova Yu V, Kychkin A, et al (2019) Sol–gel preparation of protective and decorative coatings on wood. J Sol-Gel Sci Technol 92(2):474–483

33. Makhlouf ASH, Abu-Thabit NY, Ferretiz D (2020) Shape-memory coatings, polymers, and alloys with self-healing functionality for medical and industrial applications. In: Advances in smart coatings and thin films for future industrial and biomedical engineering applications, pp 335–358
34. Makhlouf ASH, Perez A, Guerrero E (2020) Recent trends in smart polymeric coatings in biomedicine and drug delivery applications. In: Advances in smart coatings and thin films for future industrial and biomedical engineering applications. Elsevier, pp 359–381
35. Makhlouf ASH, Prado J (2020) Recent developments in smart coatings for steel alloys, their impact in the steel industry, and applications. In: Advances in smart coatings and thin films for future industrial and biomedical engineering applications, pp 39–55
36. Shang S, Zeng W (2013) Conductive nanofibres and nanocoatings for smart textiles. In: Multidisciplinary know-how for smart-textiles developers. Elsevier, pp 92–128
37. Ruys AJ, Sutton BA (2021) 9-Metal-ceramic functionally graded materials (FGMs). In: Metal reinforced ceramics, Elsevier Series on Advanced Ceramic Materials, pp 327–329
38. Shahidi S, Moazzenchi B, Ghoranneviss M (2015) A review-application of physical vapor deposition (PVD) and related methods in the textile industry. Eur Phys J Appl Phy 71(3):31302
39. Sproul WD (1996) Physical vapor deposition tool coatings. Surf Coat Technol 81(1):1–7
40. Faraji G, Kim HS, Kashi HT (2018) Severe plastic deformation: methods, processing and properties. Elsevier
41. de Damborenea J, Navas C, García J, Arenas M, Conde A (2007) Corrosion–erosion of TiN-PVD coatings in collagen and cellulose meat casing. Surf Coat Technol 201(12):5751–5757
42. Wang Y, Chen W, Wang B, Zheng Y (2014) Ultrathin ferroelectric films: growth, characterization, physics and applications. Materials 7(9):6377–6485
43. Rane AV, Kanny K, Abitha V, Thomas S (2018) Methods for synthesis of nanoparticles and fabrication of nanocomposites. In: Synthesis of inorganic nanomaterials. Elsevier, pp 121–139
44. Hagarová M (2007) Experimental methods of assessment of PVD coatings properties. J Metals, Mater Miner 17(2)
45. Sharma M, Adalati R, Kumar A, Chawla V, Chandra R (2021) Elevated performance of binder-free Co_3O_4 electrode for the supercapacitor applications. Nano Express 2(1):010002
46. Behera A, Aich S, Theivasanthi T (2022) Magnetron sputtering for development of nanostructured materials. In: Design, fabrication, and characterization of multifunctional nanomaterials. Elsevier, pp 177–199
47. Awasthi K (2021) Nanostructured zinc oxide: synthesis, properties and applications. Elsevier
48. Bonafos C, Khomenkhova L, Gourbilleau F, Talbot E, Slaoui A, Carrada M, et al (2022) Nanocomposite MO_x materials for NVMs. In: Metal oxides for non-volatile memory. Elsevier, pp 201–244
49. Kubart T, Gudmundsson JT, Lundin D (2020) Reactive high power impulse magnetron sputtering. In: High power impulse magnetron sputtering. Elsevier, pp 223–263
50. Kumar A, Malik G, Chandra R, Mulik RS (2022) Sputter-grown hierarchical nitride (TiN and h-BN) coatings on BN nanoplates reinforced Al7079 alloy with improved corrosion resistance. Surf Coat Technol 432:128061. https://doi.org/10.1016/j.surfcoat.2021.128061
51. Tripathi U, Kumar A, Kumar A, Mulik RS (2022) Electrochemical characteristics of sputter deposited ZrN nanoflowers coating for enhanced wetting and anti-corrosion properties. Surf Coat Technol 440:128466. https://doi.org/10.1016/j.surfcoat.2022.128466
52. Lundin D, Minea T, Gudmundsson JT (2019) High power impulse magnetron sputtering: fundamentals, technologies, challenges and applications
53. Mileiko ST (1997) Composite materials series, 12: metal and ceramic based components. Composite materials series
54. Mareci D, Cimpoesu N, Popa M (2012) Electrochemical and SEM characterization of NiTi alloy coated with chitosan by PLD technique. Mater Corros 63(11):985–991
55. Delmdahl R, Wiessner A (2007) Pulsed laser deposition for coating applications. J Phys: Conf Ser: IOP Publ 006
56. Best S, Marti P (2012) Mineral coatings for orthopaedic applications. In: Coatings for biomedical applications. Elsevier, pp 43–74

57. Bozovic I, Schlom D (2001) Superconducting thin films: materials, preparation, and properties. In: Encyclopedia of materials: science and technology, pp 8955–8964
58. Kumar A, Malik G, Pandey MK, Chandra R, Mulik RS (2021) Corrosion behavior of pulse laser deposited 2D nanostructured coating prepared by self-made h-BN target in salinity environment. Ceram Int 47(9):12537–12546. https://doi.org/10.1016/j.ceramint.2021.01.111
59. Bashir A, Awan TI, Tehseen A, Tahir MB, Ijaz M (2020) Interfaces and surfaces. Chem Nanomater 51–87
60. Wasa K, Kitabatake M, Adachi H (2004) Thin film materials technology: sputtering of control compound materials. Springer Science & Business Media
61. Koskinen J, Bandorf R, Sittinger V, Bräuer G, Zhang D, Guan L, et al (2014) Comprehensive materials processing, p 99
62. Mattox DM, Mattox V (2003) Vacuum coating technology. Springer
63. Eccher J, Zajaczkowski W, Faria GrC, Bock H, Von Seggern H, Pisula W, et al (2015) Thermal evaporation versus spin-coating: electrical performance in columnar liquid crystal OLEDs. ACS Appl Mater Interfaces 7(30):16374–16381
64. Wördenweber R (2011) Deposition technologies, growth and properties of high-Tc films. In: High-temperature superconductors. Elsevier, pp 3–38e
65. Jilani A, Abdel-Wahab MS, Hammad AH (2017) Advance deposition techniques for thin film and coating. In: Modern technologies for creating the thin-film systems and coatings, vol 2, no 3, pp 137–149
66. Kerdcharoen T, Wongchoosuk C (2013) Carbon nanotube and metal oxide hybrid materials for gas sensing. In: Semiconductor gas sensors. Elsevier, pp 386–407
67. Ohring M (2001) Materials science of thin films: depositon and structure. Elsevier
68. Plawsky J, Fedorov A, Garimella S, Ma H, Maroo S, Chen L et al (2014) Nano-and microstructures for thin-film evaporation—a review. Nanoscale Microscale Thermophys Eng 18(3):251–269
69. Chen Y (2015) Nanofabrication by electron beam lithography and its applications: a review. Microelectron Eng 135:57–72
70. Sarkar J (2010) Sputtering materials for VLSI and thin film devices. William Andrew
71. Bishop C (2011) Vacuum deposition onto webs, films and foils. William Andrew
72. Seshan K (2012) Handbook of thin film deposition. William Andrew
73. Ting J-M, Tsai B (2001) DC reactive sputter deposition of ZnO: Al thin film on glass. Mater Chem Phys 72(2):273–277
74. Mayrhofer P, Kunc F, Musil J, Mitterer C (2002) A comparative study on reactive and non-reactive unbalanced magnetron sputter deposition of TiN coatings. Thin Solid Films 415(1–2):151–159
75. Kadlec S, Musil J, Vyskocil H (1986) Hysteresis effect in reactive sputtering: a problem of system stability. J Phys D Appl Phys 19(9):L187
76. Hollands E, Campbell D (1968) The mechanism of reactive sputtering. J Mater Sci 3(5):544–552
77. Wang WY, Li J, Liu W, Liu Z-K (2019) Integrated computational materials engineering for advanced materials: a brief review. Comput Mater Sci 158:42–48
78. Delplancke-Ogletree M-P (2006) Cathodic arc evaporation and its applications to thin-film synthesis. In: Materials surface processing by directed energy techniques. Elsevier, pp 383–410
79. Sanchette F, Ducros C, Schmitt T, Steyer P, Billard A (2011) Nanostructured hard coatings deposited by cathodic arc deposition: from concepts to applications. Surf Coat Technol 205(23–24):5444–5453
80. Hsu C-H, Chen M-L, Lai K-L (2006) Corrosion resistance of TiN/TiAlN-coated ADI by cathodic arc deposition. Mater Sci Eng, A 421(1–2):182–190
81. Mittal M, Sardar S, Jana A (2021) Nanofabrication techniques for semiconductor chemical sensors. In: Handbook of nanomaterials for sensing applications. Elsevier, pp 119–137
82. Behera A, Mallick P, Mohapatra S (2020) Nanocoatings for anticorrosion: an introduction. In: Corrosion protection at the nanoscale. Elsevier, pp 227–243

83. Xia L (2021) Importance of nanostructured surfaces. In: Bioceramics. Elsevier, pp 5–24
84. Choy K (2003) Chemical vapour deposition of coatings. Prog Mater Sci 48(2):57–170
85. Pottathara YB, Grohens Y, Kokol V, Kalarikkal N, Thomas S (2019) Synthesis and processing of emerging two-dimensional nanomaterials. In: Nanomaterials synthesis. Elsevier, pp 1–25
86. Han GH, Rodríguez-Manzo JA, Lee C-W, Kybert NJ, Lerner MB, Qi ZJ et al (2013) Continuous growth of hexagonal graphene and boron nitride in-plane heterostructures by atmospheric pressure chemical vapor deposition. ACS Nano 7(11):10129–10138
87. Wang S, Hibino H, Suzuki S, Yamamoto H (2016) Atmospheric pressure chemical vapor deposition growth of millimeter-scale single-crystalline graphene on the copper surface with a native oxide layer. Chem Mater 28(14):4893–4900
88. Sierra-Castillo A, Haye E, Acosta S, Arenal R, Bittencourt C, Colomer J-F (2021) Atmospheric pressure chemical vapor deposition growth of vertically aligned SnS 2 and SnSe 2 nanosheets. RSC Adv 11(58):36483–36493
89. Pham PV (2018) Atmospheric pressure chemical vapor deposition of graphene. Chem Vap Depos Nanotechnol 6:115–134
90. Phuong Pham V (2018) Atmosphere pressure chemical vapor deposition of graphene. arXiv e-prints arXiv:1807.07774
91. Vlassiouk I, Fulvio P, Meyer H, Lavrik N, Dai S, Datskos P et al (2013) Large scale atmospheric pressure chemical vapor deposition of graphene. Carbon 54:58–67. https://doi.org/10.1016/j.carbon.2012.11.003
92. Kang S, Mauchauffé R, You YS, Moon SY (2018) Insights into the role of plasma in atmospheric pressure chemical vapor deposition of titanium dioxide thin films. Sci Rep 8(1):1–13
93. Mandracci P (2019) Chemical vapor deposition for nanotechnology. BoD–Books on Demand
94. Zappe HP, Tabata O, Gianchandani YB (2008) Comprehensive microsystems
95. Jeong Y, Sakuraba M, Matsuura T, Murota J (2003) Si epitaxial growth on the atomic-order nitrided Si (100) surface in SiH4 reaction. In: Rapid thermal processing for future semiconductor devices. Elsevier BV, pp 139–144
96. Miyazaki S, Hamamoto Y, Yoshida E, Ikeda M, Hirose M (2000) Control of self-assembling formation of nanometer silicon dots by low pressure chemical vapor deposition. Thin Solid Films 369(1–2):55–59
97. Chiu CC, Desu SB, Tsai CY (1993) Low pressure chemical vapor deposition (LPCVD) of β–SiC on Si (100) using MTS in a hot wall reactor. J Mater Res 8(10):2617–2626
98. Parker G (2001) Encyclopedia of materials: science and technology
99. Broadbent E, Ramiller C (1984) Selective low pressure chemical vapor deposition of tungsten. J Electrochem Soc 131(6):1427
100. Weimer R, Powell D, Lenahan P (2003) Process and technology drivers for single wafer processes in DRAM manufacturing. In: Rapid thermal processing for future semiconductor devices, pp 17–28
101. Caro J, Doudkowsky M, Figueras A, Fraxedas J, García G, Santiso J, et al (2001) Morphological and structural aspects of thin films prepared by vapor deposition. In: Handbook of surfaces and interfaces of materials. Academic Press, Cambridge, MA, USA
102. Li M, Liu D, Wei D, Song X, Wei D, Wee ATS (2016) Controllable synthesis of graphene by plasma-enhanced chemical vapor deposition and its related applications. Adv Sci 3(11):1600003
103. Guyon C, Barkallah A, Rousseau F, Giffard K, Morvan D, Tatoulian M (2011) Deposition of cobalt oxide thin films by plasma-enhanced chemical vapour deposition (PECVD) for catalytic applications. Surf Coat Technol 206(7):1673–1679
104. Martinu L, Zabeida O, Klemberg-Sapieha J (2010) Plasma-enhanced chemical vapor deposition of functional coatings. In: Handbook of deposition technologies for films and coatings, pp 392–465
105. Bo Z, Yang Y, Chen J, Yu K, Yan J, Cen K (2013) Plasma-enhanced chemical vapor deposition synthesis of vertically oriented graphene nanosheets. Nanoscale 5(12):5180–5204

106. Barankin M, Gonzalez Ii E, Ladwig A, Hicks R (2007) Plasma-enhanced chemical vapor deposition of zinc oxide at atmospheric pressure and low temperature. Sol Energy Mater Sol Cells 91(10):924–930
107. Baldacchini T (2015) Three-dimensional microfabrication using two-photon polymerization: fundamentals, technology, and applications. William Andrew
108. Alemohammad H, Toyserkani E (2010) Laser-assisted additive fabrication of micro-sized coatings. In: Advances in laser materials processing, pp 735–762
109. Pou J, Lusquiños F, Comesaña R, Boutinguiza M (2010) Production of biomaterial coatings by laser-assisted processes. In: Advances in laser materials processing. Elsevier, pp 394–425
110. Schmidt V (2012) Laser-based micro-and nano-fabrication of photonic structures. In: Laser growth and processing of photonic devices. Elsevier, pp 162–237
111. Duty C, Jean D, Lackey W (2001) Laser chemical vapour deposition: materials, modelling, and process control. Int Mater Rev 46(6):271–287
112. Ivanov BL, Beam JC, Harris AG, Wellons MS, Kozub JA, Lukehart CM (2021) 7—Confined-plume chemical deposition. In: Aliofkhazraei M, Ali N, Chipara M, Bensaada Laidani N, De Hosson JTM (eds) Handbook of modern coating technologies. Elsevier, Amsterdam, pp 181–207
113. Ivanov BL, Wellons MS, Lukehart CM (2009) Confined-plume chemical deposition: rapid synthesis of crystalline coatings of known hard or superhard materials on inorganic or organic supports by resonant IR decomposition of molecular precursors. J Am Chem Soc 131(33):11744–11750
114. Sahoo SK, Manoharan B, Sivakumar N (2018) Introduction: why perovskite and perovskite solar cells? In: Perovskite photovoltaics. Elsevier, pp 1–24
115. Mohammadzadeh A, Zadeh SKN, Saidi MH, Sharifzadeh M (2020) Mechanical engineering of solid oxide fuel cell systems: geometric design, mechanical configuration, and thermal analysis. In: Design and operation of solid oxide fuel cells. Elsevier, pp 85–130
116. Neacşu IA, Nicoară AI, Vasile OR, Vasile BŞ (2016) Inorganic micro-and nanostructured implants for tissue engineering. In: Nanobiomaterials in hard tissue engineering. Elsevier, pp 271–295
117. Decker W, Henry B (2002) Basic principles of thin film barrier coatings. In: Proceedings of the annual technical conference-society of vacuum coaters, pp 492–502
118. Reiners G, Beck U, Jehn HA (1994) Decorative optical coatings. Thin Solid Films 253(1):33–40. https://doi.org/10.1016/0040-6090(94)90290-9
119. Beck U, Reiners G, Urban I, Jehn HA, Kopacz U, Schack H (1993) Decorative hard coatings: new layer systems without allergy risk. Surf Coat Technol 61(1–3):215–222
120. Budke E, Krempel-Hesse J, Maidhof H, Schüssler H (1999) Decorative hard coatings with improved corrosion resistance. Surf Coat Technol 112(1–3):108–113
121. Wang T, Stiegel GJ (2016) Integrated gasification combined cycle (IGCC) technologies. Woodhead Publishing
122. Moroz L, Pagur P, Rudenko O, Burlaka M, Joly C (2015) Evaluation for scalability of a combined cycle using gas and bottoming sCO_2 turbines. In: ASME power conference. American Society of Mechanical Engineers, p V001T09A7
123. Soares C (2011) Gas turbines: a handbook of air, land and sea applications. Elsevier
124. Gurrappa I, Rao AS (2006) Thermal barrier coatings for enhanced efficiency of gas turbine engines. Surf Coat Technol 201(6):3016–3029
125. Ghosh S (2015) Thermal barrier ceramic coatings—a review. InTech London, UK
126. Moskal G (2009) Thermal barrier coatings: characteristics of microstructure and properties, generation and directions of development of bond. J Achiev Mater Manuf Eng 37(2):323–331
127. Randhawa H (1988) Cathodic arc plasma deposition technology. Thin Solid Films 167(1–2):175–186
128. Anders A (2002) Atomic scale heating in cathodic arc plasma deposition. Appl Phys Lett 80(6):1100–1102
129. Randhawa H, Johnson P (1987) A review of cathodic arc plasma deposition processes and their applications. Surf Coat Technol 31(4):303–318

130. Vinh PV, Ngoc TVD (2011) Deposition of TiAlSiN hard film by cathodic arc plasma evaporation using a single target combined with a shield filter
131. Kim S, Vinh P, Kim J, Ngoc T (2005) Deposition of superhard TiAlSiN thin films by cathodic arc plasma deposition. Surf Coat Technol 200(5–6):1391–1394
132. Nguyen N-T (2011) Micromixers: fundamentals, design and fabrication. William Andrew
133. Paras AK. Anti-bacterial and anti-viral polymeric coatings
134. Chen X, Hu Y, Xie Z, Wang H Materials and design of photocatalytic membranes. In: Current trends and future developments on (bio-) membranes. Elsevier, pp 71–96
135. Mandoj F, Nardis S, Di Natale C, Paolesse R (2018) Porphyrinoid thin films for chemical sensing
136. Mishra A, Bhatt N, Bajpai A (2019) Nanostructured superhydrophobic coatings for solar panel applications. In: Nanomaterials-based coatings. Elsevier, pp 397–424
137. Yilbas B, Al-Sharafi A, Ali H (2019) Self-cleaning of surfaces and water droplet mobility. Elsevier
138. Sikdar S, Menezes PV, Maccione R, Jacob T, Menezes PL (2021) Plasma electrolytic oxidation (PEO) process—processing, properties, and applications. Nanomaterials 11(6):1375
139. Walsh F, Low C, Wood R, Stevens K, Archer J, Poeton AR et al (2009) Plasma electrolytic oxidation (PEO) for production of anodised coatings on lightweight metal (Al, Mg, Ti) alloys. Trans IMF 87(3):122–135
140. Fattah-alhosseini A, Babaei K, Molaei M (2020) Plasma electrolytic oxidation (PEO) treatment of zinc and its alloys: a review. Surf Interfaces 18:100441
141. Fattah-Alhosseini A, Keshavarz MK, Molaei M, Gashti SO (2018) Plasma electrolytic oxidation (PEO) process on commercially pure Ti surface: effects of electrolyte on the microstructure and corrosion behavior of coatings. Metall Mater Trans A 49(10):4966–4979
142. Mehri Ghahfarokhi N, Shayegh Broujeny B, Hakimizad A, Doostmohammadi A (2022) Plasma electrolytic oxidation (PEO) coating to enhance in vitro corrosion resistance of AZ91 magnesium alloy coated with polydimethylsiloxane (PDMS). Appl Phys A 128(2):1–13
143. Sobolev A, Wolicki I, Kossenko A, Zinigrad M, Borodianskiy K (2018) Coating formation on Ti–6Al–4V alloy by micro arc oxidation in molten salt. Materials 11(9):1611
144. Yerokhin A, Nie X, Leyland A, Matthews A (2000) Characterisation of oxide films produced by plasma electrolytic oxidation of a Ti–6Al–4V alloy. Surf Coat Technol 130(2–3):195–206
145. Wang Y, Jiang B, Lei T, Guo L (2004) Dependence of growth features of microarc oxidation coatings of titanium alloy on control modes of alternate pulse. Mater Lett 58(12–13):1907–1911
146. Fei C, Hai Z, Chen C, Yangjian X (2009) Study on the tribological performance of ceramic coatings on titanium alloy surfaces obtained through microarc oxidation. Prog Org Coat 64(2–3):264–267
147. Lin Z, Wang T, Yu X, Sun X, Yang H (2021) Functionalization treatment of micro-arc oxidation coatings on magnesium alloys: a review. J Alloy Compd 879:160453
148. Qadir M, Li Y, Munir K, Wen C (2018) Calcium phosphate-based composite coating by micro-arc oxidation (MAO) for biomedical application: a review. Crit Rev Solid State Mater Sci 43(5):392–416
149. Yu C, Cui L-Y, Zhou Y-F, Han Z-Z, Chen X-B, Zeng R-C et al (2018) Self-degradation of micro-arc oxidation/chitosan composite coating on Mg–4Li–1Ca alloy. Surf Coat Technol 344:1–11. https://doi.org/10.1016/j.surfcoat.2018.03.007
150. Zhao L, Cui C, Wang Q, Bu S (2010) Growth characteristics and corrosion resistance of micro-arc oxidation coating on pure magnesium for biomedical applications. Corros Sci 52(7):2228–2234
151. Ray SSS, Gusain R, Kumar N (2020) Carbon nanomaterial-based adsorbents for water purification: fundamentals and applications. Elsevier
152. Hong M, Zou J, Chen Z-G (2021) Synthesis of thermoelectric materials. In: Thermoelectricity and advanced thermoelectric materials. Elsevier, pp 73–103
153. Gurrappa I, Binder L (2008) Electrodeposition of nanostructured coatings and their characterization—a review. Sci Technol Adv Mater

154. Wang R, Wu J (2017) Structure and basic properties of ternary metal oxides and their prospects for application in supercapacitors. In: Metal oxides in supercapacitors. Elsevier, pp 99–132

155. AlFalah MGK, Kamberli E, Abbar AH, Kandemirli F, Saracoglu M (2020) Corrosion performance of electrospinning nanofiber ZnO–NiO–CuO/polycaprolactone coated on mild steel in acid solution. Surf Interfaces 21:100760

156. Zhang Y, Zhang K, Lei S, Su Y, Yang W, Wang J et al (2022) Formation and oxidation behavior of TiO_2 modified Al_2O_3-Nb_2O_5/$NbAl_3$ composite coating prepared by two-step methods. Surf Coat Technol 433:128081

157. Ho SM, SA V, Ahmed G, Vidya NS (2018) A review of nanostructured thin films for gas sensing and corrosion protection. Mediterr J Chem 7(6)

158. Yin G, Zhang J (2014) Rotating electrode methods and oxygen reduction electrocatalysts. Elsevier

159. Tian B, Cheng Y (2007) Electrolytic deposition of Ni–Co–Al_2O_3 composite coating on pipe steel for corrosion/erosion resistance in oil sand slurry. Electrochim Acta 53(2):511–517

160. Yang Y, Cheng Y (2011) Electrolytic deposition of Ni–Co–SiC nano-coating for erosion-enhanced corrosion of carbon steel pipes in oilsand slurry. Surf Coat Technol 205(10):3198–3204

161. Fayomi O, Abdulwahab M, Popoola A (2013) Properties evaluation of ternary surfactant-induced Zn–Ni–Al_2O_3 films on mild steel by electrolytic chemical deposition. J Ovonic Res 9(5):123–132

162. Redondo M, Breslin CB (2007) Polypyrrole electrodeposited on copper from an aqueous phosphate solution: corrosion protection properties. Corros Sci 49(4):1765–1776

163. Zhang L, Sun H, Yu J, Yang H, Song F, Huang C (2018) Application of electrophoretic deposition to occlude dentinal tubules in vitro. J Dent 71:43–48

164. Castro Y, Ferrari B, Moreno R, Durán A (2005) Corrosion behaviour of silica hybrid coatings produced from basic catalysed particulate sols by dipping and EPD. Surf Coat Technol 191(2–3):228–235

165. Gebhardt F, Seuss S, Turhan MC, Hornberger H, Virtanen S, Boccaccini AR (2012) Characterization of electrophoretic chitosan coatings on stainless steel. Mater Lett 66(1):302–304

166. Chen X, Chen S, Liang L, Hong H, Zhang Z, Shen B (2018) Electrochemical behaviour of EPD synthesized graphene coating on titanium alloys for orthopedic implant application. Procedia Cirp 71:322–328

167. Fei J, Luo D, Zhang C, Li H, Cui Y, Huang J (2018) Friction and wear behavior of SiC particles deposited onto paper-based friction material via electrophoretic deposition. Tribol Int 119:230–238

168. Mah JC, Muchtar A, Somalu MR, Ghazali MJ (2017) Metallic interconnects for solid oxide fuel cell: a review on protective coating and deposition techniques. Int J Hydrogen Energy 42(14):9219–9229

169. Dhiflaoui H, Jaber NB, Lazar FS, Faure J, Larbi ABC, Benhayoune H (2017) Effect of annealing temperature on the structural and mechanical properties of coatings prepared by electrophoretic deposition of TiO_2 nanoparticles. Thin Solid Films 638:201–212

170. Yadav V, Sankar M, Pandey L (2020) Coating of bioactive glass on magnesium alloys to improve its degradation behavior: interfacial aspects. J Magnes Alloys 8(4):999–1015

171. Szklarska M, Łosiewicz B, Dercz G, Maszybrocka J, Rams-Baron M, Stach S (2020) Electrophoretic deposition of chitosan coatings on the $Ti_{15}Mo$ biomedical alloy from a citric acid solution. RSC Adv 10(23):13386–13393

172. Priyadarshini B, Rama M, Chetan, Vijayalakshmi U (2019) Bioactive coating as a surface modification technique for biocompatible metallic implants: a review. J Asian Ceram Soc 7(4):397–406

173. Radwan AB, Ali K, Shakoor R, Mohammed H, Alsalama T, Kahraman R et al (2018) Properties enhancement of Ni–P electrodeposited coatings by the incorporation of nanoscale Y_2O_3 particles. Appl Surf Sci 457:956–967

174. Pawlowski L (2008) The science and engineering of thermal spray coatings. John Wiley & Sons

175. Davis JR (2004) Handbook of thermal spray technology. ASM international
176. Meghwal A, Anupam A, Murty B, Berndt CC, Kottada RS, Ang ASM (2020) Thermal spray high-entropy alloy coatings: a review. J Therm Spray Technol 29(5):857–893
177. Thorpe ML, Richter HJ (1992) A pragmatic analysis and comparison of HVOF processes. J Therm Spray Technol 1(2):161–170
178. Fernández J, Gaona M, Guilemany J (2007) Effect of heat treatments on HVOF hydroxyapatite coatings. J Therm Spray Technol 16(2):220–228
179. Bolelli G, Lusvarghi L, Varis T, Turunen E, Leoni M, Scardi P et al (2008) Residual stresses in HVOF-sprayed ceramic coatings. Surf Coat Technol 202(19):4810–4819
180. Scrivani A, Ianelli S, Rossi A, Groppetti R, Casadei F, Rizzi G (2001) A contribution to the surface analysis and characterisation of HVOF coatings for petrochemical application. Wear 250(1–12):107–113
181. Stokes J, Looney L (2001) HVOF system definition to maximise the thickness of formed components. Surf Coat Technol 148(1):18–24
182. Toma D, Brandl W, Marginean G (2001) Wear and corrosion behaviour of thermally sprayed cermet coatings. Surf Coat Technol 138(2–3):149–158
183. Karthikeyan J, Berndt C, Tikkanen J, Reddy S, Herman H (1997) Plasma spray synthesis of nanomaterial powders and deposits. Mater Sci Eng, A 238(2):275–286
184. Bulloch J, Callagy A (1999) An in situ wear-corrosion study on a series of protective coatings in large induced draft fans. Wear 233:284–292
185. Knuuttila J, Ahmaniemi S, Mäntylä T (1999) Wet abrasion and slurry erosion resistance of thermally sprayed oxide coatings. Wear 232(2):207–212
186. Petrovicova E, Schadler L (2002) Thermal spraying of polymers. Int Mater Rev 47(4):169–190
187. Joukar A, Mehta J, Marks D, Goel VK (2017) Lumbar-sacral destruction fixation biome-chanics: a finite element study. Spine J 17(11):S335
188. Moridi A, Hassani-Gangaraj SM, Guagliano M, Dao M (2014) Cold spray coating: review of material systems and future perspectives. Surf Eng 30(6):369–395
189. Champagne VK (2007) The cold spray materials deposition process: fundamentals and applications
190. Li C-J, Wang H-T, Zhang Q, Yang G-J, Li W-Y, Liao H (2010) Influence of spray materials and their surface oxidation on the critical velocity in cold spraying. J Therm Spray Technol 19(1):95–101
191. Dean SW, Potter JK, Yetter RA, Eden TJ, Champagne V, Trexler M (2013) Energetic intermetallic materials formed by cold spray. Intermetallics 43:121–130
192. Kawakita J, Katanoda H, Watanabe M, Yokoyama K, Kuroda S (2008) Warm spraying: an improved spray process to deposit novel coatings. Surf Coat Technol 202(18):4369–4373
193. Kawakita J, Maruyama N, Kuroda S, Hiromoto S, Yamamoto A (2008) Fabrication and mechanical properties of composite structure by warm spraying of Zr-base metallic glass. Mater Trans 0712250284-
194. Skarvelis P, Papadimitriou G (2009) Plasma transferred arc composite coatings with self lubricating properties, based on Fe and Ti sulfides: microstructure and tribological behavior. Surf Coat Technol 203(10–11):1384–1394
195. Gedzevicius I, Valiulis A (2006) Analysis of wire arc spraying process variables on coatings properties. J Mater Process Technol 175(1–3):206–211
196. Champagne VK, Helfritch DJ (2013) A demonstration of the antimicrobial effectiveness of various copper surfaces. J Biol Eng 7(1):1–7

Chapter 4
Applications of Coating Materials: A Critical Overview

Hariome Sharan Gupta, Sushanta K. Sethi, and Akarsh Verma

Abbreviations

PA 6	Polyamide, 6
PPS	Polyphenylene sulfide
PU	Polyurethane
DTA	Diethylenetriamine
TETA	Triethylenetetramine
TEPA	Tetraethylenepentamine
MPD	m-phenylenediamine
DDS	Diamino diphenyl sulfone
MDA	Methylenedianiline
ABS	Acrylonitrile butadiene styrene
PET	Polyester polymer or polyethylene terephthalate
PVC	Polyvinyl chloride
TiO_2	Titanium dioxide
PVA	Polyvinyl acetate
PPX	Poly-p-xylylene

H. S. Gupta
Department of Polymer and Process Engineering, Indian Institute of Technology Roorkee, Saharanpur 247001, India

S. K. Sethi (✉)
Department of Metallurgical Engineering and Materials Science, Indian Institute of Technology Bombay, Mumbai, India
e-mail: ssethi@pe.iitr.ac.in

A. Verma
Department of Mechanical Engineering, University of Petroleum and Energy Studies, Dehradun 248007, India

Department of Mechanical Science and Bioengineering, Osaka University, Osaka 560-8531, Japan

PX   p-xylylene monomer
DPX   Dip-xylylene
PDMS   Polydimethylsiloxane
PCR   Polymerase chain reaction

1 Introduction

A coating material is used to cover the surface of an object and to enhance its life of the objects. The purpose of applying the coating on the objects' surface is to decorate them and protect them from external factors, i.e. light, rain, cold, humidity, air, heat, etc. On various surfaces, coatings are applied for aesthetic, protective, and functional reasons [1, 2]. The selection of coatings is done for some specific applications, such as corrosion resistance, heat resistance, insulating resistance, chemical resistance, wears resistance, scratch resistance, also for punch ability resistance, and weldability [3, 4]. The demand for coatings is increasing day by day due to furnishing and decorating objects and protecting the objects from environmental factors, and also protection for internal and external factors. The low gloss paint on the ceiling of a room serves as a good finish and decoration while also diffusing the light. In soft drink cans, the inside coating shields the cans container from the liquid; the coating on the inside of cans protects the soft drinks from outside factors. An automobile's outside coating enhances its life and also enhanced its attractiveness and keeps it from rusting and rapturing. Coatings are preventing buildings, wind turbines, rail cars, aircraft parts, offshore constructions, ballistic tanks, storage tanks, cargo tanks, petrochemical plants, ships, marine, decks, bridges, cargo holds, and some containers. Other coatings are also preventing optical fibers from abrasion, slow the formation of barnacles on ship bottoms, and so forth [3, 5–7].

For years, organic and inorganic coatings have been widely used to protect objects from environmental factors [8]. In the past, coatings evolved gradually in reaction to emerging technologies, performance standards, and pressures from the market. The difficulty in anticipating product performance was a significant factor in the slow rate of change. In order to enable quicker answers to the demands for change, there has been an increase in a study recently on understanding the fundamental links between composition and performance of coatings. Waterborne coatings are being used more frequently than solvent-borne coatings in terms of volume. For many years, latex paints have been employed in architectural coatings. The solvent content of these coatings was lower than that of conventional solvent-borne paints, but it was still relatively high. Latex paints with low and no solvent content are being introduced [9].

The application of aqueous industrial coatings has significantly increased. Although solvent levels are being decreased, solvent-borne coatings are still in use. High solids coatings have been effectively used in a variety of applications. The creation of coatings without solvents is the subject of ongoing research. The use of powder coatings for industrial purposes has been a growing industry. Utilizing powder coatings allows for the total elimination of solvent emissions in many applications. Clear coatings on heat-sensitive substrates have seen a rise in the use of radiation-curable coatings, especially UV-cured coatings. They don't require any solvents, and the energy needed to cure them is extremely minimal [10, 11]. The majority of coatings are applied as liquids that are later transformed into solid films. Powder coatings are applied in the form of solid particles that are subsequently fused with a liquid to create a solid layer. The term solid has no precise definition because almost all of the polymers used in coatings are amorphous. A solid film is one that does not flow considerably under the forces it is subjected to during testing or use, which is a suitable definition of the term. If a set of parameters are met, a film can be said to be solid by indicating the lowest viscosity at which flow can be seen within the given time frame [2, 12].

Polymer coatings provide environment-friendly coatings. The synthesis of polymer coatings required solvent-free, air purifier coatings, waterborne, and thermal-insulating coatings. It performs better scratch resistance, fingerprint resistance, corrosion resistance, heat resistance, and mar resistance better. Many sectors, i.e. automobiles, electrical, electronics, and optical applications, make use of polymer coatings. These coatings' ability to adhere to the substrate is a factor in how well they operate [13]. Coatings such as polymer work as binders with objects, enhancing the object's life and protecting it from external factors. The polymeric protective coatings may be single or multiple layers. Polymer coating techniques traditionally provide a dense barrier against the corrosive species on a metal surface. Coatings are prone to cracking, which develops deep inside the structure where detection is challenging, and rehabilitation is nearly impossible. Polymeric coatings shield a substrate from environmental exposure, and substrate corrosion is significantly accelerated when they fail. Because they are frequently thin and in direct contact with the environment, they may be contaminated by the environment to some extent. In the physical, chemical, and medicinal sciences, polymer coatings on solid materials are becoming more and more significant [14, 15].

In order to create polymer coatings, a liquid layer must first be applied to a substrate or object and then solidified. The liquid can be a monomer or oligomer with an initiator, a polymer latex suspension, or a polymer dissolved in a solvent [16]. A polymer can be dissolved in one or more solvents at the required application concentration, the coating can then be applied, and the solvent can then evaporate to create a film layer. The solvent evaporation rate in the initial stage is mostly unaffected by the presence of the polymer-coated object's surface. The rate of solvent loss depends on how quickly solvent molecules can diffuse to an object's surface as the solvent evaporates, increasing viscosity, glass transition temperature, and free volume. Somewhere polymer coatings require some additives, fillers, or pigments to prevent the objects from environmental factors. Epoxy coatings containing feldspar filler enhance the mechanical properties (impact resistance, damping hardness, and abrasion resistance) and dielectric properties of the object's surface [17].

Today, historical monuments are frequently protected with polymeric coatings, particularly those with high hydrophobicity, to slow or even halt further deterioration.

The primary causes of historical monuments being damaged and defaced, the importance of contact angle in choosing polymeric coatings, and some common types of polymeric materials used for monument protection are all covered in the section on protecting monuments. The most well-known polymers employed as protective materials are fluorinated polymers, alkoxysilanes polymers, acrylic polymers, and hybrid organic–inorganic polymer coatings [18, 19]. The historical monuments are made by some parameters such as, i.e. stone, sand, iron, marble, glass, bricks, ceramics, mud, cement, lime, and plaster. Climatic changes in nature, such as summer, winter, and rain, create harmful effects on the historical monuments. So enhancing the life of historical monuments requires coatings, and polymeric coatings are the best choice. The polymeric coatings are applied on the historical monuments by hand, spray, brush, rubbed roller, etc. Polymeric coatings also have less cost compared to other coatings [20–23].

Coatings are also used in fiber-reinforced polymer composites to improve the mechanical, physical, chemical, and thermal properties of the polymer composites, which have fiber coated by coating resins. Carbon fiber was coated by phenoxy resins after carbon fiber was dipped in phenoxy resin and coated with carbon fiber-reinforced polyamide 6 (PA 6) composites [24]. Coated carbon fabrics were placed between PA 6 films at the required temperature and pressure without special equipment. The composites were made due to the stacking of coated carbon fabrics and PA 6 and also happening chemical reaction between phenoxy resin and PA 6 polymers. Interlaminar shear strength was an increase of the coated carbon fiber-reinforced PA 6 composites due to the coating of carbon fiber by phenoxy resin [24]. Glass fiber and carbon fiber were coated with a polyimide polymer to create polymer composite laminates, which were then pressed and cured at temperatures as high as 350 °C. Occasionally, these polyamide-coated laminates can be utilized continually at temperatures between 250 and 400 °C, depending on the application. The missile, jet, aerospace, and aircraft sector has used polyamide-coated laminates, particularly in the development of supersonic aircraft [25]. The mechanical properties of carbon fiber-reinforced polymer composites increase after carbon fiber is coated by epoxy resin [26]. And also, many synthetic fibers and natural fibers have been coated with different–different polymeric coatings materials to enhance the mechanical, physical, and thermal properties of the fiber-reinforced polymer composites, i.e. natural fiber coated by epoxy, polyester, and phenolic polymer [27].

Nowadays, the self-healing polymeric coating is also available in the market, which has been used to repair objects by itself. Self-healing polymeric coating is basically a polymer that reacts with environmental species, i.e. humidity, moisture, air, etc. This emphasis on polymeric coatings results from several factors, including their practical susceptibility to scratch formation, the presence of a rigid, undamaged substrate that helps to prevent the separation of the crack faces from being reconnected, and the wide range of coating functions, including mechanical, anti-corrosive, and aesthetical functions that can be restored [28, 29]. The most popular and economical way to increase corrosion protection and, consequently, the durability of metallic structures is to apply self-healing coatings. The self-healing coating systems may efficiently safeguard a wide variety of engineering structures, including automobiles, aircraft, chemical industries, and household appliances [30]. Polyphenylene sulfide (PPS) is one of the most polymer used as self-healing polymeric coatings. To boost the self-healing properties of the objects by PPS resin, add some additive or filler which reacts with environment species and fills the crack or scratch by itself. Calcium silicate or calcium carbonate is mainly used as a filler or additive for PPS resin [31].

2 Types of Coating Materials Used and Their Applications

Many types of polymeric materials are used for coatings which have enhanced the object's life and also protect the objects from external factors, i.e. weathering (cold, summer, rain), air, heating effect, wear and abrasions factors, solubility in the solvent, etc. Coatings have also been made to furnish and decorate the objects. Some polymeric materials, which have been used for coatings purposes, are described below.

2.1 Polyurethane Coatings

A polyurethane coating is a polyurethane layer that has been placed on a substrate's surface to serve as protection. Polyurethane coatings may be in liquid form or paint form. These coatings shield substrates from various flaws, including corrosion, degrading processes, weathering, abrasion, and others [32, 33]. Polyurethane is one of the most valuable coatings, which has been mainly used for electric wire coatings. The polymer known as polyurethane is linked to a class of chemical compounds known as carbamates. Additionally, this polymer substance has a thermosetting characteristic, which means that when heated, it burns rather than melts. Customizability is another quality of polyurethane coatings. These coatings can be created in transparent, muted, glossy, or opaque forms. Sometimes it shows yellowing in nature due to photo-oxidations and irradiations of the sunlight [34].

The coating by polyurethane provides better electric wire flexibility, high electric insulation properties, better corrosion resistance, high abrasive resistance, high chemical resistance, increased toughness of electric wire, and better moisture resistance. Polyurethane coatings have become very popular due to their flexible nature and outstanding toughness and also enhanced electric wire life. Polyurethane finishes produce high-gloss, durable coatings that are resistant to chemicals. They are very helpful in heavy wear regions and have good abrasion and impact resistance [35]. Despite having good weather resistance, they lose luster in direct sunlight. If old or damaged parts are painted over, it can be challenging to recoat weathered polyurethane coatings, and topcoats won't stick without care. Polyurethane coatings are also used in military and marine coatings. The most significant two- or three-component systems of polyurethane marine coatings are available [36–38]. The polyurethane resin is present in one component, while an organic polyol is present in

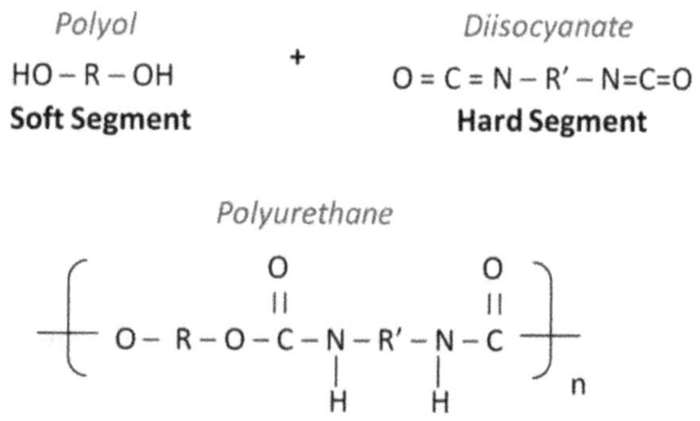

Fig. 1 Structure of
polyurethane polymer [41]

Polyol   *Diisocyanate*

$$HO - R - OH \quad + \quad O = C = N - R' - N = C = O$$

Soft Segment   **Hard Segment**

Polyurethane

$$\left\{ O - R - O - \overset{\overset{O}{\|}}{C} - \underset{\underset{H}{|}}{N} - R' - \underset{\underset{H}{|}}{N} - \overset{\overset{O}{\|}}{C} \right\}_n$$

the second component. In order to speed up the curing process, specific systems call for the addition of a third component that contains catalysts (such as metal soaps or amine compounds). Made from resins containing the isocyanate group, polyurethane coatings for marine applications are strongly reactive with substances containing hydroxyl groups (such as water and alcohol), which are frequently employed as curing agents. Coating films are created after the chemical reactions between the polyurethane resin and the curing agents and solvent evaporation. Water and polyols are common curing agents for polyurethane coatings. Moisture-cured polyurethane coatings are polyurethane coatings that cure as a result of a chemical reaction with water. They come in a single can and are air-cured by moisture in the air. This kind of coating is frequently employed as a transparent varnish or as a cosmetic coating [39, 40] (Fig. 1).

2.2 Epoxy Coatings

Epoxy coatings are one of the most valuable coatings in an electric field. Experimentally, epoxy materials were invented before 1930, but their industrial utilization started after 1950. Epoxy polymer shows a highly polar nature and surface-active nature. Epoxy coatings provide both mechanical and chemical interlocking with the surface of the objects. The hydroxyl groups are present in the epoxy polymer, making hydrogen bonding and interlock with the objects' surface during coating. Epoxy coatings provide highly abrasive resistance, good thermal stability, high impact resistance, UV-radiations resistance, good electric resistance, salt spray resistance, good weather familiarity, high-stress insulation properties, and high moisture resistance. Epoxy coatings are primarily used in protective coatings in electrical fields, housing fields, electronic fields, appliance coatings, maintenance of paints, protection of storage tanks, and also in other commercial applications [42–44].

In order to create epoxy coatings with exceptional mechanical properties, processability, high strength and modulus, and chemical resistance, as well as low creep, good electrical insulating properties, and high thermal stability, epoxy can react with

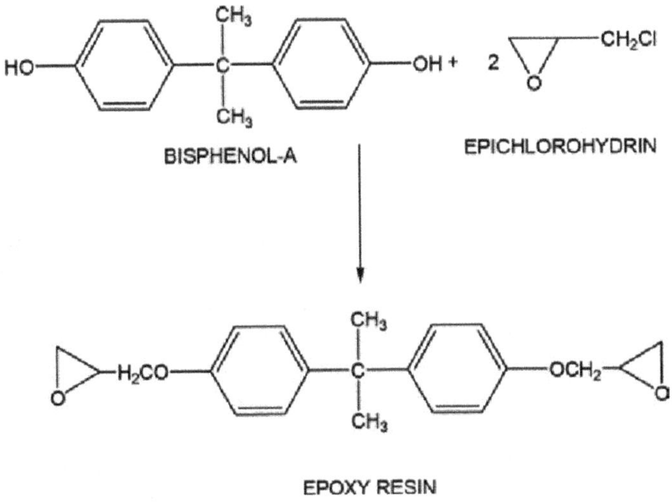

Fig. 2 Structure of epoxy polymer [47]

polyfunctional curing agents, i.e. amines, phenols, alcohols, acids, and thiols [45]. The solidification of epoxy coatings on the surface of objects requires curing agents or hardeners. Adding curing agents in epoxy polymers is solidified and makes a bridge for cross-linking. The curing agents decide the time of polymerization reaction rate and also determine the thermal, mechanical, chemical, and physical properties of cured polymer coatings. The curing agents need a suitable temperature and weather to cure polymer coatings to achieve better properties. Primary, secondary, and tertiary amines are mostly used for the curing of epoxy coatings; they are cured at room temperature for a short time, 20–30 min, and provide excellent adhesion between objects and epoxy coatings. Amides, anhydride, and boron trifluoride are also used as curing agents for epoxy coatings. Here are some of the common amines used as a curing agents for the epoxy coatings, i.e. polyamine salt, cycloaliphatic amine, ketamine, cycloaliphatic polyamine, aliphatic polyamines diethylenetriamine (DTA), triethylenetetramine (TETA), tetraethylenepentamine (TEPA), m-phenylenediamine (MPD), diamino diphenyl sulfone (DDS), and methylenedianiline (MDA) [45, 46].

Esterified solution coatings, Non-esterified solution coatings, and 100% solid coatings are the three types of coatings categories in epoxy coatings (Fig. 2).

2.3 Esterified Solution Epoxy Coatings

In this coating, the epoxy polymer is esterified by soyabean oil, castor oil, unsaturated fatty acid, and linseed oil. Alkyd and metal naphthalene are used in esterified solution epoxy coatings as curing agents. In this coating, make air-drying or baking types of coating samples, which have operated for coatings on the surfaces of objects.

When making a baking type of coating, samples required 60–180 °C temperature in 20–60 min for curing of esterified solution epoxy coatings [48, 49].

2.4   Non-esterified Solution Epoxy Coatings

Polyamides, amines, and acid anhydride are used as curing agents in non-esterified solution epoxy coatings. In this coating, the coatings may be cured at room temperature or higher temperature depending on curing agents, either reactivity between epoxy polymers and curing agents during coatings on the surface of objects. Air drying is used to remove solvents during step-by-step curing. Sometimes solvents are not permanently removed during air drying, then after air is trapped, and volatiles and solvents present are the very common problems in this coating. Add some pigments or fillers to remove all air-trapped or volatiles or solvents present during coatings on the surface of objects by non-esterified solution epoxy coatings [50, 51].

2.5   100% Solid Epoxy Coatings

In this coating, solvent or other ingredients are not used during coatings; only pure epoxy polymer is used for coatings. These coatings are mostly used in automobile sectors, space vehicles, ships, boats, submarine crew compartments, etc. Applied 100% solid epoxy coatings on the surface of objects by spray method, casting method, dip method, and also used brush. Dip method coating is mostly used in electrical instruments, motors, insulations, transformers, etc. Spray and casting method coating is mostly used in ablative coatings, circuit board coatings, insulation coatings, etc. A thicker layer achieves in a single spray or single dip after coating on the surface of objects. In this coating, very rare possibilities arise problems, i.e. air-trapped or volatiles on the coated surface of objects. Some disadvantages are also in 100% solid epoxy coatings, i.e. highly viscous, need special coating equipment, very short pot lives, poor shock resistance, and higher brittleness nature. To reduce the brittleness nature of 100% solid epoxy coatings, polyamides or low molecular weight amine or polysulfides curing agents are used. The curing agent's polyamide concentrations are increased in 100% solid epoxy coatings then flexibility is also increasing, but electrical resistance and moisture resistance will be decreased and also reduce softening points. Hence, the perfect amount of polyamide curing agents should be added to the epoxy polymer to achieve the required properties. Polyamide curing agents in epoxy polymers are mostly used in ablative coatings, which protect electronic space components and also protect electronic missile components from the excessive heat generated during the launching and reentry of missiles and rockets [45, 52, 53].

Fig. 3 Structure of phenoxy
polymer [59]

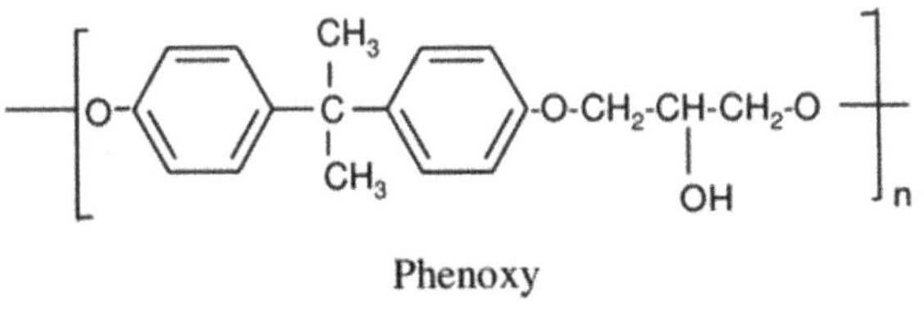

Phenoxy

2.6 Phenoxy Coatings

Phenoxy coatings are also known as polyhydroxy coatings and they are made by condensation reactions of bisphenol-A and propylene oxide [54, 55]. Phenoxy is mainly used as a primer in epoxy or acrylics, or vinyl coatings. Phenoxy coatings are used to adhere plastics in the medical and industrial field and also used in electronics applications, but they need more than 80 °C temperature to adhere the phenoxy coatings on the surface of objects [56]. Solvents are released after being coated by phenoxy and solvent entrapped on the coated surface, one of the very common problems when phenoxy coatings coat objects. Due to entrapped solvent on the surface of coated objects not properly adhering to the object's surface by phenoxy coatings reduces its physical and chemical properties. The solvents release is reduced after adding some pigments or fillers [57]. The phenoxy coatings provide good corrosion, abrasive, and wear resistance [58] (Fig. 3).

2.7 Acrylics Coatings

Acrylic polymer coatings have established a solid footing in the coatings and related industries, as well as automobile sectors, because of their superior flexibility and adherence properties. The acrylic polymer coatings are better-adhering properties compared to vinyl polymer, phenolic polymer, and styrenic polymer, as well as their affordable price. The acrylic polymer coatings provide better scratch resistance, antifogging resistance, chemical resistance, frost resistance, heat stability, and abrasion resistance [60, 61]. Acrylic polymer coatings are mostly used in bridges, storage tanks, finished furniture, kitchen area (i.e. remove the blockage by the detergent), industrial areas, the automobile sector, wood finishing, and machine maintenance. Acrylic polymers are made by bulk, solutions, suspensions, and emulsion polymerization technique but mostly use solutions and emulsions of acrylic polymer for coatings. And the acrylonitrile butadiene styrene (ABS) acrylic polymer coating is mostly used in business machines, i.e. coating in copy machine, calculator, typewriter, and analytical machine. Chemical resistance, hardness, as well as resistance to corrosion and humidity, are requirements for acrylic polymer coatings. Steel is still a widely utilized material; thus, its corrosion resistance is still a crucial need [62].

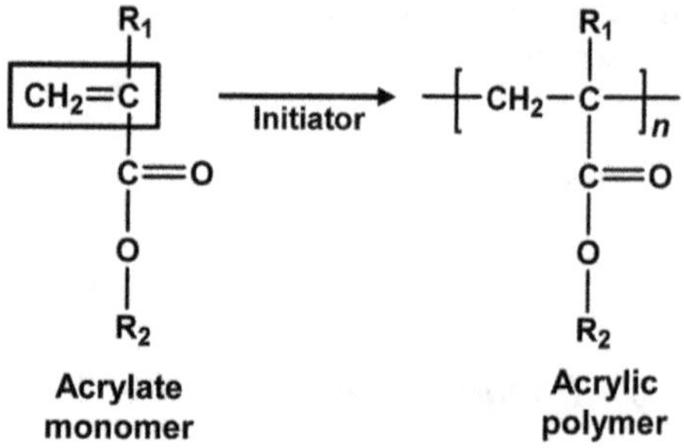

Fig. 4 Structure of acrylic
polymer [64]

A new product specification may be necessary to achieve adhesion to each substrate. Things that stick to metal might not adhere to plastic. To obtain the necessary performance, coatings for metals may be very different from those for plastic or wood. As a result, each substrate area needs to be specifically designed to accommodate its unique end-use requirements for the intended application. The use of coatings has expanded along with the use of acrylic polymer coatings in automotive applications. The acrylic polymer coatings are coated to increase exterior durability and resistance to chemicals, solvents, UV-radiation, and abrasion. The area around engine enamels, underbodies, and auto refurbishing projects are other areas where acrylic polymer coatings are used. And also, grease resistance, durability, and adherence to greasy metals are performance requirements [63] (Fig. 4).

2.8 Vinyl Ether Coatings

First, vinyl ether polymer is converted into resinous form then it is used in coatings. The vinyl ether polymer is made by bulk and solutions polymerization technique, and initially, its viscous and rubbery form. The coating enhances the qualities of the uncoated goods with vinyl ether polymers coatings, including anchoring, adhesion or tack on challenging substrates, resistance to plasticizers, and aging. In order to create surface coatings, the polyvinyl methyl ether and medium molecular weight of polyvinyl ethyl ether are combined with cellulose nitrate, chlorinated binders, and styrene copolymers. Vinyl ether polymer coatings are used for coating metallic structures to protect against corrosion, coating on plastics surface, plastic film surfaces, paper surfaces, and other flexible substrates, UV-curing coatings, and also used in antifouling paints. Vinyl ether coatings seem to be the most desirable substitute for the widely used acrylate polymer coatings in UV-curing applications [65, 66].

Vinyl ether coatings provide a barrier between objects and an aggressive environment, leading to metallic surface corrosion. To improve corrosion resistance on a metallic surface, add fluorine in vinyl ether polymer for coatings [67]. The internal surfaces of equipment (tanks, towers, pipes and fittings, and valves) that are exposed to a range of corrosive substances are coated with fluorine vinyl ether, which is widely employed in the chemical industry. Among non-inert coatings, fluorine vinyl

$$HC\!\!-\!\!\overset{+}{CH}\!\!-\!\!OCH_3$$

Fig. 5 Structure of vinyl ether [71]

ether coatings offer the best chemical and thermal resistance [67]. Fluorine vinyl ether coatings are used in industrial applications, bridge applications to protect from corrosions, marine applications, architectural applications, and also in space applications [68–70] (Fig. 5).

2.9 Polyester Coatings

The polyester polymer or polyethylene terephthalate (PET) is formed by the condensation reaction between carboxyl and hydroxyl groups containing compounds. The polyester polymer was a condensation product of glycol and phthalic anhydride. The polyester polymer is shown to have an amorphous and crystalline nature. The amorphous nature of polyester coatings is mostly used where crystalline polyester coatings reduce the permeability of the coatings. Crystalline polyester is not easily dissolved in many common solvents, but amorphous polyester is easily dissolved in a common solvent. The crystalline polymer has been dissolved in methylene chloride and aromatic hydrocarbons. Where amorphous polyester is dissolved in ester, ketones, chlorinated hydrocarbons, ether ester, and aromatic hydrocarbons, both crystalline and amorphous polyester are not dissolved in water solvents. Initially, polyester polymers were used in the paints industry due to their low molecular weight, but in today's time, polyester coatings are used in automobile sectors, packaging sectors, furniture finished areas, and also used for the protection of food items [72]. High molecular polyester coatings are used in thermoplastic powder coatings. Polyester polymer coatings provide better flexible properties, physical properties, mechanical properties, optical properties, chemical resistance, abrasive resistance, heat resistance, UV-radiation resistance, wear resistance, scratch resistance, crack formation resistance, and also provide better chipping resistance. Polyester polymer coatings are mostly used in automobile sectors as a primer, i.e. fixers for rare view mirrors, bumpers, windshield wipers, the basecoat, oil filters, clearcoat, wheel, air cleaners, engine compartment, and interior and exterior parts of automobiles [73].

For the past two decades, polyester has been used to coat on the sheet and coat for coils and cans. Polyester coatings provide an outstanding balance of physical characteristics, particularly in terms of elasticity and surface hardness. They also have excellent adhesion, anti-yellowing capabilities, and stability against most materials now packaged in coils, cans, and tubes. In solvent-borne paints, such as coil and can coatings, where a high degree of elasticity and resilience to weather or other attacks is required, polyester isocyanate-based coating systems are increasingly being used. Polyesters modified with silicone are renowned for having excellent weather resistance as well as outstanding chemical and thermal stability when the silicone

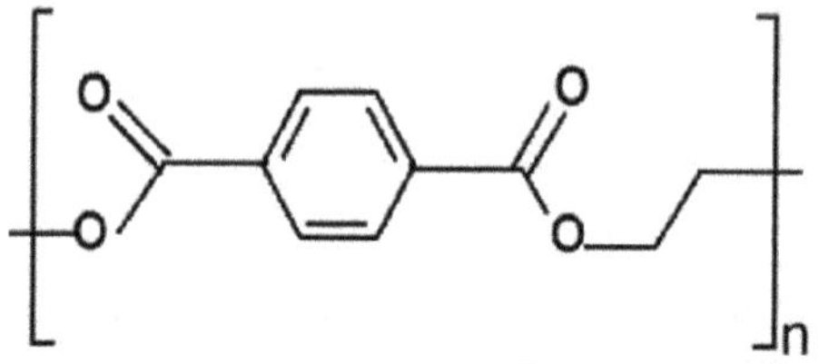

Fig. 6 Structure of polyester
polymer [82]

concentration is sufficiently high. Polyesters modified by epoxy can be cross-linked without the use of formaldehyde and in a less harmful manner by using acidic resins or acid anhydrides. And acrylate polyesters are also used as coating materials in radiation-cured coatings for adhesives, printing inks, and varnishes [74–76].

To improve adhesion, water resistance, and glassiness of the object's surface, add benzoguanamine to the polyester coating. Melamine is also used in polyester coating to enhance the compatibility between the coating and the object surface. Hexamethoxymethyl melamine is the most common resin used for polyester coating to improve compatibility. The reaction between the polyester resin and melamine resin improves the self-life and appreciably extends the object's life which the polyester has coated with melamine coatings [77, 78].

Additionally, unsaturated polyesters or cross-linked polyester also need to be specified for coatings. The unsaturated polyesters coatings feature double bonds in the main chain because they include fumaric, maleic, or other unsaturated acids. In the furniture sector, they are typically used as coatings dissolved in monomeric styrene. Office furniture, i.e. filling cabinets and still desks, require high impact resistance, so, here mostly prefer unsaturated polyester coatings. The unsaturated polyesters coatings are also can be used with other base polymers materials to improve and expand adhesion qualities without styrene [79, 80]. Isocyanate and polyurethane-cured polyester coatings are used to repair automobile parts due to cheaper cast and also protect from weathering conditions. Also, the cross-linked polyester coatings display exceptional flexibility or even elasticity and excellent impact, abrasion, stain resistance, and UV-radiation resistance. They have become indispensable in a variety of fields because of their excellent adhesive characteristics, particularly to metals, as well as high corrosion protection and weather resistance [81] (Fig. 6).

2.10 Polyvinyl Chloride (PVC) Coatings

Polyvinyl Chloride (PVC) resins are made by emulsion and suspension polymerizations technique. Improve flexibility of PVC coating using plasticizer (i.e. phthalic anhydride, sebacic acid, phosphoric acids, adipic acid, and trimellitic acid). The plasticizer has reduced the viscosity of PVC coating and also improved the finishing of coated objects. Some fillers, i.e. calcium carbonate, clay, and talc, are also used to reduce tackiness, make it easy to process, impart a dry hand, deagglomeration, good surface finish, and also control the viscosity of the PVC coatings. During PVC

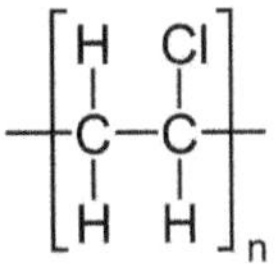

Fig. 7 Structure of PVC
polymer [89]

$$\left[\begin{array}{cc} H & Cl \\ | & | \\ C-C \\ | & | \\ H & H \end{array}\right]_n$$

coating, if any required colors, used pigments, i.e. titanium dioxide (TiO_2) are mostly used. Surfactants, i.e. polyethylene glycol, are used in PVC coating to control the viscosity and also to release air bubbles during coating on the surface of the objects. Some other additive has also been required for necessary use in PVC coatings, such as stabilizing agents (zinc oxide), blowing agents (sodium bicarbonates), flame retardants (zinc borate, magnesium hydroxide, and alumina trihydrate), UV-radiation protection agents, fungicide agents, and weathering protection agents [83, 84].

PVC coatings have provided better electrical, corrosion, abrasive, wear, chemical, weather, and UV-radiation resistance and also provide better insulation properties. PVC coatings are processed by dip-coated tool handles, spray, garden gloves, dishwasher rack, planting rack, and hot and cold dipping methods. PVC coatings by pray method required PVC diluted in solvents and also required viscous control during the spraying of PVC resin coatings. When coatings provide an irregular shape on the surface of the object, they need the spray method of PVC coatings. Hot dipping PVC coating is basically used for metal coating. In the hot dipping PVC coating method, preheated metal is dipped in PVC resin. PVC coatings are basically utilized in coatings of electric wire cable, automobile parts, appliances, medical (i.e. gloves, catheters, tubing, dialysis equipment, and blood, plasma, and fluid bags), furniture finished, housing items, inactive food pathogens, automobile kick panels, outdoor metal furniture, and also many protective applications [85–87]. PVC coatings are used in the textile field (i.e. raincoats, shoe fabrics, pocketbooks, and also in gloves) and also used in decorative places (i.e. mat casting and wall and floor covering) [88] (Fig. 7).

2.11 Polyvinyl Acetate (PVA) Coatings

Polyvinyl acetate resin (PVA) is prepared by the mineral acid catalyzed acetalizations of the vinyl alcohol. A free radical polymerization process is used to prepare a polyvinyl acetate resin from the use of equivalent vinyl acetate monomer. To improve the softness or flexibility of PVA coatings, add plasticizing agents, i.e. adipate diester, diethydiphenyl phthalates, and phosphate ester. PVA coatings provide better mechanical, physical, electrical, chemical, toughness, and thermal properties. PVA coatings are easily cross-linked by dialdehyde, amino, epoxy, isocyanates, and phenolic resins [90]. PVA coatings have provided better furnishing and optical clarity of the object surface. PVA coatings easily adhere and are mostly used for wood furnished, solder masks, ceramics, pharmaceutical applications, fusible heat wire, glass, high glass

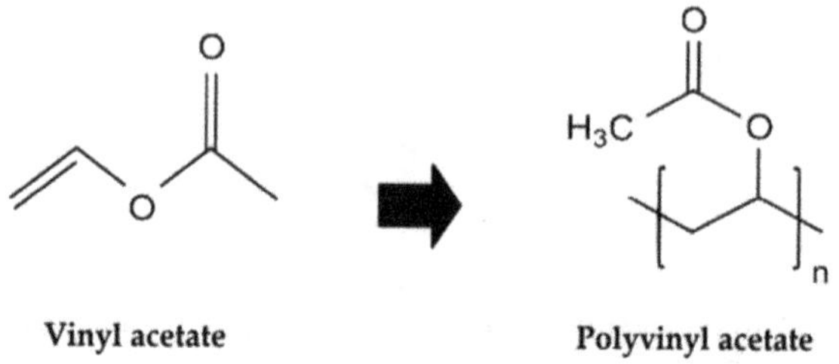

Fig. 8 Structure of PVA
polymer [108]

Vinyl acetate   Polyvinyl acetate

edible, paper, food industries, printed circuit board, photographic, electrographic, and photoconductive coatings, metals to protect by the corrosions, oil resistance insulation coatings, and also used in optical recording disk [91–93]. Paper is mostly used for packaging materials due to its cheaper cast, biodegradability, sustainability, nontoxic, flexibility, and recyclability but its poor resistance to water and poor mechanical properties. To solve the poor resistance of water and poor mechanical properties of paper, they mostly used PVA coatings. The PVA coatings have provided waterproof, bursting, and air resistance, as well as enhanced the mechanical properties of the paper. The PVA coatings have to decrease the porosity of packaging paper. It is extensively used for wood coating due to the better resistance of natural wood oils. PVA coatings are used for high glass edible coatings to form a glassy surface on apples, chew gum, citrus fruits, and chocolate candy because PVA coatings provide high permeance to oxygen and water, which have to protect the edible foods [91, 93]. A few researchers have used polydimethylsiloxane (PDMS) alongside PVA to design hydrophobic easy-clean coatings [94–107] (Fig. 8).

2.12 Polyimide Coatings

Condensation polymerization reactions between pyromellitic anhydrides and primary diamines prepare polyimide polymer resins. A polyimide's polymer chain contains –CO–NR–CO– groups as part of the ring. The polyimide polymer resin is a high-performance polymer used in high-temperature dimension stability places. Polyimide coatings are used in automobile, microelectronic device fabrications, missile wire cables, aerospace and aircraft parts, electric motors, electric circuit boards, wire insulation, enamel wire, magnetic wire, electronic, alfa particles barrier, flat flexible cable, and metal to protect from the corrosions [25, 109, 110]. Polyimide coatings are mostly used for high-temperature bearing places, and aerospace and aircraft applications need the high-temperature bearing capability. Molded polyimide coatings have been widely used for brackets in aircraft structures and the automobile industry and in load-bearing applications, i.e. struts and chassis, because of the highly flexural modulus and compressive strength of polyimide polymer resin. Polyimide coatings have provided high-temperature bearing stability, highly mechanical, physical, and electrical properties, high thermal and chemical stability, and high

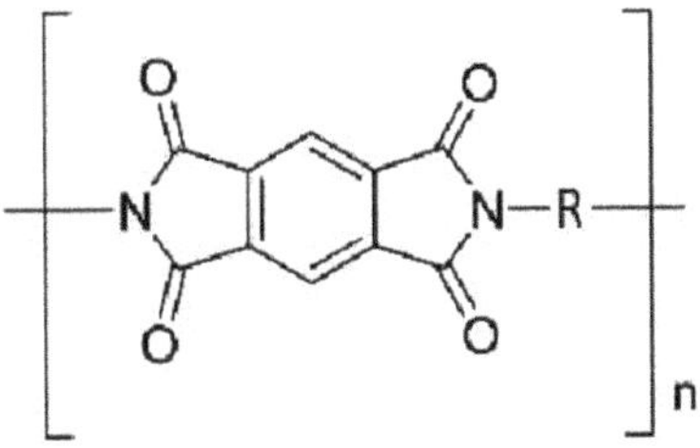

Fig. 9 Structure of polyimide polymer [25]

UV-radiation, moisture absorption, abrasive, wear, creep, and corrosion resistance. Add siloxane to enhance the thermal and oxidative thermal stability and mechanical properties of polyimide coatings. Polyimide coatings have also provided natural lubricity properties, making them ideal-bearing materials used for jet applications and appliance use [111, 112] (Fig. 9).

2.13 Perylene Coatings

The poly-p-xylylene (PPX) family of linear, high molecular weight organic polymers makes up the entire perylene coating. The gaseous p-xylylene monomer (PX) becomes converted into a solid perylene polymer coating without transitioning through a liquid stage. Pyrolytic cleavage of its dimer, dip-xylylene, conveniently produces the monomer (DPX). The production of monomers from dimers proceeds quantitatively with no byproducts. The coatings created with these dimers are known as perylene N and perylene C [113, 114]. Perylene coatings provide better barrier and dielectric properties. The perylene polymer resin is shown to be crystalline in nature. The physical strength and solvent resistance are higher due to the crystalline nature of perylene polymer resin. Although permeable, perylenes coatings are preferable as barriers to compare to other organic polymer coatings that can easily be manufactured as coatings at a given thickness. Perylene coatings are used in biomedical applications, conformal, gas impermeable, and pinhole-free coatings, circuit and insulations purpose coatings, electronic constructions, and surface growing purpose coatings of the objects. Nowadays perylene coatings are used for preservations of old books and also preservations of museum artifacts. To improve compatibility, transparency, and easy process of perylene coatings on the surface of the objects, add polydimethylsiloxane (PDMS) in perylene coatings. PDMS has also reduced the cast of perylene coatings. PDMS has been widely used in microfabricated devices for analytical chemistry, implantable devices, and cell culture. Perylene coatings are used in PDMS-based micro polymerase chain reaction (PCR) chips and also used in PDMS microdevices. Perylene coatings have prevented bubble formation and reduced the porosity and also hydrophobicity of PDMS-based micro PCR chips and PDMS microdevices [113–116] (Fig. 10).

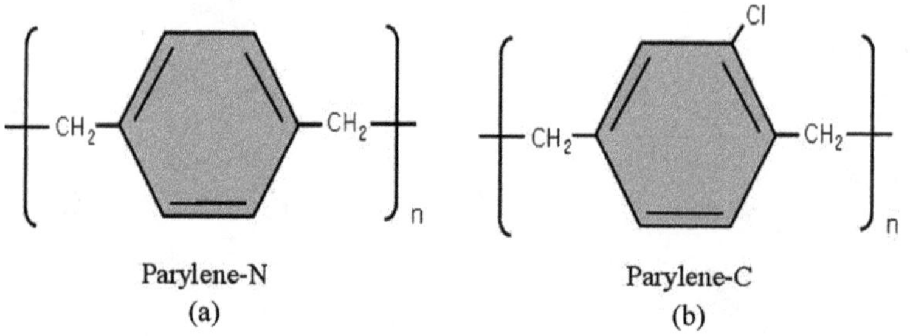

Parylene-N
(a)

Parylene-C
(b)

Fig. 10 Structure of perylene polymer [117]

3 Types of Coating Materials Used and Their Properties and Applications

Polymeric coatings have enhanced the object life by protecting objects and increasing their mechanical and physical properties. Polymeric coatings have enhanced the thermal stability and chemical resistance of objects. Some of the polymeric coatings' properties and applications are described in Table 1. Lastly, the authors have a notable experience in the polymer coatings and computational mechanics field [118–194].

Table 1 Different types of coating materials, their properties, and applications

Sr. no.	Coating materials	Properties of coating materials	Applications of coating materials	References
1.	Polyurethane (PU)	Provide better flexibility of the electric wire and high electric insulation properties, better corrosion, impact, abrasive, chemical, and moisture resistance	Used in electric wire coatings, military, and marine coatings	[34–38]
2.	Epoxy	Provide highly abrasive, impact, UV-radiations, electric, moisture, chemical and salt spray resistance, good thermal stability, good weather familiarity, and high-stress insulation properties	Used in electrical, housing, and electronic fields, appliance coatings, maintenance of paints, protections of the storage tank, and also used in other commercial applications	[42–45]

(continued)

Table 1 (continued)

Sr. no.	Coating materials	Properties of coating materials	Applications of coating materials	References
3.	Phenoxy	Provide good corrosion, abrasive, and also wear resistance	Used in the medical and industrial fields and also used in electronics applications	[56–58]
4.	Acrylic	Provide better scratch, antifogging, abrasion, chemical, grease, UV-radiation and frost resistance, and better heat stability	Used in bridges, storage tanks, finished furniture, kitchen area (i.e. remove the blockage by the detergent), industrial areas, automobile sector, wood finishing, and maintenance of the machine	[60–63]
5.	Vinyl ether	Provide better aging and corrosion resistance	Used for the coating of metallic structures to protect from corrosion, coating on the plastics surface and film surface and paper surface and other flexible substrates, UV-curing coatings and used in antifouling paints, bridge, marine, architectural, and also in space applications	[65, 66, 68–70]
6.	Polyester	Provide better flexible, physical, mechanical and optical properties; chemical, water, abrasive, heat, UV-radiation, wear, scratch, and crack formation resistance; and also provide better chipping resistance	Used in automobile sectors as (i.e. fixer for rare view mirrors, bumpers, windshield wipers, the basecoat, oil filters, clearcoat, wheel, air cleaners, engine compartment, and interior and exterior parts of automobiles), also coated for coil and cans	[73–76]
7.	Polyvinyl chloride (PVC)	Provide better electrical, corrosion, abrasive, wear, chemical, weather, and UV-radiation resistance and also provide better insulation properties	Used coating in electric wire cable, automobile parts, appliances, textile field (i.e. raincoats, shoe fabrics, pocketbooks, and also in gloves), decorative places (i.e. mat casting and wall and floor covering), medical (i.e. gloves, catheters, tubing, dialysis equipment, and blood, plasma, and fluid bags), furniture finished, housing items, inactive food pathogens, automobile kick panels, outdoor metal furniture, and also many protective applications	[85–88]

(continued)

Table 1 (continued)

Sr. no.	Coating materials	Properties of coating materials	Applications of coating materials	References
8.	Polyvinyl acetate (PVA)	Provide waterproof, bursting, and air resistance, as well as enhances the mechanical properties of the paper, improves sustainability and flexibility nature, and is also nontoxic in nature	Used for wood furnished, solder masks, ceramics, pharmaceutical applications, fusible heat wire, glass, high glass edible, paper, food industries, printed circuit board, photographic, electrographic, and photoconductive coatings, metals to protect from corrosions, oil resistance insulation coatings, and also used in optical recording disk	[91–93]
9.	Polyimides	Provide high-temperature bearing stability, highly mechanical, physical, and electrical properties and high thermal and chemical stability, and high UV-radiation, moisture absorption, abrasive, wear, creep, and corrosion resistance	Used for high-temperature bearing places and in aerospace and aircraft applications that need high-temperature handling capability	[111, 112]
10.	Perylene	Provide better barrier and dielectric properties	Used in biomedical applications, conformal, gas impermeable, and pinhole-free coatings, circuit and insulations purpose coatings, electronic constructions and surface growing purpose coatings of objects and used in preservations of old books and also preservations of museum artifacts	[113–116]

4 Conclusion

The data presented and discussed in this chapter may surely act as a guide in helping the reader/researchers to select the appropriate coating formulation. Regardless of which specific area the reader is concerned with, it is important for one to know and understand specifically which polymer, acid, or solvents will be encountered in order to select the appropriate coating formulation.

Acknowledgements Author "Akarsh Verma" is grateful to the monetary support provided by the University of Petroleum and Energy Studies (UPES)-SEED Grant program.

Conflict of Interest There are no conflicts of interest to declare by the authors.

References

1. Howarth GA, Manock HL (1997) Water-borne polyurethane dispersions and their use in functional coatings. Surf Coat Int 80(7):324–328. https://doi.org/10.1007/BF02692680
2. Xu C-N (2002) Coatings. In: Encyclopedia of smart materials. https://doi.org/10.1002/0471216275.ESM015
3. Sørensen PA, Kiil S, Dam-Johansen K, Weinell CE (2009) Anticorrosive coatings: a review. J Coat Technol Res 6:135–176. https://doi.org/10.1007/S11998-008-9144-2/FIGURES/24
4. Lindenmo M, Coombs A, Snell D (2000) Advantages, properties and types of coatings on non-oriented electrical steels. J Magn Magn Mater 215–216:79–82. https://doi.org/10.1016/S0304-8853(00)00071-8
5. Fragata F, Salai RP, Amorim C, Almeida E (2006) Compatibility and incompatibility in anticorrosive painting: the particular case of maintenance painting. Prog Org Coat 56:257–268. https://doi.org/10.1016/J.PORGCOAT.2006.01.012
6. Shipilov SA, Le May I (2006) Structural integrity of aging buried pipelines having cathodic protection. Eng Fail Anal 13:1159–1176. https://doi.org/10.1016/J.ENGFAILANAL.2005.07.008
7. Kudina EF, Barkanov E, Vinidiktova NS (2016) Use of nano-structured modifiers to improve the operational characteristics of pipelines' protective coatings. Glas Phys Chem 42(5):512–517. https://doi.org/10.1134/S1087659616050072
8. Picciotti M, Picciotti F (2006) Selecting corrosion-resistant materials. Chem Eng Prog 102:45–50
9. Weiss KD (1997) Paint and coatings: a mature industry in transition. Prog Polym Sci 22:203–245. https://doi.org/10.1016/S0079-6700(96)00019-6
10. Zhang R, Chen H, Cao H, Huang CM, Mallon PE, Li Y, He Y, Sandreczki TC, Jean YC, Suzuki R, Ohdaira T (2001) Degradation of polymer coating systems studied by positron annihilation spectroscopy. IV. Oxygen effect of UV irradiation. J Polym Sci Part B: Polym Phys 39:2035–2047. https://doi.org/10.1002/POLB.1179
11. Marchebois H, Joiret S, Savall C, Bernard J, Touzain S (2002) Characterization of zinc-rich powder coatings by EIS and Raman spectroscopy. Surf Coat Technol 157:151–161. https://doi.org/10.1016/S0257-8972(02)00147-0
12. Miller RA (1987) Current status of thermal barrier coatings—an overview. Surf Coat Technol 30:1–11. https://doi.org/10.1016/0257-8972(87)90003-X
13. Ritter JE, Lardner TJ, Rosenfeld L, Lin MR (1998) Measurement of adhesion of thin polymer coatings by indentation. J Appl Phys 66:3626. https://doi.org/10.1063/1.344071
14. Cho SH, White SR, Braun PV (2009) Self-healing polymer coatings. Adv Mater 21:645–649. https://doi.org/10.1002/ADMA.200802008
15. Wei Q, Haag R (2015) Universal polymer coatings and their representative biomedical applications. Mater Horiz 2:567–577. https://doi.org/10.1039/C5MH00089K
16. Francis LF, Mccormick AV, Vaessen DM, Payne JA. Development and measurement of stress in polymer coatings
17. Kouloumbi N, Ghivalos LG, Pantazopoulou P (2005) Determination of the performance of epoxy coatings containing feldspars filler. Pigm Resin Technol 34:148–153. https://doi.org/10.1108/03699420510597992/FULL/PDF
18. Sadat-Shojai M, Ershad-Langroudi A (2009) Polymeric coatings for protection of historic monuments: opportunities and challenges. J Appl Polym Sci 112:2535–2551. https://doi.org/10.1002/APP.29801
19. Favaro M, Mendichi R, Ossola F, Russo U, Simon S, Tomasin P, Vigato PA (2006) Evaluation of polymers for conservation treatments of outdoor exposed stone monuments. Part I: photo-oxidative weathering. Polym Degrad Stab 91:3083–3096. https://doi.org/10.1016/J.POLYMDEGRADSTAB.2006.08.012

20. Sadat-Shojai M (2009) Siloxane-based coatings as potential materials for protection of brick-made monuments superhydrophobic nanoparticles and nanocomposites view project fabrication of polymeric composites based on biodegradable polyesters and bioactive nanoparticles optimized by organic chemicals for bone tissue regeneration view project Amir Ershad-Langroudi Iran Polymer and Petrochemical Institute

21. Zielecka M, Bujnowska E (2006) Silicone-containing polymer matrices as protective coatings: properties and applications. Prog Org Coat 55:160–167. https://doi.org/10.1016/J.POR GCOAT.2005.09.012

22. De Lorenzis L, Nanni A (2004) International workshop on preservation of historical structures with FRP composites final report

23. Han JT, Zheng Y, Cho JH, Xu X, Cho K (2005) Stable superhydrophobic organic–inorganic hybrid films by electrostatic self-assembly. J Phys Chem B 109:20773–20778. https://doi.org/10.1021/JP052691X/ASSET/IMAGES/LARGE/JP052691XF00006.JPEG

24. Yi JW, Lee W, Seong DG, Won HJ, Kim SW, Um MK, Byun JH (2016) Effect of phenoxy-based coating resin for reinforcing pitch carbon fibers on the interlaminar shear strength of PA6 composites. Compos A Appl Sci Manuf 87:212–219. https://doi.org/10.1016/J.COM POSITESA.2016.04.028

25. Frącz W, Janowski G, Mikhasev G (2016) The manufacturing issues of technical products made of polyimide—carbon fibers composite by means injection moulding process. Sci Lett Univ Rzesz Technol—Mech 101–113. https://doi.org/10.7862/RM.2016.9

26. Aoki R, Yamaguchi A, Hashimoto T, Urushisaki M, Sakaguchi T, Kawabe K, Kondo K, Iyo H (2019) Preparation of carbon fibers coated with epoxy sizing agents containing degradable acetal linkages and synthesis of carbon fiber-reinforced plastics (CFRPs) for chemical recycling. Polym J 51(9):909–920. https://doi.org/10.1038/s41428-019-0202-7

27. Mylsamy G, Krishnasamy P (2022) A review on natural plant fiber epoxy and polyester composites—coating and performances. https://doi.org/10.1080/15440478.2022.2075517

28. García SJ, Fischer HR, Van Der Zwaag S (2011) A critical appraisal of the potential of self healing polymeric coatings. Prog Org Coat 72:211–221. https://doi.org/10.1016/J.POR GCOAT.2011.06.016

29. Samadzadeh M, Boura SH, Peikari M, Kasiriha SM, Ashrafi A (2010) A review on self-healing coatings based on micro/nanocapsules. Prog Org Coat 68:159–164. https://doi.org/10.1016/J.PORGCOAT.2010.01.006

30. Abu-Thabit NY, Hamdy AS (2016) Stimuli-responsive polyelectrolyte multilayers for fabrication of self-healing coatings—a review. Surf Coat Technol 303:406–424. https://doi.org/10.1016/J.SURFCOAT.2015.11.020

31. Sugama T, Gawlik K (2003) Self-repairing poly(phenylenesulfide) coatings in hydrothermal environments at 200 °C. Mater Lett 57:4282–4290. https://doi.org/10.1016/S0167-577 X(03)00304-5

32. González-García Y, González S, Souto RM (2007) Electrochemical and structural properties of a polyurethane coating on steel substrates for corrosion protection. Corros Sci 49:3514–3526. https://doi.org/10.1016/J.CORSCI.2007.03.018

33. Noreen A, Zia KM, Zuber M, Tabasum S, Zahoor AF (2016) Bio-based polyurethane: an efficient and environment friendly coating systems: a review. Prog Org Coat 91:25–32. https://doi.org/10.1016/J.PORGCOAT.2015.11.018

34. Singh RP, Tomer NS, Bhadraiah SV (2001) Photo-oxidation studies on polyurethane coating: effect of additives on yellowing of polyurethane. Polym Degrad Stab 73:443–446. https://doi.org/10.1016/S0141-3910(01)00127-6

35. Chattopadhyay DK, Raju KVSN (2007) Structural engineering of polyurethane coatings for high performance applications. Prog Polym Sci 32:352–418. https://doi.org/10.1016/J.PRO GPOLYMSCI.2006.05.003

36. Bleile H, Rodgers SD (2001) Marine coatings. In: Encyclopedia of materials: science and technology, pp 5174–5185. https://doi.org/10.1016/B0-08-043152-6/00899-8

37. Yang XF, Vang C, Tallman DE, Bierwagen GP, Croll SG, Rohlik S (2001) Weathering degradation of a polyurethane coating. Polym Degrad Stab 74:341–351. https://doi.org/10.1016/S0141-3910(01)00166-5

38. Crawford DM, Escarsega JA (2000) Dynamic mechanical analysis of novel polyurethane coating for military applications. Thermochim Acta 357–358:161–168. https://doi.org/10.1016/S0040-6031(00)00385-3

39. Papaj EA, Mills DJ, Jamali SS (2014) Effect of hardener variation on protective properties of polyurethane coating. Prog Org Coat 77:2086–2090. https://doi.org/10.1016/J.PORGCOAT.2014.08.013

40. Noreen A, Zia KM, Zuber M, Tabasum S, Saif MJ (2015) Recent trends in environmentally friendly water-borne polyurethane coatings: a review. Korean J Chem Eng 33(2):388–400. https://doi.org/10.1007/S11814-015-0241-5

41. Patti A, Costa F, Perrotti M, Barbarino D, Acierno D (2021) Polyurethane impregnation for improving the mechanical and the water resistance of polypropylene-based textiles. Materials 14. https://doi.org/10.3390/MA14081951

42. Shi X, Nguyen TA, Suo Z, Liu Y, Avci R (2009) Effect of nanoparticles on the anticorrosion and mechanical properties of epoxy coating. Surf Coat Technol 204:237–245. https://doi.org/10.1016/J.SURFCOAT.2009.06.048

43. Allahverdi A, Ehsani M, Janpour H, Ahmadi S (2012) The effect of nanosilica on mechanical, thermal and morphological properties of epoxy coating. Prog Org Coat 75:543–548. https://doi.org/10.1016/J.PORGCOAT.2012.05.013

44. Kotnarowska D (1999) Influence of ultraviolet radiation and aggressive media on epoxy coating degradation. Prog Org Coat 37:149–159. https://doi.org/10.1016/S0300-9440(99)00070-3

45. Seidi F, Jouyandeh M, Taghizadeh M, Taghizadeh A, Vahabi H, Habibzadeh S, Formela K, Saeb MR (2020) Metal-organic framework (MOF)/epoxy coatings: a review. Materials 13:2881. https://doi.org/10.3390/MA13122881

46. Licari JJ, Hughes LA (n.d.) Handbook of polymer coatings for electronics: chemistry, technology and applications—Google Books. https://books.google.co.in/books?hl=en&lr=&id=QbE7AAAAQBAJ&oi=fnd&pg=PP1&dq=types+of+polymer+coatings&ots=BQn40UFe1X&sig=ROnDBrcTZ9GHeVsX88izqYG0NSE&redir_esc=y#v=onepage&q=types%20of%20polymer%20coatings&f=false. Accessed 21 Nov 2022

47. Gul S, Kausar A, Mehmood M, Muhammad B, Jabeen S (2016) Progress on epoxy/polyamide and inorganic nanofiller-based hybrids: introduction, application, and future potential. Polym—Plast Technol Eng 55:1842–1862. https://doi.org/10.1080/03602559.2016.1185628

48. Shikha D, Kamani PK, Shukla MC (2003) Studies on synthesis of water-borne epoxy ester based on RBO fatty acids. Prog Org Coat 47:87–94. https://doi.org/10.1016/S0300-9440(02)00159-5

49. Ding Y, Liu M, Li S, Zhang S, Zhou W-F, Wang B (2001) Contributions of the side groups to the characteristics of water absorption in cured epoxy resins. Macromol Chem Phys 202:2681–2685. https://doi.org/10.1002/1521-3935(20010901)202:13

50. Croll SG (undefined 1979) Residual stress in a solventless amine-cured epoxy coating. J Coat Technol (Researchgate.Net)

51. Nazemi MK, Valix M (2016) Evaluation of acid diffusion behaviour of amine-cured epoxy coatings by accelerated permeation testing method and prediction of their service life. Prog Org Coat 97:307–312. https://doi.org/10.1016/J.PORGCOAT.2016.04.025

52. Pradhan S, Pandey P, Mohanty S, Nayak SK (2016) Insight on the chemistry of epoxy and its curing for coating applications: a detailed investigation and future perspectives. 55:862–877. https://doi.org/10.1080/03602559.2015.1103269

53. Schmidt RG, Bell JP (1986) Epoxy adhesion to metals. Adv Polym Sci 75:33–71. https://doi.org/10.1007/BFB0017914/COVER

54. Yang F, Qiu Z (2011) A comparative study of crystallization kinetics of a miscible biodegradable poly(butylene succinate-co-butylene adipate) blend with poly(hydroxyl ether biphenyl A) and its neat component. Thermochim Acta 523:200–206. https://doi.org/10.1016/J.TCA.2011.05.024

55. An J, Cao X, Ma Y, Ke Y, Yang H, Wang H, Wang F (2014) Preparation and characterization of polyamide 66/poly(hydroxyl ether of bisphenol A) blends without compatibilizer. J Appl Polym Sci 131:40437. https://doi.org/10.1002/APP.40437

56. Dharaiya D, Jana SC, Shafi A (2003) A study on the use of phenoxy resins as compatibilizers of polyamide 6 (PA6) and polybutylene terephthalate (PBT). Polym Eng Sci 43:580–595. https://doi.org/10.1002/PEN.10047

57. Wu D, Su Q, Chen L, Cui H, Zhao Z, Wu Y, Zhou H, Chen J (2022) Achieving high anti-wear and corrosion protection performance of phenoxy-resin coatings based on reinforcing with functional graphene oxide. Appl Surf Sci 601:154156. https://doi.org/10.1016/J.APSUSC. 2022.154156

58. Zhang H, Wang G, Gao T, Fu G, Wu J, Kuang H, Fu C, Zhang H, Wang G, Gao T, Fu G, Wu J, Kuang H, Fu C (2018) Preparation and cryogenic properties of cyanate ester resin blends co-modified by phenoxy resin/epoxy resin. J Aeronaut Mater 38(3):83–90. https://doi.org/10. 11868/J.ISSN.1005-5053.2017.000130

59. Singhal R (n.d.) Synthesis and characterization of phenoxy modified epoxy blends

60. Amerio E, Fabbri P, Malucelli G, Messori M, Sangermano M, Taurino R (2008) Scratch resistance of nano-silica reinforced acrylic coatings. Prog Org Coat 62:129–133. https://doi. org/10.1016/J.PORGCOAT.2007.10.003

61. Zhao J, Meyer A, Ma L, Ming W (2013) Acrylic coatings with surprising antifogging and frost-resisting properties. Chem Commun 49:11764–11766. https://doi.org/10.1039/C3CC46 561F

62. Dhole GS, Gunasekaran G, Ghorpade T, Vinjamur M (2017) Smart acrylic coatings for corrosion detection. Prog Org Coat 110:140–149. https://doi.org/10.1016/J.PORGCOAT.2017. 04.048

63. Tyagi P, Lucia LA, Hubbe MA, Pal L (2019) Nanocellulose-based multilayer barrier coatings for gas, oil, and grease resistance. Carbohyd Polym 206:281–288. https://doi.org/10.1016/J. CARBPOL.2018.10.114

64. Gaytán I, Burelo M, Loza-Tavera H (2021) Current status on the biodegradability of acrylic polymers: microorganisms, enzymes and metabolic pathways involved. Appl Microbiol Biotechnol 105:991–1006. https://doi.org/10.1007/S00253-020-11073-1/FIGURES/4

65. Decker C, Morel F, Decker D (2000) UV-radiation curing of vinyl ether-based coatings. Surf Coat Int 83(4):173–180. https://doi.org/10.1007/BF02692689

66. Brand BG, Schoen HO, Gast LE, Cowan JC (1964) Evaluation of fatty vinyl ether polymers and styrenated polymers for metal coatings. J Am Oil Chemists' Soc 41:597–599. https://doi. org/10.1007/BF02664973

67. Mirzaee M, Rezaei Abadchi M, Mehdikhani A, Riahi Noori N, Zolriasatein A (undefined 2023) Corrosion and UV resistant coatings using fluoroethylene vinyl ether polymer. Prog Color, Color Coat (Researchgate.Net). https://doi.org/10.30509/PCCC.2022.166951.1156

68. Darden W, Takayanagi T, Masuda S (2007) Fluoroethylene vinyl ether resins for applications in marine environments

69. Zhong B, Shen L, Zhang X, Li C, Bao N (2021) Reduced graphene oxide/silica nanocomposite-reinforced anticorrosive fluorocarbon coating. J Appl Polym Sci 138:49689. https://doi.org/10.1002/APP.49689

70. Shen L, Chen H, Qi C, Fu Q, Xiong Z, Sun Y, Liu Y (2021) A green and facile fabrication of rGO/FEVE nanocomposite coating for anti-corrosion application. Mater Chem Phys 263:124382. https://doi.org/10.1016/J.MATCHEMPHYS.2021.124382

71. Zacheslavskaya RK, Rappoport LY, Petrov GN, Trofimov BA (1983) Protonated forms of vinyl ethers. Bull Acad Sci USSR, Div Chem Sci 32(5):886–889. https://doi.org/10.1007/ BF00956132

72. Paseiro-Cerrato R, Noonan GO, Begley TH (2016) Evaluation of long-term migration testing from can coatings into food simulants: polyester coatings. J Agric Food Chem 64:2377–2385. https://doi.org/10.1021/ACS.JAFC.5B05880/ASSET/IMAGES/LARGE/JF-2015-058 80N_0005.JPEG

73. Moon JI, Lee YH, Kim HJ (2012) Synthesis and characterization of flexible polyester coatings for automotive pre-coated metal. Prog Org Coat 73:123–128. https://doi.org/10.1016/J.POR GCOAT.2011.09.009

74. Stojanović I, Šimunović V, Alar V, Kapor F (2018) Experimental evaluation of polyester and epoxy–polyester powder coatings in aggressive media. Coatings 8:98. https://doi.org/10.3390/COATINGS8030098

75. Shimizu T, Higashiura S, Ohguchi M, Murase H, Akitomo Y (n.d.) Water-borne polyester for inks and coatings: structural elucidation of acrylic-grafted polyester and the particle of its aqueous dispersion. https://doi.org/10.1002/(SICI)1099-1581(199907)10:7

76. Gholamiyan H, Gholampoor B, Hosseinpourpia R (2022) Application of Waterborne acrylic and solvent-borne polyester coatings on plasma-treated fir (Abies alba M.) wood. Materials 15:370. https://doi.org/10.3390/MA15010370/S1

77. Greunz T, Lowe C, Bradt E, Hild S, Strauß B, Stifter D (2018) A study on the depth distribution of melamine in polyester-melamine clear coats. Prog Org Coat 115:130–137. https://doi.org/10.1016/J.PORGCOAT.2017.11.014

78. Zhang WR, Zhu TT, Smith R, Lowe C (2010) An investigation on the melamine self-condensation in polyester/melamine organic coating. Prog Org Coat 69:376–383. https://doi.org/10.1016/J.PORGCOAT.2010.07.011

79. Atta AM, Elsaeed AM, Farag RK, El-Saeed SM (2007) Synthesis of unsaturated polyester resins based on rosin acrylic acid adduct for coating applications. React Funct Polym 67:549–563. https://doi.org/10.1016/J.REACTFUNCTPOLYM.2007.03.009

80. Karakaya AE, Karahalil B, Yilmazer M, Aygün N, Şardaş S, Burgaz S (1997) Evaluation of genotoxic potential of styrene in furniture workers using unsaturated polyester resins. Mutat Res/Genet Toxicol Environ Mutagen 392:261–268. https://doi.org/10.1016/S1383-5718(97)00080-6

81. Wang X, Soucek MD (2013) Investigation of non-isocyanate urethane dimethacrylate reactive diluents for UV-curable polyurethane coatings. Prog Org Coat 76:1057–1067. https://doi.org/10.1016/J.PORGCOAT.2013.03.001

82. Lu G, Van Driel WD, Fan X, Yazdan Mehr M, Fan J, Jansen KMB, Zhang GQ (2015) Degradation of microcellular PET reflective materials used in LED-based products. Opt Mater 49:79–84. https://doi.org/10.1016/J.OPTMAT.2015.07.026

83. Babinsky R (2006) PVC additives: a global review. Plast, Addit Compd 8:38–40. https://doi.org/10.1016/S1464-391X(06)70526-8

84. Daniels PH (2009) A brief overview of theories of PVC plasticization and methods used to evaluate PVC-plasticizer interaction. J Vinyl Addit Technol 15:219–223. https://doi.org/10.1002/VNL.20211

85. Messori M, Toselli M, Pilati F, Fabbri E, Fabbri P, Pasquali L, Nannarone S (2004) Prevention of plasticizer leaching from PVC medical devices by using organic–inorganic hybrid coatings. Polymer 45:805–813. https://doi.org/10.1016/J.POLYMER.2003.12.006

86. Zaikov GE, Gumargalieva KZ, Pokholok TV, Moiseev YV (2006) PVC wire coatings: part 1-ageing process dynamics. 39:79–125. https://doi.org/10.1080/00914039808041037

87. Li X, Xing Y, Jiang Y, Ding Y, Li W (2009) Antimicrobial activities of ZnO powder-coated PVC film to inactivate food pathogens. Int J Food Sci Technol 44:2161–2168. https://doi.org/10.1111/J.1365-2621.2009.02055.X

88. Dartman T, Shishoo R (2016) Studies of adhesion mechanisms between PVC coatings and different textile substrates. 22:317–335. https://doi.org/10.1177/152808379302200409

89. Mohamed A, Tuhaiwer A, Razzaq ZS (undefined 2016) Optical properties of polyvinyl chloride doped with DCM dye thin films. World Sci News (Bibliotekanauki.Pl)

90. Geng S, Shah FU, Liu P, Antzutkin ON, Oksman K (2017) Plasticizing and crosslinking effects of borate additives on the structure and properties of poly(vinyl acetate). RSC Adv 7:7483–7491. https://doi.org/10.1039/C6RA28574K

91. Zhang N, Liu P, Yi Y, Gibril ME, Wang S, Kong F (2021) Application of polyvinyl acetate/lignin copolymer as bio-based coating material and its effects on paper properties. Coatings 11:192. https://doi.org/10.3390/COATINGS11020192

92. Kolter K, Dashevsky A, Irfan M, Bodmeier R (2013) Polyvinyl acetate-based film coatings. Int J Pharm 457:470–479. https://doi.org/10.1016/J.IJPHARM.2013.08.077

93. Hagenmaier RD, Grohmann K (1999) Polyvinyl acetate as a high-gloss edible coating. J Food Sci 64:1064–1067. https://doi.org/10.1111/J.1365-2621.1999.TB12283.X

94. Sethi SK, Manik G (2021) A combined theoretical and experimental investigation on the wettability of MWCNT filled PVAc-g-PDMS easy-clean coating. Prog Org Coat 151:106092. https://doi.org/10.1016/j.porgcoat.2020.106092

95. Sethi SK, Manik G (2018) Recent progress in super hydrophobic/hydrophilic self-cleaning surfaces for various industrial applications: a review. Polym Plast Technol Eng 57:1932–1952. https://doi.org/10.1080/03602559.2018.1447128

96. Kadian S, Sethi SK, Manik G (2021) Recent advancements in synthesis and property control of graphene quantum dots for biomedical and optoelectronic applications. Mater Chem Front 5. https://doi.org/10.1039/d0qm00550a

97. Sethi SK, Shankar U, Manik G (2019) Fabrication and characterization of non-fluoro based transparent easy-clean coating formulations optimized from molecular dynamics simulation. Prog Org Coat 136. https://doi.org/10.1016/j.porgcoat.2019.105306

98. Sethi SK, Gogoi R, Manik G (2021) Plastics in self-cleaning applications. Ref Modul Mater Sci Mater Eng. https://doi.org/10.1016/B978-0-12-820352-1.00113-9

99. Sethi SK, Kadian S, Manik G (2022) A review of recent progress in molecular dynamics and coarse-grain simulations assisted understanding of wettability. Arch Comput Methods Eng 2021(1):1–27. https://doi.org/10.1007/S11831-021-09689-1

100. Sethi SK, Singh M, Manik G (2020) A multi-scale modeling and simulation study to investigate the effect of roughness of a surface on its self-cleaning performance. Mol Syst Des Eng 5:1277–1289. https://doi.org/10.1039/d0me00068j

101. Singh AK (2022) Surface engineering using PDMS and functionalized nanoparticles for superhydrophobic coatings: selective liquid repellence and tackling COVID-19. Prog Org Coat 171:107061

102. Sethi SK, Soni L, Manik G (2018) Component compatibility study of poly(dimethyl siloxane) with poly(vinyl acetate) of varying hydrolysis content: an atomistic and mesoscale simulation approach. J Mol Liq 272:73–83. https://doi.org/10.1016/J.MOLLIQ.2018.09.048

103. Trantidou T, Elani Y, Parsons E, Ces O (2017) Hydrophilic surface modification of PDMS for droplet microfluidics using a simple, quick, and robust method via PVA deposition. Microsyst Nanoeng 3(1):1–9

104. Sethi SK, Manik G, Sahoo SK (2019) Fundamentals of superhydrophobic surfaces. In: Superhydrophobic polymer coatings. Elsevier, pp 3–29. https://linkinghub.elsevier.com/retrieve/pii/B9780128166710000011. Accessed 29 Aug 2019

105. Sethi SK, Gogoi R, Verma A, Manik G (2022) How can the geometry of a rough surface affect its wettability?—a coarse-grained simulation analysis. Prog Org Coat 172:107062. https://doi.org/10.1016/J.PORGCOAT.2022.107062

106. Sethi SK, Soni L, Shankar U, Chauhan RP, Manik G (2020) A molecular dynamics simulation study to investigate poly(vinyl acetate)–poly(dimethyl siloxane) based easy-clean coating: an insight into the surface behavior and substrate interaction. J Mol Struct 1202:127342. https://doi.org/10.1016/j.molstruc.2019.127342

107. Sethi SK, Kadian S, Anubhav G, Chauhan RP, Manik G (2020) Fabrication and analysis of ZnO quantum dots based easy clean coating: a combined theoretical and experimental investigation. ChemistrySelect 5:8942–8950. https://doi.org/10.1002/slct.202001092

108. Sim S, Kim YM, Park YJ, Siddiqui MX, Gang Y, Lee J, Lee C, Suh HJ (2020) Determination of polyvinyl acetate in chewing gum using high performance liquid chromatography-evaporative light scattering detector and pyrolyzer-gas chromatography-mass spectrometry. Foods 9. https://doi.org/10.3390/FOODS9101473

109. Ghosh MK, Mittal KL (2018) Thermal curing in polyimide films and coatings, pp 207–248. https://doi.org/10.1201/9780203742945-8

110. Buchwalter LP (2012) Adhesion of polyimides to metal and ceramic surfaces: an overview 4:697–721. https://doi.org/10.1163/156856190X00612

111. Ghosh M (2018) Polyimides: fundamentals and applications
112. Maudgal S, St Clair TL (1984) Preparation and characterization of siloxane-containing thermoplastic polymides. Int J Adhes Adhes 4:87–90. https://doi.org/10.1016/0143-7496(84)90105-2
113. Flueckiger J, Bazargan V, Stoeber B, Cheung KC (2011) Characterization of postfabricated parylene C coatings inside PDMS microdevices. Sens Actuators, B Chem 160:864–874. https://doi.org/10.1016/J.SNB.2011.08.073
114. Golda-Cepa M, Engvall K, Hakkarainen M, Kotarba A (2020) Recent progress on parylene C polymer for biomedical applications: a review. Prog Org Coat 140:105493. https://doi.org/10.1016/J.PORGCOAT.2019.105493
115. Shin YS, Cho K, Lim SH, Chung S, Park SJ, Chung C, Han DC, Chang JK (2003) PDMS-based micro PCR chip with parylene coating. J Micromech Microeng 13:768. https://doi.org/10.1088/0960-1317/13/5/332
116. Sasaki H, Onoe H, Osaki T, Kawano R, Takeuchi S (2010) Parylene-coating in PDMS microfluidic channels prevents the absorption of fluorescent dyes. Sens Actuators, B Chem 150:478–482. https://doi.org/10.1016/J.SNB.2010.07.021
117. Sharifi H, Lahiji RR, Lin HC, Ye PD, Katehi LPB, Mohammadi S (2009) Characterization of parylene-N as flexible substrate and passivation layer for microwave and millimeter-wave integrated circuits. IEEE Trans Adv Packag 32:84–92. https://doi.org/10.1109/TADVP.2008.2006760
118. Verma A, Singh VK, Verma SK, Sharma A (2016) Human hair: a biodegradable composite fiber–a review. Int J Waste Resour 6(206):2
119. Verma A, Parashar A, Packirisamy M (2019) Effect of grain boundaries on the interfacial behaviour of graphene-polyethylene nanocomposite. Appl Surf Sci 470:1085–1092
120. Verma A, Gaur A, Singh VK (2017) Mechanical properties and microstructure of starch and sisal fiber biocomposite modified with epoxy resin. Mater Perform Charact 6(1):500–520
121. Jain N, Verma A, Singh VK (2019) Dynamic mechanical analysis and creep-recovery behaviour of polyvinyl alcohol based cross-linked biocomposite reinforced with basalt fiber. Mater Res Express 6(10):105373
122. Chaurasia A, Verma A, Parashar A, Mulik RS (2019) Experimental and computational studies to analyze the effect of h-BN nanosheets on mechanical behavior of h-BN/polyethylene nanocomposites. J Phys Chem C 123(32):20059–20070
123. Verma A, Kumar R, Parashar A (2019) Enhanced thermal transport across a bi-crystalline graphene–polymer interface: an atomistic approach. Phys Chem Chem Phys 21(11):6229–6237
124. Verma A, Negi P, Singh VK (2018) Physical and thermal characterization of chicken feather fiber and crumb rubber reformed epoxy resin hybrid composite. Adv Civ Eng Mater 7(1):538–557
125. Bharath KN, Madhu P, Gowda TG, Verma A, Sanjay MR, Siengchin S (2020) A novel approach for development of printed circuit board from biofiber based composites. Polym Compos 41(11):4550–4558
126. Verma A, Joshi K, Gaur A, Singh VK (2018) Starch-jute fiber hybrid biocomposite modified with an epoxy resin coating: fabrication and experimental characterization. J Mech Behav Mater 27(5–6)
127. Verma A, Singh C, Singh VK, Jain N (2019) Fabrication and characterization of chitosan-coated sisal fiber—Phytagel modified soy protein-based green composite. J Compos Mater 53(18):2481–2504
128. Rastogi S, Verma A, Singh VK (2020) Experimental response of nonwoven waste cellulose fabric-reinforced epoxy composites for high toughness and coating applications. Mater Perform Charact 9(1):151–172
129. Bharath KN, Madhu P, Gowda TY, Verma A, Sanjay MR, Siengchin S (2021) Mechanical and chemical properties evaluation of sheep wool fiber-reinforced vinylester and polyester composites. Mater Perform Charact 10(1):99–109

130. Verma A, Parashar A, Jain N, Singh VK, Rangappa SM, Siengchin S (2020) Surface modification techniques for the preparation of different novel biofibers for composites. In: Biofibers and biopolymers for biocomposites. Springer, Cham, pp 1–34

131. Verma A, Singh VK (2016) Experimental investigations on thermal properties of coconut shell particles in DAP solution for use in green composite applications. J Mater Sci Eng 5(3):1000242

132. Singh K, Jain N, Verma A, Singh VK, Chauhan S (2020) Functionalized graphite-reinforced cross-linked poly(vinyl alcohol) nanocomposites for vibration isolator application: morphology, mechanical, and thermal assessment. Mater Perform Charact 9(1):215–230

133. Verma A, Singh VK, Arif M (2016) Study of flame retardant and mechanical properties of coconut shell particles filled composite. Res Rev: J Mater Sci 4(3):1–5

134. Verma A, Jain N, Rastogi S, Dogra V, Sanjay SM, Siengchin S, Mansour R (2020) Mechanism, anti-corrosion protection and components of anti-corrosion polymer coatings. In: Polymer coatings. CRC Press, pp 53–66

135. Chaudhary A, Sharma S, Verma A (2022) WEDM machining of heat treated ASSAB'88 tool steel: a comprehensive experimental analysis. Mater Today: Proc 50:946–951

136. Bisht N, Verma A, Chauhan S, Singh VK (2021) Effect of functionalized silicon carbide nano-particles as additive in cross-linked PVA based composites for vibration damping application. J Vinyl Add Tech 27(4):920–932

137. Verma A, Jain N, Parashar A, Singh VK, Sanjay MR, Siengchin S (2020) Design and modeling of lightweight polymer composite structures. In: Lightweight polymer composite structures. CRC Press, pp 193–224

138. Dogra V, Kishore C, Verma A, Rana AK, Gaur A (2021) Fabrication and experimental testing of hybrid composite material having biodegradable bagasse fiber in a modified epoxy resin: evaluation of mechanical and morphological behavior. Appl Sci Eng Prog 14(4):661–667

139. Deji R, Verma A, Kaur N, Choudhary BC, Sharma RK (2022) Density functional theory study of carbon monoxide adsorption on transition metal doped armchair graphene nanoribbon. Mater Today: Proc 54:771–776

140. Verma A, Jain N, Parashar A, Gaur A, Sanjay MR, Siengchin S (2021) Lifecycle assessment of thermoplastic and thermosetting bamboo composites. In: Bamboo fiber composites. Springer, Singapore, pp 235–246

141. Deji R, Verma A, Choudhary BC, Sharma RK (2022) New insights into NO adsorption on alkali metal and transition metal doped graphene nanoribbon surface: a DFT approach. J Mol Graph Model 111:108109

142. Verma A, Jain N, Parashar A, Singh VK, Sanjay MR, Siengchin S (2020) Lightweight graphene composite materials. In: Lightweight polymer composite structures. CRC Press, pp 1–20

143. Verma A, Parashar A (2020) Characterization of 2D nanomaterials for energy storage. In: Recent advances in theoretical, applied, computational and experimental mechanics. Springer, Singapore, pp 221–226

144. Deji R, Jyoti R, Verma A, Choudhary BC, Sharma RK (2022) A theoretical study of HCN adsorption and width effect on co-doped armchair graphene nanoribbon. Comput Theor Chem 1209:113592

145. Verma A, Samant SS (2016) Inspection of hydrodynamic lubrication in infinitely long journal bearing with oscillating journal velocity. J Appl Mech Eng 5(3):1–7

146. Kataria A, Verma A, Sanjay MR, Siengchin S (2022) Molecular modeling of 2D graphene grain boundaries: mechanical and fracture aspects. Mater Today: Proc 52:2404–2408

147. Arpitha GR, Verma A, Sanjay MR, Siengchin S (2021) Preparation and experimental investigation on mechanical and tribological performance of hemp-glass fiber reinforced laminated composites for lightweight applications. Adv Civ Eng Mater 10(1):427–439

148. Verma A, Jain N, Rangappa SM, Siengchin S, Jawaid M (2021) Natural fibers based bio-phenolic composites. In: Phenolic polymers based composite materials. Springer, Singapore, pp 153–168

149. Verma A, Parashar A, Singh SK, Jain N, Sanjay SM, Siengchin S (2020) Modeling and simulation in polymer coatings. In: Polymer coatings. CRC Press, pp 309–324
150. Verma A, Singh VK. Experimental characterization of modified epoxy resin assorted with almond shell particles. ESSENCE-Int J Environ Rehabil Conserv 36
151. Deji R, Verma A, Kaur N, Choudhary BC, Sharma RK (2022) Adsorption chemistry of co-doped graphene nanoribbon and its derivatives towards carbon based gases for gas sensing applications: quantum DFT investigation. Mater Sci Semicond Process 146:106670
152. Verma A (2022) A perspective on the potential material candidate for railway sector applications: PVA based functionalized graphene reinforced composite. Appl Sci Eng Prog 15(2):5727–5727
153. Verma A, Jain N, Singh K, Singh VK, Rangappa SM, Siengchin S (2022) PVA-based blends and composites. In: Biodegradable polymers, blends and composites. Woodhead Publishing, pp 309–326
154. Raja S, Verma A, Rangappa SM, Siengchin S (2022) Development and experimental analysis of polymer based composite bipolar plate using Aquila Taguchi optimization: design of experiments. Polym Compos 43(8):5522–5533
155. Prabhakaran S, Sharma S, Verma A, Rangappa SM, Siengchin S (2022) Mechanical, thermal, and acoustical studies on natural alternative material for partition walls: a novel experimental investigation. Polym Compos
156. Verma A, Ogata S (2023) Magnesium based alloys for reinforcing biopolymer composites and coatings: a critical overview on biomedical materials. Adv Ind Eng Polym Res. https://doi.org/10.1016/j.aiepr.2023.01.002
157. Verma A, Jain N, Sethi SK (2022) Modeling and simulation of graphene-based composites. In: Innovations in graphene-based polymer composites. Woodhead Publishing, pp 167–198
158. Chaturvedi S, Verma A, Sethi SK, Rangappa SM, Siengchin S (2022) Stalk fibers (rice, wheat, barley, etc.) composites and applications. In: Plant fibers, their composites, and applications. Woodhead Publishing, pp 347–362
159. Arpitha GR, Verma A, MR S, Gorbatyuk S, Khan A, Sobahi TR, Asiri AM, Siengchin S (2022) Bio-composite film from corn starch based vetiver cellulose. J Nat Fibers 1–11
160. Thimmaiah SH, Narayanappa K, Thyavihalli Girijappa Y, Gulihonenahali Rajakumara A, Hemath M, Thiagamani SMK, Verma A (2022) An artificial neural network and Taguchi prediction on wear characteristics of Kenaf–Kevlar fabric reinforced hybrid polyester composites. Polym Compos
161. Chaturvedi S, Verma A, Singh SK, Ogata S (2022) EAM inter-atomic potential—its implication on nickel, copper, and aluminum (and their alloys). In: Forcefields for atomistic-scale simulations: materials and applications. Springer, Singapore, pp 133–156
162. Verma A, Sharma S (2022) Atomistic simulations to study thermal effects and strain rate on mechanical and fracture properties of graphene like BC3. In: Forcefields for atomistic-scale simulations: materials and applications. Springer, Singapore, pp 237–252
163. Chaturvedi S, Verma A, Sethi SK, Ogata S (2022) Defect energy calculations of nickel, copper and aluminium (and their alloys): molecular dynamics approach. In: Forcefields for atomistic-scale simulations: materials and applications. Springer, Singapore, pp 157–186
164. Kataria A, Verma A, Sethi SK, Ogata S (2022) Introduction to interatomic potentials/forcefields. In: Forcefields for atomistic-scale simulations: materials and applications. Springer, Singapore, pp 21–49
165. Shankar U, Gogoi R, Sethi SK, Verma A (2022) Introduction to materials studio software for the atomistic-scale simulations. In: Forcefields for atomistic-scale simulations: materials and applications. Springer, Singapore, pp 299–313
166. Shankar U, Sethi SK, Verma A (2022) Forcefields and modeling of polymer coatings and nanocomposites. In: Forcefields for atomistic-scale simulations: materials and applications. Springer, Singapore, pp 81–98
167. Verma A, Ogata S (2022) Computational modelling of deformation and failure of bone at molecular scale. In: Forcefields for atomistic-scale simulations: materials and applications. Springer, Singapore, pp 253–268

168. Homer ER, Verma A, Britton D, Johnson OK, Thompson GB (2022, July) Simulated migration behavior of metastable $\Sigma 3$ (11 8 5) incoherent twin grain boundaries. IOP Conf Ser: Mater Sci Eng 1249(1):012019

169. Verma A, Parashar A, van Duin AC (2022) Graphene-reinforced polymeric membranes for water desalination and gas separation/barrier applications. In: Innovations in graphene-based polymer composites. Woodhead Publishing, pp 133–165

170. Verma A, Jain N, Sanjay MR, Siengchin S Viscoelastic properties of completely biodegradable polymer-based composites. In: Vibration and damping behavior of biocomposites. CRC Press, pp 173–188

171. Verma A, Jain N, Mishra RR (2022) Applications and drawbacks of epoxy/natural fiber composites. In: Handbook of epoxy/fiber composites. Springer, Singapore, pp 1–15

172. Lila MK, Verma A, Bhurat SS (2022) Impact behaviors of epoxy/synthetic fiber composites. In: Handbook of epoxy/fiber composites. Springer, Singapore, pp 1–18

173. Verma A, Parashar A (2017) The effect of STW defects on the mechanical properties and fracture toughness of pristine and hydrogenated graphene. Phys Chem Chem Phys 19(24):16023–16037

174. Verma A, Parashar A (2018) Molecular dynamics based simulations to study failure morphology of hydroxyl and epoxide functionalised graphene. Comput Mater Sci 143:15–26

175. Verma A, Parashar A (2018) Molecular dynamics based simulations to study the fracture strength of monolayer graphene oxide. Nanotechnology 29(11):115706

176. Verma A, Parashar A (2018) Structural and chemical insights into thermal transport for strained functionalised graphene: a molecular dynamics study. Mater Res Express 5(11):115605

177. Verma A, Parashar A, Packirisamy M (2018) Tailoring the failure morphology of 2D bicrystalline graphene oxide. J Appl Phys 124(1):015102

178. Verma A, Parashar A (2018) Reactive force field based atomistic simulations to study fracture toughness of bicrystalline graphene functionalised with oxide groups. Diam Relat Mater 88:193–203

179. Singla V, Verma A, Parashar A (2018) A molecular dynamics based study to estimate the point defects formation energies in graphene containing STW defects. Mater Res Express 6(1):015606

180. Verma A, Parashar A, Packirisamy M (2019) Role of chemical adatoms in fracture mechanics of graphene nanolayer. Mater Today: Proc 11:920–924

181. Chaudhary A, Sharma S, Verma A (2022) Optimization of WEDM process parameters for machining of heat treated ASSAB'88 tool steel using Response surface methodology (RSM). Mater Today: Proc 50:917–922

182. Verma A, Zhang W, Van Duin AC (2021) ReaxFF reactive molecular dynamics simulations to study the interfacial dynamics between defective h-BN nanosheets and water nanodroplets. Phys Chem Chem Phys 23(18):10822–10834

183. Verma A, Parashar A, Packirisamy M (2018) Atomistic modeling of graphene/hexagonal boron nitride polymer nanocomposites: a review. Wiley Interdiscip Rev: Comput Mol Sci 8(3):e1346

184. Verma A, Baurai K, Sanjay MR, Siengchin S (2020) Mechanical, microstructural, and thermal characterization insights of pyrolyzed carbon black from waste tires reinforced epoxy nanocomposites for coating application. Polym Compos 41(1):338–349

185. Verma A, Budiyal L, Sanjay MR, Siengchin S (2019) Processing and characterization analysis of pyrolyzed oil rubber (from waste tires)-epoxy polymer blend composite for lightweight structures and coatings applications. Polym Eng Sci 59(10):2041–2051

186. Verma A, Negi P, Singh VK (2019) Experimental analysis on carbon residuum transformed epoxy resin: chicken feather fiber hybrid composite. Polym Compos 40(7):2690–2699

187. Verma A, Singh VK (2018) Mechanical, microstructural and thermal characterization of epoxy-based human hair-reinforced composites. J Test Eval 47(2):1193–1215

188. Jain N, Verma A, Ogata S, Sanjay MR, Siengchin S (2022) Application of machine learning in determining the mechanical properties of materials. In: Machine learning applied to composite materials. Springer, Singapore, pp 99–113

189. Kataria A, Verma A, Sanjay MR, Siengchin S, Jawaid M (2022) Physical, morphological, structural, thermal, and tensile properties of coir fibers. In: Coir fiber and its composites. Woodhead Publishing, pp 79–107

190. Kumar G, Mishra RR, Verma A (2022) Introduction to molecular dynamics simulations. In: Forcefields for atomistic-scale simulations: materials and applications. Springer, Singapore, pp 1–19. https://doi.org/10.1007/978-981-19-3092-8_1

191. Arpitha GR, Jain N, Verma A, Madhusudhan M (2022) Corncob bio-waste and boron nitride particles reinforced epoxy-based composites for lightweight applications: fabrication and characterization. Biomass Convers Biorefinery 1–8. https://doi.org/10.1007/s13399-022-037 17-1

192. Kataria A, Chaturvedi S, Chaudhary V, Verma A, Sanjay NJMR, Siengchin S (2023) Cellulose fiber-reinforced composites—history of evolution, chemistry, and structure. In: Cellulose fibre reinforced composites. Woodhead Publishing, pp 1–22. https://doi.org/10.1016/B978-0-323-90125-3.00012-4

193. Chaturvedi S, Kataria A, Chaudhary V, Verma A, Sanjay NJMR, Siengchin S (2023) Bionanocomposites reinforced with cellulose fibers and agro-industrial wastes. In: Cellulose fibre reinforced composites. Woodhead Publishing, pp 1–22. https://doi.org/10.1016/B978-0-323-90125-3.00017-3

194. Arpitha GR, Mohit H, Madhu P, Verma A (2023) Effect of sugarcane bagasse and alumina reinforcements on physical, mechanical, and thermal characteristics of epoxy composites using artificial neural networks and response surface methodology. Biomass Convers Biorefinery. https://doi.org/10.1007/s13399-023-03886-7

Chapter 5
Basics of Density Functional Theory, Molecular Dynamics, and Monte Carlo Simulation Techniques in Materials Science

Sandeep Kumar Singh, Ankur Chaurasia, and Akarsh Verma

1 Introduction

Initially, Computational science was an essential tool for physics and chemistry long ago, but in materials science, it was always taken for granted. But now, there is a slight shift in the thinking of the researchers. Computational techniques play a vital role in materials science to understand behavior and quantify various material properties [1]. Experimental techniques are always more time-consuming and more laborious. Although experimental techniques are the most accurate way to study the behavior of any engineering material. However, computational techniques are also an essential tool to provide the right direction for these experimental techniques [2]. Nowadays, computational materials science is a mainstream part of materials science, including computational techniques, which strongly involve the knowledge of physics, chemistry, and mathematics. Computational techniques are a powerful partner of experimental studies. Due to continuous increments in the power of computers and supercomputers, the computational techniques field is growing rapidly [3]. There are many computational techniques, which are used nowadays to study engineering materials.

S. K. Singh
Department of Mechanical and Materials Engineering, Queen's University, Kingston, Canada

A. Chaurasia
Department of Mechanical Engineering, Pandit Deendayal Energy University, Gandhinagar 382007, India

A. Verma (✉)
Department of Mechanical Engineering, University of Petroleum and Energy Studies, Dehradun 248007, India
e-mail: akarshverma007@gmail.com

Department of Mechanical Science and Bioengineering, Osaka University, Osaka 560-8531, Japan

A material system can be mainly defined by three different levels of scales. The electronic level consists of a nucleus and electrons, the molecular or atomic level, and the macro level consists of finite elements. The most accurate computational technique is based on electronic structure level and density functional theory (DFT). At the atomistic level, molecular dynamics (MD)-based simulations and Monte Carlo simulations come. And at the macro level, finite element analysis (FEA) is utilized very often.

2   Density Functional Theory

As mentioned in the previous section, the DFT technique is the most accurate computer simulation technique. The governing laws of DFT are based on small particle interactions, known as quantum mechanics. This method is entirely dependent on quantum mechanics, so there is not any fitting parameter used unlike molecular dynamics(MD) methods [4].

This method is dependent on the Schrodinger wave equation, given by the following equation:-

$$\widehat{H}\Psi = E\Psi \tag{1}$$

Here, $\widehat{H}$ is the Hamiltonian operator, Ψ is the wave function, and E is the energy of the system.

However, its computational cost is very high for many atom systems. So, it is not feasible to use DFT in the many-body system for dynamic studies.

3   Molecular Dynamics Simulations

To overcome the computational cost issue of DFT simulations, MD simulation is a great substitution for both static and dynamic kinds of simulations. Molecular Dynamics Simulation can be used to study the material properties of metals, alloys, nanomaterials, polymers, etc. MD simulations use the laws of motion, energy, and forces to model interactions between all the atoms in a system. Through this, researchers can study the behavior of materials under different conditions, including temperature, pressure, and alloy composition. Furthermore, the results of MD simulations can be compared to experimental data, which will provide a better understanding of the system's behavior. In molecular dynamics simulations, Newton's law of motion is integrated with respect to time evolution [5, 6]. There are the following steps involved in typical molecular dynamics simulation:

- Initially, the position and initial velocities of each atom in the system are defined. And the forces per atom can be calculated by the provided interatomic potential.

- By using this default knowledge about the system, the initial positions are further changed to lower energy state positions, by using a small time interval during minimization, which assigns new positions and velocity to the atoms, which results in a change in various properties function of position and velocity as shown in Fig. 1. For time integration, different integration schemes can be used, and the velocity Verlet algorithm [7] is one of them. Energy minimization occurs at 0 K and various minimization techniques can be used to perform minimization [8–10]. (Conjugate gradient method [11], Steepest descend method, etc.)
- This lower energy trend is followed till the equilibration is reached (Defined by no change in system properties with respect to time evolution).
- The raw data during minimization and equilibration is recorded in dump files (configuration files) which include the position of atoms, temperature, energy, velocity, force, etc. [12–14]. These files are used for visualization purposes at various timesteps with the help of Open visualization tool (OVITO) software. Some properties and analyses can be directly done from OVITO itself, like diffusion, grain boundary study, radial distribution function (RDF), crack growth, shock wave propagation, irradiation damage, etc. [15].

As compared to the DFT method, MD is significantly fast and computationally cheap. However, MD simulations have some limitations:

- It is well known that the accuracy of MD simulations is entirely dependent on empirical interatomic potentials. And these empirical potentials are available in very limited quantity, especially for multi-component systems. Additionally, there is always a question about the accuracy of empirical potential.
- The length scale of molecular dynamics simulation is still not macroscopic that's why again accuracy can be questioned. In addition, the timescale is also up to

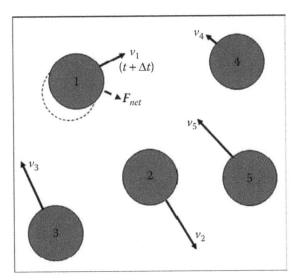

Fig. 1 Atom movement in five-atom system during minimization (From dotted position to solid position) [Adapted from Ref. 1]

nanoseconds only. So, dynamic simulations can be performed for a very less time duration.

3.1 Interatomic Potential for Molecular Dynamics Simulations

As discussed earlier, the accuracy of any molecular dynamics simulation entirely depends on the chosen interatomic potential. In molecular dynamics simulations, atoms are considered as the sphere. It is assumed that there is not any nucleus and electrons inside the atom [1]. The force interaction among these atoms was given by empirical relation, developed by first principle calculations [16]. These relations are known as interatomic potentials. In molecular dynamics, the electrostatic potential energy of an atom or molecule can be expressed in terms of the attractive or repulsive forces between its constituent atoms or molecules [17]. An entity with a positive electrostatic potential energy is attracted to an entity with a negative electrostatic potential energy, while an entity with a zero electrostatic potential energy is unaffected by any other entity's electrostatic potential energy. The electrostatic potential energy of a molecule is determined by the number of protons in the molecule, the strength of the covalent bonds between the atoms, and the electron degeneracy of the molecule. In general, molecules with more protons have higher electrostatic potential energy or free energy, while molecules with fewer protons have lower electrostatic potential energy. Likewise, molecules with stronger covalent bonds have higher free energy (potential energy), while molecules with weaker covalent bonds have lower electrostatic potential energy. Finally, molecules with more electron degeneracy have higher electrostatic potential energy, while molecules with less electron degeneracy have lower electrostatic potential energy. This electrostatic potential energy can be formulated by some empirical relations, which are known as interatomic potentials. There are many types of potentials used in molecular dynamics simulations.

For solid–solid interaction (Metals and alloys), mainly embedded atom method (EAM) and modified embedded atom method potentials (MEAM) can be used.

$$E_i = F_\alpha \left(\sum_{j \neq i} \rho_\beta \left(r_{ij} \right) \right) + \frac{1}{2} \sum_{j \neq i} \phi_{\alpha\beta} \left(r_{ij} \right) \tag{2}$$

Equation 2 consists of empirical relation of energy as a function of embedding function F_α and pair potential $\phi_{\alpha\beta}$. Embedding function F_α is further function of electron density ρ_β. For polymer interactions reactive force field (reaxFF), Tersoff, etc. potentials are used. For fluid interaction L-J, Morse, etc. potentials are mainly used.

Nowadays, machine learning-based [18] and deep learning-based [19] interatomic potentials are also developed. The data used for training for various structures (energy and force) was obtained from DFT calculations. The size of the structure is kept at a

Table 1 Various ensembles used in molecular dynamics and Monte Carlo simulations

Sr. No	Ensemble	Fixed variables	Remarks
1	Microcanonical ensemble (NVE)	N, V, E	Isolated system, Very common in MD $S = k \ln\_NVE$
2	Canonical ensemble (NVT)	N, V, T	Very common in MC, Common in MD $F = -kT \ln\_NVT$
3	Isobaric-isothermal ensemble (NPT)	N, P, T	$G = -kT \ln\_NPT$
4	Grand canonical ensemble (μVT)	μ, V, T	Rarely in MD, more in MC $\mu = -kT \ln\_NPT/N$
5	Semi grand canonical ensemble	P, T	Hybrid MCMD
6	Variance controlled semi grand canonical ensemble		Hybrid MCMD

tractable level, as DFT is significantly costly from the computational point of view. For the development of deep learning-based potential, DeePMD kit [20] can be used.

3.2 Ensembles Used in Molecular Dynamics Simulations

Various ensembles are used in MD and Monte Carlo simulations. These ensembles provide the required simulation environment for any simulation. For instance, for applying external force (tensile, compressive, shear loading case) on a solid structure, maximum time NPT ensemble can be used. For studying the effect of impact shock loading, the NVE ensemble can be used, as the shock process is adiabatic. The ensembles used in typical molecular dynamics and Monte Carlo simulations are shown in Table 1.

4 Monte Carlo Simulations

Monte Carlo simulations word came into existence from the city of Monte Carlo, which is famous for its casino. Here there was a need for a probabilistic model which can increase the winning probability of any person. Monte Carlo simulation is a common technique used in materials science for simulating the behavior of materials on the atomic level. By using Monte Carlo simulation, materials scientists can predict the properties of complex materials and optimize the design of new materials. Monte Carlo simulation can also provide insight into the physical and chemical processes that take place within materials, allowing for a deeper understanding and more accurate prediction of material behavior. Monte Carlo simulation can be used to simulate the behavior of a variety of materials, such as metals [21], alloys, ceramics [22], and polymers [23]. Materials scientists can use this technique to study the structure and properties of a material, such as its elasticity, electrical and thermal properties, permeability, fracture toughness, and corrosion resistance. Monte Carlo simulations can also be used to optimize the design of a material, by searching for optimal conditions that maximize a chosen material property. Optimization through simulation can lead to new, improved materials with enhanced performance. Monte Carlo simulations can also be used to study the relationship between a material's microstructure and its macroscopic properties. By simulating the microstructure at various levels of detail, scientists can better understand which microstructural features are the most important to consider when designing a material. Simulations can also be used to determine the effects of different processing steps, like heat treatments, on a material's properties. Finally, Monte Carlo simulations can be used to predict the behavior of complex material systems such as multi-layer composites, thin films, and functionally graded materials. Monte Carlo simulations can be combined with powerful optimization algorithms to identify the material parameters that lead to a desired macroscopic property or behavior. For example, simulations can be used to determine the best combination of parameters that lead to the highest strength or the lowest heat conductivity. Through the optimization process, scientists can discover optimal combinations of parameters that were previously unknown, providing powerful tools for material design. Monte Carlo simulations can also be used to predict the failure of a material under certain load conditions, providing insight into the safety and reliability of components.

Monte Carlo simulations are not limited to determining material properties. They are also used to study complex systems such as biochemical networks, weather patterns, and economic trends. By simulating thousands or even millions of scenarios, researchers can gain insight into the behavior of these systems. The results of the simulations can often be used to design or improve existing systems, providing a powerful tool for designing intelligent systems. Furthermore, Monte Carlo simulations are increasingly being used to analyze and improve decision-making processes, helping to make better decisions in a variety of fields. The schematic diagram for a typical Monte Carlo simulation model is shown in Fig. 2.

5 Hybrid Monte Carlo/Molecular Dynamics Simulations

Monte Carlo Simulations (MCS) and Molecular Dynamics (MD) are powerful computational techniques that allow scientists to simulate a variety of complex material systems. High entropy alloys (HEAs) [3, 25–28] are a particular type of material that have generated significant interest in recent years due to their unique combination of physical and chemical properties. MCS and MD have been used to study a range of properties in these materials, including structural, thermodynamic, and mechanical properties. Recent advances in computational capabilities have enabled researchers to gain further insights into the behavior of HEAs through MCS and MD simulations.

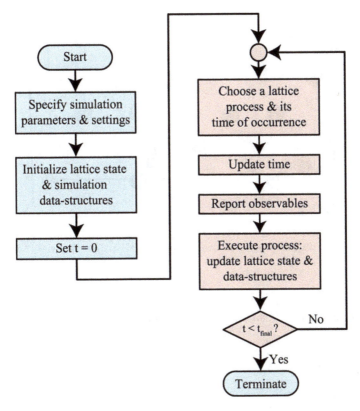

Fig. 2 Schematic diagram of typical Monte Carlo simulation [Adapted from reference 24]

MCS and MD simulations are invaluable tools for studying the properties of high entropy alloys (HEAs). By standalone MD simulations, the structure obtained will be of random type by using random numbers. After proceeding further, with this random alloy we can get the results of only one set pattern of atoms. But using MCS, we can get the most stable configuration of HEA, by repetitively swapping the place of parent elements, as shown in Fig. 3. By combining MCS with MD simulations, the results obtained from MD simulations can be improved. These simulations allow researchers to study a range of parameters, such as temperature, pressure, and alloy composition, to determine how they affect the material's properties. By comparing the output of the simulations to experimental data, researchers can gain a better understanding of the behavior of particular material. Additionally, these simulations can be used to optimize the properties of the alloy for specific applications and to study the mechanistic effects of phase transitions and microstructural evolution in various materials, especially in HEAs [29–31]. To study the effect of chemical short range order [29, 32–34] on various properties of random alloys, hybrid Monte Carlo molecular dynamics simulation is an essential tool. By swapping the parent atoms and calculating the chemical potential difference between two entities, chemical short

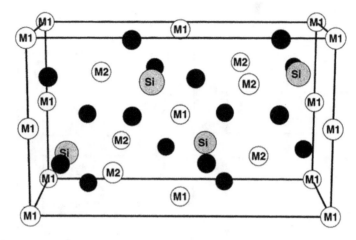

Fig. 3 Schematic diagram showing the two possible sites M1 and M2 in olivine [Adapted from Ref. 36]

range order can be calculated [35, 36]. Lastly, the authors have a vast experience in the field of molecular dynamics and polymer composites [37–112].

6 Summary

Computational modeling and simulation techniques are an important part of materials science. To study engineering materials experimentally, it is essential to find the right direction of research by using computational techniques. To opt, which computational technique is better for a particular study, it is always necessary to understand the required accuracy and availability of computational equipment. This chapter would help in making an understanding of nanoscale simulation techniques for various applications.

Acknowledgements Corresponding Author "Akarsh Verma" is grateful to the monetary support provided by the University of Petroleum and Energy Studies (UPES)-SEED Grant program.

References

1. Lee JG (2016) Computational materials science, 2nd edn. CRC Press, Taylor & Francis, Boca Raton
2. Singh SK, Parashar A (2021) Defect dynamics and uniaxial tensile deformation of equi and non-equi-atomic configuration of multi-elemental alloys. Mater Chem Phys 266:124549
3. Murty BS, Yeh JW, Ranganathan S (2014) Alloy Design in the twenty-first century. High Entropy Alloys. Elsevier, pp 57–76
4. Zhao W, Li W, Li X, Gong S, Vitos L, Sun Z (2019) Thermo-mechanical properties of Ni-Mo solid solutions: a first-principles study. Comput Mater Sci 158:140–148
5. Sharma AK, Sharma SS, Singh SK, Parashar A (2021) Atomistic simulations to study the effect of helium nanobubble on the shear deformation of nickel crystal. J Nucl Mater 557:153245
6. Singh SK, Parashar A (2022) Effect of lattice distortion and grain size on the crack tip behaviour in Co-Cr-Cu-Fe-Ni under mode-I and mode-II loading. Eng Fract Mech 274:108809
7. Grubmüller H, Heller H, Windemuth A, Schulten K (1991) Generalized Verlet algorithm for efficient molecular dynamics simulations with long-range interactions. Mol Simul 6:121–142
8. Chaturvedi S, Verma A, Singh SK, Ogata S (2022) EAM inter-atomic potential—its implication on nickel, copper, and aluminum (and their alloys). In: Forcefields for atomistic-scale simulations: materials and applications. Springer, Singapore, pp 133–156
9. Verma A, Parashar A, Singh SK, Jain N, Sanjay SM, Siengchin S (2020) Modeling and simulation in polymer coatings. In: Polymer coatings. CRC Press, pp 309–324
10. Kumar Singh S, Chaurasia A, Parashar A (2022) Atomistic simulations to study shock and ultrashort pulse response of high entropy alloy. Mater. Today Proc
11. Grujicic M, Zhou XW (1995) Atomistic simulation of thermally activated glide of dislocations in FeNiCrN austenite. 190
12. Kumar Singh S, Parashar A (2021) Atomistic simulations to study crack tip behaviour in multi-elemental alloys. Eng Fract Mech 243:107536
13. Singh SK, Parashar A (2022) Shock resistance capability of multi-principal elemental alloys as a function of lattice distortion and grain size. J Appl Phys 132:095903
14. Chaurasia A, Kumar Singh S, Parashar A (2022) Atomistic scale insight to investigate the strain rate effect on mechanical response of boron nitride nanosheet reinforced nanocomposites. Mater. Today Proc
15. Kumar Singh S, Parashar A (2022) Effect of lattice distortion and nanovoids on the shock compression behavior of (Co-Cr-Cu-Fe-Ni) high entropy alloy. Comput Mater Sci 209:111402
16. Plimpton S (1995) Fast parallel algorithms for short-range molecular dynamics. J Comput Phys 117:1–19
17. Koči L, Bringa EM, Ivanov DS, Hawreliak J, McNaney J, Higginbotham A, Zhigilei LV, Belonoshko AB, Remington BA, Ahuja R (2006) Simulation of shock-induced melting of Ni using molecular dynamics coupled to a two-temperature model. Phys Rev B Condens Matter Mater Phys 74:012101
18. Byggmästar J, Nordlund K, Djurabekova F (2022) Simple machine-learned interatomic potentials for complex alloys. Phys Rev Mater 6:083801
19. Zhang L, Qian K, Schuller BW, Shibuta Y (2021) Prediction on mechanical properties of non-equiatomic high-entropy alloy by atomistic simulation and machine learning. Metals (Basel) 11
20. Wang H, Zhang L, Han J, Weinan E (2018) DeePMD-kit: a deep learning package for many-body potential energy representation and molecular dynamics. Comput Phys Commun 228:178–184
21. Esfandiarpour A, Alvarez-Donado R, Papanikolaou S, Alava M (2022) Atomistic simulations of dislocation plasticity in concentrated VCoNi medium entropy alloys: effects of lattice distortion and short range order. Front Mater 9
22. Liu S, Tian Z, Shen L, Qiu M, Gao X (2021) Monte Carlo simulation of ceramic grain growth during laser ablation processing. Optik (Stuttg) 227:165569

23. Gervasio M, Lu K 2019 Monte Carlo simulation modeling of nanoparticle–polymer cosuspensions. Langmuir 35:61–70
24. Pineda M, Stamatakis M (2022) Kinetic Monte Carlo simulations for heterogeneous catalysis: fundamentals, current status, and challenges J Chem Phys 156:120902
25. Wu Y, Bönisch M, Alkan S, Abuzaid W, Sehitoglu H (2018) Experimental determination of latent hardening coefficients in FeMnNiCoCr. Int J Plast 105:239–260
26. Shaginyan LR, Britun VF, Krapivka NA, Firstov SA, Kotko AV, Gorban VF (2018) The properties of Cr–Co–Cu–Fe–Ni Alloy films deposited by magnetron sputtering powder. Metall Met Ceram 57:293–300
27. Yeh JW (2016) Physical metallurgy. High-entropy alloys: fundamentals and applications. pp 51–113
28. Divinski SV, Pokoev AV, Esakkiraja N, Paul A (2018) A Mystery of "sluggish diffusion" in high-entropy alloys: the truth or a myth? Diffus Found 17:69–104
29. Zhao S, Stocks GM, Zhang Y (2016) Defect energetics of concentrated solid-solution alloys from ab initio calculations: Ni0.5Co0.5, Ni0.5Fe0.5, Ni0.8Fe0.2 and Ni0.8Cr0.2. Phys Chem Chem Phys 18:24043–24056
30. Huang X, Liu L, Duan X, Liao W, Huang J, Sun H, Yu C (2021) Atomistic simulation of chemical short-range order in HfNbTaZr high entropy alloy based on a newly-developed interatomic potential. Mater Des 202:109560
31. Lokman Ali M, Shinzato S, Wang V, Shen Z, Du JP, Ogata S (2020) An atomistic modeling study of the relationship between critical resolved shear stress and atomic structure distortion in FcC high entropy alloys ® relationship in random solid solution and chemical-short-range-order alloys ®. Mater Trans 61:605–609
32. Jian WR, Wang L, Bi W, Xu S, Beyerlein IJ (2021) Role of local chemical fluctuations in the melting of medium entropy alloy CoCrNi Appl. Phys Lett 119:121904
33. Xie Z, Jian WR, Xu S, Beyerlein IJ, Zhang X, Wang Z, Yao X (2021) Role of local chemical fluctuations in the shock dynamics of medium entropy alloy. CoCrNi Acta Mater 221:117380
34. Liu D, Wang Q, Wang J, Chen XF, Jiang P, Yuan FP, Cheng ZY, Ma E, Wu XL (2021) Chemical short-range order in Fe50Mn30Co10Cr10 high-entropy alloy. Mater Today Nano 16:100139
35. Najafabadi R, Wang HY, Srolovitz DJ, LeSar R (1991) A new method for the simulation of alloys: application to interfacial segregation, vol 39
36. Purton JA, Barrera GD, Allan NL, Blundy JD (1998) Monte Carlo and hybrid Monte Carlo/molecular dynamics approaches to order−disorder in alloys, oxides, and silicates. J Phys Chem B 102:5202–5207
37. Deji R, Verma A, Kaur N, Choudhary BC, Sharma RK (2022) Density functional theory study of carbon monoxide adsorption on transition metal doped armchair graphene nanoribbon. Mater Today: Proc 54:771–776
38. Deji R, Verma A, Choudhary BC, Sharma RK (2022) New insights into NO adsorption on alkali metal and transition metal doped graphene nanoribbon surface: a DFT approach. J Mol Graph Model 111:108109
39. Verma A, Jain N, Parashar A, Singh VK, Sanjay MR, Siengchin S (2020) Lightweight graphene composite materials. In: Lightweight polymer composite structures. CRC Press, pp 1–20
40. Verma A, Parashar A (2020) Characterization of 2D nanomaterials for energy storage. In: Recent advances in theoretical, applied, computational and experimental mechanics. Springer, Singapore, pp 221–226
41. Deji R, Jyoti R, Verma A, Choudhary BC, Sharma RK (2022) A theoretical study of HCN adsorption and width effect on co-doped armchair graphene nanoribbon. Comput Theor Chem 1209:113592
42. Verma A, Parashar A, Packirisamy M (2019) Effect of grain boundaries on the interfacial behaviour of graphene-polyethylene nanocomposite. Appl Surf Sci 470:1085–1092
43. Deji R, Verma A, Kaur N, Choudhary BC, Sharma RK (2022) Adsorption chemistry of co-doped graphene nanoribbon and its derivatives towards carbon based gases for gas sensing applications: quantum DFT investigation. Mater Sci Semicond Process 146:106670

44. Verma A, Jain N, Sethi SK (2022) Modeling and simulation of graphene-based composites. In: Innovations in graphene-based polymer composites. Woodhead Publishing, pp 167–198
45. Verma A, Parashar A, van Duin AC (2022) Graphene-reinforced polymeric membranes for water desalination and gas separation/barrier applications. In: Innovations in graphene-based polymer composites. Woodhead Publishing, pp. 133–165
46. Verma A, Parashar A (2017) The effect of STW defects on the mechanical properties and fracture toughness of pristine and hydrogenated graphene. Phys Chem Chem Phys 19(24):16023–16037
47. Verma A, Parashar A (2018) Molecular dynamics based simulations to study failure morphology of hydroxyl and epoxide functionalised graphene. Comput Mater Sci 143:15–26
48. Verma A, Parashar A (2018) Molecular dynamics based simulations to study the fracture strength of monolayer graphene oxide. Nanotechnology 29(11):115706
49. Verma A, Parashar A (2018) Structural and chemical insights into thermal transport for strained functionalised graphene: a molecular dynamics study. Mater Res Express 5(11):115605
50. Verma A, Parashar A, Packirisamy M (2018) Tailoring the failure morphology of 2D bicrystalline graphene oxide. J Appl Phys 124(1):015102
51. Verma A, Parashar A (2018) Reactive force field based atomistic simulations to study fracture toughness of bicrystalline graphene functionalised with oxide groups. Diam Relat Mater 88:193–203
52. Singla V, Verma A, Parashar A (2018) A molecular dynamics based study to estimate the point defects formation energies in graphene containing STW defects. Mater Res Express 6(1):015606
53. Verma A, Parashar A, Packirisamy M (2019) Role of chemical adatoms in fracture mechanics of graphene nanolayer. Mater Today: Proc 11:920–924
54. Verma A, Parashar A, Packirisamy M (2018) Atomistic modeling of graphene/hexagonal boron nitride polymer nanocomposites: a review. Wiley Interdiscip Rev: Comput Mol Sci 8(3):e1346
55. Verma A, Kumar R, Parashar A (2019) Enhanced thermal transport across a bi-crystalline graphene–polymer interface: an atomistic approach. Phys Chem Chem Phys 21(11):6229–6237
56. Verma A, GaurA, Singh VK (2017) Mechanical properties and microstructure of starch and sisal fiber biocomposite modified with epoxy resin mechanical properties and microstructure modified with epoxy resin. 6(1):500–520
57. Verma A, Parashar A, Jain N, Singh VK, Rangappa SM, Siengchin S (2020) Surface modification techniques for the preparation of different novel biofibers for composites. In: Biofibers and biopolymers for biocomposites. Springer, Cham, pp 1–34
58. Verma A, Singh VK (2016) Experimental investigations on thermal properties of coconut shell particles in DAP solution for use in green composite applications. J Mater Sci Eng 5(3):1000242
59. Singh K, Jain N, Verma A, Singh VK, Chauhan S (2020) Functionalized graphite–reinforced cross-linked poly (vinyl alcohol) nanocomposites for vibration isolator application: Morphology, mechanical, and thermal assessment. Mater Perform Charact 9(1):215–230
60. Verma A, Singh VK, Arif M (2016) Study of flame retardant and mechanical properties of coconut shell particles filled composite. Res Rev J Mater Sci 4(3):1–5
61. Verma A, Jain N, Rastogi S, Dogra V, Sanjay SM, Siengchin S, Mansour R (2020) Mechanism, anti-corrosion protection and components of anti-corrosion polymer coatings. In: Polymer coatings. CRC Press, pp 53–66
62. Chaudhary A, Sharma S, Verma A (2022) WEDM machining of heat treated ASSAB'88 tool steel: a comprehensive experimental analysis. Mater Today: Proc 50:946–951
63. Bisht N, Verma A, Chauhan S, Singh VK (2021) Effect of functionalized silicon carbide nano-particles as additive in cross-linked PVA based composites for vibration damping application. J Vinyl Add Tech 27(4):920–932
64. Verma A, Jain N, Parashar A, Singh VK, Sanjay MR, Siengchin S (2020) Design and modeling of lightweight polymer composite structures. In: Lightweight Polymer composite structures. CRC Press, pp 193–224

65. Dogra V, Kishore C, Verma A, Rana AK, Gaur A (2021) Fabrication and experimental testing of hybrid composite material having biodegradable bagasse fiber in a modified epoxy resin: evaluation of mechanical and morphological behavior. Appl Sci Eng Progress 14(4):661–667

66. Verma A, Baurai K, Sanjay MR, Siengchin S (2020) Mechanical, microstructural, and thermal characterization insights of pyrolyzed carbon black from waste tires reinforced epoxy nanocomposites for coating application. Polym Compos 41(1):338–349

67. Verma A, Budiyal L, Sanjay MR, Siengchin S (2019) Processing and characterization analysis of pyrolyzed oil rubber (from waste tires)-epoxy polymer blend composite for lightweight structures and coatings applications. Polym Eng Sci 59(10):2041–2051

68. Verma A, Negi P, Singh VK (2019) Experimental analysis on carbon residuum transformed epoxy resin: chicken feather fiber hybrid composite. Polym Compos 40(7):2690–2699

69. Verma A, Singh VK (2018) Mechanical, microstructural and thermal characterization of epoxy-based human hair–reinforced composites. J Test Eval 47(2):1193–1215

70. Jain N, Verma A, Ogata S, Sanjay MR, Siengchin S (2022) Application of machine learning in determining the mechanical properties of materials. In: Machine learning applied to composite materials. Springer, Singapore, pp 99–113

71. Verma A, Singh VK, Verma SK, Sharma A (2016) Human hair: a biodegradable composite fiber–a review. Int J Waste Resour 6(206):1000206

72. Kataria A, Verma A, Sanjay MR, Siengchin S, Jawaid M (2022) Physical, morphological, structural, thermal, and tensile properties of coir fibers. In: Coir fiber and its composites. Woodhead Publishing, pp 79–107

73. Jain N, Verma A, Singh VK (2019) Dynamic mechanical analysis and creep-recovery behaviour of polyvinyl alcohol based cross-linked biocomposite reinforced with basalt fiber. Mater Res Express 6(10):105373

74. Chaurasia A, Verma A, Parashar A, Mulik RS (2019) Experimental and computational studies to analyze the effect of h-BN nanosheets on mechanical behavior of h-BN/polyethylene nanocomposites. J Phys Chem C 123(32):20059–20070

75. Verma A, Negi P, Singh VK (2018) Physical and thermal characterization of chicken feather fiber and crumb rubber reformed epoxy resin hybrid composite. Adv Civ Eng Mater 7(1):538–557

76. Bharath KN, Madhu P, Gowda TG, Verma A, Sanjay MR, Siengchin S (2020) A novel approach for development of printed circuit board from biofiber based composites. Polym Compos 41(11):4550–4558

77. Verma A, Joshi K, Gaur A, Singh VK (2018) Starch-jute fiber hybrid biocomposite modified with an epoxy resin coating: fabrication and experimental characterization. J Mech Behav Mater 27(5–6):20182006

78. Verma A, Singh C, Singh VK, Jain N (2019) Fabrication and characterization of chitosan-coated sisal fiber–Phytagel modified soy protein-based green composite. J Compos Mater 53(18):2481–2504

79. Rastogi S, Verma A, Singh VK (2020) Experimental response of nonwoven waste cellulose fabric–reinforced epoxy composites for high toughness and coating applications. Mater Perform Charact 9(1):151–172

80. Verma A, Zhang W, Van Duin AC (2021) ReaxFF reactive molecular dynamics simulations to study the interfacial dynamics between defective h-BN nanosheets and water nanodroplets. Phys Chem Chem Phys 23(18):10822–10834

81. Verma A, Jain N, Parashar A, Gaur A, Sanjay MR, Siengchin S (2021) Lifecycle assessment of thermoplastic and thermosetting bamboo composites. In: Bamboo fiber composites. Springer, Singapore, pp 235–246

82. Arpitha GR, Verma A, Sanjay MR, Siengchin S (2021) Preparation and experimental investigation on mechanical and tribological performance of hemp-glass fiber reinforced laminated composites for lightweight applications. Adv Civil Eng Mater 10(1):427–439

83. Verma A, Samant SS (2016) Inspection of hydrodynamic lubrication in infinitely long journal bearing with oscillating journal velocity. J Appl Mech Eng 5(3):1–7

84. Verma A, Jain N, Rangappa SM, Siengchin S, Jawaid M (2021) Natural fibers based bio-phenolic composites. In: Phenolic polymers based composite materials. Springer, Singapore, pp 153–168
85. Arpitha, G.R., Jain, N., Verma, A. and Madhusudhan, M., 2022. Corncob bio-waste and boron nitride particles reinforced epoxy-based composites for lightweight applications: fabrication and characterization. Biomass Convers Biorefinery 1–8. https://doi.org/10.1007/s13399-022-03717-1
86. Verma A, Singh VK Experimental characterization of modified epoxy resin assorted with almond shell particles. ESSENCE-Int J Environ Rehabil Conserv 7(1):36–44
87. Verma A (2022) A perspective on the potential material candidate for railway sector applications: PVA based functionalized graphene reinforced composite. Appl Sci Eng Prog 15(2):5727–5727
88. Verma A, Jain N, Singh K, Singh VK, Rangappa SM, Siengchin S (2022) PVA-based blends and composites. In: Biodegradable polymers, blends and composites. Woodhead Publishing, pp 309–326
89. Raja S, Verma A, Rangappa SM, Siengchin S (2022) Development and experimental analysis of polymer based composite bipolar plate using Aquila Taguchi optimization: design of experiments. Polym Compos 43(8):5522–5533
90. Prabhakaran S, Sharma S, Verma A, Rangappa SM, Siengchin S (2022) Mechanical, thermal, and acoustical studies on natural alternative material for partition walls: a novel experimental investigation. Polym Compos 43(7):4711–4720
91. Sethi SK, Gogoi R, Verma A, Manik G (2022) How can the geometry of a rough surface affect its wettability?-a coarse-grained simulation analysis. Prog Org Coat 172:107062
92. Chaudhary A, Sharma S, Verma A (2022) Optimization of WEDM process parameters for machining of heat treated ASSAB'88 tool steel using response surface methodology (RSM). Mater Today: Proc 50:917–922
93. Verma A, Jain N, Sanjay MR, Siengchin S Viscoelastic properties of completely biodegradable polymer-based composites. In: Vibration and damping behavior of biocomposites. CRC Press, pp 173–188
94. Verma A, Jain N, Mishra RR (2022) Applications and drawbacks of epoxy/natural fiber composites. In: Handbook of epoxy/fiber composites. Singapore, Springer Singapore, pp 1–15
95. Lila MK, Verma A, Bhurat SS (2022) Impact behaviors of epoxy/synthetic fiber composites. In: Handbook of epoxy/fiber composites. Singapore, Springer Singapore, pp. 1–18
96. Kumar G, Mishra RR, Verma A (2022) Introduction to molecular dynamics simulations. In: Forcefields for atomistic-scale simulations: materials and applications. Springer, Singapore, pp 1–19. https://doi.org/10.1007/978-981-19-3092-8_1
97. Verma A, Sharma S (2022) Atomistic simulations to study thermal effects and strain rate on mechanical and fracture properties of graphene like BC3. In: Forcefields for atomistic-scale simulations: materials and applications. Springer, Singapore, pp 237–252
98. Chaturvedi S, Verma A, Sethi SK, Ogata S (2022) Defect energy calculations of nickel, copper and aluminium (and their alloys): molecular dynamics approach. In: Forcefields for atomistic-scale simulations: materials and applications. Springer, Singapore, pp 157–186
99. Shankar U, Gogoi R, Sethi SK, Verma A (2022) Introduction to materials studio software for the atomistic-scale simulations. In: Forcefields for atomistic-scale simulations: materials and applications. Springer, Singapore, pp 299–313
100. Shankar U, Sethi SK, Verma A (2022) Forcefields and modeling of polymer coatings and nanocomposites. In: Forcefields for atomistic-scale simulations: materials and applications. Springer, Singapore, pp 81–98
101. Verma A, Ogata S (2022) Computational modelling of deformation and failure of bone at molecular scale. In: Forcefields for atomistic-scale simulations: materials and applications. Springer, Singapore, pp 253–268
102. Homer ER, Verma A, Britton D, Johnson OK, Thompson GB (2022) Simulated migration behavior of metastable $\Sigma 3$ (11 8 5) incoherent twin grain boundaries. IOP Conf Ser Mater Sci Eng 1249(1):012019

103. Chaturvedi S, Verma A, Sethi SK, Rangappa SM, Siengchin S (2022) Stalk fibers (rice, wheat, barley, etc.) composites and applications. In: Plant fibers, their composites, and applications. Woodhead Publishing, pp 347–362

104. Arpitha GR, Verma A, Sanjay MR, Gorbatyuk S, Khan A, Sobahi TR, Asiri AM, Siengchin S (2022) Bio-composite film from corn starch based vetiver cellulose. J Nat Fibers 19(16):14634–14644

105. Thimmaiah SH, Narayanappa K, Thyavihalli Girijappa Y, Gulihonenahali Rajakumara A, Hemath M, Thiagamani SMK, Verma A (2022) An artificial neural network and Taguchi prediction on wear characteristics of Kenaf-Kevlar fabric reinforced hybrid polyester composites. Polym Compos 44(1):261–273

106. Kataria A, Verma A, Sethi SK, Ogata S (2022) Introduction to interatomic potentials/forcefields. In: Forcefields for atomistic-scale simulations: materials and applications. Springer, Singapore, pp 21–49

107. Kataria A, Verma A, Sanjay MR, Siengchin S (2022) Molecular modeling of 2D graphene grain boundaries: mechanical and fracture aspects. Mater Today Proc 52:2404–2408

108. Verma A, Ogata S (2023) Magnesium based alloys for reinforcing biopolymer composites and coatings: a critical overview on biomedical materials. Adv Ind Eng Polym Res. https://doi.org/10.1016/j.aiepr.2023.01.002

109. Bharath KN, Madhu P, Gowda TY, Verma A, Sanjay MR, Siengchin S (2021) Mechanical and chemical properties evaluation of sheep wool fiber–reinforced vinylester and polyester composites. Mater Perform Charact 10(1):99–109

110. Kataria A, Chaturvedi S, Chaudhary V, Verma A, Jain N, Rangappa SM, Siengchin S (2023) Cellulose fiber-reinforced composites—history of evolution, chemistry, and structure. In: Cellulose fibre reinforced composites. Woodhead Publishing, pp 1–22. https://doi.org/10.1016/B978-0-323-90125-3.00012-4

111. Chaturvedi S, Kataria A, Chaudhary V, Verma A, Jain N, Rangappa SM, Siengchin S (2023) Bionanocomposites reinforced with cellulose fibers and agro-industrial wastes. In: Cellulose fibre reinforced composites. Woodhead Publishing, pp 1–22. https://doi.org/10.1016/B978-0-323-90125-3.00017-3

112. Arpitha GR, Mohit H, Madhu P, Verma* A (2023) Effect of sugarcane bagasse and alumina reinforcements on physical, mechanical, and thermal characteristics of epoxy composites using artificial neural networks and response surface methodology. Biomass Convers Biorefinery. https://doi.org/10.1007/s13399-023-03886-7

Chapter 6
Molecular Dynamics Simulations in Coatings

Aditya Kataria, Suhaib Zafar, Akarsh Verma, and Shigenobu Ogata

1 Introduction

There has been a growing interest in studying the intricate working of coating materials on a molecular level. Such interests, when pursued experimentally, end up costing a lot in expenses and are time-consuming. Moreover, there is not much control over the environment or on the working of the coating reaction itself. Molecular dynamics simulations provide a good balance of being inexpensive while allowing us to control most of the undesired variables. Nano coatings have the ability to shape the properties of a material to the users' liking, be it increased hardness, anti-stain surface finish, cavitation control, improved current flow to even healing surface defects. The simulations have been done using various kinds of interatomic potentials and forcefields [1].

A. Kataria
Department of Mechanical Engineering, University of Minnesota, Twin Cities, Minneapolis, USA

S. Zafar
Stellantis, Chrysler Technology Center, 800 Chrysler Drive, Auburn Hills, MI 48326, USA

A. Verma (✉)
Department of Mechanical Engineering, University of Petroleum and Energy Studies,
Dehradun 248007, India
e-mail: akarshverma007@gmail.com

A. Verma · S. Ogata
Department of Mechanical Science and Bioengineering, Osaka University, Osaka 560-8531, Japan

S. Ogata
Center for Elements Strategy Initiative for Structural Materials, Kyoto University,
Kyoto 606-8501, Japan

2 Types of Coatings

2.1 Graphene

The most versatile nanomaterial, Graphene, has itself been used for a variety of tasks like exhibiting super capacitive behavior [2] for photocatalysis and catalytic reduction [3], or for sensing of dopamine using fluorescence [4]. The molecular modeling of 2D graphene grain boundaries has been studied, and further research has been done to understand their impact on coatings [5]. Peng et al. in 2019 [6] examined the strengthening processes of graphene coatings on Cu film substrate using molecular dynamics simulations of nanoindentation. After applying graphene coatings, it is discovered that the load-bearing capacity of Cu substrate may be visibly increased, and this improvement is related to the number of graphene layers used, ranging from monolayer to trilayer.

In order to study the strengthening mechanisms, a molecular dynamics simulation of nanoindentation for single-crystal Cu films with multilayer graphene coatings was used in this work. Dislocation initiation and progression during the entire indentation process were discussed in detail. The large-scale atomic/molecular massively parallel simulator (LAMMPS) is used for all simulations in this work [7]. The first model consists of a monocrystalline Cu substrate that is defect-free, 1–3 layers of graphene coatings, and an almost spherical indenter. The X, Y, and Z axes of the Cu substrate's crystal orientation are [100], [010], and [001], respectively.

With a = 0.3615 nm as the lattice constant, the size of the Cu substrate is kept as $73.75a \times 73.75a \times 85a$. 1,872,792 atoms make up the Cu substrate, which is split into a border zone and a Newtonian zone. The Velocity Verlet algorithm is used to integrate the atomic motion in the Newtonian zone, which complies with the original Newton's second law [8]. CuGr1, CuGr2, and CuGr3 are the designations for the composite samples after monolayer, bilayer, and trilayer graphene coatings have been applied to a Cu substrate, respectively. The adaptive intermolecular reactive empirical bond order (AIREBO) potential is used to characterize the C–C contact force in the graphene layer [9].

According to Mishin et al. in 2001 [10], the interactions between Cu atoms define the embedded atom model potential (EAM). To explain the interactions between Cu and C atoms, the Morse potential [11] is used, and the parameters [12] are set to $D_0 = 0.1$ eV, a = 1.7 Å^{-1}, and $r_0 = 2.2$ Å, where D_0 is the cohesive energy, a is the potential's breadth, and r_0 is the interatomic equilibrium distance.

LJ potential is used to represent the interactions between graphene layers, and the parameters are $\varepsilon_{C-C} = 2.967$ meV and $\sigma_{C-C} = 3.407$ Å [13]. For the simulated system, the periodic boundary condition is applied in the X and Y directions. Prior to indentation, the system is relaxed using energy minimization, and then the stress in the X, Y, and Z directions is released to zero at a temperature of 0.1 K using a time step of 1 fs in an NPT ensemble for 100 ps.

The Nose–Hoover thermostat is then enforced as part of an NVT ensemble to further optimize the system's energy use for the next 100 ps. To perform nanoindentation, a virtual spherical indenter is positioned above the model for displacement control at 3 nm indentation depths. The virtual indenter has a radius of 6 nm, and a repulsive potential that simulates it has the formula:

$$F = AH(r)(R - r)^2$$

where r is the distance from an atom to the center of the spherical indenter and A is a constant force with a value of 10 eVÅ^{-2}. H(r) is used as a step function.

NVT ensemble is used during the indentation process to regulate the temperature of the graphene and Newtonian zone layers. With the NVE ensemble, atoms in the boundary zone are also fixed. In order to discover dislocation flaws, the dislocation extraction algorithm (DXA) [14] converts all recognized dislocations into continuous lines and fully automatically calculates corresponding Burger vectors. After applying a graphene coating, the Cu substrate's ability to support loads can be increased. Additionally, when graphene layers go from monolayer to trilayer, capacity increases. From the indentation response of the Cu/graphene system, the indentation process may be split into elastic and plastic stages.

Prismatic loops and surface-sliding dislocations are examples of the plastic stage. Graphene coatings have various strengthening processes at various phases. Due to the ultra-high elastic modulus of graphene, the stress homogenization effect of the graphene contact is the primary cause of the increase in load-bearing capacity in the elastic stage. The primary strengthening mechanism during the plastic stage is the interaction between dislocations and the graphene interface. An important factor in strengthening is the rise in surface image force acting on prismatic dislocation loops as they form. The strengthening mechanism throughout this process is the graphene coating, which prevents the dislocation from slipping over the surface and increases the load force for the motion of the dislocations.

Tang et al. in 2019 [15] investigated the impact of graphene coating on nanofilm boiling on an atomically smooth surface involves running molecular dynamics (MD) simulations, on both bare and surfaces coated with graphene, and boiling modes that transition from regular evaporation to explosive boiling. The graphene-coated surface's beginning temperature of explosive boiling is roughly 170 K, which is quite close to the value of roughly 160 K for the naked surface.

Classic molecular dynamics (MD) have been extensively acknowledged as a tool to examine the micro-nanoscale heat transfer because they are able to take advantage of the atomic-sized level and sub-femtosecond timescale comprehension of phenomena [16–26]. One of the most effective ways to increase the CHF of pool boiling is nanoscale surface engineering [27–29]. Graphene coating, which has a 2D planar structure and an extremely high thermal conductivity, has received a lot of attention due to its tremendous potential for enhancing pool boiling [30–34].

The development of three simulation systems includes graphene-coated Cu systems, hydrophilic Cu systems, and hydrophobic Cu systems. The three systems are first set up as cuboid confinements with the same dimensions of 8.39 nm (x)*9.68 nm (y)*100 nm (z) and consist of solid, liquid, and vapor sections. The 39,150 atom arrangement is identical in both the hydrophilic and hydrophobic Cu systems. The face-centered cubic (FCC) lattice with a lattice constant of 0.35 nm, which corresponds to its density at 90 K, generates the bottom Cu group, which has 11 layers and 15,170 atoms.

The liquid argon group, which has a thickness of 5 nm and a density of 8640 argon (Ar) atoms at the saturation state of 1.33 atm and 90 K, is positioned over the FCC (111) Cu surface. The geometric mirror of the lower Cu group is the upper Cu group. 170 vapor argon atoms are used to fill the remaining space in the box. A graphene monolayer containing 3200 carbon atoms is superimposed on the bottom Cu group of the hydrophilic Cu systems to create the graphene-coated Cu system. The use of the LJ potential for Cu–C interactions is justified by the weak contact of graphene with Cu [35, 36].

The interatomic interactions except C–C atom pairs are described by the LJ potential as follows:

$$\phi(r_{ij}) = \begin{cases} 4\varepsilon\left[(\sigma/r_{ij})^{12} - (\sigma/r_{ij})^{6} \right], & r \leq r_{cut} \\ 0, & r > r_{cut} \end{cases}$$

where ε and σ are the energy and distance parameters, and r_{ij} is the distance between atoms i and j. The distance and energy parameters for Ar–Cu, Ar–C, and Cu–C interactions are obtained according to the Lorentz–Berthelot combining rules, as follows:

$$\sigma_{A-B} = (\sigma_{A-A} + \sigma_{B-B})/2$$

$$\varepsilon_{A-B} = \sqrt{\varepsilon_{A-A}\varepsilon_{B-B}}$$

where the subscripts A and B represent different atom types.

The three sections that make up a simulation are the NPT, NVT, and NVE-Langevin sections. Three systems are equilibrated for 0.2 ns at the saturation state of 90 K at 1.33 atm in the NPT segment. At this portion, a 3D periodic boundary condition is used. The atoms' positions and velocities are then revised for the Nose/Hoover NVT segment. In this section, the x- and y-axis dimensions use periodic boundary conditions, whereas the z-axis dimension uses fixed boundary conditions at the top and bottom surfaces of the simulation domain.

The z-axis dimension boundaries are made up of three layers, or 4144 Cu atoms, and are locked at each position to prevent translational movement and atom penetration. To reach the predetermined temperature of 90 K, the remaining atoms equilibrate for 0.2 ns. In the NPT and NVT sections, the Nose/Hoover thermostat and barostat's timescale is empirically set to 100 timesteps, or 0.05 ps. The NPT (as well as the NVT) segment uses the same parameters for all three systems and runs once for each system. The table below describes the LJ potentials used (Table 1).

Table 1 Interaction parameters for the LJ potential used

Atom pairs	ε (meV)	σ (nm)
Cu–Cu	409.33	0.2338
Cu–C	31.28	0.2842
Ar–Ar	10.33	0.3405
Ar–C	4.97	0.3775
Ar–Cu (hydrophilic)	65.03	0.2872
Ar–Cu (hydrophobic)	4.97	0.3775

In the final NVE-Langevin segment, the Langevin thermostat is employed in conjunction with the NVE ensemble to mimic the boiling of the nanofilm. The Langevin thermostat merely alters forces to achieve thermostating; it does not do time integration [37]. Furthermore, as the atoms are heated (or cooled) by the Langevin thermostat, cumulative thermal energy is added (or withdrawn) from them. According to the NVE-Langevin simulation, the combined force acting on each atom is given by

$$F = F_c + F_f + F_r$$

where the terms conservative, frictional, and solvent forces, respectively, are denoted by F_c, F_f, and F_r. The conservative force F_C is computed using atom interactions. The frictional term Ff is performed as

$$F_f = -\frac{m}{t_{\text{damp}}} v$$

where t_{damp} is the Langevin thermostat's timescale and m and v are the atom's mass and velocity, respectively. The size of the solvent term Fr is given by the fluctuation–dissipation hypothesis to be

$$F_r \propto \sqrt{\frac{m k_B T}{t_{\text{damp}} \delta t}}$$

where t is the length of the timestep, T is the desired temperature, and kB is the Boltzmann constant. An appropriate timescale for LJ atoms is

$$t_{\text{damp}} = \sqrt{\frac{m \sigma^2}{\varepsilon}}$$

An abrupt temperature increase of the specific solid atoms at the bottom area causes the nanofilm to boil.

The VMD code's atom distribution animation vividly shows how the nanofilm's boiling mode shifts from routine evaporation to explosive boiling. The initial superheated temperatures for the hydrophilic copper, graphene-coated copper, and hydrophobic copper systems are 160, 170, and 280 K, respectively, in the explosive boiling mode.

The bottom liquid progressively turns to vapor in circumstances of normal evaporation, which can be exhibited by a steady reduction in film thickness. The sudden departure and swift motion of the liquid clusters, in contrast, point to an explosive boiling mode. The boiling mode transition's superheated temperature dependence is consistent with the findings of earlier research [17–38]. In addition, absorption of a nonevaporating layer always takes place at the superheated solid surface in the given hydrophilic Cu system. The graphene layer also marginally lowers the CHF of nanofilm boiling and the heat flux density. Therefore, the graphene covering is detrimental to the improvement of heat transfer of the boiling nanofilm.

Ma et al. in 2021 [39] determined the cooling performance of a graphene-coated heated silicon substrate, for which a suitable manner of heat transmission must be chosen. High heat fluxes (1 kW/cm^2) and low thermal resistances (0.1 °C/W) can be dissipated by subcooled liquid convection cooling; however, this method frequently has a coefficient of performance that is found to be low because a high pumping pressure is needed for the generation of a solid convective current. In comparison to being cooled by liquid, two-phase cooling that is based on boiling and/or evaporation offers an improved solution since it allows for extremely high heat dissipation capacities due to the fluid's high latent heat of vaporization. Additionally, the heat source's temperature rise lowers considerably once the coolant's temperature stays almost constant throughout the entire process of phase change.

A liquid nanofilm's evaporation differs significantly from the evaporation of a larger liquid pool in several ways. For macroscale evaporative heat transfer, vapor transport resistance and conduction resistance, which are both in the magnitude of 10^{-2} m^2 K/W or more, are what essentially influence the evaporative heat flux. The Kapitza resistance, which is the thermal resistance across solid–liquid interfaces, which is in the magnitude of 10^{-7} m^2 K/W [40–44], becomes comparable with the conduction resistance and is influential in the overall heat transfer performance if the thickness of the liquid film is only a few microns thick.

The vacuum spaces among two touching solid surfaces cause contact resistance, whereas the Kapitza resistance occurs due to the uneven vibrational frequencies among the molecules in two distinct mediums [45].

As a result, the intensity of the intermolecular contacts at the solid–liquid boundary, such as the bonding strength and the vibrational coupling among the solid and liquid, that directly affect surface wettability, greatly influence the size of the Kapitza resistance. On both sides along the z-axis between two ends of the replication region, two solid silicon (100) structures were built. The two silicon substrates were then used to build two identical liquid argon cuboids.

At 85 K in the saturated state, individual liquid layers are 30 Å wide and comprise 568 argon atoms, giving it a density of 1.40 g/cm^3 [46]. The simulation domain has dimensions of 220 mm in height and 30 mm on each side. These figures were chosen in accordance with other literature reviews [38, 47]. It is important to keep in mind that if the two liquid layers are in each other's vicinity, the transfer due to evaporation will be hindered by the high frequency of vapor molecule collisions in the middle region. Except for the z direction, which is restricted by a fixed solid wall, periodic boundary conditions were used in all other directions.

The coordinates of the highest and lowest layers of silicon atoms were specified in the simulation in order to guarantee that the silicon structure remains locked to its place during the evaporation process. The forcefields created by the stationary atoms in the simulation nonetheless have an impact on the remaining atoms, particularly the ones that are near the outermost layer. The hot and cold layers of silicon atoms were then subjected to an isothermal boundary condition. In particular, after each cycle, the mean temperature of these atoms was continuously resized to the desired levels.

This technique replicates an isothermal boundary condition simply and successfully [47, 48]. Under the influence of intermolecular interactions, all atoms that were left could move around freely.

A 12-6 Lennard–Jones (LJ) potential was used in the simulation to determine the non-bonded interaction between various atoms [49] as

$$E_{nonbonded} = 4\varepsilon \left[\left(\frac{\sigma}{r} \right)^{12} - \left(\frac{\sigma}{r} \right)^{6} \right]$$

where r is the distance between the molecules, ε is the depth of the potential well, and σ is the distance between particles at which the intermolecular potential drops to zero. The associated cutoff distance was established at 12 Å [38]. The following harmonic equations were used to simulate all of the bound molecule interactions:

$$E_{bonded} = k_s(r - r_0)^2 + k_b(\theta - \theta_0)^2 + k_t\{1 + d_t\cos(n_t\varphi_t)\} + k_0\{1 + d_0\cos(n_0\varphi_0)\}$$

where k_s, k_b, k_t, and k_0 are the forcefield coefficients used to describe the potential energy change brought on by stretching, bending, torsion, and out-of-plane distortion of the bond, respectively; r_0 is the equilibrium bonding length; θ_0 is the equilibrium angle formed between two bonds; d_t and d_0 are the potential sign parameters; n_t and n_0 are the potential multiplier factor; r is the actual distance between two bonded atoms. φ_t and φ_0 are the torsion and distortion angles, respectively.

For non-bonded intermolecular potentials between different types of atoms, the LJ parameters shown are calculated based on the Lorentz–Berthelot mixing rule [50]:

$$\sigma_{ij} = \frac{\sigma_{ii} + \sigma_{ij}}{2}$$

$$\epsilon_{ij} = \sqrt{\epsilon_{ii} * \epsilon_{jj}}$$

A Nosé–Hoover thermostat (NVT ensemble) was then used to increase the system temperature to 85 K over a period of 50 ps while maintaining a constant full atom count, system temperature, and domain volume. After updating the trajectory and velocity of the molecules by resolving Newton's equation with no limitations on the temperature and pressure, a microcanonical (NVE) ensemble is then put on the argon molecules for a duration of 100 ps.

The silicon molecules remained static in the NVT ensemble throughout this time to maintain a steady temperature. Near the termination of the integration cycles, a reasonably constant temperature in both the liquid and solid domains confirmed an equilibrium state for every stage along the given time step. The results show that, despite its good 2D planar structure and thermal conductivity, using graphene covering in thin-film evaporation degrades heat transfer functioning because of the predominance of resistance of the interfacial thermal.

The large Kapitza resistance that exists among the graphene layer and liquid argon, in particular, influences the pace of the total heat transfer while hindering evaporative mass motion. The resistance at the interfacial thermal resistance and the wetting properties are equally significantly influenced by the strength of the intermolecular potential amid solid and liquid atoms, so even though the results were obtained using silicon and graphene surfaces, they're anticipated to be generally valid to other such systems. However, a stronger intermolecular connection also ends up resulting in a bigger disjoining pressure on the liquid molecules, which can dramatically slow down evaporative transport because the film thickness is low.

Al and Cu are both very good electrical conductors. Many scientists believe that Al and Cu materials are current collectors based on these results. Due to their different material compositions, current collectors might be difficult to combine during the battery pack production process [51, 52]. The disparity in melting points between the materials is a significant issue when trying to combine them. Brittle intermetallic compounds (IMC), fracture growth, unbonded surfaces, metallurgical flaws, etc. are additional problems [53–55]. These flaws cause batteries to lose energy and occasionally even malfunction.

Dissimilar metals display poor fusion weldability as a result of their various melting points and chemical characteristics. According to the authors, solid-state welding is preferable to previous fusion welding procedures because it causes fewer thermal flaws, less cracking, and less brittle IMC formation when combining aluminum and copper [56–58]. Some solid-state welding techniques include friction stir spot welding, diffusion welding, and ultrasonic welding (FSSW).

Mypati et al. in 2022 [59] used large-scale atomic/molecular massively parallel simulator (LAMMPS) tools; MD simulations are carried out for the samples without and with GL.

An embedded atom method potential (EAM) is used to specify the interatomic interaction between Al and Cu, Airebo potentials are used to specify the interaction between the carbon (C) atoms, and a 12-6-Lennard–Jones (LJ) potential is used to specify the interaction between Al, C, and Cu [60, 61]. The simulation box sizes are 51.6 Å*108.46 Å* 108.46 Å and 27.94 Å* 18.68 Å* 51.18 Å, respectively, with and without GL.

In order to identify the atom group and optimize the simulation time, the simulation box size is determined arbitrarily. In the entire MD simulation, the x-y-z planes are taken into account as a periodic boundary condition. The heat applied in each simulation is done using a Nose–Hoover thermostat at a particular temperature of

300 K and a time step of 1 fs during the solid–solid interface in the FSSW process. For the atoms of Al, Cu, and C, the initial velocity is set to follow the Gaussian distribution. The mole (N), volume (V), and temperature (T) have been employed in a canonical ensemble NVT.

Compressive strain rates are calculated using the experiment's axial load. By dividing the axial load by the tool's surface area on the sample, one can calculate the axial stress. LAMMPS software is used to determine the mean square displacement (MSD) of the Al–Cu and C atoms. Atomic motion during FSSW circumstances has been measured using this technique. To calculate the mutual atom diffusion at the interface, MSD data is gathered every 1000 steps. The MSD is computed using the following equations:

$$\text{MSD} = \langle \left| X_i\left(t + t'\right) - X_i(t) \right|^2 \rangle$$

where $X_i(t)$ is the instantaneous particle at the ith position, t is the lag time, and $\langle - \rangle$ is the ensemble average.

The Einstein equation is used to determine Al and Cu diffusion coefficients (D) based on their locations over time as follows:

$$D = \frac{1}{6} \lim_{t \to \infty} \frac{d\text{MSD}}{dt}$$

Al and Cu atoms with FCC structures are initially stacked one on top of the other, and the FSSW condition is then applied to the system.

Then, the relationships between Al–Cu–C are further examined. In general, there is a straight proportionality between temperature and the binding energy. As observed, sample SGL-1500 generates less heat than sample S1500. The result from the MD simulation shows a rise in time, or cooling time.

Twin nucleation is also increased for FCC alloys at lower temperatures and higher strain rates. Additionally, temperature and the rate of deformation strain affect twin growth [62, 63].

When two dislocations cross-slip onto successful (001) and (003) planes, respectively, for the samples with and without GL, twin growth occurs. In experiments, the presence of graphene causes a rise in nucleation, which raises deformation strain and strengthens joints. As a result, sample S1500 shows less tensile load.

The number of twin grains, however, rises as the deformation strain rises and eventually reaches saturation during the welding process after the strain exceeds an accumulated capacity. In sample SGL-1500, the high dislocation density inhibits twin formation, resulting in narrower twins on the Cu side of the weld contact [64–67]. Additionally, the Al–Cu metal matrix's dislocations prevent twins from expanding, which is why sample SGL-1500's twin thickness is lower. As a result, it can be said that sample SGL-1500 has a high twin boundary density, which raises the tensile load [63].

2.2 Polymers

Smith et al. in 2005 [68] examined, using computer models, how the aspect ratio of nanoparticle fillers in a polymer coating impacts the film's capacity to repair nanoscale flaws in the substrate beneath it.

Standard "Lennard–Jones + FENE" model simulations of molecular dynamics are carried out using the LAMMPS program [7, 69].

Atoms i and j interact noncovalently in the Lennard–Jones form, with the hardcore diameter r_c acting as a shift:

$$U_{ij}^{LJ} = \begin{cases} 4\epsilon \left[\left(\frac{\sigma}{r_{ij} - r_c} \right)^{12} - \left(\frac{\sigma}{r_{ij} - r_c} \right)^6 \right], & r_{ij} - r_c \le 2^{1/6}\sigma \\ 0, & r_{ij} - r_c > 2^{1/6}\sigma \end{cases}$$

using $\epsilon_{mm} = 1$ mm and $\sigma_{mm} = 1$ mm as the units of energy and length for interactions between monomers, respectively. The monomers, $r_{c,mm} = 0$, do not have a hardcore. The finite extensibility nonlinear elastic (FENE) potential can be written as

$$U_{ij}^{FENE} = \begin{cases} -0.5kR_0^2 \ln \left[1 - \left(r_{ij}/R_0 \right)^2 \right], & r_{ij} \le r_0 \\ \infty, & r_{ij} > r_0 \end{cases}$$

where $R_0 = 1.5\,\sigma$ and $k = 30\epsilon/\sigma^2$ represents bonds between monomers.

The model for rod-like particles is a collection of LJ beads that have been "glued" together and are therefore restricted to move as a stiff body. (For clarity, the complete stiff rod-like object is referred to as a "particle" and its individual parts as "beads". Of course, a single bead constitutes a spherical particle.) The quantity of beads in the particle equals the particle aspect ratio, a. According to these results, it was found that there were more rods than spheres in the notch. Additionally, longer rods have a propensity to align in the notch's direction, filling the notch more densely packed.

Zee et al. in 2015 [70] demonstrated that a system bonded by van der Waals (Lennard–Jones) forces that is also crosslinked with strong covalent (FENE) bonds to form three or six functional solid networks also contains nano-sized cavitation gaps. Previous molecular dynamics investigations have shown that when there is just one sort of force, typically van der Waals, connecting particles, cavitation forms pores in simulated liquids, including metals (and the resulting solids).

In a few trials, systems containing 1,428,000 beads (125 times larger) were used to test for finite-size effects and validate the outcomes of the smaller networks. In order to prevent border effects, periodic boundary conditions and a rectangular parallelepiped simulation cell design were adopted.

With a cutoff at $r_c = 2.5\,\sigma$, the Lennard–Jones (LJ) potential is used by all non-bonded beads to interact [71]:

$$U(r) = \begin{cases} U_{LJ}(r) - U_{LJ}(r_c), & \text{for } r \le r_c \\ 0, & \text{for } r > r_c \end{cases}$$

where

$$U_{LJ}(r) = 4\varepsilon \left[\left(\frac{\sigma}{r} \right)^{12} - \left(\frac{\sigma}{r} \right)^{6} \right]$$

Here, r is the distance between bead centers, and is the energy parameter. This is the length scale in the LJ potential. This potential simulates the van der Waals forces of attraction between each bead and the potently repelling core that establishes the bead's boundaries. Using a potential that forbids chain crossing, "covalent" links between beads that were already present or created during a process are described [71].

This bond potential is made up of a finite-extensible nonlinear elastic (FENE) attractive potential and a purely repulsive LJ interaction with a cutoff at $2^{1/6}\ \sigma$ (the minimum of the LJ potential) [72]:

$$U_{ij}^{FENE} = \begin{cases} -0.5kR_0^2 \ln\left[1 - \left(r_{ij}/R_0 \right)^2 \right], & r_{ij} \le r_0 \\ \infty, & r_{ij} > r_0 \end{cases}$$

where $R_0 = 1.5\ \sigma$ and $k = 30\epsilon/\sigma^2$ represents bonds between monomers.

The maximal bond extension in this case is R_0, and k governs how strong the spring (bond) potential is. Because the characteristics can be chosen of an amorphous material in a study, which introduces incompatible length scales to the inter- and intra-polymer interactions that prevent crystallization, this decision also ensures the absence of chain crossing and relatively strong connections. For these parameter selections, the average bond length is tightly clustered around $0.961\ \sigma$.

Using the large-scale atomic/molecular massively parallel simulator (LAMMPS) [7], the simulations are carried out.

As the network cools off within its rubbery to its glassy low-temperature state, it is discovered that void formation and growth begin above the glass transition temperature of the network. It is a little unexpected that these cavitation voids manifest in crosslinked polymer simulations when the network is held together by both powerful directional FENE (covalent) connections and far weaker non-directional Lennard–Jones (van der Waals) interactions.

In simulations of liquids and amorphous solids using solely Lennard–Jones forces, at remarkably identical pressure, temperature, and density settings, such voids have been observed and explored. These gaps are not generated by a loss of volatiles or consolidation of "free volume," but rather by cavitation, which occurs when the stresses of solidification and cooling surpass the material's local tensile strength. The rubbery condition of the voids first manifests itself while the network is highly mobile. The reported discovery, that voids develop while the creation of crosslinked networks is a novel finding with implications for the final characteristics of such materials as well as the long-term performance.

These results are very consistent with earlier simulations of liquids, but because the timescales of all such simulations are so short, it's possible that the pores in actual liquids and solids will vanish if they are much warmer than their glass transition temperature and are not restrained by adhesion to or between more rigid material bodies. The simulations demonstrate that internal pressure, network connectedness, and network strength are all diverse. As the network expands toward the gel point, clusters of reacted material are produced by polymers generated from mobile reactive precursors. It follows that there may be nearby weak spots where voids can form to lessen the internal tensile strain brought on by the solidification/cooling process. These weak spots may be adjacent regions that are not connected by covalent bonds.

Sun et al. in 2021 [73] examined adding reactive microcapsules to the coating matrix to create self-healing anti-corrosion coatings. Materials that can mend themselves can be stimulated outside of themselves. In the last ten years, a lot of laboratory research has been done on these smart materials in the areas of coatings, nanofluid, microfluid, statistical modeling, and surface treatment [74–77]. When the reactants are released from the capsules and react after minor damage to the coated surface, a cross-network that is crucial for improved cracking is established [78–81].

As an application strategy adopted by other researchers, Wang et al. [77] placed the chemical dicyclopentadiene in a formaldehyde shell. The easy and inexpensive in situ polymerization method of an oil-in-water emulsion was employed by researchers to create microcapsules. Polymeric nano-capsules are coated on samples utilizing the electrospray (ELS) technique while they are in an ambient environment (under sand and dirt); Kouhi et al. [78] analyzed the effects of corrosion and how to prevent them. The findings agreed with the previous researcher's findings. Based on formaldehyde microcapsules, Kouhi et al. [73] developed self-healing micro/nano smart coatings, used oil as a catalyst and repair agent, and improved it to stop corrosion.

Wei et al.'s studies [79] of self-healing spherical nano-capsules in various sizes between 40 and 400 m, with a speed between 500 and 1500 rpm, and a thickness of 2–17 m demonstrated that the sphericity of nano-capsules leads to a more uniform distribution. They contained isocyanate with a polymer coating (polyurethane) for use in humid environments.

The following steps have been taken in order to put the MD approach into practice; first, a simulation box was created in which the chemical makeup of primary and secondary components was originally evaluated. Then, box production was energy-minimized as a component of the energy-minimization process. Then, at a temperature of 300 K, NVT is applied to MDs, such as a simulation box. This curve aims to increase the energy of the system so that the atoms have sufficient energy to move in the direction of equilibrium.

In other words, NVT is used to relieve the molecular structure of the initial internal stresses that were imposed during the simulation box's formation. Then, to allow molecules and atoms to travel to the ideal state, the initial density of the system in gr/cm^3 is considered to be 0.9 at this stage. The simulation lasted for 50 PS. To get the system density closer to the actual density, a molecular dynamics simulation is run as NPT at the system pressure of 1 atm and at 300 K.

The NPT may also lower any leftover pressures on the system with a simulation time of 50 PS. Additionally, there is a significant risk of over-energizing the system due to the energy loss caused by the changes made to the links. With the aid of the optimization method, the system energy is subsequently reduced to a minimum level by shifting the atoms about. It should be stressed that the ratios (0, 20, 40, and 50%) must be determined first, and the number of molecules must then be computed using the atomic number in order to simulate this covering of nanocomposites.

Since these microcapsules display higher corrosion resistance than the sample without them, they can be offered as a viable substitute in the oil and gas industry. The obtained results go on to show how the proper microcapsule coating is applied to coatings with smooth substrates. Since coatings are commercially viable and have desirable qualities like hardness or abrasion resistance, the hardness has increased by 40% from 36 to 68 in the sample. This is due to the presence of microcapsules carrying resin-hardener.

2.3 Anti-stain/Hydrophobic

Scientists from all over the world have been fascinated by the synthesis and characterization of anti-stain coatings for a very long time. The stain often develops on a surface as a result of the deposition of molecules that resemble oil. Due to this important problem, oleophobic coatings have recently gained increased relevance, and great effort has been made to understand the fundamental principles behind how these coatings work.

Kumar et al. in 2016 [82] researched the uses of molecular simulations to forecast the proposed coating's surface characteristics, including adhesion energy to the substrate, surface energy, and contact angle. Researchers working to enhance comparable coating compositions in the future may find this study useful. Energy is created and independently minimized for molecules of Perfluorooctane (PFO) and chains of Polyvinyl acetate (PVAc) with isotactic stereochemical structures using molecular mechanics (MM).

Different amorphous cells are built using the PVAc (10 chains) and PFO (in various weight percent) energy-minimized structures. Realistic 3D periodic structures were first created using Material Studio's Amorphous Cell module. By reducing close interactions between atoms and guaranteeing a realistic distribution of torsion angles for the given force-field, the module constructs molecules in a cell in a Monte Carlo manner. Periodic boundary conditions (PBC), in which the center cell repeats itself in three dimensions (x, y, and z), have been integrated to prevent limited cell size effects.

The original high-energy periodic amorphous cells (with varied PFO wt percent) are subjected to energy relaxing using the Smart Minimizer module for 5000 MM iterations. The energy-minimized structures are then put through equilibration via NVT (constant number of particles, volume, and temperature) MD simulations to produce structures with realistic density and low potential energy. For this, the system's initial configuration bias was eliminated by running a faster NVT dynamics at a high temperature (750 K) for about 50 ps.

The generated frames are then saved every 0.1 ps while NVT simulations are run at 298 K for 300 ps with a time step of 1 fs. The lowest potential energy frame is chosen from the set of 3000 frames produced, and it is decreased even more using MM to a convergence of 0.01 kcal/mol/Å. For 300 ps at 298 K and 0.0001 GPa (1 atm), NPT dynamics are applied to the MM-generated frame to remove local system stressors and rebalance the density.

After making an estimate of the value of cohesive energy density (CED), the solubility parameters were computed. When all of an $E_{cohesive}$ polymer's intermolecular tensions are removed in a unit molar volume, CED states the rise in energy per mole that results. So, if V_{molar} is the molar volume of the polymer, then CED can be written as follows:

$$CED = (E_{cohesive} / V_{molar})$$

The following equation relates the Hildebrand solubility parameter (δ) to CED:

$$\delta = CED^{1/2} = (E_{cohesive} / V_{molar})^{1/2}$$

The literature estimates for pure PVAc [83–87] and PFO [88] are contrasted with the simulated δ values. The comparison reveals that the simulated values and reported values closely match each other, verifying the simulation technique and its outcomes.

Due to the surface aggregation seen with rising PFO content, it has been determined that the surface energy of the simulated structures decreases ideally with increasing PFO content. However, as the PFO concentration was increased, the interaction energy of the formed PVAc-PFO structures with the Al substrate (which was hypothetically used as the base coating substrate) dropped. The PFO content in PVAc is restricted to approximately 35.2–45.8 wt% due to the opposing behavior displayed by surface energy and interaction energy. On the basis of the aforementioned rationale, a coating formulation containing 35.2 wt% PFO in PVAc is suggested.

Additionally, the contact angle estimations demonstrate that when PFO content increased, the surface hydrophobicity increased practically linearly, whereas oleophobicity increased generally. The acetate moieties of PVAc may be partially hydrolyzed to hydroxyl groups if it is intended to create a self-cleaning coating (a coating that cleans accumulated dirt if rinsed with water).

Many applications, such as self-cleaning surfaces [89], anti-fogging surfaces [90], oil/water separation surfaces [91], anti-icing surfaces [92], anti-corrosion surfaces [93], and drag reduction surfaces [94], all hold great promise for the future of superhydrophobic surfaces. Innovations in the fields of building, environmental protection, maritime engineering, energy saving, smart gadgets, microfluidics, and other fields are anticipated. As a result, the creation of superhydrophobic surfaces has received extensive research in recent years. The ability of superhydrophobic surfaces to reject water with a water contact angle (WCA) greater than 150° is their distinctive characteristic [95]. Water droplets may easily flow across and even wipe the surface of the lotus leaf, one of the most superhydrophobic surfaces known to nature [96].

The high roughness and low free energy of the surface are essential elements for replicating and creating a superhydrophobic surface, according to studies on the wetting property of this plant [97].

Daneshmand et al. [98] in 2021 applied MD computations to examine the water droplet adsorption on surfaces made of glass, alumina, and alumina treated with stearic acid (SA). The work conducted by Soganoma et al. [99] was used to achieve this goal.

The amorphous cell module with 1.0 g/cm^3 density and cube periodic condition produced 500 water molecules. The cube lattice was then removed, and the nanocluster module made the cube spherical. Finally, the canonical NVT ensemble at 500 ps, 1.0 fs of integration time, and 300 K of Nose thermostat was used to optimize the nano-water droplet. As a result, the nano-water droplet with a radius of 25.0 Å, which is adequate to measure a WCA, was constructed for wetting calculations.

Glass and Alumina lattices were first downloaded from Materials Studio's structural database for use in the simulation procedure. The (0 2 0) crystallographic orientation and a repeat unit of alumina were both broken apart. The surface was then covered with a vacuum slab whose lattice size along the z-axis was 150 Å. The final cell was built with 95*95*150 Å^3 dimensions. Additionally, an amorphous cell containing 300 SA molecules and a density of 0.94 g/cm^3 was employed to create the majority of the SA.

A SA molecule was employed that had been tuned for this procedure and was acquired from the DFT simulation. Additionally, the Materials Studio's build layers module was used to prepare multilayer bulk material. An approach called conjugate gradient was used to optimize the structures. Afterward, the Nose thermostat was set to 300 K and the canonical NVT and NPT ensembles were implemented to optimize all surfaces at 200 ps. The structures were first optimized at the standard NVT and NPT ensembles at the examined temperature (100–400 °C, respectively), in order to study the impact of different temperatures on the wetting characteristics of SA surfaces. The conjugate gradient approach was then used to optimize SA surfaces before wetting simulations.

The MD simulation was employed to demonstrate the process of SA adsorption on the surface of alumina. SA begins the adsorption process by being adsorbed vertically. This is because SA contains an alkyl chain and a carboxyl group. The carboxyl group of SA is entirely adsorbed at the start of the adsorption process because of the hydrophilicity and polarity of the Alumina surface. Due to the flexibility and hydrophobicity of the alkyl chain, SA flexes to the surface in the meantime. After 200 ps, SA finally reaches equilibrium with the alumina surface, just bending to the surface and not totally absorbing it. Figures 1 and 2 show the mechanism of layer formation based on the temperatures.

The adsorption mechanism investigated by MD modeling is likely the basis for the minimal amount of SA adsorbed on the Alumina surface. The outcome of the MD simulation showed that SA's strong hydrophobicity of the alkyl chain prevented it from being completely absorbed by the alumina surface. The approach created in this study can be used with ease on various substrate sizes and has the necessary mechanical durability.

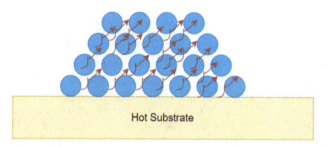

Fig. 1 The mechanism schematic of forming layer based on the low temperature

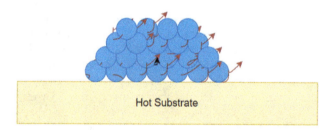

Fig. 2 The mechanism schematic of forming layer based on the high temperature

Due to their cheap maintenance costs, great durability, and wide range of potential uses, such as providing resistance to fouling [100], icing, smear, corrosion [101], and oil–water separation [102], easy-clean coatings are gaining popularity [103–105]. This served as inspiration for other authors, who used a variety of experimental procedures to create numerous easy-clean coatings [103].

A coating material's phobicity or philicity toward testing fluids (water and oil) may be adequate to assess how easily it may be cleaned. Sethi et al. [106] in a study conducted in 2020, the hydrophobicity and oleophobicity were assessed using the Forcite module of MS by calculating fluid contact angles using blend structures that had already been constructed and stabilized. Above the equilibrated amorphous structures, a 100 Å vacuum layer was constructed to prevent the contact of the polymer with its own image (which may occur due to the periodic boundary condition).

In order to allow the fluid droplet to form and sustain itself comfortably on the polymer system, the layer was next subjected to the construction of super cells by lengthening the x and y directions to 4 times that of the layers built earlier. Then, utilizing clever algorithm-based geometry optimization with ultra-fine quality and energy and force convergence tolerances of $2*10^5$ kcal/mol/Å and 0.001 kcal/mol/Å, the various produced models were equilibrated.

Using the SPC/E model [107], water droplet construction was carried out. They used about 1000 water molecules to construct their simulation cell, and then they optimized the geometry to get rid of any potential stress that would have been created at particular local areas. The simulated cell was then annealed for structural optimization and equilibrated cells using 10 annealing cycles and 5 heating ramps, each cycle from 0 to 298 K (assumed to be room temperature).

After the cell had been removed for density equilibration at 298 K, the software's Nanocluster construct function was used to create distinct droplet forms with varied diameters (d = 15, 20, 25, and 30 Å). It was observed that post-annealing water droplets over PDMS did not exhibit any symptoms of water molecules flying away after 100 ps of NVT dynamics, in contrast to non-annealed droplets that did.

Upon observing water droplets of various diameters that have undergone structural optimization and annealing, all of them appear to be very highly spherical. A cluster of 300 decane (oil-like) molecules was similarly equilibrated and transformed into a sphere with varied diameters (d = 15, 20, and 25 Å) using the nanocluster build function in order to create an oil droplet. Additionally, in order to allow the water and oil droplets to fall and interact on the surface, they were injected 5 Å above the previously constructed super cells of various surfaces in both cases.

On its base Al surface, Al typically oxidizes and forms a thin Al_2O_3 coating. Therefore, it was thought to be crucial to compare the interaction energy results to those for Al_2O_3. Similar to how interaction with Al_2O_3 was estimated, Materials Studio's structural database was used to import the Al_2O_3 lattice. The surface was then split at its lowest energy surface (0001) and a vacuum slab was put above it, resulting in a 400 Å lattice along the z-axis.

The polymer/blend layer and Al_2O_3 layer were then combined to create a 2-layer system. The following equation was used to calculate the interaction energy ($E_{Interaction}$) of the polymers/blends with Al or Al_2O_3:

$$E_{Interaction} = \frac{1}{A}\left(E_{total} - \left(E_{Al/Al2O3} + E_{polymer}\right)\right)$$

where E_{Total} is the combined energy of the Al and polymer/blend system, E_{Al}, $E_{Polymer}$ are the energies of the separately simulated amorphous cells of Al/Al_2O_3 and polymer/blends, respectively, and A is the surface area of the polymer/blends.

In addition, the simulated estimates of WCA and transparency are in good agreement with the experimental results, validating the accuracy of the simulation methods used. The parameters of pure PDMS, PVAc, and PVOH, such as surface energy and RI, were in line with the previously published values. Up to 20 weight percent of hyd. PVAc has been observed to boost the energy of interaction with the Al surface. It starts to decrease after this crucial content because of potential phase separation. The simulated water contact angles are close to characterization estimations made using a contact angle analyzer as well as some values that have already been published in the literature. The findings of the water and oil contact angles demonstrate that the hydrophobicity and oleophobicity of the prepared coating materials are unfavorably decreased as PVAc content (all hydrolysis grades) is increased.

2.4 Electroosmotic Flow

An essential form of electrokinetic transfer is electroosmotic flow [108]. An electrical double layer (EDL) with a total charge forms close to the exterior when an ionic solution comes into touch with a charged surface. If an external electrical field is applied in a direction tangential to the surface of an EDL, the fluid will be drawn by the moving ions, resulting in an electroosmotic flow. Because the driving force for the flow only exists in the thin EDL, the velocity profile of an electroosmotic flow is flat in most sections of the channel unless the channel width is equal to the thickness of the EDL (usually a few angstroms to tens of nanometers in thickness).

Due to this, electroosmotic flow has a far smaller dispersion than pressure-driven flow, which is frequently used at macroscales, and scales more favorably [109]. Electroosmotic flow has thus gained popularity as a method of transporting fluid in micro- and nanofluidic systems [110].

Qiao in 2006 [111] made a system that comprises a solid wall covered in KCl electrolyte and grafted with polymers. The wall's surface number density is 7.82/nm^2, and its cross dimensions are 4.29*4.29 nm^2. Every atom is connected to the lattice site with a spring, and the structure is made up of 3 layers of atoms organized in a square lattice. The top wall layer is anchored by sixteen charge groups, each of which has a silicon atom that is neutral and an oxygen atom that has a $-e$ charge. This results in a surface charge density (σ_s) of -0.14 C/m^2.

An illustrative surface A-B that reflects back ions crossing the surface prevents ions from entering the water-vacuum boundary. The thickness of the KCl solution layer is set so that the concentration of K $+$ and Cl$-$ ions near surface A-B is the same. The bulk concentration of the KCl solution is 0.95 ± 0.05 M. This device uses a bulk electrolyte concentration that is higher than that utilized in most microfluidic systems. This results from the restricted length scale that an MD simulation may investigate. The system size for an MD simulation of electrokinetic transport must be significantly bigger than the Debye length.

A big simulation system is necessary when using a low bulk concentration. For instance, the characteristic length of the system must be at least tens of nanometers long for a bulk concentration of 0.001 M (Debye length around 10 nm), which is beyond the current computing capacity. The conclusions reached here, however, are applicable to low bulk concentration scenarios because the counterion concentration near the charged surface, the study's main emphasis, is insensitive to the bulk concentration. To create an electroosmotic flow that can be extracted from MD simulations, a strong external electric field of 0.24 V/nm is supplied along the solid wall.

As the EDL (and hence the driving force for flow) only exists near the polymer grafted wall in such a setup, the area below surface A–B replicates electroosmotic flow in a broad microchannel. There are several orders of magnitude more electric fields like this one than are routinely utilized in research. However, the results reported here should be applicable to laboratory settings because the system is still in the linear response domain for the field strength taken into consideration [112]. Three polymer grafting densities were used to investigate the electroosmotic fluxes (0, 11, and 33%). The $(CH_2)_{17}$-COOH polymer is the one chosen. It is an example of an extensive class of amphiphilic polymers.

A sizable layer of water that extends into the electrolyte solution by about 1 nm can be observed. Numerous MD simulations of fluid–solid interfaces have reported the presence of such a density oscillation, which is brought on by the solid wall [112, 113]. The K+ ion concentration reaches its maximum ($c_K^+ = 7.4$ M) at $z = 0.53$ nm, and the Cl-ions are essentially depleted in the region; at $z > 1.19$ nm, where their concentrations are almost identical. The EDL is around 1.0 nm thick, as determined by the thickness of the area where K + and Cl-ion concentrations differ. The electroosmotic flow for the $(CH_2)_{17}$COOH polymer under study is modestly increased at low grafting densities (11%), but is greatly reduced at high densities (33%).

Though the initial observation is infrequently recorded (because polymers are normally coated tightly onto the surface), it can take place locally on a surface with an uneven coating. The latter finding is consistent with past research showing that nonionic polymers placed on silica surfaces can dramatically lessen electroosmotic flow [114, 115]. Although identifying the underlying mechanisms from experiments is challenging, MD simulations can shed light on these discoveries. In simulations, it was discovered that the presence of hydrophobic polymers close to the charged surface considerably alters the electrical double layer, and the modulation of electroosmotic flow greatly depends on those physical processes.

By altering the local dielectric constant and ion hydration, the interfacial polymers have an impact on the electrical double layer. The hydration effects take over at low grafting densities, and the ion concentrations close to the surface decrease. Electroosmotic flow somewhat increases as the electrical double layer becomes more diffuse. Strong counterion adsorption takes place at high grafting densities, probably as a result of a noticeably decreased dielectric constant close to the surface. In this scenario, the electro friction between the ion and surface as well as the drag of the polymer on the fluid and ions both work to suppress the electroosmotic flow.

Cao et al. in 2010 [116] examined for athermal, good, and poor solvent scenarios, the effects of the grafting density and the electric field strength on electroosmotic flow velocity, counterion distribution, and structural properties of grafted chains. Even though the physics of electroosmosis is well established, there are still a lot of new problems that need to be solved for researchers, such as the effects of roughness on channel walls and two-liquid electroosmotic flow [117–121].

In this study, a model is developed in which the electroosmotic flow is influenced by polymer coatings by using a coarse-grained molecular dynamics technique. Previous molecular simulations have not taken into account the flow rate of electroosmosis, which is predicted to be pertinent to the intensity of interactions between solvent particles and polymer monomers.

Two walls that each have two layers of solid atoms organized to form a (1 1 1) plane of a fcc crystal are used to hold fluid particles. Polymer chains made of N monomers are fixed at one end and arranged in a square lattice with a spacing of $d = \rho_g^{-1/2}$, where ρ_g is the number of end-grafted chains per square inch. The first layer's wall atoms are chosen to form the surface-charged particles. The wall atoms in the first layer and the grafted monomers of polymer chains initially share the same plane.

They can be thought of as fixed ghost particles that do not communicate with other particles. Ideal harmonic springs with a $400\varepsilon_{LJ}/\sigma^2$ spring constant hold all wall particles to their starting places. The truncated and shifted Lennard–Jones (LJ) potential simulates the short-range interaction between any two particles separated by a distance r, where r and ε_{LJ} are the Lennard–Jones parameters,

$$U_{LJ}(r) = 4\varepsilon_{LJ}\left[(\sigma/r)^{12} - (\sigma/r)^6 - (\sigma/r_c)^{12} + (\sigma/r_c)^6\right], r < r_c$$

It should be noted that the chosen cutoff radius, $r_c = 2^{1/6}\,\sigma$ corresponds to an interaction between the particles that are solely repulsive. All other units, including the time unit $(m\sigma^2/\varepsilon_{LJ})^{1/2}$ and the temperature unit ε_{LJ}/k_B, are built upon these fundamental units (k_B is Boltzmann constant). The polymer chains are simulated using a widely known coarse-grained bead-spring model.

By a finitely extendable nonlinear elastic (FENE) potential with the formula

$$U_{bond}(r) = -\left(kR_0^2/2\right)\ln\left(1 - r^2/R_0^2\right)$$

where the maximal bond length is $R_0 = 1.5\sigma$ and the spring constant is provided by $k = 30\varepsilon_{LJ}/\sigma^2$, the beads are connected, according to Kremer and Grest [69].

The outcome of choosing the parameter for the average bond length is $a = 0.98\,\sigma$. Combining LJ and FENE potentials prevents the constituent chains from traveling through one another. The simulation box has the following measurements: L_x*L_y*h, where $L_x = L_y = 22.2\,\sigma$ and $h = 24.3\sigma$. The x and y directions use periodic boundary conditions.

The simulation results reveal that the maximum velocity u_{max} in the channel center region significantly depends on the grafting density and solvent quality for a fixed electric field strength. However, there is a minimal value of u_{max} at the intermediate grafting regime for both excellent and poor solvent conditions. In the range of the grafting density tested, u_{max} in the athermal solvent case falls with an increase in grafting density. The three essential parts of these characteristics are counterion distribution, polymer coverage, and viscous drag a polymer exerts on a fluid. It should be underlined that counterion distribution, particularly in good and bad solvent scenarios, will have a considerable impact on flow velocity as grafting density increases.

Polymer–solvent attraction can cause substantial viscous friction between fluid particles and polymer chains in a suitable solvent at modest grafting densities. It is also discovered that electroosmotic flow is markedly improved in the situation of inadequate solvent at high grafting densities, and u_{max} even exceeds that in the case without polymer covering. Counterions travel toward the polymer layer and fluid interface as the electric field strength increases, weakening the polymer's ability to screen against the electroosmotic flow. The athermal, good, and poor solvent scenarios, respectively, are represented by u_{max} at a given grafting density, which is independent of the applied electric field strength. In the end, the authors have a vast experience in the field of molecular dynamics and polymer composites which proves the worth of writing this chapter [122–197].

Conclusions

The chapter has tried to provide an overview of the various kinds of molecular dynamics simulations done on coating. The four types of coatings covered are Graphene, Polymers, Anti-stain or hydrophobic, and electro-osmotic flow. Various uses of these coatings were studied like increase in load-bearing strength of copper, improved nanofilm boiling, repairing of nanoscale flaws for anti-corrosion coatings, and reducing cavitation, among many other uses. All of these simulations have some kind of Atomic Potential and creation of forcefields which the chapter also delves deep upon.

Acknowledgements Academic support from Osaka University (Japan) and monetary support from Japan Society for the Promotion of Science (JSPS) is highly appreciable (Grant Number: P21355).

Conflict of Interest There are no conflicts of interest to declare by the authors.

References

1. Kataria A, Verma A, Sethi SK, Ogata S (2022) Introduction to interatomic potentials/forcefields. In: Forcefields for atomistic-scale simulations: materials and applications. Springer, Singapore, pp 21–49
2. Goswami T, Kamboj N, Bheemaraju A, Kataria A, Dey RS (2021) Supercapacitive behaviour of a novel nanocomposite of 3,4,9,10-perylenetetracarboxylic acid incorporated captopril-Ag nanocluster decorated on graphene nanosheets. Mater Adv 2(4):1358–1368. https://doi.org/10.1039/D0MA00527D
3. Goswami T et al (2022) TiO2 nanoparticles and Nb2O5 nanorods immobilized rGO for efficient visible-light photocatalysis and catalytic reduction. Catal Lett 1:1–17. https://doi.org/10.1007/S10562-022-04000-8/SCHEMES/3

4. Goswami T et al (2021) Highly fluorescent water-soluble PTCA incorporated silver nano-cluster for sensing of dopamine. Mater Chem Phys 259:124086. https://doi.org/10.1016/J.MATCHEMPHYS.2020.124086
5. Kataria A, Verma A, Sanjay MR, Siengchin S (2022) Molecular modeling of 2D graphene grain boundaries: mechanical and fracture aspects. Mater Today: Proc 52:2404–2408
6. Peng W, Sun K, Abdullah R, Zhang M, Chen J, Shi J (2019) Strengthening mechanisms of graphene coatings on Cu film under nanoindentation: a molecular dynamics simulation. Appl Surf Sci 487:22–31. https://doi.org/10.1016/J.APSUSC.2019.04.256
7. Plimpton S (1995) Fast parallel algorithms for short-range molecular dynamics. J Comput Phys 117(1):1–19. https://doi.org/10.1006/JCPH.1995.1039
8. Verlet L (1967) Computer 'experiments' on classical fluids. I. Thermodynamical properties of lennard-jones molecules. Phys Rev 159(1):98. https://doi.org/10.1103/PhysRev.159.98
9. Stuart SJ, Tutein AB, Harrison JA (2000) A reactive potential for hydrocarbons with intermolecular interactions. J Chem Phys 112(14):6472–6486. https://doi.org/10.1063/1.481208
10. Mishin Y, Mehl MJ, Papaconstantopoulos DA, Voter AF, Kress JD (2001) Structural stability and lattice defects in copper: *Ab initio*, tight-binding, and embedded-atom calculations. Phys Rev B 63(22):224106. https://doi.org/10.1103/PhysRevB.63.224106
11. Morse PM, Stueckelberg ECG (1929) Diatomic molecules according to the wave mechanics I: Electronic levels of the hydrogen molecular ion. Phys Rev 33(6):932–947. https://doi.org/10.1103/PHYSREV.33.932
12. Li L, Song W, Xu M, Ovcharenko A, Zhang G (2015) Atomistic insights into the loading—unloading of an adhesive contact: a rigid sphere indenting a copper substrate. Comput Mater Sci 98:105–111. https://doi.org/10.1016/J.COMMATSCI.2014.10.064
13. Vodenitcharova T, Zhang LC (2004) Mechanism of bending with kinking of a single-walled carbon nanotube. Phys Rev B 69(11):115410. https://doi.org/10.1103/PhysRevB.69.115410
14. Stukowski A, Bulatov VV, Arsenlis A (2012) Automated identification and indexing of dislocations in crystal interfaces. Model Simul Mat Sci Eng 20(8):085007. https://doi.org/10.1088/0965-0393/20/8/085007
15. Tang YZ, Zhang XG, Lin Y, Xue J, He Y, Ma LX (2019) Molecular dynamics simulation of nanofilm boiling on graphene-coated surface. Adv Theory Simul 2(8):1900065. https://doi.org/10.1002/ADTS.201900065
16. Carlborg Junichiro CFS, Maruyama S (2008) Thermal boundary resistance between single-walled carbon nanotubes and surrounding matrices. Phys Rev B 78(20):205406-NA. https://doi.org/10.1103/physrevb.78.205406
17. Yi Dimos PP, Walther JH, Yadigaroglu G (2002) Molecular dynamics simulation of vaporization of an ultra-thin liquid argon layer on a surface. Int J Heat Mass Transf 45(10):2087–2100. https://doi.org/10.1016/s0017-9310(01)00310-6
18. Nagayama Takaharu GT, Cheng P (2006) Molecular dynamics simulation on bubble formation in a nanochannel. Int J Heat Mass Transf 49(23):4437–4443. https://doi.org/10.1016/j.ijheatmasstransfer.2006.04.030
19. Mao Yuwen YZ (2014) Molecular dynamics simulation on rapid boiling of water on a hot copper plate. Appl Therm Eng 62(2):607–612. https://doi.org/10.1016/j.applthermaleng.2013.10.032
20. Hens Rahul AA, Biswas G (2014) Nanoscale study of boiling and evaporation in a liquid Ar film on a Pt heater using molecular dynamics simulation. Int J Heat Mass Transf 71(NA):303–312. https://doi.org/10.1016/j.ijheatmasstransfer.2013.12.032
21. Shavik Mohammad Nasim MH, Monjur Morshed AKM (2016) Molecular dynamics study on explosive boiling of thin liquid argon film on nanostructured surface under different wetting conditions. J Electron Packag 138(1):010904-NA. https://doi.org/10.1115/1.4032463
22. Diaz Zhixiong RG (2017) Molecular dynamics study of wettability and pitch effects on maximum critical heat flux in evaporation and pool boiling heat transfer. Numeri Heat Transf A Appl 72(12):891–903. https://doi.org/10.1080/10407782.2017.1412710

23. Wang SY, Lu G, Wang X-D (2018) Explosive boiling of nano-liquid argon films on high temperature platinum walls: effects of surface wettability and film thickness. Int J Therm Sci 132(NA):610–617. https://doi.org/10.1016/j.ijthermalsci.2018.07.007

24. Liu H, Zhang CH, Zhao M, Zhu Y, Wang W (2017) Influence of interface wettability on normal and explosive boiling of ultra-thin liquid films on a heated substrate in nanoscale: a molecular dynamics study. Micro Nano Lett 12(11):843–848. https://doi.org/10.1049/mnl.2017.0425

25. Inaoka HI, Nobuyasu RS (2013) Numerical simulation of pool boiling of a Lennard-Jones liquid. Physica A Statistical Mechanics and its Applications, 392(18):3863–3868. https://doi.org/10.1016/j.physa.2013.05.002

26. Diaz R, Guo Z (2016) A molecular dynamics study of phobic/philic nano-patterning on pool boiling heat transfer. Heat Mass Trans 53(3):1061–1071. https://doi.org/10.1007/s00231-016-1878-2

27. Dong LQ, Xiao J, Cheng P (2014) An experimental investigation of enhanced pool boiling heat transfer from surfaces with micro/nano-structures. Int J Heat Mass Transf 71(71):189–196. https://doi.org/10.1016/j.ijheatmasstransfer.2013.11.068

28. Prakash CGJ, Prasanth R (2018) Enhanced boiling heat transfer by nano structured surfaces and nanofluids. Renewable and Sustainable Energy Reviews 82(NA):4028–4043. https://doi.org/10.1016/j.rser.2017.10.069

29. Wang Q, Chen R (2018) Ultrahigh flux thin film boiling heat transfer through nanoporous membranes. Nano Lett 18(5):3096–3103. https://doi.org/10.1021/acs.nanolett.8b00648

30. Balandin AA, Suchismita G, Bao W, Calizo I, Teweldebrhan D, Miao F, Lau CN (2008) Superior thermal conductivity of single-layer graphene. Nano Lett 8(3):902–907. https://doi.org/10.1021/nl0731872

31. Shahil KMF, Balandin AA (2012) Thermal properties of graphene and multilayer graphene: applications in thermal interface materials. Solid State Commun 152(15):1331–134. https://doi.org/10.1016/j.ssc.2012.04.034

32. Chang S-W Nair AK, Buehler MJ (2012) Geometry and temperature effects of the interfacial thermal conductance in copper- and nickel-graphene nanocomposites. J Phys Condens Matter 24(24):245301-NA. https://doi.org/10.1088/0953-8984/24/24/245301.

33. Xu Z, Buehler MJ (2012) Heat dissipation at a graphene–substrate interface. J Phys Condens Matter 24(47):475305-NA. https://doi.org/10.1088/0953-8984/24/47/475305.

34. Hong Y, Lei L, Zeng XC, Zhang J (2015) Tuning thermal contact conductance at graphene-copper interface via surface nanoengineering. Nanoscale 7(14):6286–6294. https://doi.org/10.1039/c5nr00564g

35. Giovannetti G, Khomyakov P, Brocks G, Karpan V, van den Brink JE, Kelly Paul J (2008) Doping graphene with metal contacts. Phys Rev Lett 101(2):026803-NA. https://doi.org/10.1103/physrevlett.101.026803

36. Xu Z, Buehler MJ (2010) Interface structure and mechanics between graphene and metal substrates: a first-principles study. J Phys Condens Matter 22(48):485301-NA. https://doi.org/10.1088/0953-8984/22/48/485301

37. Schneider T, Stoll E (1978) Molecular-dynamics study of a three-dimensional one-component model for distortive phase transitions. Phys Rev B 17(3):302–1322. https://doi.org/10.1103/physrevb.17.1302

38. Yu Hao JW (2012) A molecular dynamics investigation on evaporation of thin liquid films. Int J Heat Mass Transf 55(4):1218–1225. https://doi.org/10.1016/j.ijheatmasstransfer.2011.09.035

39. Ma B, Guye K, Dogruoz B, Agonafer D (2021) Molecular dynamics simulations of thin-film evaporation: the influence of interfacial thermal resistance on a graphene-coated heated silicon substrate. Appl Therm Eng 195:117142. https://doi.org/10.1016/J.APPLTHERMALENG.2021.117142

40. Nagayama G, Kawagoe M, Tokunaga A, Tsuruta T (2010) On the evaporation rate of ultra-thin liquid film at the nanostructured surface: A molecular dynamics study. Int J Therm Sci 49(1):59–66. https://doi.org/10.1016/j.ijthermalsci.2009.06.001

41. Pham A, Barisik M, Kim B (2014) Molecular dynamics simulations of Kapitza length for argon-silicon and water-silicon interfaces. Int J Precis Eng Manuf 15(2):323–329. https://doi.org/10.1007/s12541-014-0341-x

42. Kim B, Beskok A, Cagin T (2008) Molecular dynamics simulations of thermal resistance at the liquid-solid interface. J Chem Phys 129(17):174701. https://doi.org/10.1063/1.3001926

43. Han H, Schlawitschek C, Katyal N, Stephan P, Gambaryan-Roisman T, Leroy F, Müller-Plathe F (2017) Solid–liquid interface thermal resistance affects the evaporation rate of droplets from a surface: a study of perfluorohexane on chromium using molecular dynamics and continuum theory. Langmuir 33(21):5336–5343. https://doi.org/10.1021/acs.langmuir.7b01410

44. Ramos-Alvarado B, Kumar S, Peterson GP (2016) Solid-liquid thermal transport and its relationship with wettability and the interfacial liquid structure. J Phys Chem Lett 7(17):3497–3501. https://doi.org/10.1021/acs.jpclett.6b01605

45. Mahan GD (2008) Kapitza resistance at a solid-fluid interface. Nanoscale Microscale Thermophys Eng 12(4):294–310. https://doi.org/10.1080/15567260802591944

46. van Itterbeek A, Verbeke O (1960) Density of liquid nitrogen and argon as a function of pressure and temperature. Physica 26(11):931–938. https://doi.org/10.1016/0031-8914(60)90042-2

47. Liang Z, Biben T, Keblinski P (2017) Molecular simulation of steady-state evaporation and condensation: Validity of the Schrage relationships. Int J Heat Mass Transf 114:105–114. https://doi.org/10.1016/j.ijheatmasstransfer.2017.06.025

48. Ramos-Alvarado B, Kumar S, Peterson GP (2016) On the wettability transparency of graphene-coated silicon surfaces. J Chem Phys 144(1):14701. https://doi.org/10.1063/1.4938499

49. Lennard-Jones JE (1931) Cohesion. Proc Phys Soc 43(5):461–482. https://doi.org/10.1088/0959-5309/43/5/301

50. Lorentz HA (1881) Ueber die Anwendung des Satzes vom Virial in der kinetischen Theorie der Gase. Ann Phys 248(1):127–136. https://doi.org/10.1002/andp.18812480110

51. Mypati O, Mishra D, Sahu S, Pal SK, Srirangam P (2019) A study on electrical and electrochemical characteristics of friction stir welded lithium-ion battery tabs for electric vehicles. J Electron Mater 49(1):72–87. https://doi.org/10.1007/S11664-019-07711-8

52. Das A, Li D, Williams D, Greenwood D (2018) Joining technologies for automotive battery systems manufacturing. World Electr Veh J 9(2):22. https://doi.org/10.3390/WEVJ9020022

53. Mahto RP, Anishetty S, Sarkar A, Mypati O, Pal SK, Majumdar JD (2019) Interfacial microstructural and corrosion characterizations of friction stir welded AA6061-T6 and AISI304 materials. Met Mater Int 25(3):752–767. https://doi.org/10.1007/S12540-018-00222-X/FIGURES/20

54. Mypati O et al (2020) Enhancement of joint strength in friction stir lap welding between AA6061 and AISI 304 by adding diffusive coating agents. Proc Inst Mech Eng B J Eng Manuf 234(1–2):204–217. https://doi.org/10.1177/0954405419838379/ASSET/IMAGES/LARGE/10.1177_0954405419838379-FIG2.JPEG

55. Sahu S, Mypati O, Pal SK, Shome M, Srirangam P (2021) Effect of weld parameters on joint quality in friction stir welding of Mg alloy to DP steel dissimilar materials. CIRP J Manuf Sci Technol 35:502–516. https://doi.org/10.1016/J.CIRPJ.2021.06.012

56. Asemabadi M, Sedighi M, Honarpisheh M (2012) Investigation of cold rolling influence on the mechanical properties of explosive-welded Al/Cu bimetal. Mater Sci Eng: A 558:144–149. https://doi.org/10.1016/j.msea.2012.07.102

57. Shen N, Samanta A, Ding H, Cai WW (2016) Simulating microstructure evolution of battery tabs during ultrasonic welding. J Manuf Process 23:306–31. https://doi.org/10.1016/j.jmapro.2016.04.005

58. Heideman R, Johnson C, Kou S (2010) Metallurgical analysis of Al/Cu friction stir spot welding. Sci Technol Weld Join 15(7):597–604. https://doi.org/10.1179/136217110x12785889549985

59. Mypati O, Kumar PP, Pal SK, Srirangam P (2022) TEM analysis and molecular dynamics simulation of graphene coated Al-Cu micro joints. Carbon Trends 9:100223. https://doi.org/10.1016/J.CARTRE.2022.100223

60. Heinz H, Vaia RA, Farmer Barry L, Naik Rajesh R (2008) Accurate simulation of surfaces and interfaces of face-centered cubic metals using $12-6$ and $9-6$ lennard-jones potentials. J Phys Chem C 112(44):17281–17290. https://doi.org/10.1021/jp801931d

61. Kumar S (2017) Effect of applied force and atomic organization of copper on its adhesion to a graphene substrate. RSC Adv 7(40):25118–25131. https://doi.org/10.1039/c7ra01873h

62. Borovikov VM, Mendelev MI, King AH, LeSar R (2015) Effect of Stacking Fault Energy on Mechanism of Plastic Deformation in Nanotwinned FCC Metals. Model Simul Mat Sci Eng 23(5):055003-NA. https://doi.org/10.1088/0965-0393/23/5/055003

63. Lu L, Shen Y, Chen X, Qian L, Lu K (2004) Ultrahigh strength and high electrical conductivity in copper. Science 304(5669):422–426. https://doi.org/10.1126/science.1092905

64. Lloyd JT (2018) A dislocation-based model for twin growth within and across grains. Proc R Soc A: Math, Phys Eng Sci 474(2210). https://doi.org/10.1098/RSPA.2017.0709

65. Zhao W, Tao N, Guo JY, Lu Q, Lu K (2005) High density nano-scale twins in Cu induced by dynamic plastic deformation. Scr Mater 53(6):745–749. https://doi.org/10.1016/j.scriptamat.2005.05.022

66. Liu C, Shanthraj P, Robson JD, Diehl M, Dong S, Dong J, Ding W, Raabe D (2019) On the interaction of precipitates and tensile twins in magnesium alloys. Acta Mater 178:146–162. https://doi.org/10.1016/j.actamat.2019.07.046

67. Gosling JH, Makarovsky O, Wang F, Cottam ND, Greenaway M, Patanè A, Wildman RD, Tuck C, Turyanska L, Fromhold TM (2021) Universal mobility characteristics of graphene originating from charge scattering by ionised impurities. Commun Phys 4(1):30. https://doi.org/10.1038/s42005-021-00518-2

68. Smith KA, Tyagi S, Balazs AC (2005) Healing surface defects with nanoparticle-filled polymer coatings: effect of particle geometry. Macromolecules 38(24):10138–10147. https://doi.org/10.1021/MA0515127

69. Kremer K, Grest GS (1990) Dynamics of entangled linear polymer melts: A molecular-dynamics simulation. J Chem Phys 92(8):5057–5086. https://doi.org/10.1063/1.458541

70. Zee M, Feickert AJ, Kroll DM, Croll SG (2015) Cavitation in crosslinked polymers: molecular dynamics simulations of network formation. Prog Org Coat 83:55–63. https://doi.org/10.1016/J.PORGCOAT.2015.01.022

71. Beveridge DL (1997) Monte Carlo and Molecular Dynamics Simulations in Polymer Science Edited by Kurt Binder (Johannes-Gutenberg University). Oxford University Press, New York, Oxford, 1995. xiv + 587 pp $95.00. ISBN 0-19-509438-7. J Am Chem Soc 119(41):9938–9938. https://doi.org/10.1021/JA965637Q

72. Grest GS, Kremer K (1990) Statistical properties of random cross-linked rubbers. Macromolecules 23(23):4994–5000. https://doi.org/10.1021/ma00225a020

73. Sun C et al (2021) Self-healing polymers using electrosprayed microcapsules containing oil: molecular dynamics simulation and experimental studies. J Mol Liq 325:115182. https://doi.org/10.1016/J.MOLLIQ.2020.115182

74. Khodadadi H, Toghraie D, Karimipour A (2019) Effects of nanoparticles to present a statistical model for the viscosity of MgO-Water nanofluid. Powder Technol 342:166–180. https://doi.org/10.1016/j.powtec.2018.09.076

75. He W, Toghraie D, Lotfipour A, Pourfattah F, Karimipour A, Afrand M (2020) Effect of twisted-tape inserts and nanofluid on flow field and heat transfer characteristics in a tube. Int Commun Heat Mass Transf 110:104440-NA. https://doi.org/10.1016/j.icheatmasstransfer.2019.104440

76. Varzaneh AA, Toghraie D, Karimipour A (2019) Comprehensive simulation of nanofluid flow and heat transfer in straight ribbed microtube using single-phase and two-phase models for choosing the best conditions. J Therm Anal Calorim 139(1):701–720. https://doi.org/10.1007/s10973-019-08381-8

77. Wang X, Xu X, Choi SUS (1999) Thermal conductivity of nanoparticle -fluid mixture. J Thermophys Heat Trans 13(4):474–480. https://doi.org/10.2514/2.6486

78. Kouhi M, Mohebbi A, Mirzaei M (2012) Evaluation of the corrosion inhibition effect of micro/nanocapsulated polymeric coatings: a comparative study by use of EIS and Tafel experiments

and the area under the Bode plot. Res Chem Intermed 39(5):2049–2062. https://doi.org/10. 1007/s11164-012-0736-1

79. Wei H, Wang Y, Guo J, Shen NZ, Jiang D, Zhang X, Yan X, Zhu J, Wang Q, Shao L, Lin H, Wei S, Guo Z (2015) Advanced micro/nanocapsules for self-healing smart anticorrosion coatings. J Mater Chem A Mater 3(2):469–480. https://doi.org/10.1039/c4ta04791e

80. Alipour P, Toghraie D, Karimipour A, Hajian M (2019) Modeling different structures in perturbed Poiseuille flow in a nanochannel by using of molecular dynamics simulation: Study the equilibrium. Phys A: Stat Mech Appl 515:13–30. https://doi.org/10.1016/j.physa.2018. 09.177

81. Alipour P, Toghraie D, Karimipour A, Hajian M (2019) Molecular dynamics simulation of fluid flow passing through a nanochannel: effects of geometric shape of roughnesses. J Mol Liq 275:192–203. https://doi.org/10.1016/j.molliq.2018.11.057

82. Kumar N, Manik G (2016) Molecular dynamics simulations of polyvinyl acetate-perfluorooctane based anti-stain coatings. Polymer (Guildf) 100:194–205. https://doi.org/10. 1016/J.POLYMER.2016.08.019

83. Small PA (2007) Some factors affecting the solubility of polymers. J Appl Chem 3(2):71–80. https://doi.org/10.1002/jctb.5010030205

84. Daoust H, Rinfret M (1952) Solubility of polymethyl methacrylate and polyvinyl acetate. J Colloid Sci 7(1):11–19. https://doi.org/10.1016/0095-8522(52)90016-0

85. Hariharan SS, Meenakshi A (1977) A new water-soluble initiator for vinyl polymerization. Journal of Polymer Science: Polymer Letters Edition15(1):1–7. : https://doi.org/10.1002/pol. 1977.130150101.

86. DiPaola-Baranyi G, Guillet JE, Klein J, Jeberien H-E (1978) Estimation of solubility parameters for poly(vinyl acetate) by inverse gas chromatography. J Chromatogr A 166(2):349–356.https://doi.org/10.1016/s0021-9673(00)95616-4

87. Barton AFM (1990) CRC handbook of polymer-liquid interaction parameter s and solubility parameters

88. Hougham G, Cassidy, PE, Johns K, Davidson T (eds) (2002) Fluoropolymers 2. https://doi. org/10.1007/B114560

89. Siddiqui AR, Li W, Wang F, Ou J, Amirfazli A (2021) One-step fabrication of transparent superhydrophobic surface. Appl Surf Sci 542:148534. https://doi.org/10.1016/j.apsusc.2020. 148534

90. Lai YT, Yu X, Gong J, Gong D, Chi L, Lin C, Chen Z (2012) Transparent superhydrophobic/ superhydrophilic TiO2-based coatings for self-cleaning and anti-fogging. J Mater Chem 22(15):7420–7426. https://doi.org/10.1039/c2jm16298a

91. Rasouli S, Rezaei N, Hamedi H, Zendehboudi S, Duan X (2021) Superhydrophobic and superoleophilic membranes for oil-water separation application: a comprehensive review. Mater Des 204:109599. https://doi.org/10.1016/J.MATDES.2021.109599

92. Wang P, Li Z, Xie Q, Duan W, Zhang X, Han H (2021) A passive anti-icing strategy based on a superhydrophobic mesh with extremely low ice adhesion strength. J Bionic Eng 18(1):55–64. https://doi.org/10.1007/s42235-021-0012-4

93. Xu S, Wang Q, Wang N (2020) Eco-friendly fabrication of superhydrophobic surface with anti-corrosion by transferring dendrite-like structures to aluminum substrate. Colloids Surf A Physicochem Eng Asp 595:124719. https://doi.org/10.1016/j.colsurfa.2020.124719

94. Zhu Y, Yang F, Guo Z (2021) Bioinspired surfaces with special micro-structures and wettability for drag reduction: which surface design will be a better choice?. Nanoscale 13(6):3463–3482. https://doi.org/10.1039/d0nr07664c

95. Wang C, Tanf F, Li Q, Zhang Y, Wang X (2017) Spray-coated superhydrophobic surfaces with wear-resistance, drag-reduction and anti-corrosion properties. Colloids Surf A Physicochem Eng Asp 514:236–242. https://doi.org/10.1016/j.colsurfa.2016.11.059

96. Jiang Y, Choi CH (2020) Droplet retention on superhydrophobic surfaces: a critical review. Adv Mater Interfaces 8(2):2001, 205. https://doi.org/10.1002/admi.202001205

97. Samaha MA, Tafreshi HV, Gad-el-Hak M (2012) Superhydrophobic surfaces: From the lotus leaf to the submarine. Comptes Rendus Mécanique 340(1):18–34. https://doi.org/10.1016/j. crme.2011.11.002

98. Daneshmand H, Sazgar A, Araghchi M (2021) Fabrication of robust and versatile super-hydrophobic coating by two-step spray method: an experimental and molecular dynamics simulation study. Appl Surf Sci 567:150825. https://doi.org/10.1016/J.APSUSC.2021.150825

99. Suganuma Y, Yamamoto S, Kinjo T, Mitsuoka T, Umemoto K (2017) Wettability of Al2O3 surface by organic molecules: insights from molecular dynamics simulation. J Phys Chem B 121(42):9929–9935. https://doi.org/10.1021/acs.jpcb.7b07062

100. Fyrner T, Lee HH, Mangone A, Ekblad T, Pettitt ME, Callow ME, Callow JA, Conlan SL, Mutton R, Clare AS, Konradsson P, Liedberg B, Ederth T (2011) Saccharide-functionalized alkanethiols for fouling-resistant self-assembled monolayers: synthesis, monolayer properties, and antifouling behavior. Langmuir 27(24):15034–15047. https://doi.org/10.1021/la202774e

101. Messali M, Larouj M, Lgaz H, Rezki N, Al-blewi FF, Aouad MR, Chaouiki A, Salghi R, Chung IM (2018) A new schiff base derivative as an effective corrosion inhibitor for mild steel in acidic media: experimental and computer simulations studies. J Mol Struct 1168:39–48. https://doi.org/10.1016/j.molstruc.2018.05.018

102. Luo ZY, Chen KX, Wang JH, Mo DC, Lyu SS (2016) Hierarchical nanoparticle-induced superhydrophilic and under-water superoleophobic Cu foam with ultrahigh water permeability for effective oil/water separation. J Mater Chem A Mater 4(27):10566–10574. https://doi.org/10.1039/c6ta04487e

103. Sethi SK, Manik G (2018) Recent progress in super hydrophobic/hydrophilic self-cleaning surfaces for various industrial applications: a review. Polym Plast Technol Eng 57(18):1932–1952. https://doi.org/10.1080/03602559.2018.1447128

104. Sethi SK, Manik G, Sahoo SK (2019) Fundamentals of superhydrophobic surfaces. In: Superhydrophobic polymer coatings. pp 3–29. https://doi.org/10.1016/b978-0-12-816671-0.00001-1

105. Sethi SK, Shankar U, Manik G (2019) Fabrication and characterization of non-fluoro based transparent easy-clean coating formulations optimized from molecular dynamics simulation. Prog Oat 136:105306. https://doi.org/10.1016/j.porgcoat.2019.105306

106. Sethi SK, Soni L, Shankar U, Chauhan RP, Manik G (2020) A molecular dynamics simulation study to investigate poly(vinyl acetate)-poly(dimethyl siloxane) based easy-clean coating: An insight into the surface behavior and substrate interaction. J Mol Struct 1202. https://doi.org/10.1016/J.MOLSTRUC.2019.127342

107. Berendsen HJC, Grigera JR, Straatsma TP (1987) The missing term in effective pair potentials. J Phys Chem 91(24):6269–6271. https://doi.org/10.1021/j100308a038

108. Lyklema J Fundamentals of interface and colloid science. Volume III, Liquid-fluid interfaces

109. Probstein RF (1994) Physicochemical hydrodynamics. https://doi.org/10.1002/0471725137

110. Vilkner T, Janasek D, Manz A (2004) Micro total analysis systems. Recent developments. Anal Chem 76(12):3373–3385. https://doi.org/10.1021/ac040063q

111. Qiao R (2006) Control of electroosmotic flow by polymer coating: effects of the electrical double layer. https://doi.org/10.1021/la060883t

112. Qiao R, Aluru NR (2003) Ion concentrations and velocity profiles in nanochannel electroosmotic flows. J Chem Phys 118(10):4692–4701. https://doi.org/10.1063/1.1543140

113. Qiao R, Aluru NR (2005) Atomistic simulation of KCl transport in charged silicon nanochannels: interfacial effects. Colloids Surf A Physicochem Eng Asp 267(1):103–109. https://doi.org/10.1016/j.colsurfa.2005.06.067

114. Bruin GC, Chang JP, Kuhlman RH, Zegers K, Kraak JC, Poppe H (1989) Capillary zone electrophoretic separations of proteins in polyethylene glycol-modified capillaries. J Chromatogr A 471:429–436. https://doi.org/10.1016/s0021-9673(00)94190-6

115. Belder D, Jörg W (2001) Electrokinetic effects in poly(ethylene glycol)-coated capillaries induced by specific adsorption of cations. Langmuir 17(16):4962–4966. https://doi.org/10.1021/la010115w

116. Cao Q, Zuo C, Li L, Ma Y, Li N (2010) Electroosmotic flow in a nanofluidic channel coated with neutral polymers. Microfluid Nanofluidics 9(6):1051–1062. https://doi.org/10.1007/S10404-010-0620-5/FIGURES/15

117. Y. W. Gao Teck Neng; Yang Chun; Ooi Kim Tiow, "Two-fluid electroosmotic flow in microchannels.," *J Colloid Interface Sci*, vol. 284, no. 1, pp. 306–314, 2005, doi: https://doi.org/10.1016/j.jcis.2004.10.011.

118. Lee JSH, Li D (2006) Electroosmotic flow at a liquid–air interface. Microfluid Nanofluidics 2(4):361–365. https://doi.org/10.1007/s10404-006-0084-9

119. Hu Y, Werner C, Li D (2003) Electrokinetic transport through rough microchannels. Anal Chem 75(21):5747–5758. https://doi.org/10.1021/ac0347157

120. Qiao R, He P (2007) Modulation of electroosmotic flow by neutral polymers. Langmuir 23(10):5810–5816. https://doi.org/10.1021/la063042v

121. Yang D, Liu Y (2008) Numerical simulation of electroosmotic flow in microchannels with sinusoidal roughness. Colloids Surf A Physicochem Eng Asp 328(1):28–33 https://doi.org/10.1016/j.colsurfa.2008.06.029

122. Bharath KN, Madhu P, Gowda TY, Verma A, Sanjay MR, Siengchin S (2021) Mechanical and chemical properties evaluation of sheep wool fiber–reinforced vinylester and polyester composites. Mater Perform Charact 10(1):99–109

123. Verma A, Parashar A, Jain N, Singh VK, Rangappa SM, Siengchin S (2020) Surface modification techniques for the preparation of different novel biofibers for composites. In: Biofibers and biopolymers for biocomposites. Springer, Cham, pp 1–34

124. Verma A, Singh VK (2016) Experimental investigations on thermal properties of coconut shell particles in DAP solution for use in green composite applications. J Mater Sci Eng 5(3):1000242

125. Singh K, Jain N, Verma A, Singh VK, Chauhan S (2020) Functionalized graphite–reinforced cross-linked poly (vinyl alcohol) nanocomposites for vibration isolator application: morphology, mechanical, and thermal assessment. Mater Perform Charact 9(1):215–230

126. Verma A, Singh VK, Arif M (2016) Study of flame retardant and mechanical properties of coconut shell particles filled composite. Res Rev: J Mater Sci 4(3):1–5

127. Verma A, Jain N, Rastogi S, Dogra V, Sanjay SM, Siengchin S, Mansour R (2020) Mechanism, anti-corrosion protection and components of anti-corrosion polymer coatings. In: Polymer coatings. CRC Press, pp 53–66

128. Chaudhary A, Sharma S, Verma A (2022) WEDM machining of heat treated ASSAB'88 tool steel: a comprehensive experimental analysis. Mater Today: Proc 50:946–951

129. Bisht N, Verma A, Chauhan S, Singh VK (2021) Effect of functionalized silicon carbide nano-particles as additive in cross-linked PVA based composites for vibration damping application. J Vinyl Add Tech 27(4):920–932

130. Verma A, Jain N, Parashar A, Singh VK, Sanjay MR, Siengchin S (2020) Design and modeling of lightweight polymer composite structures. In: Lightweight polymer composite structures. CRC Press, pp 193–224

131. Dogra V, Kishore C, Verma A, Rana AK, Gaur A (2021) Fabrication and experimental testing of hybrid composite material having biodegradable bagasse fiber in a modified epoxy resin: evaluation of mechanical and morphological behavior. Appl Sci Eng Prog 14(4):661–667

132. Deji R, Verma A, Kaur N, Choudhary BC, Sharma RK (2022) Density functional theory study of carbon monoxide adsorption on transition metal doped armchair graphene nanoribbon. Mater Today: Proc 54:771–776

133. Verma A, Jain N, Parashar A, Gaur A, Sanjay MR, Siengchin S (2021) Lifecycle assessment of thermoplastic and thermosetting bamboo composites. In: Bamboo fiber composites. Springer, Singapore, pp 235–246

134. Deji R, Verma A, Choudhary BC, Sharma RK (2022) New insights into NO adsorption on alkali metal and transition metal doped graphene nanoribbon surface: a DFT approach. J Mol Graph Model 111:108109

135. Verma A, Jain N, Parashar A, Singh VK, Sanjay MR, Siengchin S (2020) Lightweight graphene composite materials. In: Lightweight polymer composite structures. CRC Press, pp 1–20

136. Verma A, Parashar A (2020) Characterization of 2D nanomaterials for energy storage. In: Recent advances in theoretical, applied, computational and experimental mechanics. Springer, Singapore, pp 221–226

137. Deji R, Jyoti R, Verma A, Choudhary BC, Sharma RK (2022) A theoretical study of HCN adsorption and width effect on co-doped armchair graphene nanoribbon. Comput Theor Chem 1209:113592
138. Verma A, Samant SS (2016) Inspection of hydrodynamic lubrication in infinitely long journal bearing with oscillating journal velocity. J Appl Mech Eng 5(3):1–7
139. Verma A, Parashar A, Packirisamy M (2019) Effect of grain boundaries on the interfacial behaviour of graphene-polyethylene nanocomposite. Appl Surf Sci 470:1085–1092
140. Arpitha GR, Verma A, Sanjay MR, Siengchin S (2021) Preparation and experimental investigation on mechanical and tribological performance of hemp-glass fiber reinforced laminated composites for lightweight applications. Adv Civil Eng Mater 10(1):427–439
141. Verma A, Jain N, Rangappa SM, Siengchin S, Jawaid M (2021) Natural fibers based bio-phenolic composites. In: Phenolic polymers based composite materials. Springer, Singapore, pp 153–168
142. Verma A, Parashar A, Singh SK, Jain N, Sanjay SM, Siengchin S (2020) Modelling and simulation in polymer coatings. In: Polymer coatings. CRC Press, pp 309–324
143. Verma A, Singh VK Experimental characterization of modified epoxy resin assorted with almond shell particles. ESSENCE-Int J Environ Rehabil Conserv 7(1):36–44
144. Deji R, Verma A, Kaur N, Choudhary BC, Sharma RK (2022) Adsorption chemistry of co-doped graphene nanoribbon and its derivatives towards carbon based gases for gas sensing applications: quantum DFT investigation. Mater Sci Semicond Process 146:106670
145. Verma A (2022) A perspective on the potential material candidate for railway sector applications: PVA based functionalized graphene reinforced composite. Appl Sci Eng Prog 15(2):5727–5727
146. Verma A, Jain N, Singh K, Singh VK, Rangappa SM, Siengchin S (2022) PVA-based blends and composites. In: Biodegradable polymers, blends and composites. Woodhead Publishing, pp 309–326
147. Raja S, Verma A, Rangappa SM, Siengchin S (2022) Development and experimental analysis of polymer based composite bipolar plate using Aquila Taguchi optimization: design of experiments. Polym Compos 43(8):5522–5533
148. Prabhakaran S, Sharma S, Verma A, Rangappa SM, Siengchin S (2022) Mechanical, thermal, and acoustical studies on natural alternative material for partition walls: a novel experimental investigation. Polym Compos 43(7):4711–4720
149. Sethi SK, Gogoi R, Verma A, Manik G (2022) How can the geometry of a rough surface affect its wettability?-a coarse-grained simulation analysis. Prog Org Coat 172:107062
150. Verma A, Jain N, Sethi SK (2022) Modeling and simulation of graphene-based composites. In: Innovations in graphene-based polymer composites. Woodhead Publishing, pp 167–198
151. Chaturvedi S, Verma A, Sethi SK, Rangappa SM, Siengchin S (2022) Stalk fibers (rice, wheat, barley, etc.) composites and applications. In: Plant fibers, their composites, and applications. Woodhead Publishing, pp 347–362
152. Arpitha GR, Verma A, Sanjay MR, Gorbatyuk S, Khan A, Sobahi TR, Asiri AM, Siengchin S (2022O Bio-composite film from corn starch based vetiver cellulose. J Nat Fibers 19(16):14634–14644
153. Thimmaiah SH, Narayanappa K, Thyavihalli Girijappa Y, Gulihonenahali Rajakumara A, Hemath M, Thiagamani SMK, Verma A (2022) An artificial neural network and Taguchi prediction on wear characteristics of Kenaf-Kevlar fabric reinforced hybrid polyester composites. Polym Compos 44(1):261–273
154. Chaturvedi S, Verma A, Singh SK, Ogata S (2022) EAM inter-atomic potential—its implication on nickel, copper, and aluminum (and their alloys). In: Forcefields for atomistic-scale simulations: materials and applications. Springer, Singapore, pp. 133–156
155. Verma A, Sharma S (2022) Atomistic simulations to study thermal effects and strain rate on mechanical and fracture properties of graphene like BC3. In: Forcefields for atomistic-scale simulations: materials and applications. Springer, Singapore, pp. 237–252
156. Chaturved S, Verma A, Sethi SK, Ogata S (2022) Defect energy calculations of nickel, copper and aluminium (and their alloys): molecular dynamics approach. In: Forcefields for atomistic-scale simulations: materials and applications. Springer, Singapore, pp 157–186

157. Verma A, Gaur A, Singh VK (2017) Mechanical properties and microstructure of starch and sisal fiber biocomposite modified with epoxy resin. Mater Perform Charact 6(1):500–520

158. Shankar U, Gogoi R, Sethi SK, Verma A (2022) Introduction to materials studio software for the atomistic-scale simulations. In: Forcefields for atomistic-scale simulations: materials and applications. Springer, Singapore, pp 299–313

159. Shankar U, Sethi SK, Verma A (2022) Forcefields and modeling of polymer coatings and nanocomposites. In: Forcefields for atomistic-scale simulations: materials and applications. Springer, Singapore, pp 81–98

160. Verma A, Ogata S (2022) Computational modelling of deformation and failure of bone at molecular scale. In: Forcefields for atomistic-scale simulations: materials and applications. Springer, Singapore, pp 253–268

161. Homer ER, Verma A, Britton D, Johnson OK, Thompson GB (2022) Simulated migration behavior of metastable Σ3 (11 8 5) incoherent twin grain boundaries. IOP Conf Ser: Mater Sci Eng 1249(1):012019

162. Verma A, Parashar A, van Duin AC (2022) Graphene-reinforced polymeric membranes for water desalination and gas separation/barrier applications. In Innovations in graphene-based polymer composites. Woodhead Publishing, pp 133–165

163. Verma A, Jain N, Sanjay MR, Siengchin S Viscoelastic Properties of completely biodegradable polymer-based composites. In: Vibration and damping behavior of biocomposites. CRC Press, pp 173–188

164. Verma A, Jain N, Mishra RR (2022) Applications and drawbacks of epoxy/natural fiber composites. In: Handbook of epoxy/fiber composites. Singapore, Springer Singapore, pp 1–15

165. Lila MK, Verma A, Bhurat SS (2022) Impact behaviors of epoxy/synthetic fiber composites. In: Handbook of epoxy/fiber composites. Singapore, Springer Singapore, pp 1–18

166. Verma A, Parashar A (2017) The effect of STW defects on the mechanical properties and fracture toughness of pristine and hydrogenated graphene. Phys Chem Chem Phys 19(24):16023–16037

167. Verma A, Parashar A (2018) Molecular dynamics based simulations to study failure morphology of hydroxyl and epoxide functionalised graphene. Comput Mater Sci 143:15–26

168. Verma A, Parashar A (2018) Molecular dynamics based simulations to study the fracture strength of monolayer graphene oxide. Nanotechnology 29(11):115706

169. Verma A, Parashar A (2018) Structural and chemical insights into thermal transport for strained functionalised graphene: a molecular dynamics study. Mater Res Express 5(11):115605

170. Verma A, Parashar A, Packirisamy M (2018) Tailoring the failure morphology of 2D bicrystalline graphene oxide. J Appl Phys 124(1):015102

171. Verma A, Parashar A (2018) Reactive force field based atomistic simulations to study fracture toughness of bicrystalline graphene functionalised with oxide groups. Diam Relat Mater 88:193–203

172. Singla V, Verma A, Parashar A (2018) A molecular dynamics based study to estimate the point defects formation energies in graphene containing STW defects. Mater Res Express 6(1):015606

173. Verma A, Parashar A, Packirisamy M (2019) Role of chemical adatoms in fracture mechanics of graphene nanolayer. Mater Today: Proc 11:920–924

174. Chaudhary A, Sharma S, Verma A (2022) Optimization of WEDM process parameters for machining of heat treated ASSAB'88 tool steel using Response surface methodology (RSM). Mater Today: Proc 50:917–922

175. Verma A, Zhang W, Van Duin AC (2021) ReaxFF reactive molecular dynamics simulations to study the interfacial dynamics between defective h-BN nanosheets and water nanodroplets. Phys Chem Chem Phys 23(18):10822–10834

176. Verma A, Parashar A, Packirisamy M (2018) Atomistic modeling of graphene/hexagonal boron nitride polymer nanocomposites: a review. Wiley Interdiscip Rev: Comput Mol Sci 8(3):e1346

177. Verma A, Baurai K, Sanjay MR, Siengchin S (2020) Mechanical, microstructural, and thermal characterization insights of pyrolyzed carbon black from waste tires reinforced epoxy nanocomposites for coating application. Polym Compos 41(1):338–349
178. Verma A, Budiyal L, Sanjay MR, Siengchin S (2019) Processing and characterization analysis of pyrolyzed oil rubber (from waste tires)-epoxy polymer blend composite for lightweight structures and coatings applications. Polym Eng Sci 59(10):2041–2051
179. Verma A, Negi P, Singh VK (2019) Experimental analysis on carbon residuum transformed epoxy resin: chicken feather fiber hybrid composite. Polym Compos 40(7):2690–2699
180. Verma A, Singh VK (2018) Mechanical, microstructural and thermal characterization of epoxy-based human hair–reinforced composites. J Test Eval 47(2):1193–1215
181. Jain N, Verma A, Ogata S, Sanjay MR, Siengchin S (2022) Application of machine learning in determining the mechanical properties of materials. In: Machine learning applied to composite materials. Springer, Singapore, pp 99–113
182. Verma A, Singh VK, Verma SK, Sharma A (2016) Human hair: a biodegradable composite fiber–a review. Int J Waste Resour 6(206):2
183. Kataria A, Verma A, Sanjay MR, Siengchin S, Jawaid M (2022) Physical, morphological, structural, thermal, and tensile properties of coir fibers. In: Coir fiber and its composites. Woodhead Publishing, pp 79–107
184. Jain N, Verma A, Singh VK (2019) Dynamic mechanical analysis and creep-recovery behaviour of polyvinyl alcohol based cross-linked biocomposite reinforced with basalt fiber. Mater Res Express 6(10):105373
185. Chaurasia A, Verma A, Parashar A, Mulik RS (2019) Experimental and computational studies to analyze the effect of h-BN nanosheets on mechanical behavior of h-BN/polyethylene nanocomposites. J Phys Chem C 123(32):20059–20070
186. Verma A, Kumar R, Parashar A (2019) Enhanced thermal transport across a bi-crystalline graphene–polymer interface: an atomistic approach. Phys Chem Chem Phys 21(11):6229–6237
187. Verma A, Negi P, Singh VK (2018) Physical and thermal characterization of chicken feather fiber and crumb rubber reformed epoxy resin hybrid composite. Adv Civ Eng Mater 7(1):538–557
188. Bharath KN, Madhu P, Gowda TG, Verma A, Sanjay MR, Siengchin S (2020) A novel approach for development of printed circuit board from biofiber based composites. Polym Compos 41(11):4550–4558
189. Verma A, Joshi K, Gaur A, Singh VK (2018) Starch-jute fiber hybrid biocomposite modified with an epoxy resin coating: fabrication and experimental characterization. J Mech Behav Mater 27(5–6)
190. Verma A, Singh C, Singh VK, Jain N (2019) Fabrication and characterization of chitosan-coated sisal fiber–Phytagel modified soy protein-based green composite. J Compos Mater 53(18):2481–2504
191. Rastogi S, Verma A, Singh VK (2020) Experimental response of nonwoven waste cellulose fabric–reinforced epoxy composites for high toughness and coating applications. Mater Perform Charact 9(1):151–172
192. Verma A, Ogata S (2023) Magnesium based alloys for reinforcing biopolymer composites and coatings: A critical overview on biomedical materials. Adv Ind Eng Polym Res. https://doi.org/10.1016/j.aiepr.2023.01.002
193. Kumar G, Mishra RR, Verma A (2022) Introduction to Molecular Dynamics Simulations. In: Forcefields for Atomistic-scale simulations: materials and applications. Springer, Singapore, pp 1–19. https://doi.org/10.1007/978-981-19-3092-8_1
194. Arpitha GR, Jain N, Verma A, Madhusudhan M (2022). Corncob bio-waste and boron nitride particles reinfrced epoxy-based composites for lightweight applications: fabrication and characterization. Biomass Convers Biorefinery. 1–8. https://doi.org/10.1007/s13399-022-03717-1
195. Kataria A, Chaturvedi S, Chaudhary V, Verma A, Jain N, Sanjay MR, Siengchin S (2023) Cellulose fiber-reinforced composites—history of evolution, chemistry, and structure. In:

Cellulose fibre reinforced composites. Woodhead Publishing, pp 1–22. https://doi.org/10.1016/B978-0-323-90125-3.00012-4

196. Kataria A, Chaturvedi S, Chaudhary V, Verma A, Jain N, Sanjay MR, Siengchin S (2023) Bionanocomposites reinforced with cellulose fibers and agro-industrial wastes. In: Cellulose fibre reinforced composites. Woodhead Publishing, pp 1–22. https://doi.org/10.1016/B978-0-323-90125-3.00017-3

197. Arpitha GR, Mohit H, Madhu P, Verma* A (2023) Effect of sugarcane bagasse and alumina reinforcements on physical, mechanical, and thermal characteristics of epoxy composites using artificial neural networks and response surface methodology. Biomass Convers Biorefinery.https://doi.org/10.1007/s13399-023-03886-7

Chapter 7
Molecular Dynamics Based Simulations as an Effective Tool for Studying Coating Materials on Metals

Divya Singh

1 Introduction

A wide range of coating materials, including metals, organic coatings as well as inhibitors areavailable for metals mainly for anti-corrosion applications. Metals and inorganic coatings on metal surfaces are a class of permanent coatings, whereas inhibitors are a class of temporary coatings.

A process involving deposition of coatings on metal surfaces like sputtering and film thin deposition have a lot of process parameters that can affect the structural and functional properties of coatings. The development and improvement of coating materials and performance by experimental methods require a lot of money and time. With the growth of computational hardware and software, corrosion scientists have turned towards using powerful tools like molecular dynamics based simulations to understand the underlying mechanisms of coating and surface interactions as well as process parameters. The results from simulations can be used to select appropriate coatings as well as improve coating surface interactions.

In this chapter, many aspects of coatings on metals have been recorded. This includes MD based studies on the organic coatings, metal coatings as well as the effect of inhibitors on metal surfaces. In addition, some effort has been made to compile MD based studies on the process parameters of coating process on metal surfaces.

D. Singh (✉)
Engineering Department, Utah Tech University, St. George, Utah 84780, USA
e-mail: divya.singh@utahtech.edu

2 Simulation Methodology

2.1 Molecular Dynamics Simulations

MD based approach falls in the category of the classical mechanic-based techniques. The basic equation solved in any MD simulation is given by Eq. 1:

$$m_\alpha a_\alpha = F_\alpha = -\left(\frac{\partial E}{\partial r_\alpha}\right) \tag{1}$$

where $\alpha = 1, 2......N$

Here, N is the total number of atoms in the system, m_α, r_α and F_α are the mass, position and time dependent external force acting on each atom, respectively. The energy E is the total potential energy that has two components; E_i and E_{ext}. E_i is the total internal energy of system, due to the interaction of the atoms among themselves. E_{ext} accounts for the external fields and constraints. Velocity vectors and atom positions are the factors of energy and forces in an MD based simulation. Macroscopic properties such as pressure, stress, temperature, strains etc. are calculated as time averages utilizing statistical mechanics approach.

Numerical integration schemes such as Verlet, Velocity Verlet [1], Leapfrog [2] and Beeman's [3] algorithm are employed in MD based calculations to update the position and velocity vector after each time step. The most widely used integration scheme is the velocity verlet algorithm, which is an extension of the verlet algorithm. The position vector 'r' at any given time t and after an increment Δt during the simulation, for any atom in the system can be expanded as Taylor series expansion as given by Eq. 2

$$r(t + \Delta t) \approx r(t) + \dot{r}(t)\Delta t + \frac{\ddot{r}t\Delta t^2}{2} + \ldots \tag{2}$$

The updated velocity vector at the new current time step '$t + \Delta t$' is calculated by using the velocity-verlet equation as:

$$v(t + \Delta t) \approx v(t) + \left(\frac{F(t + \Delta t) + F(t)}{2m}\right)\Delta t \tag{3}$$

MD simulations can be carried out under different types of ensembles that includes most extensively used NPT, NVT and NVE. The NPT ensemble stands for Nose Hoover Pressure barostat and temperature thermostat. It is an isobaric-isothermal ensemble. The canonical NVT is a fix volume thermostat kind of an ensemble, whereas the micro-canonical NVE is a constant volume-constant energy ensemble.

As the MD simulations are carried out at the atomic level, the continuum level principles are not applicable. Hence, the stresses in MD simulations cannot be expressed by the Cauchy stress tensor. An analogous stress at the atomic level referred as virial stress was used to simulate the tensile behavior of any atomistic system. Virial stress constitutes of two major energy components: one component is a factor of velocities or kinetic energies of the atoms while the other component is dependent upon the position and interatomic forces between the atoms. Mathematically, the combination of these stresses can be expressed in the virial stress equation as:

$$\sigma_{ij}^{\alpha} = \frac{1}{\varphi^{\alpha}}\left(\frac{1}{2}m^{\alpha}v_i^{\alpha}v_j^{\alpha} + \sum_{\beta=1,n} r_{\alpha\beta}^j f_{\alpha\beta}^i\right) \tag{4}$$

where i and j denote indices in Cartesian coordinate system; α and β are the atomic indices; m_{α} and v_{α} are mass and velocity of atom α; $r_{\alpha\beta}$ is the distance between atoms α and β; and φ_{α} is the atomic volume of atom α.

The accuracy of any MD based simulation entirely depends on the type of interatomic potential employed for estimating interatomic forces. The interatomic potential is, thus, a constitutive part of any MD simulation. The potential energy estimated for any atomic system by the interatomic potential can be generalized as:

$$U_{total} = \frac{1}{2}\sum_{\alpha=\beta=1}^{N}\sum_{\beta=1}^{N}\phi\left(r_{\alpha\beta}\right) \tag{5}$$

where, $r_{\alpha\beta}$ is the distance between particles α and β, $\varnothing(r_{\alpha\beta})$ is the potential energy between particles α and β.

These interatomic potentials are semi empirically derived mathematical expressions and authenticated against experimental observation or verified with the help of higher fidelity numerical simulations, such as quantum mechanics based DFT approach. For metals, the most commonly employed interatomic potentials are EAM (embedded atom method), MEAM (modified embedded atom method) and ADP (angular dependent potential). The most widely used potential for hcp metals are EAM and MEAM potential.

2.2 Force Fields

Embedded Atom Method (EAM) Potential

The energy of an atomic system as estimated by a many-body EAM potential can be expressed mathematically by Eq. 6:

$$E = F_{\alpha}\left(\sum_{j\neq i}\rho\beta r_{ij}\right) + \frac{1}{2}\sum_{j\neq i}\phi_{\alpha\beta}\left(r_{ij}\right) \tag{6}$$

where F_α is the embedding energy which is a function of atomic electron density ρ, α and β are element types of atoms i and j, and $\emptyset$ is a pair potential interaction.

The EAM potential is employed for studying the properties of metals and alloys.

COMPASS: An ab Initio Force-Field Optimized for Condensed-Phase Applications.

COMPASS is a general all-atom force field developed for atomistic simulation of common organic molecules, inorganic small molecules, and polymers. It developed by implementing an ab initio and empirical parametrization techniques. This force field is validated against twenty eight molecular classes. These can be classified broadly as isolated molecules, liquids and crystals. The accuracy of the COMPASS force field was established by replicating accurate and concurrent estimates of structural, conformational, vibrational, and thermophysical properties for the above-mentioned range of molecules in both isolations as well as in condensed phases. It has been established through many atomistic-based MD studies that the interaction of inorganic coatings or corrosion inhibitors and metal surfaces can be accurately studied by employing this force field.

2.3   Non-metal Coatings on Metal Surfaces

MD simulations can provide important information on metal polymer coating interaction which are difficult to measure experimentally. Polyvinyl Chloride is used as a coating material on Aluminum given its chemical and thermal stability and barrier properties from the C–Cl polar bond. Interaction between PVC and Al_2O_3 is studied using MD simulations at various temperatures in order to uncover the influence of the temperature on adhesion [4]. The measured interaction energy revealed that the adhesion between PVC coating and Al_2O_3 increases as temperature increases.

Adhesion properties of anti-corrosive coating of DGEBA-polyaminoamide on steel were studied using MD [5]. The computational results were key to understanding the type of bonding and mechanism of interaction between the coating and steel substrate. It was revealed through computational analysis that the DGBEA-polyaminoamide had a strong tendency to stick to the steel surface and this is what helps to stop the corrosive dissolution of steel surface in a simulated marine environment. Obtaining such intricate information from experiments would have been very time and resource consuming.

The most common wear occurring on the surface of fluid dynamic system components, like water hydraulic valves, causing the failure of metal parts is cavitation erosion and corrosion. MD based simulations were used as a powerful tool to study the effects of PTFE chain length on Al_2O_3 in order to understand the influence of chain length on the wear and corrosion resistance of coating materials used on Al_2O_3 composite hydraulic valve elements [6]. This is done by measuring their mechanical properties as well performing studies on binding energy, cohesive energy density and solubility. The calculated values of cohesive energy density and solubility are indicators of the water's superior hydrophobic effect as a function of chain length of coating material on the composite surface.

Green coating materials have caught attention in recent years owing to their environmentally friendly nature. MD based simulations act as a quick tool to assess the effectiveness of such novel emerging materials as anti-corrosive coating on metal surfaces. For instance, results on the corrosion properties of lemon seeds obtained through MD simulations have revealed that it could be a prominent green material used for anti-corrosive coatings on mild steel [7].

Many interesting insights into Carbon and its effects on the oxidation of Aluminum nanoparticles have been revealed through atomistic simulations [8]. Properties such as full dissociation curve of Al-C bond, the angle of distortion energies of the Al/C/H/O clusters, binding energies of hydrocarbon species on Al(111) surfaces, and reaction kinetics of systems related to C coating process were studied through MD simulations. These results revealed the atomistic factors that influence the growth of coating layers. It was also discovered that the oxidation of Carbon-coated Aluminum nanoparticles increased due to the removal of the coating layer as a result of the formation of H_2O, H_2, CO and CO_2.

2.4 Metal Coatings on Metal Surfaces

The reasons behind the instability of Al coatings of Fe have been studied through MD based simulations [9]. Many atomistic phenomena, which are hard to capture experimentally, but actually influence the instability of Al coatings on Fe were figured out. Knowledge of such phenomenon is crucial for improving Al coating stability and can be achieved through MD simulations with much ease and accuracy. For instance, it was discovered that a process of Al film spreading was responsible for initializing diffusion of Fe atoms into Al. Information regarding such phenomenon previously did not exist which restricted the work on the improvement of adhesion of Al coatings on Fe. MD simulations are also capable of catching atomistic MD simulations enabled to generate a relation between temperature and diffusion coefficient of Al atoms in an iron crystal. Phenomena at high temperatures provided accurate and appropriate potentials are employed for the study. The instability of Al atoms at 873 K were found to be linked with the penetration of Al atom into the Fe matrix through such a simulation.

Sputtering is one of the most widely used methods to deposit metal coatings. Temperature, particle fluxes and composition strongly influence the surface properties of coatings. The precursor incident during sputtering is also a big influencer on the properties of coatings. Since these processes induce changes at atomistic level and occur at very small time scales, it is hard to capture the microstructure of coatings and surface morphological evolution experimentally. MD based simulations, on the other hand, can be utilized to understand the effect of sputtering variables on surface properties of coatings. The sputtering depth of the Au atoms, the particle distribution of the Au/Cu coating system, radial distribution functions of particles in the coatings, the mean square displacement of Cu atoms in the substrate and roughness of the coatings were analyzed by sputtering Cu substrate by Au atoms at different incident energies in MD simulation environment [10]. The results revealed that the incident energy had almost no influence on the sputtering depth of Au atoms. The study made a very critical conclusion in weening out the incident energy as a potential factor influencing surface roughness of the coating. Such powerful information could be drawn with simple MD simulations where experimental set up would have taken up a lot of time.

The microscopic nucleation and growth during early stages of jet electrodeposition of metal coatings directly influence the consistency and reliability of the coating structure. A coarse-grained MD model based on nucleation diffusion and growth analysis is developed in order to gain a deeper understanding of different growth environments on the microscopic nucleation growth of coating structure [11]. Properties like ion dynamic diffusion and nucleation kinetic mechanism under various electrodeposition conditions as well as the physical structure of the surface was obtained through atomistic simulations. These findings provided a theoretical framework for competent preparation of nickel coatings.

Friction and wear behaviors of Fe-based amorphous coatings were studied by MD simulations [12]. The simulations indicated that the ceramic phase was capable of improving the plasticity of composite coatings and reducing wear. It was also established that at 15 wt% of Al_2O_3, the stress concentration between ceramic and amorphous phases causes a viscosity flow in the amorphous phase that improved plastic deformation. Such atomistic insights can help to achieve coatings with best wear resistance. An atomistic experimentation in MD simulation environment also indicated that a high sliding speed resulted in a reduction of the coatings' wear resistance.

2.5   Temporary Corrosion Inhibitors on Metal Surfaces

Ferrous metals are studied extensively for their interaction with corrosion inhibitors using MD based simulations in order to understand the atomistic mechanism of corrosion inhibition in conjunction with experimental procedures. Computational methods have proved themselves as an effective tool in studying metal inhibition being more time, cost and environmental efficient. Additionally, conventional methods are also incapable of providing notable insight into metal surface-inhibitor interaction. Conventional experimental procedures for estimating the performance of corrosion inhibitors such as weight loss method, electrochemical impedance spectroscopy, potentiodynamic polarization and cyclic voltammetry are effective but on the other hand, have time and resources constraints. In comparison, MD based studies of metal corrosion inhibitors can help capture mechanisms involved in surface and inhibitor interaction at a low cost. The inhibition of mild steel corrosion in 15% HCl by four main monomers of polysaccharide: glucose, mannose, galactose and fructose were studied using MD simulations. The computed binding energies indicated that fructose had the highest adsorption on the surface [13].

Computer simulations can reveal the modes of action of inhibitor molecules on metal surfaces in the simulated solvent environment. MD results on the patterns of different inhibitors on the iron surface have been compared in order to make quicker and wiser decisions on choosing an appropriate inhibitor [14].

The adsorption process of ant-corrosion additives BTA on Cu surface was studied in terms of adsorption dynamics and the charge transfer on a solid Cu surface [15]. It was found through MD simulations that BTA molecules formed on the adsorbed layer are parallel to the surface and surface diffusion leads to aggregation of molecules. A selective adsorption phenomenon was discovered by performing MD adsorption studies on the surface of a hybrid slab consisting of both a Cu area and a Cu_2O area. It was found that BTA molecules had five times greater affinity in the Cu area when compared to Cu2O area. The study established that this selective nature of adsorption onto a newly formed metal surface due to charge transfer from the metal surface is the reason behind the ineffectiveness of most anti-corrosion additives. Such atomistic information can be crucial in devising chemical methodologies to improve inhibitor performance.

2.6 Coating Deposition on Nanostructures

Molecular dynamics simulations have been used for the structural characterization of metal cluster deposition on surfaces [16]. The effect of adsorption coverage, the impact energy, and the properties of materials in contact with the structural characteristics in the adlayer has been studied using MD simulations. The atomistic simulation results showed that a small single cluster was reconstructed post deposition because of the interface stress that matched the surface structure. The study also revealed a competing relationship between interfacial energy minimization and intrinsic stress energy that led to the formation of a nanostructure formation of Au clusters on a Cu (001) surface as coverage increased. Such details from MD studies prove valuable in controlling the surface morphology synthesized by cluster deposition.

MD simulations have been utilized to investigate the deposition and sputtering mechanisms at the coating surface in nanofibers [17]. Such simulations provide vital information regarding sputtering and sticking probabilities that prove beneficial in estimating the parameters for the continuum transport model accurately. These continuum transport models define an evolution equation which can be further used to control the coating morphology by changing operating conditions.

A coarse-grained MD model based on the nucleation diffusion and growth analysis is developed to understand how changes in growth environments affect the microscopic nucleation growth of the coating structure [18]. The dynamic deposition process which is hard to study experimentally can be easily simulated with the 2D view in an MD environment. The MD results on the mechanism of nucleation and growth reveal that the reaction rate in jet electrodeposition is higher than that of traditional electrodeposition. Such information proves vital in improving the coating process and making it more efficient.

3 Conclusion

In conclusion, MD based simulations provide a simpler, cost-effective and time effective way to understand the atomistic phenomena behind interactions between permanent as well as temporary coatings on metals. Not only this but the process of coating deposition which occurs at very small length and time scales and are hard to capture experimentally can be studied with MD set up with relative ease when compared to experimental methods. Such information helps to accelerate the selection of the right coating materials for metals as well as improve deposition processes to improve the surface properties of coating materials against wear and roughness.

Conflict of Interest   "There are no conflicts of interest to declare by the author."

References

1. Swope WC, Andersen HC, Berens PH, Wilson KR (1982) A computer simulation method for the calculation of equilibrium constants for the formation of physical clusters of molecules: application to small water clusters. J Chem Phys 76:637
2. Hockney RW, Eastwood JW (1981) Computer simulation using particles. McGraw-Hill, New York
3. Beeman D (1976) Some multistep methods for use in MD calculations. J Comput Phys 20:130–139
4. Sharma A, Sharma S (2022) A molecular dynamics study of adhesion of polyvinyl-chloride coatings to the aluminum surface, OP Conference series: materials science and engineering 1248, 012062
5. Dagdag O, Berisha A, Safi Z, Hamed O, Jodeh S, Verma C, Ebenso E, El Harfi A (2020) DGEBA-polyaminoamide as effective anti-corrosive material for15CDV6 steel in NaCl medium: computational and experimental studies. J Appl Polym Sci2020. https://doi.org/10.1002/APP.48402
6. Mlela MK, Xu H, Wang H (2020) Molecular dynamics study of anti-wear erosion and corrosion protection of PTFE/Al2O3 (010) coating composite in water hydraulic valves. Coatings 10:1214. https://doi.org/10.3390/coatings10121214
7. Pal S, Ji G, Lgaz H, Chung I-M, Prakash R (2020) Lemon seeds as green coating material for mitigation of mild steel corrosion in acid media: Molecular dynamics simulations, quantum chemical calculations and electrochemical studies. J Mol Liq 316:113797
8. Hong S, van Duin ACT (2016) Atomistic-scale analysis of carbon coating and its effect on the oxidation of aluminum nanoparticles by reaxff-molecular dynamics simulations. J Phys Chem C 120 (17):9464–9474
9. Galashev AYe, Rakhmanova OR, Kovrov VA, Zaikov YuP (2019) Molecular dynamics study of the stability of aluminium coatings on iron. Lett Mater 9(4):436–441

10. Zhang L, Tian S, Peng T (2019) Molecular simulations of sputtering preparation and transformation of surface properties of Au/Cu alloy coatings under different incident energies. Metals 9:259. https://doi.org/10.3390/met9020259
11. Zhang‡ a F, Liu‡b S, Wang F (2022) Nucleation and growth mechanism in the early stages of nickel coating in jet electrodeposition: a coarse-grained molecular simulation and experimental study. RSC Adv 12:11052–11059
12. Chu Z, Zhou Y, Xu1 F, Xu J, Zheng X, Luo X, Shu Y, Zhang Z, Hu Q, Study of friction and wear behaviors of Fe-based amorphous coatings by MD simulations. Front Mater. https://doi.org/10.3389/fmats.2022.1048443
13. Haris NIN, Sobri S, Yusof YA, Kassim NK (2021) An overview of molecular dynamic simulation for corrosion inhibition of ferrous metals. Metals 11:46
14. Chaouiki A, Chafiq M, Rbaa M, Lgaz H, Salghi R, Lakhrissi B, Ali IH, Masroor S, Cho Y (2020) New 8-hydroxyquinoline-bearing quinoxaline derivatives as effective corrosion inhibitors for mild steel in HCL: electrochemical and computational investigations. Coatings 10:811
15. Nishikawa K, Akiyama H, Yagishita K, Washizu H, Molecular dynamics analysis of adsorption process of anti-copper-corrosion additives to the copper surface. arXiv:1812.10647
16. Wang YX, Pan ZY, Xu Y, Huang Z, Du AJ, Ho YK (2002) Molecular dynamics simulation of structural characteristics in metal cluster deposition on surfaces. Surface Coatings Technol, 158–159, 263–268
17. Buldum A, Clemons CB, Dill LH, Kreider KL, Young GW, Zheng X, Evans EA, Zhang G, Hariharan SI (2005) Multiscale modeling, simulations, and experiments of coating growth on nanofibers. Part II. DepositionJAP 98(4):044304
18. Zhang F, Liu S, Wang F (2022) Nucleation and growth mechanism in the early stages of nickel coating in jet electrodeposition: a coarse-grained molecular simulation and experimental study. RSC Adv 12:11052–11059

Chapter 8
Computational Aspects: Self-clean Coatings, Plastics and Polymers in Coatings

Hariome Sharan Gupta, Uday Shankar, Akarsh Verma, Rupam Gogoi, and Sushanta K. Sethi

1 Introduction

The coating industry represents one of the largest sectors in the world [1]. The market is diverse and overlaps with industrial paints. For this, many experimental scientists have adopted different procedures such as, blending, grafting, filler functionalization, surface treatments, etc. Many times, such experimental techniques involve the manipulation of variables, time consuming and sometimes are most expensive.

Computational techniques have been performed as a direct approach to fundamentally understand and to address various interactions of a coating material with water, oil, fillers, substrates or any other materials [2–12]. Various simulations like atomistic and coarse-grained (CG) simulations are garnering considerable attentions among researchers for their ability to model polymer structures and further their simulations.

H. S. Gupta · U. Shankar
Department of Polymer and Process Engineering, Indian Institute of Technology Roorkee, Saharanpur 247001, India

A. Verma
Department of Mechanical Engineering, University of Petroleum and Energy Studies, Dehradun 248007, India

Department of Mechanical Science and Bioengineering, Osaka University, Osaka 560-8531, Japan

R. Gogoi
Department of Chemistry, Imperial College London, London SW7 2BX, UK

S. K. Sethi (✉)
Department of Metallurgical Engineering and Materials Science, Indian Institute of Technology, Bombay, India
e-mail: ssethi@pe.iitr.ac.in

In this chapter, the applications of different computational techniques in various types of coatings have been addressed. We highlight the role of different molecular dynamics (MD) simulation techniques in discovering of different types of coatings. Specific simulation details have not been covered as it may unnecessarily increase the content and also may lose the main theme of this subject. The readers may refer to the original article cited in this chapter for more specific clarity.

1.1 Self-clean Coatings

Self-clean coatings have found a broad application in almost all sectors like aircraft, automobiles, kitchen utensils, structural glasses, etc. to design novel self-clean coating materials, MD simulations have played a crucial role [7, 13–25]. In this section, we summarize the MD simulation techniques applied to develop novel self-clean coating materials and highlight their importance.

Sethi et al. [7, 24] have considered PVAc and PDMS for the fabrication of easy-clean coatings. Since PVAc is polar and PDMS is nonpolar, hence the authors have first studied the blend compatibility among them. Authors have performed dissipative particle dynamics (DPD), mesoscopic dynamics (Mesodyn) and glass transition temperatures (T_g) to understand the compatibility of PVAc with PDMS. Some PVAc: PDMS compositions displayed multiple T_g values thereby demonstrating the incompatibility among them. Keeping the incompatibility issues in mind, they have then synthesized PVAc-g-PDMS [15] and considered the same system in MD simulations to study the performance behavior such as water/oil contact angle. The complete procedure adopted by Sethi et al. [13] for making a two-layered system out of a simulation cell for the water contact angle estimation is shown in Fig. 1. Authors have considered the Hautman and Klein technique [26] to predict the contact angles through MD simulations. The selected system demonstrated good adhesion ability on one side with substrate and a decent water contact angle on the other. Since the self-clean coating materials should demonstrate phobic nature of >150° against water; hence the authors then incorporated different fillers into PVAc-g-PDMS. First, they separately incorporated MWCNT [13], and ZnO QDs [25] into PVAc-g-PDMS and investigated the performance properties. They observed almost similar performance properties in both cases, thus suggesting that the bio-safe ZnO QDs filler is a better replacement for MWCNT.

Generally, the fillers impart roughness to the surface; thereby, the rough grooves can beneath air inside it and help to enhance the phobic nature against water. In this direction, Sethi et al. [19, 23] investigated the impact of rough surfaces and its shape on its wettability. Authors have considered different shapes of filler such as cylinder, sphere, nail, square and rectangle and demonstrated which shape is better for considering hydrophilic or hydrophobic surfaces. Water molecules penetrating inside the rough grooves demonstrated by different shape fillers are shown in Fig. 2. Here is a comprehensive study on the use of MD and CG simulations for the creation of self-cleaning coating materials [17]. Chen et al. [16] have demonstrated a one-step

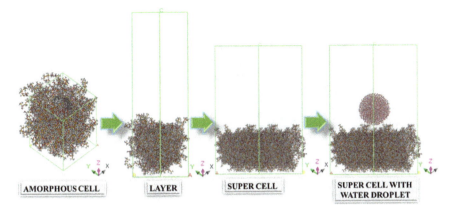

Fig. 1 Modeling of two-layered system consisting of water droplet and coating surface for the simulation and prediction of water contact angle. Refurbished with permission from [13]

fabrication technique for superhydrophobic self-clean surface using combined experimental and computational methods. Using the MD simulations, they observed the hydrophilic interactions among water droplets and ZnO and vdW interactions among water droplets and alkyl chains of the surface. The pre-prepared surface displayed a ~164° water contact angle. Similarly, Daneshmand et al. [20] have fabricated a superhydrophobic surface by employing two-step spraying technique. The adsorption of water molecules on surfaces made of alumina, glass and alumina treated with stearic acid has been studied by authors using MD simulations. The obtained contact angles were observed to be 160°, which is in line with experimental characterizations. Moradi et al. [27] have studied the hydrophobic nature of PP and PP/graphene oxide composite by using MD simulations. They found that the graphene oxide contributes to enhance the water repellence nature from 110° to 160°.

1.2 Fire-Retardant Coatings

Reactive MD simulations with a ReaxxFF forcefield were used by Vaari et al. [28] to examine how aluminum trihydroxide affects the heat breakdown of polyethylene (PE). The endothermic breakdown of aluminum trihydroxide into water and alumina has been replicated. Additionally, they have reasonably accurately observed the flame retardancy mechanism, such as the filler and its residue absorbing heat. The increase of the flame-retardant properties of polyvinyl acetate (PVAc) and attapulgite (ATP) ores was also the subject of research by Qiong et al. [29]. The authors have illustrated the interface interaction of the PVAc/ATP composite and clarified the degree of Mg ion diffusion within the PVAc, which contributes to the material's enhanced flame retardant properties. They have proposed that the ATP ores as an optimized additive, which can be applicable to future organic materials to achieve enhanced

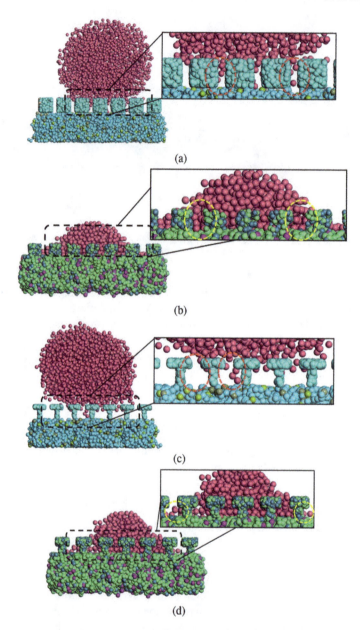

Fig. 2 Simulated water droplets on the hydrophobic **a** Square, **b** Nail, hydrophilic **c** Square and **d** Nail-shaped corrugated surfaces. Refurbished with permission from [23]

flame retardancy. The obtained outcomes are well compared with their experimental findings.

In fire modeling, one of the limiting factors is the solid pyrolysis of polymeric materials. This presents additional difficulties such as characterizing the parent fuel for combustion in a gas phase chemical reaction process. In this vein, Chen et al. [30] methodology for using MD simulations to examine the thermal degradation of polymer composites was developed. High-density polyethylene (HDPE), high-impact polystyrene (HIPS) and polymethyl methacrylate are the three polymers that the authors have chosen (PMMA). They noticed that all of the materials under consideration produced fuel gas containing alkane groups, especially C1-C3, which served as the source of the fire. By examining the buildup of pure carbon chain molecules, MD simulations can also be used to forecast the composition of char deposits. The dynamic bond-breaking behavior and bond reformation during the pyrolysis of the polymers were described in this experiment using the ReaxFF forcefield. To replicate the pyrolysis process, the chosen polymers were heated to temperatures ranging from 1300 to 2400 K to 2800 K to 3200 K to 3600 K to 3800 K, and to 4000 K and simulations were run for 300 ps at each temperature. But for 1300 K, the authors have performed the simulation for 2000 ps (2 ns) as the reaction occurs very slowly at lower temperatures. According to this study, the thermal decomposition process involves numerous multiphase interactions, such as charring, melting, internal heat and mass transfer, gas migration through a porous zone that is molten or charring, and volatile gas emission.

The reactive MD simulation is a potential technique for researching established and novel flame-retardant concepts as well as the pyrolysis chemistry of fire-retardant materials, according to the literature that has been studied. However, several research have also used a few other forcefields to study various properties of the chosen materials, which is primarily applicable for fire-retardant applications. For example, Sangkhawasi et al. [31] have investigated and compared the structural dynamics of Polyethylene vanillic (PEV) using OPLS forcefield. They have studied the flexibility of PEV single-chain polymer model consisting of 100 monomer units and found that its flexibility was higher than that of polyethylene terephthalate (PET). Saeb et al. [32] have investigated the intramolecular interactions among epoxy resin molecules and β-cyclodextrin (CD) surface modifiers using ab initio quantum mechanics (QM). Further they have performed MD simulations and observed that a strong adhesion between cure moieties and β-CD.

1.3 Anti-corrosion Coatings

In order to examine the atomic level mechanism of corrosion inhibitors in minimizing corrosion, MD simulations are widely employed in the development of innovative anti-corrosion coatings [33–35]. A detailed report on the application of MD simulations for corrosion inhibition of Ferrous material can be found here [33].

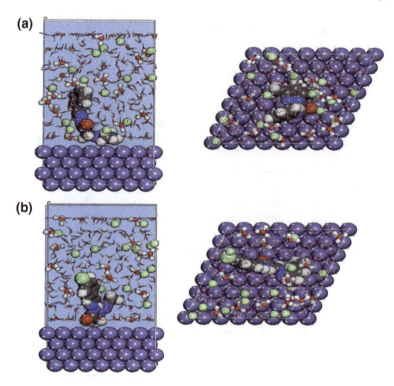

Fig. 3 Illustration of **a** N'—benzylidene—2–(3, 5—dimethyl −1H—pyrazol—1-yl) acetohydrazide (DPP) and **b** N'—(4—chlorobenzylidene)—2—(3, 5—dimethyl—1H—pyrazol—1—yl) acetohydrazide (4—CP) inhibitor in the simulated corrosions media on steel (Fe) (1 1 0) surface. Refurbished with permission from [41]

Arrouji et al. [41] studied the corrosion inhibition of two pyrazole derivates. They have performed density functional theory and MD simulations to unravel the interfacial interaction of corrosive molecules with N'—benzylidene—2—(3, 5— dimethyl—1H—pyrazol—1—yl) acetohydrazide (DPP) and N'—(4—chlorobenzylidene) —2—(3, 5—dimethyl—1H—pyrazol—1—yl) acetohydrazide (4—CP) as shown in Fig. 3. The figure displays the lowest energy configurations of both DPP and 4—CP on Fe surface in simulated corrosive media. Shen et al. [36] have fabricated a robust SiO_2-PDMS coating composite using a two-step spraying technique. To do this, the authors simulated and examined the anti-corrosive mechanism of the suggested coating material using an MD simulation program. According to the computer work, superhydrophobic coatings have extremely low diffusion coefficients and contact forces for corrosive particles (Cl ions). Keshmiri et al. [37] have fabricated a covalent organic framework (COFs) based on terepthalaldehyde and melamine. The adsorption propensity of the chosen structure onto oxidized CNTbased nanostructures and the interactions of epoxy polymer with these nanostructures were quantitatively analyzed by the authors using DFT and MD simulations. Moradi et al. [38] created a PP-based superhydrophobic anti-corrosion coating for mild steel surfaces to prevent corrosive chemicals from penetrating the substrate. They found a more compact structure at higher temperature thereby demonstrated the thermal resistance of the proposed coating material. In another investigation [27], the same authors have developed a PP-based composite coating including graphene oxide for applicable on steel. The authors found that the graphene oxide acting as an effective barrier against penetrating agents from MD simulation study. Similar to this, Dagdag et al. [39] have described a polyaminoamide anti-corrosive coating made of diglycidyl ether (DGEBA). They have carried out both computational and experimental characterizations to investigate the stability and nature of interactions between the suggested material and the high-strength low alloy steel surface 15CDV6. They used DFT-based quantum chemical computations, Monte Carlo (MC), and MD simulations to estimate the type of bonding and the mechanism of interactions. They discovered that the polar groups of DGEBA—polyaminoamide were firmly attached to the surface of Fe(110) by both chemical and physical connections. Similarly, Bahlakeh et al. [40] created a brand-new, environmentally safe chemical conversion coating that is suitable for mild steel surfaces and based on neodymium (Nd) oxide. In both dry and wet conditions, the computational analyses of their work predicted that the Nd-treated steel would form better adsorption over it than the untreated steel, which is also in good agreement with the results of the experiments.

1.4 Anti-icing Coatings

The presence of ice and snow has several adverse effects on human activities. Therefore, the dynamic mechanical of the icing and de-icing process is quite important. In this direction, the MD simulation has emerged as a powerful instrument for developing different anti-icing coatings. Sun et al. [42] have employed MD simulation, studied the underpinning physics and visualized the thermal de-icing process on a smooth surface. They studied the effect of ice thickness, wettability and surface temperature on the thermal de-icing process. They observed that the ice started melting from its mantle and proceeded inwards. Consumption of energy required for the thermal de-icing process is a bilinear function of surface temperature and ice thickness. They suggested that a hydrophobic surface, also considered as an ice-phobic can prevent ice accretion; more time and constant energy is required to melt the ice continuously. Such a study guides other researchers to design and develop several de-icing techniques and devices for engineering applications. One of the most important methods to lessen the dangers brought on by ice accretion, according to Zhuo et al. [43], is to delay the production of ice and frost. The authors have created Ionogel surfaces to control the growth of ice as well as the process of ice nucleation. Through an integrated experimental and computational investigation, they demonstrated that the ion gel surface might effectively produce a liquid interfacial layer. Such a layer is necessary to lessen ice adhesion and stop frost formation. The created surface exhibits anti-frost properties under cold, humid conditions, importing gas at 60% RH at 20 °C, i.e., below 20 °C.

Generally, the fabrication of surface energy surfaces is a passive strategy to develop anti-icing surfaces. Considering this, using MD simulations, Zhang et al. [44] explored the fundamental connection between ice adhesion and water contact angle on a graphene surface. When they took into account surface material deformations and a number of other experimental parameters, they discovered a substantial relationship based on the theoretical predictions that suggest the theoretical relation holds for ice and water. In the same line, Bao et al. [48] fabricated surfaces with single wall carbon nanotubes (SWNT) that display very low adhesion strength with ice. They have performed MD simulations and discovered that the ratio of the ice-surface interaction energy and the temperature of the ice are the two main determinants of the detaching strength of ice. The simulated snapshots of the sequence of ice detaching from a CNT surface are illustrated in Fig. 4. Xiao et al. [45] utilized MD simulations and studied the ice adhesion. To explore the determinants of the adhesion process at atomistic level descriptions, they applied force to detach and shear nano-sized ice cubes. Additionally, they investigated the mechanical impact of an aqueous water layer sandwiched between ice and surfaces. The authors discovered that a high interfacial energy inhibits mobility and raises the likelihood of a strong adhesion. They also proposed that an aqueous water layer could reduce the strength of ice adhesion by up to 60%. Jiang et al. [46] demonstrated a novel method for preparing slippery lubricant-infused porous surface. The lubricant effects on wettability were studied by employing MD simulations. They considered a copper plate, droplet of water and a lubricant oil layer in MD simulations. They proposed that the conventional solid-water contact model may be changed into a water–oil-solid contact model, which might partially inhibit icing propagation. Li et al. [47] studied active and passive anti-icing behavior together. They explored the coupled mechanism of a solute and a hydrophilic surface by utilizing MD simulations. They described the microscopic mechanism of solution freezing on hydrophilic and hydrophobic surfaces based on their observations. They discovered, in essence, that solute particle adsorption to ice crystals raises the nucleation energy barrier and lowers the local freezing temperature. As a result, it can withstand the formation and expansion of ice to a certain degree.

1.5 Anti-viral Coatings

MD simulations have also played a prime role in designing anti-viral coatings. Researchers are working globally in developing and improving novel computational strategies. Strategies employed in MD simulations to develop anti-viral coatings include the ligand-based approach that different known active compounds to extrapolate biological activities like classical quantitative structure–activity relationship

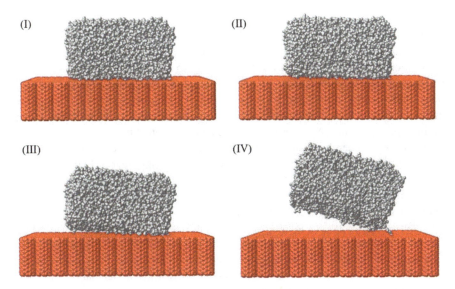

Fig. 4 Computational modeling and simulation of carbon nanotube array system during the detachment process of ice cube system. Refurbished with permission from [48]

(QSAR), structure-based techniques depending on experimentally obtained 3D structures of the targets, like molecular docking, and universal methods, which can be applicable to structure or ligand-based modeling like 3D QSAR.

Generally, the natural products are of prime interest to provide anti-viral activity. Newman and Cragg [49] have investigated several drugs approved between 19 and 2006 and showed that the anti-effective area mostly depends on natural products and their structures. In this line, the computer-assisted approaches have found entrance since last two decades in this natural product research. The MD simulations are particularly useful to investigate the drug resistance development, which is a prime concern in anti-AIDS therapies. The molecular docking led to new structural insights and helped to further understand fundamentally about the structure–property relationship. In the cultivation of agricultural cops, the bud blight disease is a serious issue which caused by bud necrosis virus (GBNV). Sangeetha et al. [50] have investigated to identify suitable organic anti-viral agents by employing MD simulations and molecular dockings of nucleocapsid Coat Protein of GBNV. Their study suggested the anti-viral activity of Squalene by fitting into binding sites of the coat protein of GBNV effectively.

In the recent pandemic of COVID-19, the MD simulations have acted as a very powerful tool to develop novel anti-viral coatings that can easily rupture the viruses instantly and effectively. In the direction of anti-viral activity of SARS COV2, Shivanika et al. [51] have evaluated the potency of ~200 natural anti-viral phytocompounds against the active sites of SARS COV2 using Autodock. Authors have reported the top 10 compounds with strong binding energies. Further, they found the drugs such as atazanavir, darunavir and saquinavir providing inhibiting the protease much effectively. In addition to this, the pharmacokinetics and toxicity of the selected natural compounds and drugs against SARS COV2 were examined. Sahihi et al. [52] have studied the interactions of fully glycosylated spike proteins of SARS COV2 with different metal surfaces. Authors have investigated the mechanism, molecular and atomic level insights of the absorption of viruses on different soft and hard materials. They found that the hydrophobic and vdW interactions of metal substrates are the main driving forces to absorb the glycan groups-based proteins. When tested against spike proteins an increase in surface area was found to provide more strong interactions and thereby strong absorbance with metal substrates. Till date several new strains of SARS COV2 were found, which has bought fear to humankind. To understand the structure, Singh et al. [53] have investigated several properties such as root mean square fluctuation (RMSF), root mean square displacement (RMSD), radius of gyration and hydrogen bonding for the main protease using MD simulation. This study helped to identify the promising molecules for clinical trials and further to combat COVID-19 transmission. In the same line, Kushwaha et al. [54] have explored the COVID-19 druggable target inhibition efficiency of phytochemicals from the Indian medical plants. Authors have performed molecular docking and MD simulations combinedly and screened ~130 phytochemicals against COVID-19. Lakhera et al. [8] have explored the antiviral activity of feverfew plant for the treatment of SARS COV2 virus using MD simulations and molecular docking. They found excellent binding affinity and reactivity responses and proposed that the Parthenolide could be a promising drug candidate against SARS COV2. To obtain the structure, dynamics and function of the COVID-19 spike protein, Casalino et al. [55] used MD simulations to combine, improve and extend the experimental database that was already available. In Fig. 5, the simulated snapshot is displayed.

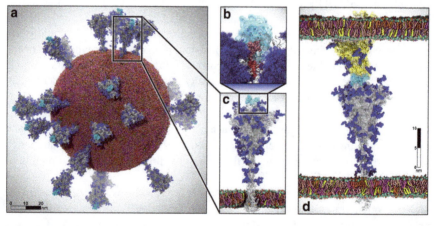

Fig. 5 Computational modelling of **a** Covid-19 virus including 24 spike proteins, **b** Full length modeling of Covid-19 spike protein embedded into a lipid bilayer membrane, **c** Glycan shield shown by overlaying multiple conformations for each glycan collected, and **d** Two parallel membrane system of spike Angiotensin-converting enzyme-2 (ACE-2) complex. Reprinted with permission from [56]

At the end, all the authors possess a vast experience in the areas of polymer composites and computational based atomistic-scale modeling that proves the worth of writing this chapter [57–128].

2 Summary and Conclusion

It is encouraging to mention that the MD simulations have been used widely in different sectors of specialized coatings to understand the atomic level descriptions of various behaviors. Some recent works on application of MD simulations on different types of coatings such as self-clean, anti-corrosion, anti-viral and fire-retardant have been collected and summarized in this chapter. The insights discussed in this chapter may surely assist the new researchers in developing novel and improved coating materials.

Acknowledgements Author "Akarsh Verma" is grateful to the monetary support provided by the University of Petroleum and Energy Studies (UPES)-SEED Grant program.

Conflict of Interest "There are no conflicts of interest to declare by the authors."

References

1. Matthews A, Leyland A, Stevenson P (1996) Widening the market for advanced PVD coatings. J Mater Process Technol 56:757–764. https://doi.org/10.1016/0924-0136(95)01889-1
2. Maurya AK, Gogoi R, Sethi SK, Manik G (2021) A combined theoretical and experimental investigation of the valorization of mechanical and thermal properties of the fly ash-reinforced polypropylene hybrid composites. J Mater Sci 56:16976–16998. https://doi.org/10.1007/S10 853-021-06383-2/FIGURES/10
3. Agrawal G, Samal SK, Sethi SK, Manik G, Agrawal R (2019) Microgel/silica hybrid colloids: bioinspired synthesis and controlled release application. Polymer (Guildf) 178:121599. https://doi.org/10.1016/j.polymer.2019.121599
4. Deji R, Jyoti R, Verma A, Choudhary BC, Sharma RK (2022) A theoretical study of HCN adsorption and width effect on co-doped armchair graphene nanoribbon. Comput Theor Chem 1209:113592
5. Shankar U, Sethi SK, Singh BP, Kumar A, Manik G, Bandyopadhyay A (2021) Optically transparent and lightweight nanocomposite substrate of poly(methyl methacrylate-co-acrylonitrile)/MWCNT for optoelectronic applications: an experimental and theoretical insight. J Mater Sci 56:17040–17061. https://doi.org/10.1007/S10853-021-06390-3/FIGURES/17
6. Saini A, Yadav C, Sethi SK, Xue BL, Xia Y, Li K, Manik G, Li X (2021) Microdesigned nanocellulose-based flexible antibacterial aerogel architectures impregnated with bioactive cinnamomum cassia. ACS Appl Mater Interfaces 13:4874–4885. https://doi.org/10.1021/acsami.0c20258
7. Sethi SK, Soni L, Manik G (2018) Component compatibility study of poly(dimethyl siloxane) with poly(vinyl acetate) of varying hydrolysis content: an atomistic and mesoscale simulation approach. J Mol Liq 272:73–83. https://doi.org/10.1016/J.MOLLIQ.2018.09.048

8. Lakhera S, Devlal K, Ghosh A, Chowdhury P, Rana M (2022) Modelling the DFT structural and reactivity study of feverfew and evaluation of its potential antiviral activity against COVID-19 using molecular docking and MD simulations. Chem Pap 76:2759–2776. https://doi.org/10.1007/S11696-022-02067-6/FIGURES/15

9. Sureshkumar B, Mary YS, Resmi KS, Suma S, Armaković S, Armaković SJ, van Alsenoy C, Narayana B, Sobhana D (2018) Spectroscopic characterization of hydroxyquinoline derivatives with bromine and iodine atoms and theoretical investigation by DFT calculations, MD simulations and molecular docking studies. J Mol Struct 1167:95–106. https://doi.org/10.1016/J.MOLSTRUC.2018.04.077

10. Dhiman A, Gupta A, Sethi SK, Manik G, Agrawal G (2022) Encapsulation of wax in complete silica microcapsules. J Mater Res 2022:1–14. https://doi.org/10.1557/S43578-022-00865-Y

11. Verma A, Jain N, Sethi SK (2022) Modeling and simulation of graphene-based composites. Innov Graphene-Based Polymer Compos, 167–198. https://doi.org/10.1016/B978-0-12-823789-2.00001-7

12. Kataria A, Verma A, Sethi SK, Ogata S (2022) Introduction to interatomic potentials/forcefields. In: Forcefields for atomistic-scale simulations: materials and applications (pp 21–49). Springer, Singapore

13. Sethi SK, Manik G (2021) A combined theoretical and experimental investigation on the wettability of MWCNT filled PVAc-g-PDMS easy-clean coating. Prog Org Coat 151:106092. https://doi.org/10.1016/j.porgcoat.2020.106092

14. Sethi SK, Manik G (2018) Recent progress in super hydrophobic/hydrophilic self-cleaning surfaces for various industrial applications: a review. Polym Plast Technol Eng 57:1932–1952. https://doi.org/10.1080/03602559.2018.1447128

15. Sethi SK, Shankar U, Manik G (2019) Fabrication and characterization of non-fluoro based transparent easy-clean coating formulations optimized from molecular dynamics simulation. Prog Org Coat 136. https://doi.org/10.1016/j.porgcoat.2019.105306

16. Chen X, Wang M, Xin Y, Huang Y (2022) One-step fabrication of self-cleaning superhydrophobic surfaces: a combined experimental and molecular dynamics study. Surfaces and Interfaces. 31:102022. https://doi.org/10.1016/J.SURFIN.2022.102022

17. Sethi SK, Kadian S, Manik G (2022) A review of recent progress in molecular dynamics and coarse-grain simulations assisted understanding of wettability. Arch Comput Methods Eng 2021(1):1–27. https://doi.org/10.1007/S11831-021-09689-1

18. Sethi SK, Singh M, Manik G (2020) A multi-scale modeling and simulation study to investigate the effect of roughness of a surface on its self-cleaning performance. Mol Syst Des Eng 5:1277–1289. https://doi.org/10.1039/d0me00068j

19. Kumar G, Mishra RR, Verma A (2022) Introduction to molecular dynamics simulations. In: Forcefields for atomistic-scale simulations: materials and applications (pp 1–19). Springer, Singapore. https://doi.org/10.1007/978-981-19-3092-8_1

20. Daneshmand H, Sazgar A, Araghchi M (2021) Fabrication of robust and versatile superhydrophobic coating by two-step spray method: An experimental and molecular dynamics simulation study. Appl Surf Sci 567:150825. https://doi.org/10.1016/J.APSUSC.2021.150825

21. Sethi SK, Shankar U, Manik G (2019) Fabrication and characterization of non-fluoro based transparent easy-clean coating formulations optimized from molecular dynamics simulation. Prog Org Coat 136:105306. https://doi.org/10.1016/j.porgcoat.2019.105306

22. Sethi SK, Manik G, Sahoo SK (2019) Fundamentals of superhydrophobic surfaces. In: Superhydrophobic polymer coatings, Elsevier, pp 3–29. https://linkinghub.elsevier.com/retrieve/pii/B9780128166710000011. Accessed August 29, 2019

23. Sethi SK, Gogoi R, Verma A, Manik G (2022) How can the geometry of a rough surface affect its wettability?—a coarse-grained simulation analysis. Prog Org Coat 172:107062. https://doi.org/10.1016/J.PORGCOAT.2022.107062

24. Sethi SK, Soni L, Shankar U, Chauhan RP, Manik G (2020) A molecular dynamics simulation study to investigate poly(vinyl acetate)-poly(dimethyl siloxane) based easy-clean coating: an insight into the surface behavior and substrate interaction. J Mol Struct 1202:127342. https://doi.org/10.1016/j.molstruc.2019.127342

25. Sethi SK, Kadian S, Goel A, Chauhan RP, Manik G (2020) Fabrication and analysis of ZnO quantum dots based easy clean coating: a combined theoretical and experimental investigation. ChemistrySelect. 5: 8942–8950. https://doi.org/10.1002/slct.202001092

26. Hautman J, Klein ML (1991) Microscopic wetting phenomena. Phys Rev Lett 67:1763–1766. https://doi.org/10.1103/PhysRevLett.67.1763

27. Moradi M, Rezaei M (2022) Long-term experimental evaluation and molecular dynamics simulation of polypropylene/graphene oxide nanocomposite coating in 3.5 wt% NaCl solution. Prog Org Coat. 164:106718. https://doi.org/10.1016/J.PORGCOAT.2022.106718

28. Vaari J, Paajanen A (2018) Evaluation of the reactive molecular dynamics method for research on flame retardants: ATH-filled polyethylene. Comput Mater Sci 153:103–112. https://doi.org/10.1016/j.commatsci.2018.06.032

29. Qiong S, Yanbin W, Li L, Junxi L (2019) Simulation investigation on flame retardancy of the PVAc/ATP nanocomposite. J Chem 2019. https://doi.org/10.1155/2019/3041097

30. Chen TBY, Yuen ACY, Lin B, Liu L, Lo ALP, Chan QN, Zhang J, Cheung SCP, Yeoh GH (2020) Characterisation of pyrolysis kinetics and detailed gas species formations of engineering polymers via reactive molecular dynamics (ReaxFF). J Anal Appl Pyrolysis, 153. https://doi.org/10.1016/j.jaap.2020.104931

31. Sangkhawasi M, Remsungnen T, Vangnai AS, Poo-Arporn RP, Rungrotmongkol T (2022) All-atom molecular dynamics simulations on a single chain of PET and PEV polymers. Polymers (Basel). 14. https://doi.org/10.3390/polym14061161

32. Saeb MR, Rastin H, Shabanian M, Ghaffari M, Bahlakeh G (2017) Cure kinetics of epoxy/ β-cyclodextrin-functionalized Fe3O4 nanocomposites: experimental analysis, mathematical modeling, and molecular dynamics simulation. Prog Org Coat 110:172–181. https://doi.org/10.1016/j.porgcoat.2017.05.007

33. Haris NIN, Sobri S, Yusof YA, Kassim NK (2021) An overview of molecular dynamic simulation for corrosion inhibition of ferrous metals. Metals 11:46. 11 (2020) 46. https://doi.org/10.3390/MET11010046

34. Douche D, Elmsellem H, Anouar EH, Guo L, Hafez B, Tüzün B, el Louzi A, Bougrin K, Karrouchi K, Himmi B (2020) Anti-corrosion performance of 8-hydroxyquinoline derivatives for mild steel in acidic medium: Gravimetric, electrochemical. DFT Molecular Dyn Simul Invest J Mol Liq 308:113042. https://doi.org/10.1016/J.MOLLIQ.2020.113042

35. Madani A, Sibous L, Hellal A, Kaabi I, Bentouhami E (2021) Synthesis, density functional theory study, molecular dynamics simulation and anti-corrosion performance of two benzidine Schiff bases. J Mol Struct 1235:130224. https://doi.org/10.1016/J.MOLSTRUC.2021.130224

36. Shen Y, Li K, Chen H, Wu Z, Wang Z (2021) Superhydrophobic F-SiO2@PDMS composite coatings prepared by a two-step spraying method for the interface erosion mechanism and anti-corrosive applications. Chem Eng J 413:127455. https://doi.org/10.1016/J.CEJ.2020.127455

37. Keshmiri N, Najmi P, Ramezanzadeh M, Ramezanzadeh B, Bahlakeh G (2022) Ultra-stable porous covalent organic framework assembled carbon nanotube as a novel nanocontainer for anti-corrosion coatings: experimental and computational studies. ACS Appl Mater Interfaces 14:19958–19974. https://doi.org/10.1021/ACSAMI.1C24185/ASSET/IMAGES/LARGE/AM1C24185_0009.JPEG

38. Moradi M, Rezaei M (2022) Construction of highly anti-corrosion and super-hydrophobic polypropylene/graphene oxide nanocomposite coatings on carbon steel: experimental, electrochemical and molecular dynamics studies. Constr Build Mater 317:126136. https://doi.org/10.1016/J.CONBUILDMAT.2021.126136

39. Dagdag O, Berisha A, Safi Z, Hamed O, Jodeh S, Verma C, Ebenso EE, el Harfi A (2020) DGEBA-polyaminoamide as effective anti-corrosive material for 15CDV6 steel in NaCl medium: computational and experimental studies. J Appl Polym Sci 137:48402. https://doi.org/10.1002/APP.48402

40. Bahlakeh G, Ramezanzadeh B, Ramezanzadeh M (2019) The role of chrome and zinc free-based neodymium oxide nanofilm on adhesion and corrosion protection properties of polyester/melamine coating on mild steel: experimental and molecular dynamics simulation study. J Clean Prod 210:872–886. https://doi.org/10.1016/J.JCLEPRO.2018.11.089

41. el Arrouji S, Karrouchi K, Berisha A, Ismaily Alaoui K, Warad I, Rais Z, Radi S, Taleb M, Ansar M, Zarrouk A (2020) New pyrazole derivatives as effective corrosion inhibitors on steel-electrolyte interface in 1 M HCl: Electrochemical, surface morphological (SEM) and computational analysis. Colloids Surf A Physicochem Eng Asp. 604, 125325. https://doi.org/10.1016/J.COLSURFA.2020.125325

42. Sun Q, Zhao Y, Choi KS, Mao X (2021) Molecular dynamics simulation of thermal de-icing on a flat surface. Appl Therm Eng 189:116701. https://doi.org/10.1016/J.APPLTHERMALENG.2021.116701

43. Zhuo Y, Xiao S, Håkonsen V, He J, Zhang Z (2020) Anti-icing Ionogel surfaces: inhibiting ice nucleation, growth, and adhesion, ACS. Mater Lett 2:616–623. https://doi.org/10.1021/ACSMATERIALSLETT.0C00094/SUPPL_FILE/TZ0C00094_SI_006.MP4

44. Rønneberg S, Xiao S, He J, Zhang Z (2020) nanoscale correlations of ice adhesion strength and water contact angle. Coatings 10:379. 10, 379. https://doi.org/10.3390/COATINGS1004 0379

45. Xiao S, He J, Zhang Z (2016) Nanoscale deicing by molecular dynamics simulation. Nanoscale 8:14625–14632. https://doi.org/10.1039/C6NR02398C

46. Jiang J, Sheng Q, Tang GH, Yang MY, Guo L (2022) Anti-icing propagation and icephobicity of slippery liquid-infused porous surface for condensation frosting. Int J Heat Mass Transf 190:122730. https://doi.org/10.1016/J.IJHEATMASSTRANSFER.2022.122730

47. Li N, Jiang J, Yang MY, Wang H, Ma Y, Li Z, Tang GH (2023) Anti-icing mechanism of combined active ethanol spraying and passive surface wettability. Appl Therm Eng 220:119805. https://doi.org/10.1016/J.APPLTHERMALENG.2022.119805

48. Bao L, Huang Z, Priezjev Nv, Chen S, Luo K, Hu H (2018) A significant reduction of ice adhesion on nanostructured surfaces that consist of an array of single-walled carbon nanotubes: a molecular dynamics simulation study. Appl Surf Sci 437:202–208. https://doi.org/10.1016/J.APSUSC.2017.12.096

49. Newman DJ, Cragg GM (2007) Natural products as sources of new drugs over the last 25 years. J Nat Prod 70:461–477. https://doi.org/10.1021/NP068054V/SUPPL_FILE/NP0680 54V_SA.PDF

50. Sangeetha B, Krishnamoorthy AS, Sharmila DJS, Renukadevi P, Malathi VG, Amirtham D (2021) Molecular modelling of coat protein of the Groundnut bud necrosis tospovirus and its binding with Squalene as an antiviral agent: In vitro and in silico docking investigations. Int J Biol Macromol 189:618–634. https://doi.org/10.1016/J.IJBIOMAC.2021.08.143

51. Shivanika C, Deepak Kumar S, Ragunathan V, Tiwari P, Sumitha A, Brindha Devi P (2020) Molecular docking, validation, dynamics simulations, and pharmacokinetic prediction of natural compounds against the SARS-CoV-2 main-protease. J Biomol Struct Dyn 40:1. https://doi.org/10.1080/07391102.2020.1815584

52. Sahihi M, Faraudo J (2022) Computer simulation of the interaction between SARS-CoV-2 spike protein and the surface of coinage metals, langmuir.https://doi.org/10.1021/acs.langmuir.2c02120

53. Singh MB, Sharma R, Kumar D, Khanna P, Mansi, Khanna L, Kumar V, Kumari K, Gupta A, Chaudhary P, Kaushik N, Choi EH, Kaushik NK, Singh P (2022) An understanding of coronavirus and exploring the molecular dynamics simulations to find promising candidates against the Mpro of nCoV to combat the COVID-19: a systematic review. J Infect Public Health 15:1326–1349. https://doi.org/10.1016/J.JIPH.2022.10.013

54. Kushwaha PP, Singh AK, Bansal T, Yadav A, Prajapati KS, Shuaib M, Kumar S (2021) Identification of natural inhibitors against SARS-CoV-2 drugable targets using molecular docking. Molecular Dyn Simul MM-PBSA Approach Front Cell Infect Microbiol 11:728. https://doi.org/10.3389/FCIMB.2021.730288/BIBTEX

55. Casalino L, Dommer AC, Gaieb Z, Barros EP, Sztain T, Ahn SH, Trifan A, Brace A, Bogetti AT, Clyde A, Ma H, Lee H, Turilli M, Khalid S, Chong LT, Simmerling C, Hardy DJ, Maia JDC, Phillips JC, Kurth T, Stern AC, Huang L, McCalpin JD, Tatineni M, Gibbs T, Stone JE, Jha S, Ramanathan A, Amaro RE (2021) AI-driven multiscale simulations illuminate mechanisms of SARS-CoV-2 spike dynamics. Int J High Perform Comput

Appl 35:432–451. https://doi.org/10.1177/10943420211006452/ASSET/IMAGES/LARGE/10.1177_10943420211006452-FIG2.JPEG

56. Kataria A, Verma A, Sanjay MR, Siengchin S (2022) Molecular modeling of 2D graphene grain boundaries: mechanical and fracture aspects. Materials Today: Proc 52:2404–2408

57. Bharath KN, Madhu P, Gowda TY, Verma A, Sanjay MR, Siengchin S (2021) Mechanical and chemical properties evaluation of sheep wool fiber–reinforced vinylester and polyester composites. Materials Performance Characterization 10(1):99–109

58. Verma A, Parashar A, Jain N, Singh VK, Rangappa SM, Siengchin S (2020) Surface modification techniques for the preparation of different novel biofibers for composites. In: Biofibers and biopolymers for biocomposites (pp 1–34). Springer, Cham

59. Verma A, Singh VK (2016) Experimental investigations on thermal properties of coconut shell particles in DAP solution for use in green composite applications. J Mater Sci Eng 5(3):1000242

60. Singh K, Jain N, Verma A, Singh VK, Chauhan S (2020) Functionalized graphite–reinforced cross-linked poly (vinyl alcohol) nanocomposites for vibration isolator application: Morphology, mechanical, and thermal assessment. Mater Performance Characterization 9(1):215–230

61. Verma A, Singh VK, Arif M (2016) Study of flame retardant and mechanical properties of coconut shell particles filled composite. Res Rev J Mater Sci 4(3):1–5

62. Verma A, Jain N, Rastogi S, Dogra V, Sanjay SM, Siengchin S, Mansour R (2020) Mechanism, anti-corrosion protection and components of anti-corrosion polymer coatings. In: Polymer coatings (pp 53–66). CRC Press

63. Chaudhary A, Sharma S, Verma A (2022) WEDM machining of heat treated ASSAB'88 tool steel: a comprehensive experimental analysis. Mater Today: Proc 50:946–951

64. Bisht N, Verma A, Chauhan S, Singh VK (2021) Effect of functionalized silicon carbide nanoparticles as additive in cross-linked PVA based composites for vibration damping application. J Vinyl Add Tech 27(4):920–932

65. Verma A, Jain N, Parashar A, Singh VK, Sanjay MR, Siengchin S (2020) Design and modeling of lightweight polymer composite structures. In: Lightweight polymer composite structures (pp 193–224). CRC Press

66. Dogra V, Kishore C, Verma A, Rana AK, Gaur A (2021) Fabrication and experimental testing of hybrid composite material having biodegradable bagasse fiber in a modified epoxy resin: evaluation of mechanical and morphological behavior. Appl Sci Eng Progress 14(4):661–667

67. Deji R, Verma A, Kaur N, Choudhary BC, Sharma RK (2022) Density functional theory study of carbon monoxide adsorption on transition metal doped armchair graphene nanoribbon. Mater Today: Proc 54:771–776

68. Verma A, Jain N, Parashar A, Gaur A, Sanjay MR, Siengchin S (2021) Lifecycle assessment of thermoplastic and thermosetting bamboo composites. In: Bamboo fiber composites (pp 235–246). Springer, Singapore

69. Deji R, Verma A, Choudhary BC, Sharma RK (2022) New insights into NO adsorption on alkali metal and transition metal doped graphene nanoribbon surface: a DFT approach. J Mol Graph Model 111:108109

70. Verma A, Jain N, Parashar A, Singh VK, Sanjay MR, Siengchin S (2020) Lightweight graphene composite materials. In: Lightweight polymer composite structures (pp 1–20). CRC Press

71. Verma A, Parashar A (2020) Characterization of 2D nanomaterials for energy storage. In: Recent advances in theoretical, applied, computational and experimental mechanics (pp 221–226). Springer, Singapore

72. Arpitha GR, Jain N, Verma A, Madhusudhan M (2022) Corncob bio-waste and boron nitride particles reinforced epoxy-based composites for lightweight applications: fabrication and characterization. Biomass Conversion Biorefinery, pp1–8. https://doi.org/10.1007/s13399-022-03717-1

73. Verma A, Samant SS (2016) Inspection of hydrodynamic lubrication in infinitely long journal bearing with oscillating journal velocity. J Appl Mech Eng 5(3):1–7

74. Verma A, Parashar A, Packirisamy M (2019) Effect of grain boundaries on the interfacial behaviour of graphene-polyethylene nanocomposite. Appl Surf Sci 470:1085–1092
75. Arpitha GR, Verma A, Sanjay MR, Siengchin S (2021) Preparation and experimental investigation on mechanical and tribological performance of hemp-glass fiber reinforced laminated composites for lightweight applications. Adv Civil Eng Mater 10(1):427–439
76. Verma A, Jain N, Rangappa SM, Siengchin S, Jawaid M (2021) Natural fibers based bio-phenolic composites. In: Phenolic polymers based composite materials(pp 153–168). Springer, Singapore
77. Verma A, Parashar A, Singh SK, Jain N, Sanjay SM, Siengchin S (2020) Modeling and simulation in polymer coatings. In: Polymer coatings (pp 309–324). CRC Press
78. Verma A, Singh VK, Experimental characterization of modified epoxy resin assorted with almond shell particles. ESSENCE-Int J Environ Rehab Conserv 7(1):36–44
79. Deji R, Verma A, Kaur N, Choudhary BC, Sharma RK (2022) Adsorption chemistry of co-doped graphene nanoribbon and its derivatives towards carbon based gases for gas sensing applications: quantum DFT investigation. Mater Sci Semicond Process 146:106670
80. Verma A (2022) A perspective on the potential material candidate for railway sector applications: PVA based functionalized graphene reinforced composite. Appl Sci Eng Progress 15(2):5727–5727
81. Verma A, Jain N, Singh K, Singh VK, Rangappa SM, Siengchin S (2022) PVA-based blends and composites. In: Biodegradable polymers, blends and composites (pp 309–326). Woodhead Publishing
82. Raja S, Verma A, Rangappa SM, Siengchin S (2022) Development and experimental analysis of polymer based composite bipolar plate using Aquila Taguchi optimization: design of experiments. Polym Compos 43(8):5522–5533
83. Prabhakaran S, Sharma S, Verma A, Rangappa SM, Siengchin S (2022) Mechanical, thermal, and acoustical studies on natural alternative material for partition walls: a novel experimental investigation. Polym Compos 43(7):4711–4720
84. Chaturvedi S, Verma A, Sethi SK, Rangappa SM, Siengchin S (2022) Stalk fibers (rice, wheat, barley, etc.) composites and applications. In: Plant fibers, their composites, and applications (pp 347–362). Woodhead Publishing
85. Arpitha GR, Verma A, MRS, Gorbatyuk S, Khan A, Sobahi TR, Asiri AM, Siengchin S (2022) Bio-composite film from corn starch based vetiver cellulose. J Natural Fibers 19(16):14634–14644
86. Thimmaiah SH, Narayanappa K, Thyavihalli Girijappa Y, Gulihonenahali Rajakumara A, Hemath M, Thiagamani SMK, Verma A (2022) An artificial neural network and Taguchi prediction on wear characteristics of Kenaf-Kevlar fabric reinforced hybrid polyester composites. Polym Compos 44(1):261–273
87. Chaturvedi S, Verma A, Singh SK, Ogata S (2022) EAM inter-atomic potential—its implication on nickel, copper, and aluminum (and Their Alloys). In: Forcefields for atomistic-scale simulations: materials and applications (pp 133–156). Springer, Singapore
88. Verma A, Sharma S (2022) atomistic simulations to study thermal effects and strain rate on mechanical and fracture properties of graphene like BC3. In: forcefields for atomistic-scale simulations: materials and applications (pp 237–252). Springer, Singapore
89. Chaturvedi S, Verma A, Sethi SK, Ogata S (2022) Defect energy calculations of nickel, copper and aluminium (and Their Alloys): Molecular dynamics approach. In: Forcefields for atomistic-scale simulations: materials and applications (pp 157–186). Springer, Singapore
90. Verma A, Gaur A, Singh VK (2017) Mechanical properties and microstructure of starch and sisal fiber biocomposite modified with epoxy resin. Mater Performance Characterization 6(1):500–520
91. Shankar U, Gogoi R, Sethi SK, Verma A (2022) Introduction to materials studio software for the atomistic-scale simulations. In: Forcefields for atomistic-scale simulations: materials and applications (pp 299–313). Springer, Singapore
92. Shankar U, Sethi SK, Verma A (2022) Forcefields and modeling of polymer coatings and nanocomposites. In: Forcefields for atomistic-scale simulations: materials and applications (pp 81–98). Springer, Singapore

93. Verma A, Ogata S (2022) Computational modelling of deformation and failure of bone at molecular scale. In: Forcefields for atomistic-scale simulations: materials and applications (pp 253–268). Springer, Singapore
94. Homer ER, Verma A, Britton D, Johnson OK, Thompson GB (2022) Simulated migration behavior of metastable Σ3 (11 8 5) incoherent twin grain boundaries. In: IOP conference series: materials science and engineering (Vol 1249, No 1, p 012019)
95. Verma A, Parashar A, van Duin AC (2022) Graphene-reinforced polymeric membranes for water desalination and gas separation/barrier applications. In: Innovations in graphene-based polymer composites (pp 133–165). Woodhead Publishing
96. Verma A, Jain N, Sanjay MR, Siengchin S, Viscoelastic properties of completely biodegradable polymer-based composites. In: Vibration and damping behavior of biocomposites (pp 173–188). CRC Press
97. Verma A, Jain N, Mishra RR (2022) Applications and drawbacks of epoxy/natural fiber composites. In: Handbook of epoxy/fiber composites (pp 1–15). Singapore: Springer Singapore
98. Lila MK, Verma A, Bhurat SS (2022) Impact behaviors of epoxy/synthetic fiber composites. In: Handbook of epoxy/fiber composites (pp 1–18). Singapore: Springer Singapore
99. Verma A, Parashar A (2017) The effect of STW defects on the mechanical properties and fracture toughness of pristine and hydrogenated graphene. Phys Chem Chem Phys 19(24):16023–16037
100. Verma A, Parashar A (2018) Molecular dynamics based simulations to study failure morphology of hydroxyl and epoxide functionalised graphene. Comput Mater Sci 143:15–26
101. Verma A, Parashar A (2018) Molecular dynamics based simulations to study the fracture strength of monolayer graphene oxide. Nanotechnology 29(11):115706
102. Verma A, Parashar A (2018) Structural and chemical insights into thermal transport for strained functionalised graphene: a molecular dynamics study. Materials Res Express 5(11):115605
103. Verma A, Parashar A, Packirisamy M (2018) Tailoring the failure morphology of 2D bicrystalline graphene oxide. J Appl Phys 124(1):015102
104. Verma A, Parashar A (2018) Reactive force field based atomistic simulations to study fracture toughness of bicrystalline graphene functionalised with oxide groups. Diam Relat Mater 88:193–203
105. Singla V, Verma A, Parashar A (2018) A molecular dynamics based study to estimate the point defects formation energies in graphene containing STW defects. Materials Res Express 6(1):015606
106. Verma A, Parashar A, Packirisamy M (2019) Role of chemical adatoms in fracture mechanics of graphene nanolayer. Materials Today: Proc 11:920–924
107. Chaudhary A, Sharma S, Verma A (2022) Optimization of WEDM process parameters for machining of heat treated ASSAB'88 tool steel using Response surface methodology (RSM). Materials Today: Proc 50:917–922
108. Verma A, Zhang W, Van Duin AC (2021) ReaxFF reactive molecular dynamics simulations to study the interfacial dynamics between defective h-BN nanosheets and water nanodroplets. Phys Chem Chem Phys 23(18):10822–10834
109. Verma A, Parashar A, Packirisamy M (2018) Atomistic modeling of graphene/hexagonal boron nitride polymer nanocomposites: a review. Wiley Interdisciplinary Rev Comput Molecular Sci 8(3):e1346
110. Verma A, Baurai K, Sanjay MR, Siengchin S (2020) Mechanical, microstructural, and thermal characterization insights of pyrolyzed carbon black from waste tires reinforced epoxy nanocomposites for coating application. Polym Compos 41(1):338–349
111. Verma A, Budiyal L, Sanjay MR, Siengchin S (2019) Processing and characterization analysis of pyrolyzed oil rubber (from waste tires)-epoxy polymer blend composite for lightweight structures and coatings applications. Polym Eng Sci 59(10):2041–2051
112. Verma A, Negi P, Singh VK (2019) Experimental analysis on carbon residuum transformed epoxy resin: Chicken feather fiber hybrid composite. Polym Compos 40(7):2690–2699

113. Verma A, Singh VK (2018) Mechanical, microstructural and thermal characterization of epoxy-based human hair–reinforced composites. J Test Eval 47(2):1193–1215

114. Jain N, Verma A, Ogata S, Sanjay MR, Siengchin S (2022) Application of machine learning in determining the mechanical properties of materials. In: Machine learning applied to composite materials (pp 99–113). Springer, Singapore

115. Verma A, Singh VK, Verma SK, Sharma A (2016) Human hair: a biodegradable composite fiber–a review. Int J Waste Res 6(206):2

116. Kataria A, Verma A, Sanjay MR, Siengchin S, Jawaid M (2022) Physical, morphological, structural, thermal, and tensile properties of coir fibers. In: Coir fiber and its composites (pp 79–107). Woodhead Publishing

117. Jain N, Verma A, Singh VK (2019) Dynamic mechanical analysis and creep-recovery behaviour of polyvinyl alcohol based cross-linked biocomposite reinforced with basalt fiber. Materials Res Express 6(10):105373

118. Chaurasia A, Verma A, Parashar A, Mulik RS (2019) Experimental and computational studies to analyze the effect of h-BN nanosheets on mechanical behavior of h-BN/polyethylene nanocomposites. J Phys Chem C 123(32):20059–20070

119. Verma A, Kumar R, Parashar A (2019) Enhanced thermal transport across a bi-crystalline graphene–polymer interface: an atomistic approach. Phys Chem Chem Phys 21(11):6229–6237

120. Verma A, Negi P, Singh VK (2018) Physical and thermal characterization of chicken feather fiber and crumb rubber reformed epoxy resin hybrid composite. Adv Civil Eng Materials 7(1):538–557

121. Bharath KN, Madhu P, Gowda TG, Verma A, Sanjay MR, Siengchin S (2020) A novel approach for development of printed circuit board from biofiber based composites. Polym Compos 41(11):4550–4558

122. Verma A, Joshi K, Gaur A, Singh VK (2018) Starch-jute fiber hybrid biocomposite modified with an epoxy resin coating: fabrication and experimental characterization. J Mech Behav Materials 27(5–6)

123. Verma A, Singh C, Singh VK, Jain N (2019) Fabrication and characterization of chitosan-coated sisal fiber–Phytagel modified soy protein-based green composite. J Compos Mater 53(18):2481–2504

124. Rastogi S, Verma A, Singh VK (2020) Experimental response of nonwoven waste cellulose fabric–reinforced epoxy composites for high toughness and coating applications. Materials Performance Characterization 9(1):151–172

125. Verma A, Ogata S (2023) Magnesium based alloys for reinforcing biopolymer composites and coatings: a critical overview on biomedical materials. Adv Indus Eng Polymer Res.https://doi.org/10.1016/j.aiepr.2023.01.002

126. Kataria A, , Vaibhav Chaudhary SC, Verma A, Mavinkere Rangappa Sanjay NJ, Siengchin S (2023) Cellulose fiber-reinforced composites—History of evolution, chemistry, and structure. In: Cellulose fibre reinforced composites (pp 1–22). Woodhead Publishing. https://doi.org/10.1016/B978-0-323-90125-3.00012-4

127. Chaturvedi S, Kataria A, Chaudhary V, Verma A, Mavinkere Rangappa Sanjay NJ, Siengchin S (2023) Bionanocomposites reinforced with cellulose fibers and agro-industrial wastes. In: Cellulose fibre reinforced composites (pp 1–22). Woodhead Publishing. https://doi.org/10.1016/B978-0-323-90125-3.00017-3

128. Arpitha GR, Mohit H, Madhu P, Verma* A (2023) Effect of sugarcane bagasse and alumina reinforcements on physical, mechanical, and thermal characteristics of epoxy composites using artificial neural networks and response surface methodology. Biomass Conver Biorefinery.https://doi.org/10.1007/s13399-023-03886-7

Chapter 9
Continuum Mechanics-Based Simulations in Coatings

Suhaib Zafar and Akarsh Verma

1 Introduction

The transportation of people and goods is a cornerstone of any economy in today's world of globalization. Consequently, it is a primary contributor to greenhouse gas emissions, directly impacting climate change. It is no secret that climate change is a major challenge for countries across the globe: Goal #13 on the Sustainable Development Goals (SDGs) from the United Nations (UN) 2030 Agenda is taking urgent action to combat climate change. In this context, the electrification of automobiles is a crucial step forward, which will reduce our dependence on internal combustion engines. However, there are serious barriers yet to be overcome, and it is not entirely clear if the reduction in emissions will be significant, given that production of batteries for electric vehicles (EVs) can produce considerable negative environmental impacts. Additionally, EVs need electric power to recharge their batteries, furthering the negative effect on the environment, assuming the grid is *conventional*.

For aviation, arguably the primary transportation sector, especially for the global movement of goods and services, the contribution to carbon dioxide emissions is 12% of the entire transportation sector, and 3% of the global CO_2 emissions. Apart from the obvious environmental impact of aviation is fuel costs for airlines in particular, accounting for around 40% of the expenses, which translates into millions of dollars annually [1]. Improving the aerodynamic efficiency of aircraft wings is an important area of research because of these numbers, since it translates into lower fuel costs

S. Zafar (✉)
Stellantis, Chrysler Technology Center, 800 Chrysler Drive, Auburn Hills, MI 48326, USA
e-mail: suhaibzafar91@gmail.com

A. Verma
Department of Mechanical Engineering, University of Petroleum and Energy Studies, Dehradun 248007, India

Department of Mechanical Science and Bioengineering, Osaka University, Osaka 560-8531, Japan

© The Author(s), under exclusive license to Springer Nature Singapore Pte Ltd. 2023
A. Verma et al. (eds.), *Coating Materials*, Materials Horizons: From Nature to Nanomaterials, https://doi.org/10.1007/978-981-99-3549-9_9

and therefore reduced emissions. The other active research domain is finding ways to improve the thermodynamic and aerodynamic efficiency of gas turbines (GTs) that power modern airliners. However, aeronautical propulsion is not the only application of GTs, since they are also heavily utilized for power generation applications and will be for the foreseeable future. According to one estimate, the cumulative sales of GTs in the 2017–2031 period will amount to approximately 2 trillion US dollars, highlighting their importance to the aerospace and power generation sectors [2]. It is also noteworthy that GTs can play a crucial role in decarbonizing the energy and transportation sectors by using net-zero emission fuels or compensating for the intermittency of renewables [3].

Any reader well-informed on the fundamentals of thermodynamics of GTs is aware that their thermodynamic efficiency is directly proportional to the compressor pressure ratio and the turbine inlet temperature. Therefore, improvements in efficiency have mainly come from increasing both in tandem, resulting in a better thrust-to-weight ratio and improved fuel economy alongside a reduction in harmful emissions. While the increase in compression ratios is not the focal point of this discussion, the increase in turbine working temperature is a result of utilizing cooling mechanisms and thermal barrier coatings (TBCs), allowing temperatures of up to 1500 °C. However, even better designs that can reduce the cooling air requirement by just 1% can result in GT efficiency gains of around 0.4%, which translates to hundreds of millions of dollars in fuel burn savings [3].

For the past few decades, Ni-based superalloys with TBCs have been widely utilized for the hot sections in GT engines. However, the increase in the working temperature of GT engines is limited by the melting point of these superalloys, and silicon carbide (SiC)-fiber-reinforced SiC ceramic matrix composites (SiC/SiC CMCs) have increasingly filled the void. Not only do CMCs provide an increased service temperature and superior strength when compared with Ni-based superalloys, but they also do so while offering significant weight savings of up to 66% [4]. As a result of this innovation, the coating design priority has shifted from thermal insulation to environmental shielding, referred to as environmental barrier coatings (EBCs). The primary reason for developing EBCs are the harsh working conditions in gas turbines that cause thermo-chemo-mechanical degradation. For instance, SiC/SiC CMCs suffer from rapid surface recession when the protective silica reacts with water vapor (a product of combustion) or corrosive species such as calcium-aluminum-magnesium silicates (CMAS) deposits [2, 4].

Although the above discussion on T/EBCs primarily applies to GTs used in aerospace propulsion, similar issues exist for land-based power generation GTs, despite different requirements and concerns. As a result, several commonalities exist in the development of TBCs for both types of engines [5]. In a similar fashion, automobile engines benefit tremendously from the application of TBCs to various engine components including pistons, rings and liners by enhancing high temperature durability via the reduction of heat transfer and decreasing the temperature of the metal underneath [6].

2 The Finite Element Method (FEM): A Primer

The finite element method (FEM), also referred to as finite element analysis (FEA) by theorists and practitioners, is a numerical method utilized to solve problems in various disciplines of engineering. Applications of the FEM include deformation and stress analysis of structures such as aircraft and automotive, heat transfer such as TBCs, and fluid mechanics including aerodynamics and acoustics. All these applications involve the solution of a set of partial differential equations (PDEs) that are mathematical models of the underlying physical process(es), such as heat transfer in a solid (governed by the heat equation).

While the FEM has a variety of applications, the method was originally developed in the context of the analysis of aircraft structures. A survey of texts published on FEM suggests that the concept of the FEM was first described by Turner et al. [7] in 1956, who derived stiffness matrices for trusses, beams, and other similar elements to analyze aircraft. However, the term *finite element* itself was coined in 1960 by Clough [8]. Earlier still, in the 1940s, Hrennikof [9] published a paper that presented a solution to elasticity problems using the "framework method", while Courant's paper [10] used piecewise polynomials for interpolation over triangular subregions to model torsion. Similar developments continued into the late 1960s and early 1970s. The first book on FEM was published in 1967 by the authors Zeinkiewicz and Cheung [11].

With the rise of powerful computing, starting in the 1980s, the FEM gained popularity and commercial software packages were introduced to aid engineers in design and analysis. The method has continued to gain popularity ever since and is now routinely used in industrial applications, especially structural engineering. As is the case with any tool, the FEM comes with several advantages as well as downsides. The major advantage of the FEM is the ability to analyze complex geometries, even if the boundaries are irregular. It is important to understand how crucial this aspect is: solving PDEs even for a simple geometry using analytical methods is practically impossible, despite significant advances on the theoretical end (primarily pure mathematics). Other advantages include relative ease of analysis when the loads are complex (point loads, inertial forces, thermal loading, and fluid–structure interactions etc.), the ability to test various configurations including different materials, nonlinearities, and a variety of boundary conditions (more on this later). All these advantages combined allow engineers to identify optimal structural designs with significant levels of confidence.

The disadvantages of FEM are obvious, especially if one is aware of the nature of numerical methods and the use of computer simulations to solve engineering problems in the first place. First, FEM results are *approximate*, as are all other numerical methods. The major reason for this fact is the use of *discretization*, which is dividing a given geometry into subdomains known as *elements*. Since the number of elements is finite, the numerical solution can never be the same in magnitude as the *exact* (also referred to as *analytical*) solution. This discrepancy is usually countered by increasing the number of elements used to model the geometry, which results in increased computational expense. As mentioned before, with the increasing power of computing hardware, it is now possible to use elements numbering in the millions, however, the discrepancy would theoretically remain.

The number of elements alone does not dictate the accuracy of the FEM solution, however, and other factors such as the type of element being utilized, and the assumptions used in modeling also affect the simulation results. Since these choices affect the accuracy of the solution, this feeds into the second disadvantage of FEM: the human using this method. A less experienced professional may use the wrong type of elements (discussed later), incorrectly input the boundary conditions in the solver, or worst (not uncommon) of all, make errors in units for physical quantities involved. The latter is usually a mismatch of units, leading to horrible mistakes in the final calculation, as both authors can attest based on their teaching experience.

Finally, the FEM has *inherent* errors, which fundamentally means that the geometry representing the structure to be analyzed is itself not accurate, there are large differences in stiffness of the elements being used leading to difficulties in solving the system numerically, the limitations of computing hardware being used, and so on. As the name implies, these errors cannot be overcome in general, and practitioners should be cognizant of them when using the FEM to analyze a given problem.

Before discussing the mathematics behind the FEM, it is worthwhile to summarize the steps involved in FEA, listed below:

1. **Discretization**: The geometry or structure is divided into finite elements. This step has a major impact on the accuracy of the solution, as discussed before.
2. **Define element properties**: The next step is to define element properties and the type of finite elements (FEs) to be used in the analysis.
3. **Assemble the element stiffness matrices**: Each element has a stiffness matrix containing coefficients that can be derived from equilibrium equations governing the behavior of the element. The stiffness matrix relates the nodal displacements to the forces acting on each node. The terms in each matrix depend on the element type, its geometry, and the material properties. The stiffness matrix is always a *square* matrix, meaning that the number of rows and columns are equal.
4. **Combine the element stiffness matrices into one global stiffness matrix**: All element stiffness matrices can be combined into one *global* stiffness matrix $[K]$ based on how the elements are connected to one another.
5. **Apply the loads**: At this stage, the externally applied loads are provided. These often include concentrated or uniform forces, moments, or a combination thereof.
6. **Apply the boundary conditions**: The boundary conditions (BCs) are an essential step before the FE model is solved. This step involves providing the values of several nodal displacements, typically at the supports.
7. **Solve the system of linear algebraic equations**: The global FEM equation $\{F\} = [K]\{U\}$ is solved using direct or *iterative* methods, typically the latter. The unknown quantities are the nodal displacements, which is the vector $\{U\}$.
8. **Compute stresses and other relevant quantities**: The final stage is to compute stresses, strains and other relevant quantities using the nodal displacements.

It is often the case that undergraduates in engineering are asked to do the above steps by hand on a simple geometry or write computer code in a programming language such as MATLAB™ or Python. This exercise allows students to fully grasp the fundamentals of the FEM, while keeping number crunching to a minimum. However, the above steps become increasingly complicated as the number of elements increases, or if the problem is not one-dimensional (or some other simplification, such as plane stress or plane strain). In this case (which is the norm), commercial FEA packages such as Abaqus or ANSYS are used by students to get practical experience, since in-house solvers are rarely used in industrial applications.

The solution workflow of commercial FEA packages is different from the steps outlined above and can be broadly categorized into three stages: *pre-processing*, *solution*, and *post-processing*. Each of these steps is discussed below:

1. *Pre-processing*: This is the fundamental step in setting up an FE simulation in a solver. Typically, the geometry is imported into the solver from a CAD/CAE program (such as SolidWorks). In some cases, the user can develop the geometry inside the FEA package. Once the geometry is defined, data preparation takes place. This involves selecting the element type, the material properties, discretization of the structure (meshing), application of loads, and specifying the boundary conditions. Using this information, the computer program (solver) prepares the equations for each element that needs to be solved (the user can select the type of solver).

2. *Solution*: This is the second and major step in the process, since the computer program solves the algebraic equations for the unknown values, such as nodal displacements. The computed values are then utilized to calculate secondary values such as stresses and strains.

3. *Post-processing*: This is the final stage, where the user can develop all sorts of plots, do follow-up calculations, display relevant quantities in a textual or tabular format, and so on. Figure 1 shows a 3D plot displaying the variation of the von Mises stress in the given geometry.

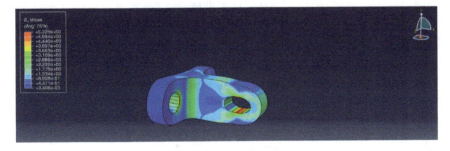

Fig. 1 3D plot of the von Mises stress in Abaqus

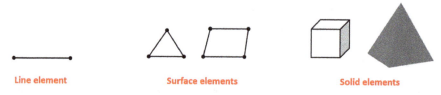

Fig. 2 Types of elements used in FEA

Before concluding this section, we would like to make a mention of the types of elements used in FEA, which usually fit into the following three categories:

1. **Line elements**: These elements are used to model one-dimensional structures such as pipes.
2. **Surface elements**: Surface elements are best suited for modeling thin structures such as shells.
3. **Solid elements**: For 3D FEA, solid elements must be utilized.

It is important to note that the categorization listed above is not the only way to categorize elements used in FEA (Fig. 2), since other distinctions such as *linear* (note the difference from *line*) and *quadratic* types of elements also exist. Such differences will become clear to the reader in the next section when we discuss the mathematical foundations of the FEM.

3 Mathematical Foundations of the FEM

Before we begin our discussion in this section, we will take a moment here to emphasize that the expectation from our readers is a decent understanding of the fundamentals of linear algebra, vector calculus and other relevant subjects such as solid mechanics. The aim of this section is to provide a quick overview of the mathematical underpinnings of the FEM and interested readers can refer to specialized texts on solid mechanics and the FEM for detailed treatment of the topics covered in this section.

To begin, we note that in continuum mechanics (which deals with materials and their mechanical behavior with the assumption of *continuous* mass rather than discrete particles), the Cauchy stress tensor (denoted by σ_{ij}) comprises of nine components, which can be represented by a 3×3 matrix. These components completely define the state of stress inside a material when it is deformed. However, this representation can be reduced to six components thanks to symmetry, and therefore the stress can be written as:

$$\boldsymbol{\sigma} = [\sigma_x, \sigma_y, \sigma_z, \tau_{yz}, \tau_{xz}, \tau_{xy}]^{\mathrm{T}}$$

The first three components are normal stresses, and the latter three are shear stresses. Using the principle of conservation of linear momentum, we can derive the following set of equations known as the Cauchy momentum equation (tensor notation allows for a single equation to be written down):

$$\frac{\partial \sigma_x}{\partial x} + \frac{\partial \tau_{xy}}{\partial y} + \frac{\partial \tau_{xz}}{\partial z} + f_x = 0$$

$$\frac{\partial \tau_{xy}}{\partial x} + \frac{\partial \sigma_y}{\partial y} + \frac{\partial \tau_{yz}}{\partial z} + f_y = 0$$

$$\frac{\partial \tau_{xz}}{\partial x} + \frac{\partial \tau_{yz}}{\partial y} + \frac{\partial \sigma_z}{\partial z} + f_z = 0$$

These equations are the *equilibrium* equations (implying the assumption of static equilibrium), and therefore a special case of the Cauchy momentum equation. They can be derived by considering an elemental volume $dV = dxdydz$, and utilizing $\sum F_x = \sum F_y = \sum F_z = 0$ for each face of the volume. For example, the first equation can be derived as follows:

$$\left(\sigma_x + \frac{\partial \sigma_x}{\partial x} dx \right) dydz - \sigma_x dydz + \left(\tau_{xy} + \frac{\partial \tau_{xy}}{\partial y} dy \right) dxdz$$

$$- \tau_{xy} dxdz + \left(\tau_{xz} + \frac{\partial \tau_{xz}}{\partial z} dz \right) dxdy$$

$$- \tau_{xz} dxdy + f_x dxdydz = 0$$

$$\frac{\partial \sigma_x}{\partial x} dxdydz + \frac{\partial \tau_{xy}}{\partial y} dxdydz + \frac{\partial \tau_{xz}}{\partial z} dxdydz + f_x dxdydz = 0$$

$$\frac{\partial \sigma_x}{\partial x} + \frac{\partial \tau_{xy}}{\partial y} + \frac{\partial \tau_{xz}}{\partial z} + f_x = 0$$

In a similar manner, the heat equation for conduction in a solid can be derived using the principle of conservation of energy, by considering an elemental volume $dV = dxdydz$ as we did for the equilibrium equations. Note however, that both sets of equations are valid for Cartesian coordinates only, and the mathematics required to derive these equations in polar or spherical coordinates differs significantly (and is more complicated).

For the sake of brevity, we will consider the one-dimensional case. To derive the heat equation, we need to utilize Fourier's law of heat conduction. This equation relates the heat flux (heat transfer per unit area) to the temperature gradient in the direction of the heat flow in the body. That is,

$$q_n'' = -k \frac{\partial T}{\partial n}$$

In the above equation, k is a material constant known as *thermal conductivity*.

Combining the conservation of energy with Fourier's law of heat conduction, we can do the following steps to get the heat equation:

$$q_x'' dydz + q_{gen} dxdydz = q_{xdx}'' dydz + \rho c \frac{\partial T}{\partial t} dxdydz$$

The first term in this equation on the left-hand side (LHS) is the heat flux *input* to the volume, while the second term is the heat generation *within* the volume. These two terms combined are equal to the sum of the two terms on the right-hand side (RHS) of the equation: the heat flux *output* and the change in internal energy of the volume (body). We can now proceed to re-arrange the terms, and introduce Fourier's law to get the following:

$$\left(q_x'' - q_{x+dx}''\right) dydz + q_{gen} dxdydz = \rho c \frac{\partial T}{\partial t} dxdydz$$

$$-\frac{\partial q_x''}{\partial x} dxdydz + q_{gen} dxdydz = \rho c \frac{\partial T}{\partial t} dxdydz$$

$$-\frac{\partial}{\partial x}\left(-k \frac{\partial T}{\partial x}\right) dxdydz + q_{gen} dxdydz = \rho c \frac{\partial T}{\partial t} dxdydz$$

Thus, the heat equation in the one-dimensional case is,

$$\frac{\partial}{\partial x}\left(k \frac{\partial T}{\partial x}\right) + q_{gen} = \rho c \frac{\partial T}{\partial t}$$

Further simplifications may be applied as necessary. For instance, if there is no heat generation ($q_{gen} = 0$), and the temperature does not vary with time (steady state), the heat equation reduces to the following form:

$$\frac{\partial}{\partial x}\left(k \frac{\partial T}{\partial x}\right) = 0$$

The thermal conductivity may or may not be a function of temperature or vary spatially. In that case, the equation reduces to the following form:

$$\frac{d^2 T}{dx^2} = 0$$

Thus, the temperature variation in a body with one-dimensional heat transfer, no internal heat generation and constant thermal conductivity, at a steady state, is *linear*. The above steps can be generalized to get the heat equation in two and three dimensions. For an *orthotropic* material, the thermal conductivity varies in every direction. Typically, most texts on heat transfer will assume the material is *isotropic*, and state the heat equation as follows:

$$\frac{\partial^2 T}{\partial x^2} + \frac{\partial^2 T}{\partial y^2} + \frac{\partial^2 T}{\partial z^2} + q_{gen} = \rho c \frac{\partial T}{\partial t}$$

Assuming no internal heat generation and steady state heat transfer, the heat equation reduces to the familiar Laplace's equation:

$$\frac{\partial^2 T}{\partial x^2} + \frac{\partial^2 T}{\partial y^2} + \frac{\partial^2 T}{\partial z^2} = \nabla^2 T = 0$$

This is a PDE that solves for temperature *field* in a geometry, and like structural analysis where nodal displacements are the unknowns, the nodal temperature is the quantity of interest when applying the FEM to heat transfer.

4 Stress–Strain Relations

Like stress σ, the (nodal) displacement can be represented by a vector. That is,

$$\boldsymbol{u} = [u, v, w]^T$$

Other physical quantities such as the applied loads and surface traction can be represented in a similar fashion. The displacement vector is used to compute strains, denoted by ϵ. The strain vector is written as,

$$\boldsymbol{\epsilon} = [\epsilon_x, \epsilon_y, \epsilon_z, \gamma_{yz}, \gamma_{xz}, \gamma_{xy}]^T$$

The first three components are normal strains, and the latter three are the engineering shear strains. The strain ϵ is related to displacements as follows:

$$\epsilon = \left[\frac{\partial u}{\partial x}\; \frac{\partial v}{\partial y}\; \frac{\partial w}{\partial z}\; \frac{\partial v}{\partial z} + \frac{\partial w}{\partial y}\; \frac{\partial u}{\partial z} + \frac{\partial w}{\partial x}\; \frac{\partial u}{\partial y} + \frac{\partial v}{\partial x} \right]^T$$

For *linear elastic* materials, Hooke's law dictates the stress–strain relations. The material properties governing these relations (assuming the material is *isotropic*) are the modulus of elasticity E and Poisson's ratio ν. The shear modulus G can be computed using E and ν. The stress–strain relations are as follows:

$$\epsilon_x = \frac{\sigma_x}{E} - \nu \frac{\sigma_y}{E} - \nu \frac{\sigma_z}{E}$$

$$\epsilon_y = \frac{\sigma_y}{E} - \nu \frac{\sigma_x}{E} - \nu \frac{\sigma_z}{E}$$

$$\epsilon_z = \frac{\sigma_z}{E} - \nu\frac{\sigma_x}{E} - \nu\frac{\sigma_y}{E}$$

$$\gamma_{yz} = \frac{\tau_{yz}}{G}$$

$$\gamma_{xz} = \frac{\tau_{xz}}{G}$$

$$\gamma_{xy} = \frac{\tau_{xy}}{G}$$

The shear modulus G is computed as,

$$G = \frac{E}{2(1+\nu)}$$

These relations can be manipulated to obtain *inverse* relations, that is, equations that express σ in terms of ϵ. For starters, note that

$$\epsilon_x + \epsilon_y + \epsilon_z = \frac{1-2\nu}{E}(\sigma_x + \sigma_y + \sigma_z)$$

To obtain a relation for σ_x, we do the following steps:

$$\sigma_x + \sigma_y + \sigma_z = \frac{E}{1-2\nu}(\epsilon_x + \epsilon_y + \epsilon_z)$$

$$\sigma_x + \left(\frac{\sigma_x - E\epsilon_x}{\nu}\right) = \frac{E}{1-2\nu}(\epsilon_x + \epsilon_y + \epsilon_z)$$

The above step can be done because,

$$\epsilon_x = \frac{\sigma_x}{E} - \nu\frac{\sigma_y}{E} - \nu\frac{\sigma_z}{E}$$

$$E\epsilon_x = \sigma_x - \nu(\sigma_y + \sigma_z)$$

$$\sigma_y + \sigma_z = \frac{\sigma_x - E\epsilon_x}{\nu}$$

Consequently,

$$\sigma_x + \frac{\sigma_x}{\nu} = \frac{E}{1-2\nu}(\epsilon_x + \epsilon_y + \epsilon_z) + \frac{E\epsilon_x}{\nu}$$

$$\sigma_x(1+\nu) = \left(\frac{E\nu}{1-2\nu} + E\right)\epsilon_x + \frac{E\nu}{1-2\nu}(\epsilon_y + \epsilon_z)$$

$$\sigma_x = \frac{E(1-v)}{(1+v)(1-2v)}\epsilon_x + \frac{Ev}{(1+v)(1-2v)}\left(\epsilon_y + \epsilon_z\right)$$

$$\sigma_x = \frac{E}{(1+v)(1-2v)}\left((1-v)\epsilon_x + v\epsilon_y + v\epsilon_z\right)$$

Similar steps can be done to derive the relationship between other stresses and strains, to get the following matrix relationship:

$$\sigma = D\epsilon$$

D is the symmetric 6×6 material matrix.

5 Special Cases

It is often possible to (and even required in some cases) reduce the problem to one or two dimensions to decrease the computational expense associated with running a FEM simulation.

1. *One dimension*: For one-dimensional problems, the stress–strain relationship can be expressed as,

$$\sigma = E\epsilon$$

2. *Two dimensions*: For two-dimensional problems, two variations exist. These are the commonly known **plane stress** and **plane strain** configurations. In the former, the stresses in the third dimension are set to zero, meaning that $\sigma_z = \tau_{yz} = \tau_{xz} = 0$. This configuration is usually applicable to thin, planar bodies with in-plane loading. The latter case implies that $\epsilon_z = \gamma_{yz} = \gamma_{xz} = 0$. This configuration usually applies to long bodies that are subjected to transverse loading along their length. Crucially, the stress σ_z may **not** be zero in this case.

6 Methods to Derive the Stiffness Matrix

In problems related to structural mechanics, our fundamental aim is to solve the *equilibrium* equations:

$$\frac{\partial \sigma_x}{\partial x} + \frac{\partial \tau_{xy}}{\partial y} + \frac{\partial \tau_{xz}}{\partial z} + f_x = 0$$

$$\frac{\partial \tau_{xy}}{\partial x} + \frac{\partial \sigma_y}{\partial y} + \frac{\partial \tau_{yz}}{\partial z} + f_y = 0$$

Fig. 3 A spring element

$$\frac{\partial \tau_{xz}}{\partial x} + \frac{\partial \tau_{yz}}{\partial y} + \frac{\partial \sigma_z}{\partial z} + f_z = 0$$

The quantity of interest is the nodal displacement u, which is related to strains, and in turn related to stresses, which appear in the equilibrium equations. The fundamental principle of the FEM is to obtain the *stiffness matrix*, which can be obtained using various methods. These methods fall under one of three categories in general:

1. *Direct method*: This method utilizes force equilibrium at the nodes and force–deformation relationships. It is best suited to simple structural elements such as springs, bars and trusses. For a demonstration of this method, consider the spring shown in Fig. 3.

The mechanical behavior of a spring is governed by Hooke's law:

$$F = kx$$

Recognize that,

$$-F + f_2 = 0$$

$$F + f_1 = 0$$

Also,

$$F = k(u_2 - u_1)$$

Therefore,

$$f_1 = -k(u_2 - u_1) = k(u_1 - u_2)$$

$$f_2 = k(u_2 - u_1)$$

In matrix form, this relationship is,

$$\begin{bmatrix} f_1 \\ f_2 \end{bmatrix} = \begin{bmatrix} k & -k \\ -k & k \end{bmatrix} \begin{bmatrix} u_1 \\ u_2 \end{bmatrix}$$

Therefore, the stiffness matrix for a spring element is,

$$[k] = \begin{bmatrix} k & -k \\ -k & k \end{bmatrix}$$

The stiffness matrix for a beam or a truss element can be derived in a similar way.

2. *Principle of Minimum Potential Energy (MPE)*: The total potential energy Π_p of an elastic body is defined to be the sum of the strain energy U of the stresses causing elastic deformation and the work potential P of the external loads such as body forces.

$$\Pi_p = U + P$$

For *conservative* systems, where the work potential is *independent* of the path taken, and therefore only depends on the initial and final configurations, the *principle of minimum potential energy (MPE)* applies. Since elastic solids and structures are an example of a conservative system, we can utilize the MPE to our advantage. The fundamental idea behind the MPE is that of all possible states of a system, those satisfying equilibrium *extremize* the total potential energy. If the extremum is a *minimum*, the equilibrium state of the system is *stable.*

Mathematically, the equilibrium of a conservative system with n degrees of freedom is established when,

$$\frac{\partial \Pi_p}{\partial q_i} = 0, i = 1, 2, \ldots..n$$

For the spring element in Fig. 3,

$$U = \frac{1}{2}k(u_2 - u_1)^2$$

$$P = -(f_1 u_1 + f_2 u_2)$$

Therefore,

$$\Pi_p = \frac{1}{2}k(u_2 - u_1)^2 - (f_1 u_1 + f_2 u_2)$$

Applying the principle of MPE yields,

$$\frac{\partial \Pi_p}{\partial u_1} = -ku_2 + ku_1 - f_1 = 0$$

$$\frac{\partial \Pi_p}{\partial u_2} = ku_2 - ku_1 - f_2 = 0$$

Therefore,

$$f_1 = k(u_1 - u_2)$$

$$f_2 = k(u_2 - u_1)$$

In matrix form, this relationship is,

$$\begin{bmatrix} f_1 \\ f_2 \end{bmatrix} = \begin{bmatrix} k & -k \\ -k & k \end{bmatrix} \begin{bmatrix} u_1 \\ u_2 \end{bmatrix}$$

Therefore, the stiffness matrix for a spring element is,

$$[k] = \begin{bmatrix} k & -k \\ -k & k \end{bmatrix}$$

This is the same result we obtained using the direct method (equilibrium). For systems involving multiple springs, the principle of MPE is much more convenient to use than the direct method. We demonstrate this with the aid of the following example.

Consider the spring system delineated in Fig. 4. The task is to determine the nodal displacements u_1 and u_2. The total potential energy Π_p for this system is,

$$\Pi_p = \frac{1}{2}k_1u_1^2 + \frac{1}{2}k_2u_1^2 + \frac{1}{2}k_3(u_2 - u_1)^2 - F_1u_1 - F_2u_2$$

Applying the principle of MPE yields,

$$\frac{\partial \Pi_p}{\partial u_1} = k_1u_1 + k_2u_1 + k_3(u_1 - u_2) - F_1 = 0$$

$$\frac{\partial \Pi_p}{\partial u_2} = k_3(u_2 - u_1) - F_2 = 0$$

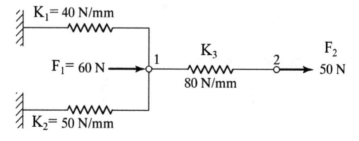

Fig. 4 Spring system with unknown nodal displacements

Substitution of numerical values gives us the following two equations,

$$170u_1 - 80u_2 = 60$$

$$80u_2 - 80u_1 = 50$$

Thus, $u_1 = 1.22mm$ and $u_2 = 1.85mm$. Note that the stiffness matrix for this system is,

$$[k] = \begin{bmatrix} k_1 + k_2 + k_3 & -k_3 \\ -k_3 & k_3 \end{bmatrix} = \begin{bmatrix} 170 & -80 \\ -80 & 80 \end{bmatrix}$$

However, this matrix is termed as the *global stiffness matrix*, since this is the stiffness matrix for the entire spring system, and not a single element. The dimensions of the global stiffness matrix are dictated by the number of degrees of freedom of a given system. The system in Fig. 4 has two degrees of freedom, hence the dimensions of the global stiffness matrix are 2×2.

3. *Weighted Residual methods*: The weighted residual methods are a class of methods that involve the solution of differential equations with the aid of test functions (scaled with appropriate coefficients) that are assumed to be an approximation of the exact solution. There are several methods that fall under this category, but the Galerkin's method is perhaps the most popular technique among this class.

The Galerkin's method is mathematically intensive, so we shall resort to demonstrating its applicability to the solution of a first-order ordinary differential equation (ODE). Many texts on the FEM usually start with an example of some sort, and then proceed to derive the stiffness matrix by applying the Galerkin's method, which we avoid explicating here for the sake of brevity and avoiding unnecessary complexity.
Consider the following ODE,

$$\frac{du}{dx} + 2u = 1, 0 \le x \le 1$$

The boundary condition for this ODE is $u(0) = 0$.
For the *exact* solution to this problem (i.e., analytical), we use the standard technique of finding an integrating factor. The integrating factor for this ODE is,

$$e^{\int 2dx} = e^{2x}$$

Multiplying both sides by the integrating factor and simplifying yields,

$$\frac{d(e^{2x}u)}{dx} = e^{2x}$$

$$e^{2x} u = \frac{1}{2} e^{2x} + C$$

Applying the boundary condition gives us,

$$C = -\frac{1}{2}$$

Therefore, the solution to this ODE is,

$$u = \frac{1}{2} \left(1 - e^{-2x} \right)$$

To compare this solution with the numerical solution, we evaluate u at $x = 0.5$. This gives us $u = 0.316$.

Next, we apply the Galerkin's method. The fundamental idea underpinning the weighted residual methods is to set the *residual* relative to a weighting function W_i, equal to zero. The *residual* is the error that arises when we seek an approximate solution $\tilde{u}$ to a differential equation. That is, if the differential equation is,

$$Lu = P$$

Then, the residual ϵ is

$$\epsilon = L\tilde{u} - P$$

The weighted residual, set to zero, can therefore be expressed as,

$$\int_V W_i (L\tilde{u} - P) dV = 0 \quad i = 1, 2, \ldots n$$

In the Galerkin's method, the weighting functions W_i are chosen to be the basis functions used to construct $\tilde{u}$. That is, if

$$\tilde{u} = \sum_{i=1}^{n} a_i \phi_i$$

Then,

$$W_i = \phi_i (i = 1, 2, \ldots n)$$

For the ODE under consideration, we let

$$\tilde{u} = a_1 + a_2 x + a_3 x^2$$

The assumed solution has to satisfy the boundary condition. Therefore, $a_1 = 0$. As a result, we are left with

$$\tilde{u} = a_2 x + a_3 x^2$$

The residual ϵ is

$$\epsilon = \frac{d\tilde{u}}{dx} + 2\tilde{u} - 1 = a_2(1 + 2x) + a_3(2x + x^2) - 1$$

The Galerkin's method requires that we compute the following integrals:

$$\int_0^1 x\big(a_2(1 + 2x) + a_3(2x + x^2) - 1\big)dx = 0$$

$$\int_0^1 x^2\big(a_2(1 + 2x) + a_3(2x + x^2) - 1\big)dx = 0$$

Solving these integrals gives us the following set of equations:

$$\frac{7}{6}a_2 + \frac{11}{12}a_3 = \frac{1}{2}$$

$$\frac{5}{6}a_2 + 0.7a_3 = \frac{1}{3}$$

Therefore, $a_2 = 0.8421$ and $a_3 = -0.5263$. This means that our approximate solution is,

$$\tilde{u} = 0.8421x - 0.5263x^2$$

To compare the numerical solution with the exact solution, we compute $\tilde{u}(0.5) = 0.289$ and observe that $u(0.5) = 0.316$. This is an error of 8.5% and could be improved by utilizing a cubic polynomial, which would increase the computational expense.

7 Von Mises Stress

The von Mises stress is used as a criterion for determining the onset of failure in ductile materials. The criterion is that the von Mises stress σ_{VM} should be less than the yield stress σ_Y of the material under consideration. Mathematically,

$$\sigma_{VM} \le \sigma_Y$$

The von Mises stress is given by the equation,

$$\sigma_{VM} = \sqrt{I_1^2 - 3I_2}$$

In the above equation, I_1 and I_2 are the first two invariants of the stress tensor, which we defined earlier in the chapter,

$$\boldsymbol{\sigma} = [\sigma_x, \sigma_y, \sigma_z, \tau_{yz}, \tau_{xz}, \tau_{xy}]^T$$

In terms of the *principal* stresses, denoted by σ_1, σ_2, and σ_3, I_1 and I_2 can be expressed as,

$$I_1 = \sigma_1 + \sigma_2 + \sigma_3$$

$$I_2 = \sigma_1\sigma_2 + \sigma_2\sigma_3 + \sigma_3\sigma_1$$

Using the above equations, we can derive the equation for the von Mises stress in terms of the three principal stresses,

$$\sigma_{VM} = \sqrt{(\sigma_1 + \sigma_2 + \sigma_3)^2 - 3(\sigma_1\sigma_2 + \sigma_2\sigma_3 + \sigma_3\sigma_1)}$$

However, the formulation found in most texts on solid mechanics and the FEM is given by

$$\sigma_{VM} = \frac{1}{\sqrt{2}}\sqrt{(\sigma_1 - \sigma_2)^2 + (\sigma_2 - \sigma_3)^2 + (\sigma_3 - \sigma_1)^2}$$

To get to the above formulation, we begin with the following step,

$$(\sigma_1 + \sigma_2 + \sigma_3)^2 = (\sigma_1 + \sigma_2)^2 + 2\sigma_3(\sigma_1 + \sigma_2) + \sigma_3^2$$

Subsequently,

$$(\sigma_1 + \sigma_2 + \sigma_3)^2 = \sigma_1^2 + 2\sigma_1\sigma_2 + \sigma_2^2 + 2\sigma_3\sigma_1 + 2\sigma_3\sigma_2 + \sigma_3^2$$

$$(\sigma_1 + \sigma_2 + \sigma_3)^2 - 3(\sigma_1\sigma_2 + \sigma_2\sigma_3 + \sigma_3\sigma_1) = \sigma_1^2 + 2\sigma_1\sigma_2 + \sigma_2^2$$
$$+ 2\sigma_3\sigma_1 + 2\sigma_3\sigma_2 + \sigma_3^2$$
$$- 3\sigma_1\sigma_2 - 3\sigma_2\sigma_3 - 3\sigma_3\sigma_1$$

Therefore,

$$\sigma_{VM} = \sqrt{\frac{2}{2}(\sigma_1^2 + 2\sigma_1\sigma_2 + \sigma_2^2 + 2\sigma_3\sigma_1 + 2\sigma_3\sigma_2 + \sigma_3^2 - 3\sigma_1\sigma_2 - 3\sigma_2\sigma_3 - 3\sigma_3\sigma_1)}$$

$$\sigma_{VM} = \frac{1}{\sqrt{2}}\sqrt{2(\sigma_1^2 + \sigma_2^2 + \sigma_3^2) - 2(\sigma_1\sigma_2 + \sigma_2\sigma_3 + \sigma_3\sigma_1)}$$

Finally,

$$\sigma_{VM} = \frac{1}{\sqrt{2}}\sqrt{(\sigma_1^2 - 2\sigma_1\sigma_2 + \sigma_2^2) + (\sigma_2^2 - 2\sigma_2\sigma_3 + \sigma_3^2) + (\sigma_3^2 - 2\sigma_3\sigma_1 + \sigma_1^2)}$$

$$\sigma_{VM} = \frac{1}{\sqrt{2}}\sqrt{(\sigma_1 - \sigma_2)^2 + (\sigma_2 - \sigma_3)^2 + (\sigma_3 - \sigma_1)^2}$$

8 Shape Functions and Local Coordinates

In this section, we introduce the concept of *shape functions* and *local coordinates*. Figure 5 shows piecewise linear basis functions $\phi_i(\mathbf{x})$, $i = 1$ to 5.

The functions are defined as,

$$\phi_i(\mathbf{x}) = \begin{bmatrix} 1 & at\ node\ i \\ 0 & if\ node\ j \neq i \end{bmatrix}$$

Utilizing these functions, we can write

$$u = \sum_{i=1}^{5} a_i(\mathbf{x})\phi_i(\mathbf{x})$$

These basis functions are referred to as *global shape functions*, and they represent a constant or a linear function exactly. Additionally, they satisfy the following property,

$$\sum_i \phi_i(\mathbf{x}) = 1$$

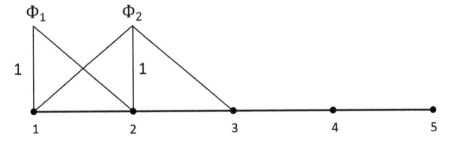

Fig. 5 Global shape functions (not all shown)

This is known as the *partition of unity* property. It is an extremely important concept, particularly when dealing with methods such as the extended finite element method (XFEM), which we will discuss later in this chapter.

It is often more convenient to analyze these basis functions from the perspective of a single element since the integration of potential energy is the sum of the potential energy over each element. For element 1, the basis functions ϕ_1 and ϕ_2 are *active* and these parts can be understood as the *shape functions* for this element.

For a general element e, if the left node is labeled 1 and the right node is labeled 2, the shape functions are N_1 and N_2, respectively. Since the shape functions are *linear* in this case, such elements are termed *linear finite elements*. The shape functions allow for linear interpolation of the displacement (or say, temperature) field *within* an element, as shown in Fig. 6.

The shape functions are defined on an interval $(-1, 1)$, with a mapping defined for each element as,

$$\zeta = \frac{2}{x_2 - x_1}(x - x_1) - 1$$

From the above relation, we see that when $x = x_1$, $\zeta = -1$. When $x = x_2$, $\zeta = +1$. We call ζ the *natural* or *intrinsic* coordinate. The shape functions are defined in terms of this coordinate system,

$$N_1(\zeta) = \frac{1 - \zeta}{2}$$

$$N_2(\zeta) = \frac{1 + \zeta}{2}$$

Using the shape functions as defined above, the *linear* displacement field *within* the element (or any physical quantity of interest for that matter) can be written in terms of nodal displacements q_1 and q_2 as,

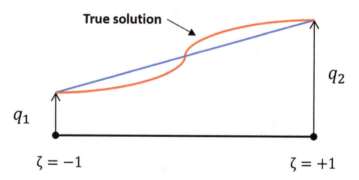

Fig. 6 Linear interpolation of the displacement field within an element

$$u = N_1 q_1 + N_2 q_2$$

In matrix form, this relationship is written as

$$u = \mathbf{Nq}$$

Here,

$$\mathbf{N} = \begin{bmatrix} N_1 & N_2 \end{bmatrix}$$

$$\mathbf{q} = \begin{bmatrix} q_1 & q_2 \end{bmatrix}^{\mathrm{T}}$$

Similar to the nodal displacement u, the transformation from x to ζ can be written as,

$$x = N_1 x_1 + N_2 x_2$$

This is because,

$$\zeta = \frac{2}{x_2 - x_1}(x - x_1) - 1$$

Therefore,

$$\frac{\zeta + 1}{2} = \frac{x - x_1}{x_2 - x_1}$$

$$x = x_1 + \frac{\zeta + 1}{2}(x_2 - x_1)$$

$$x = x_1 \left(1 - \frac{\zeta + 1}{2}\right) + \left(\frac{\zeta + 1}{2}\right) x_2$$

Finally,

$$x = \left(\frac{1 - \zeta}{2}\right) x_1 + \left(\frac{\zeta + 1}{2}\right) x_2 = N_1 x_1 + N_2 x_2$$

Since the nodal displacement u and the coordinate x are interpolated within the element using the *same* shape functions N_1 and N_2, this is known as the *isoparametric* formulation.

Additionally, since

$$u = \left(\frac{1 - \zeta}{2}\right) q_1 + \left(\frac{\zeta + 1}{2}\right) q_2$$

And,

$$\frac{du}{d\zeta} = \frac{-q_1 + q_2}{2}$$

Therefore,

$$\epsilon = \frac{du}{dx} = \frac{du}{d\zeta}\frac{d\zeta}{dx} = \left(\frac{-q_1 + q_2}{2}\right)\left(\frac{2}{x_2 - x_1}\right) = \frac{1}{x_2 - x_1}(-q_1 + q_2)$$

The relationship between strain ϵ and nodal displacements q can be written as,

$$\epsilon = \mathbf{Bq}$$

The matrix $\mathbf{B}$ is the *element strain–displacement matrix*,

$$\mathbf{B} = \frac{1}{x_2 - x_1}[-1\ 1]$$

To add to our discussion above, consider the following problem (Fig. 7):

For this problem, $q_1 = 0.02m$ and $q_1 = 0.025m$. The displacement at point P can be determined as follows,

$$\zeta = \frac{2}{x_2 - x_1}(x - x_1) - 1 = \frac{2}{23 - 15}(20 - 15) - 1 = 0.25$$

$$N_1(\zeta) = \frac{1 - \zeta}{2} = 0.375$$

$$N_2(\zeta) = \frac{1 + \zeta}{2} = 0.625$$

$$u = N_1 q_1 + N_2 q_2 = [0.375\ 0.625]\begin{bmatrix} 0.02 \\ 0.025 \end{bmatrix} = 23.125mm$$

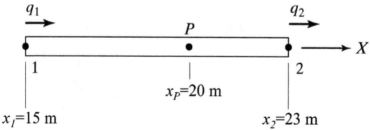

Fig. 7 A bar with known nodal displacements q

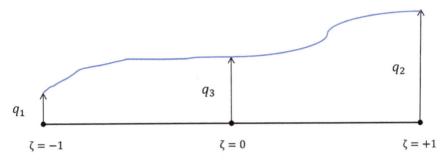

Fig. 8 Interpolation using quadratic shape functions

Note that $N_1 + N_2 = 1$. This holds true for **all** points within the element. To compute the strain in the element, utilize

$$\epsilon = \frac{1}{x_2 - x_1}(-q_1 + q_2) = \frac{1}{23 - 15}(-0.02 + 0.025) = 6.25 \times 10^{-4}$$

This concludes our discussion on *linear* elements. The use of quadratic interpolation obviously results in a higher accuracy when utilizing the FEM. Such elements are known as *quadratic finite elements*. A schematic is shown in Fig. 8.

A quadratic element has three nodes, with node 3 being in the middle of nodes 1 and 2 and therefore called an *internal node*. The coordinate transformation in this case is,

$$\zeta = \frac{2(x - x_3)}{x_2 - x_1}$$

The shape functions are again defined in terms of this coordinate system,

$$N_1(\zeta) = \frac{-1}{2}\zeta(1 - \zeta)$$

$$N_2(\zeta) = \frac{1}{2}\zeta(1 + \zeta)$$

$$N_3(\zeta) = (1 + \zeta)(1 - \zeta)$$

Similar to the linear shape functions, each of these functions has a value of 1 to its corresponding node, and zero otherwise. For instance, $N_1 = 1$ at node 1, and zero at nodes 2 and 3. The displacement field within the element can be written as,

$$u = N_1q_1 + N_2q_2 + N_3q_3$$

In matrix form, this relationship is written as

$$u = \mathbf{N}\mathbf{q}$$

Here,

$$\mathbf{N} = \begin{bmatrix} N_1 & N_2 & N_3 \end{bmatrix}$$

$$\mathbf{q} = \begin{bmatrix} q_1 & q_2 & q_3 \end{bmatrix}^{\mathrm{T}}$$

Similarly, the strain can be derived from

$$\epsilon = \frac{du}{dx} = \frac{du}{d\zeta}\frac{d\zeta}{dx} = \left(\frac{2}{x_2 - x_1}\right)\left[\frac{-1 + 2\zeta}{2}q_1 + \frac{1 + 2\zeta}{2}q_2 - 2\zeta q_3\right]$$

The relationship between strain ϵ and nodal displacements q can be written as,

$$\epsilon = \mathbf{B}\mathbf{q}$$

For quadratic shape functions, the matrix $\mathbf{B}$ is

$$\mathbf{B} = \frac{2}{x_2 - x_1}\left[\frac{-1+2\zeta}{2} \quad \frac{1+2\zeta}{2} \quad -2\zeta\right]$$

Comparing with the matrix $\mathbf{B}$ for linear shape functions,

$$\mathbf{B} = \frac{1}{x_2 - x_1}\begin{bmatrix} -1 & 1 \end{bmatrix}$$

We observe that, while the strain is **constant** within *linear finite elements*, it is **linear** within *quadratic finite elements*. By extension, these observations also apply to the stress within the element for each case.

This discussion can be extended to two and three-dimensional cases. In the two-dimensional case, it is common to study the *constant strain triangle* (CST) finite element. Since the aim of this chapter is to elucidate the fundamental concepts of the FEM in a succinct manner, we will conclude the discussion on the mathematical foundations of the FEM on this note.

9 Application of the FEM to T/EBCs

As discussed in the beginning of this chapter, TBCs are used in aircraft engines and stationary GTs for the protection of materials in the combustion chamber from extreme temperatures. Despite being successful, the potential of TBCs is hampered by their limited lifetime. The primary cause of this degradation is oxygen diffusing through the ceramic TBC and reacting with the aluminum in the bond coat to form an oxide layer, known as *thermally grown oxide* (TGO) in academic literature.

Coatings usually spill off when the TGO exceeds a critical thickness of the order of 1–10 μm due to the large stresses that develop within the TGO during growth and cooling. Consequently, predicting the lifetime of TBC systems is of paramount importance. In this context, a detailed overview of the application of the FEM to the study of TBCs by researchers Bäker and Seiler [12] is a good starting point for neophytes in the field. Therefore, we highlight the points made in this study that we believe are useful for readers to be aware of.

The article is divided into four sections, each of which discusses an aspect of the FEM as applied to TBC systems. In the introduction, Bäker and Seiler observe that finite element (FE) simulations of TBCs have allowed researchers to understand how stresses evolve in a coating system. The two main causes of stress in a TBC system are the coefficient of thermal expansion (CTE) mismatch and the growth of the TGO.

However, the evolution of the stress state is affected by an additional set of factors, such as sintering, crack propagation, the kinetics of TGO growth and the coating microstructure. New coating materials or coating strategies also require FE simulations for researchers to understand their stress evolution behavior and failure mechanisms. Thanks to the complex interplay of a multitude of factors influencing the stress state of a TBC system, practitioners have to make a number of modeling decisions when setting up a FE simulation. These include, but are not limited to, the dimensionality of the model, choice of material parameters and implementation of crack propagation.

Model dimensionality is an important parameter because FE simulations of TBC systems necessitate the use of a large number of elements, since the scale of the interface region is typically much smaller than the specimen thickness. (In atmospheric plasma-spraying (APS) coatings, a rough interface is required to ensure coating adhesion by mechanical interlocking.) To minimize computational costs, two-dimensional FE simulations are frequently used. However, crack propagation simulation shows different results in three-dimensional models than in two-dimensional cases.

Before discussing how crack propagation is simulated, it is worthwhile to highlight the process of *verification* and *validation* of FE models. Validation of a FE model can be done by making comparisons with analytical calculations (whenever possible). Verification of a FE simulation of a TBC system is much more complicated because of the actual conditions in a gas turbine. To circumvent this issue, comparisons with model experiments are usually performed. However, even using experimental results for verifying a FE model is difficult because of the complexity of the microstructure and uncertainties in the material behavior. If crack propagation is simulated, verification is done by comparisons with experimentally produced cracks. In general, experimentally verifying a FE model in the strict sense of the term would require the determination of all unknown parameters from one set of experiments and then predicting the outcomes for a different set of experiments under different conditions. The large number of parameters involved, however, makes this practically infeasible and therefore verification is done indirectly in almost all case studies.

When it comes to crack propagation simulation in TBC systems, Bäker and Seiler [12] and Wang et al. [13] quote three methods: (i) the extended finite element method (XFEM); (ii) the virtual crack closure technique (VCCT); and (iii) the cohesive zone model (CZM). Each of these methods is briefly discussed, primarily based on the review article by Wang et al. [13] along with a broad survey of literature on the topic.

First, we elucidate the XFEM. The XFEM falls under the purview of *partition of unity* methods (PUM), which allows for the use of enrichment functions, which if chosen to be discontinuous, allow for the simulation of discontinuities such as cracks. XFEM is useful because it avoids using a mesh coinciding with the discontinuous geometry. Additionally, remeshing is not required during the evolution of discontinuities [14]. The displacement approximation within XFEM is given by [15],

$$u(x) = \sum_{i \subset \Omega} N_i(x)u_i + \sum_{i \subset \Omega^1} \breve{N}_j \big(H(x) - H(x_j)\big)a_j$$
$$+ \sum_{k \subset \Omega^2} \breve{N}_k \sum_{\gamma} \big(B_\gamma(x) - B_\gamma(x_k)\big)b_{\gamma k}$$

In the above equation, N_i, $\breve{N}_j$ denote shape functions, u_i are nodal displacements, and a_j and $b_{\gamma k}$ are additional degrees of freedom for the displacement approximation. $H(x)$ is the Heaviside function, which takes a value of $+1$ at one side of the crack and -1 at the opposite side of the crack. $B_\gamma(x)$ are support functions to replicate the asymptotic field ahead of the crack tip.

The VCCT is a well-established technique that is used in the analysis of composite structures. It computes the strain energy release rate (SERR) at a crack tip. Fundamentally, it is linear elastic fracture mechanics (LEFM) analysis, and for a given fracture mode,

$$G = \frac{Fv}{2bd}$$

In the above equation, G is the SERR, F is the nodal force(s) at the crack tip, v is the nodal displacement(s) behind the crack tip, b is the width and d is the element length. The simulated crack can be located in a material layer or along the interface of two different materials or two adjacent layers. Multiple cracks can be modeled using the VCCT and can propagate simultaneously and independently from each other. Additionally, in special cases, some cracks can also merge into a single crack when propagating along the same interface. Due to these properties, the VCCT is well-suited to modeling the interface cracking behavior of TBCs [13].

The CZM is a numerical method that assumes the development of a cohesive damage zone at the crack tip when fracture initiates and begins to propagate. This technique employs a zero-thickness cohesive element with a traction–separation law, which depends on the application. The simplest case is the bilinear model, which assumes a linear elastic behavior before the damage onset and a linear softening once the degradation process evolves. There are other models reported in the literature, such as the exponential model. Similar to the VCCT, the crack plane needs to be defined as a damageable interface [16]. The major advantage of the CZM is that the mesh refinement does not need to take place during the simulation process, however, the computational cost is high, and the model requires many parameters to be tuned [13].

To conclude this discussion, modeling of TBC systems is a complex task, and several questions remain regarding the modeling of cracks and crack propagation along with the need for more experimental input regarding the material properties of individual layers and the behavior of coating systems for a better understanding of the stress evolution and failure mechanisms of TBCs. To maintain conciseness, we only provide a cursory overview of technical literature on this topic. Since the authors have considerable experience in the field of computational modeling and polymer-based materials, a relevant list of publications dealing with similar topics has been cited for interested readers [17–94].

10 Summary

In this chapter, we discussed the application of the finite element method (FEM) to the study of thermal/environmental barrier coatings (T/EBCs), in particular modeling crack propagation using the XFEM, the VCCT and the CZM. To prepare the reader for this discussion, we begin with an introduction to the topic of T/EBCs and their importance in improving the effectiveness and thermal efficiency of gas turbines, as utilized in the aerospace and power generation applications. Subsequently, we provide an overview of the FEM and discuss the mathematical concepts underpinning this widely utilized and highly effective numerical method. In the final section of this chapter, we take a cursory glance at the technical progress in the field and the challenges faced by researchers and practitioners in the FE modeling of TBC systems. We encourage interested readers to utilize this text as a springboard for further investigation into this fascinating research topic.

Acknowledgements Author "Akarsh Verma" is grateful for the academic support and funding provided by the University of Petroleum and Energy Studies (UPES)-SEED Grant program.

Conflict of Interest "There are no conflicts of interest to declare by the authors."

References

1. Vinuesa R, Lehmkuhl O, Lozano-Durán A, Rabault J (2022) Flow control in wings and discovery of novel approaches via deep reinforcement learning. Fluids 7:62. https://doi.org/10.3390/fluids7020062

2. Tejero-Martin D, Bennett C, Hussain T (2021) A review on environmental barrier coatings: history, current state of the art and future developments. J Eur Ceram Soc 41(3):1747–1768. https://doi.org/10.1016/j.jeurceramsoc.2020.10.057
3. Sandberg RD, Michelassi V (2022) Fluid dynamics of axial turbomachinery: blade-and stage-level simulations and models. Ann Rev Fluid Mech 54:255–285. https://doi.org/10.1146/ann urev-fluid-031221-105530
4. Lee KN (2020) Environmental barrier coatings. Coatings 10(6):512. https://doi.org/10.3390/coatings10060512
5. Padture NP (2016) Advanced structural ceramics in aerospace propulsion. Nat Mater 15(8):804–809. https://doi.org/10.1038/nmat4687
6. Durat M, Kapsiz M, Nart E, Ficici F, Parlak A (2012) The effects of coating materials in spark ignition engine design. Mater Des (1980–2015) 36:540–545. https://doi.org/10.1016/j.matdes.2011.11.053
7. Turner MJ, Clough RW, Martin HC, Topp LJ (1956) Stiffness and deflection analysis of complex structures. J Aeronaut Sci 23(9):805–824. https://doi.org/10.2514/8.3664
8. Clough RW (1960) The finite element method in plane stress analysis. Proceedings american society of civil engineers, 2nd conference on electronic computation, Pittsburgh, PA, 23, pp 345–378
9. Hrennikof A (1941) Solution of problems in elasticity by the framework method. J Appl Mech Trans ASME 8:169–175. https://doi.org/10.1115/1.4009129
10. Courant R (1943) Variational methods for the solution of problems of equilibrium and vibrations. Bull Am Math Soc 49:1–23
11. Zienkiewicz OC, Cheung YK (1967) The finite element method in structural and continuum mechanics. McGraw-Hill, London
12. Bäker M, Seiler P (2017) A guide to finite element simulations of thermal barrier coatings. J Therm Spray Tech 26:1146–1160. https://doi.org/10.1007/s11666-017-0592-z
13. Wang L, Li DC, Yang JS, Shao F, Zhong XH, Zhao HY, Yang K, Tao SY, Wang Y (2016) Modeling of thermal properties and failure of thermal barrier coatings with the use of finite element methods: a review. J Eur Ceram Soc 36(6):1313–1331. https://doi.org/10.1016/j.jeu rceramsoc.2015.12.038
14. Fan XL, Zhang WX, Wang TJ, Sun Q (2012) The effect of thermally grown oxide on multiple surface cracking in air plasma sprayed thermal barrier coating system. Surf Coat Technol 208:7–13. https://doi.org/10.1016/j.surfcoat.2012.06.074
15. Curiel-Sosa JL, Brighenti R, Moreno MS, Barbieri E (2015) Computational techniques for simulation of damage and failure in composite materials. In: Structural integrity and durability of advanced composites. Woodhead Publishing, pp 199–219. https://doi.org/10.1016/B978-0-08-100137-0.00008-0
16. Wang CH, Duong CN (2016) Damage tolerance and fatigue durability of scarf joints. Bonded Joints and Repairs to Composite Airframe Structures; Elsevier: Amsterdam, The Netherlands, 141–172. https://doi.org/10.1016/B978-0-12-417153-4.00006-2
17. Verma A, Parashar A (2018) Structural and chemical insights into thermal transport for strained functionalised graphene: a molecular dynamics study. Mater Res Express 5(11):115605
18. Verma A, Parashar A, Packirisamy M (2018) Tailoring the failure morphology of 2D bicrystalline graphene oxide. J Appl Phys 124(1):015102
19. Verma A, Parashar A (2018) Reactive force field based atomistic simulations to study fracture toughness of bicrystalline graphene functionalised with oxide groups. Diam Relat Mater 88:193–203
20. Singla V, Verma A, Parashar A (2018) A molecular dynamics based study to estimate the point defects formation energies in graphene containing STW defects. Mater Res Express 6(1):015606
21. Verma A, Parashar A, Packirisamy M (2019) Role of chemical adatoms in fracture mechanics of graphene nanolayer. Materials Today: Proceedings 11:920–924
22. Verma A, Parashar A, Packirisamy M (2018) Atomistic modeling of graphene/hexagonal boron nitride polymer nanocomposites: a review. Wiley Interdisciplinary Rev Comput Molecular Sci 8(3):e1346

23. Verma A, Kumar R, Parashar A (2019) Enhanced thermal transport across a bi-crystalline graphene–polymer interface: an atomistic approach. Phys Chem Chem Phys 21(11):6229–6237
24. Verma A, Gaur A, Singh VK (2017) Mechanical properties and microstructure of starch and sisal fiber biocomposite modified with epoxy resin mechanical properties and microstructure modified with epoxy resin 6(1):500–520
25. Deji R, Verma A, Kaur N, Choudhary BC, Sharma RK (2022) Density functional theory study of carbon monoxide adsorption on transition metal doped armchair graphene nanoribbon. Materials Today: Proc 54:771–776
26. Deji R, Verma A, Choudhary BC, Sharma RK (2022) New insights into NO adsorption on alkali metal and transition metal doped graphene nanoribbon surface: a DFT approach. J Mol Graph Model 111:108109
27. Verma A, Jain N, Parashar A, Singh VK, Sanjay MR, Siengchin S (2020) Lightweight graphene composite materials. In: Lightweight polymer composite structures, pp 1–20. CRC Press
28. Verma A, Parashar A (2020) Characterization of 2D nanomaterials for energy storage. In: Recent advances in theoretical, applied, computational and experimental mechanics, pp 221–226. Springer, Singapore
29. Deji R, Jyoti R, Verma A, Choudhary BC, Sharma RK (2022) A theoretical study of HCN adsorption and width effect on co-doped armchair graphene nanoribbon. Comput Theor Chem 1209:113592
30. Verma A, Parashar A, Packirisamy M (2019) Effect of grain boundaries on the interfacial behaviour of graphene-polyethylene nanocomposite. Appl Surf Sci 470:1085–1092
31. Deji R, Verma A, Kaur N, Choudhary BC, Sharma RK (2022) Adsorption chemistry of co-doped graphene nanoribbon and its derivatives towards carbon based gases for gas sensing applications: Quantum DFT investigation. Mater Sci Semicond Process 146:106670
32. Verma A, Jain N, Sethi SK (2022) Modeling and simulation of graphene-based composites. In: Innovations in graphene-based polymer composites, pp 167–198. Woodhead Publishing
33. Verma A, Parashar A, van Duin AC (2022) Graphene-reinforced polymeric membranes for water desalination and gas separation/barrier applications. In: Innovations in graphene-based polymer composites, pp 133–165. Woodhead Publishing
34. Verma A, Parashar A (2017) The effect of STW defects on the mechanical properties and fracture toughness of pristine and hydrogenated graphene. Phys Chem Chem Phys 19(24):16023–16037
35. Verma A, Parashar A (2018) Molecular dynamics based simulations to study failure morphology of hydroxyl and epoxide functionalised graphene. Comput Mater Sci 143:15–26
36. Verma A, Parashar A (2018) Molecular dynamics based simulations to study the fracture strength of monolayer graphene oxide. Nanotechnology 29(11):115706
37. Bharath KN, Madhu P, Gowda TY, Verma A, Sanjay MR, Siengchin S (2021) Mechanical and chemical properties evaluation of sheep wool fiber–reinforced vinylester and polyester composites. Mater Performance Characterization 10(1):99–109
38. Verma A, Parashar A, Jain N, Singh VK, Rangappa SM, Siengchin S (2020) Surface modification techniques for the preparation of different novel biofibers for composites. In: Biofibers and biopolymers for biocomposites, pp 1–34. Springer, Cham
39. Verma A, Singh VK (2016) Experimental investigations on thermal properties of coconut shell particles in DAP solution for use in green composite applications. J Mater Sci Eng 5(3):1000242
40. Singh K, Jain N, Verma A, Singh VK, Chauhan S (2020) Functionalized graphite–reinforced cross-linked poly (vinyl alcohol) nanocomposites for vibration isolator application: Morphology, mechanical, and thermal assessment. Mater Performance Characterization 9(1):215–230
41. Chaudhary A, Sharma S, Verma A (2022) Optimization of WEDM process parameters for machining of heat treated ASSAB'88 tool steel using Response surface methodology (RSM). Mater Today: Proc 50:917–922
42. Verma A, Jain N, Sanjay MR, Siengchin S, Viscoelastic properties of completely biodegradable polymer-based composites. In: Vibration and damping behavior of biocomposites, pp 173–188. CRC Press

43. Verma A, Jain N, Mishra RR (2022) Applications and drawbacks of epoxy/natural fiber composites. In: Handbook of epoxy/fiber composites, pp 1–15. Singapore: Springer Singapore

44. Lila MK, Verma A, Bhurat SS (2022) Impact behaviors of epoxy/synthetic fiber composites. In: Handbook of epoxy/fiber composites (pp 1–18). Singapore: Springer Singapore

45. Chaturvedi S, Verma A, Singh SK, Ogata S (2022) EAM inter-atomic potential—its implication on nickel, copper, and aluminum (and their alloys). In: Forcefields for atomistic-scale simulations: materials and applications (pp 133–156). Springer, Singapore

46. Verma A, Sharma S (2022) Atomistic simulations to study thermal effects and strain rate on mechanical and fracture properties of graphene like BC3. In: Forcefields for atomistic-scale simulations: materials and applications (pp 237–252). Springer, Singapore

47. Chaturvedi S, Verma A, Sethi SK, Ogata S (2022) Defect energy calculations of nickel, copper and aluminium (and their alloys): molecular dynamics approach. In: Forcefields for atomistic-scale simulations: materials and applications (pp 157–186). Springer, Singapore

48. Shankar U, Gogoi R, Sethi SK, Verma A (2022) Introduction to materials studio software for the atomistic-scale simulations. In: Forcefields for atomistic-scale simulations: materials and applications (pp 299–313). Springer, Singapore

49. Shankar U, Sethi SK, Verma A (2022) Forcefields and modeling of polymer coatings and nanocomposites. In: Forcefields for atomistic-scale simulations: materials and applications (pp 81–98). Springer, Singapore

50. Verma A, Ogata S (2022) computational modelling of deformation and failure of bone at molecular scale. In: Forcefields for atomistic-scale simulations: materials and applications (pp 253–268). Springer, Singapore

51. Homer ER, Verma A, Britton D, Johnson OK, Thompson GB (2022) Simulated migration behavior of metastable $\Sigma 3$ (11 8 5) incoherent twin grain boundaries. In: IOP conference series: materials science and engineering (Vol 1249, No 1, p. 012019)

52. Chaturvedi S, Verma A, Sethi SK, Rangappa SM, Siengchin S (2022) Stalk fibers (rice, wheat, barley, etc.) composites and applications. In: Plant fibers, their composites, and applications (pp 347–362). Woodhead Publishing

53. Arpitha GR, Verma A, MRS, Gorbatyuk S, Khan A, Sobahi TR, Asiri AM, Siengchin S (2022). Bio-composite film from corn starch based vetiver cellulose. J Nat Fibers 19(16):14634–14644

54. Thimmaiah SH, Narayanappa K, Thyavihalli Girijappa Y, Gulihonenahali Rajakumara A, Hemath M, Thiagamani SMK, Verma A (2022) An artificial neural network and Taguchi prediction on wear characteristics of Kenaf-Kevlar fabric reinforced hybrid polyester composites. Polym Compos 44(1):261–273

55. Kataria A, Verma A, Sethi SK, Ogata S (2022) Introduction to interatomic potentials/forcefields. In: Forcefields for atomistic-scale simulations: materials and applications (pp 21–49). Springer, Singapore

56. Kataria A, Verma A, Sanjay MR, Siengchin S (2022) Molecular modeling of 2D graphene grain boundaries: mechanical and fracture aspects. Mater Today Proc 52:2404–2408

57. Verma A, Ogata S (2023) Magnesium based alloys for reinforcing biopolymer composites and coatings: a critical overview on biomedical materials. Adv Indus Eng Polymer Res. https://doi.org/10.1016/j.aiepr.2023.01.002

58. Arpitha GR, Jain N, Verma A, Madhusudhan M (2022) Corncob bio-waste and boron nitride particles reinforced epoxy-based composites for lightweight applications: fabrication and characterization. Biomass Conver nd Biorefinery, pp 1–8. https://doi.org/10.1007/s13399-022-03717-1

59. Kumar G, Mishra RR, Verma A (2022) Introduction to molecular dynamics simulations. In: Forcefields for atomistic-scale simulations: materials and applications (pp 1–19). Springer, Singapore. https://doi.org/10.1007/978-981-19-3092-8_1

60. Kataria A, Chaturvedi S, Chaudhary V, Verma A, Rangappa Sanjay NJM, Siengchin S (2023) Cellulose fiber-reinforced composites—History of evolution, chemistry, and structure. In Cellulose fibre reinforced composites (pp 1–22). Woodhead Publishing. https://doi.org/10.1016/B978-0-323-90125-3.00012-4

61. Verma A, Jain N, Rastogi S, Dogra V, Sanjay SM, Siengchin S, Mansour R (2020) Mechanism, anti-corrosion protection and components of anti-corrosion polymer coatings. In: Polymer coatings (pp 53–66). CRC Press

62. Chaudhary A, Sharma S, Verma A (2022) WEDM machining of heat treated ASSAB'88 tool steel: a comprehensive experimental analysis. Mater Today: Proc 50:946–951

63. Bisht N, Verma A, Chauhan S, Singh VK (2021) Effect of functionalized silicon carbide nanoparticles as additive in cross-linked PVA based composites for vibration damping application. J Vinyl Add Tech 27(4):920–932

64. Verma A, Jain N, Parashar A, Singh VK, Sanjay MR, Siengchin S (2020) Design and modeling of lightweight polymer composite structures. In: Lightweight polymer composite structures (pp 193–224). CRC Press

65. Dogra V, Kishore C, Verma A, Rana AK, Gaur A (2021) Fabrication and experimental testing of hybrid composite material having biodegradable bagasse fiber in a modified epoxy resin: evaluation of mechanical and morphological behavior. Appl Sci Eng Progress 14(4):661–667

66. Verma A, Baurai K, Sanjay MR, Siengchin S (2020) Mechanical, microstructural, and thermal characterization insights of pyrolyzed carbon black from waste tires reinforced epoxy nanocomposites for coating application. Polym Compos 41(1):338–349

67. Verma A, Budiyal L, Sanjay MR, Siengchin S (2019) Processing and characterization analysis of pyrolyzed oil rubber (from waste tires)-epoxy polymer blend composite for lightweight structures and coatings applications. Polym Eng Sci 59(10):2041–2051

68. Verma A, Negi P, Singh VK (2019) Experimental analysis on carbon residuum transformed epoxy resin: Chicken feather fiber hybrid composite. Polym Compos 40(7):2690–2699

69. Verma A, Singh VK (2018) Mechanical, microstructural and thermal characterization of epoxy-based human hair–reinforced composites. J Test Eval 47(2):1193–1215

70. Jain N, Verma A, Ogata S, Sanjay MR, Siengchin S (2022) Application of machine learning in determining the mechanical properties of materials. In: Machine learning applied to composite materials (pp 99–113). Springer, Singapore

71. Verma A, Singh VK, Verma SK, Sharma A (2016) Human hair: a biodegradable composite fiber–a review. Int J Waste Resour 6(206):1000206

72. Kataria A, Verma A, Sanjay MR, Siengchin S, Jawaid M (2022) Physical, morphological, structural, thermal, and tensile properties of coir fibers. In: Coir fiber and its composites (pp 79–107). Woodhead Publishing

73. Jain N, Verma A, Singh VK (2019) Dynamic mechanical analysis and creep-recovery behaviour of polyvinyl alcohol based cross-linked biocomposite reinforced with basalt fiber. Mater Res Express 6(10):105373

74. Chaurasia A, Verma A, Parashar A, Mulik RS (2019) Experimental and computational studies to analyze the effect of h-BN nanosheets on mechanical behavior of h-BN/polyethylene nanocomposites. J Phys Chem C 123(32):20059–20070

75. Verma A, Negi P, Singh VK (2018) Physical and thermal characterization of chicken feather fiber and crumb rubber reformed epoxy resin hybrid composite. Adv Civil Eng Mater 7(1):538–557

76. Bharath KN, Madhu P, Gowda TG, Verma A, Sanjay MR, Siengchin S (2020) A novel approach for development of printed circuit board from biofiber based composites. Polym Compos 41(11):4550–4558

77. Verma A, Joshi K, Gaur A, Singh VK (2018) Starch-jute fiber hybrid biocomposite modified with an epoxy resin coating: fabrication and experimental characterization. J Mech Behav Mater 27(5–6):20182006

78. Verma A, Singh C, Singh VK, Jain N (2019) Fabrication and characterization of chitosan-coated sisal fiber–Phytagel modified soy protein-based green composite. J Compos Mater 53(18):2481–2504

79. Rastogi S, Verma A, Singh VK (2020) Experimental response of nonwoven waste cellulose fabric–reinforced epoxy composites for high toughness and coating applications. Mater Performance Characterization 9(1):151–172

80. Verma A, Zhang W, Van Duin AC (2021) ReaxFF reactive molecular dynamics simulations to study the interfacial dynamics between defective h-BN nanosheets and water nanodroplets. Phys Chem Chem Phys 23(18):10822–10834
81. Verma A, Jain N, Parashar A, Gaur A, Sanjay MR, Siengchin S (2021) Lifecycle assessment of thermoplastic and thermosetting bamboo composites. In: Bamboo fiber composites (pp 235–246). Springer, Singapore
82. Arpitha GR, Verma A, Sanjay MR, Siengchin S (2021) Preparation and experimental investigation on mechanical and tribological performance of hemp-glass fiber reinforced laminated composites for lightweight applications. Adv Civil Eng Mater 10(1):427–439
83. Verma A, Samant SS (2016) Inspection of hydrodynamic lubrication in infinitely long journal bearing with oscillating journal velocity. J Appl Mech Eng 5(3):1–7
84. Verma A, Jain N, Rangappa SM, Siengchin S, Jawaid M (2021) Natural fibers based bio-phenolic composites. In: Phenolic polymers based composite materials (pp 153–168). Springer, Singapore
85. Verma A, Parashar A, Singh SK, Jain N, Sanjay SM, Siengchin S (2020) Modeling and simulation in polymer coatings. In: Polymer coatings (pp 309–324). CRC Press
86. Verma A, Singh VK, Experimental characterization of modified epoxy resin assorted with almond shell particles. ESSENCE Int J Environ Rehab Conserv 7(1):36–44
87. Verma A (2022) A Perspective on the potential material candidate for railway sector applications: PVA based functionalized graphene reinforced composite. Appl Sci Eng Progress 15(2):5727–5727
88. Verma A, Jain N, Singh K, Singh VK, Rangappa SM, Siengchin S (2022) PVA-based blends and composites. In: Biodegradable polymers, blends and composites (pp 309–326). Woodhead Publishing
89. Raja S, Verma A, Rangappa SM, Siengchin S (2022) Development and experimental analysis of polymer based composite bipolar plate using Aquila Taguchi optimization: Design of experiments. Polym Compos 43(8):5522–5533
90. Prabhakaran S, Sharma S, Verma A, Rangappa SM, Siengchin S (2022) Mechanical, thermal, and acoustical studies on natural alternative material for partition walls: a novel experimental investigation. Polym Compos 43(7):4711–4720
91. Sethi SK, Gogoi R, Verma A, Manik G (2022) How can the geometry of a rough surface affect its wettability?-A coarse-grained simulation analysis. Prog Org Coat 172:107062
92. Chaturvedi S, Kataria A, Chaudhary V, Verma A, Mavinkere Rangappa Sanjay NJ, Siengchin S (2023) Bionanocomposites reinforced with cellulose fibers and agro-industrial wastes. In: Cellulose fibre reinforced composites (pp 1–22). Woodhead Publishing. https://doi.org/10.1016/B978-0-323-90125-3.00017-3
93. Verma A, Singh VK, Arif M (2016) Study of flame retardant and mechanical properties of coconut shell particles filled composite. Res Rev J Mater Sci 4(3):1–5
94. Arpitha GR, Mohit H, Madhu P, Verma* A (2023) Effect of sugarcane bagasse and alumina reinforcements on physical, mechanical, and thermal characteristics of epoxy composites using artificial neural networks and response surface methodology. Biomass Conv Biorefinery. https://doi.org/10.1007/s13399-023-03886-7

Chapter 10
Recent Progress in Computational Techniques in Various Coating Materials

Pankaj Kumar

1 Introduction

A substance applied to the substrate or materials is called a coating. It is used to change the surface properties, such as resistance to wear or chemical attack, color, gloss, and permeability, without changing the substrate or material properties. It is applied to most of the products across the industry: from precision engineering through electronics, construction to machinery, buildings, medical tools, and even food packaging. Coating materials may be protective, decorative, wear-resistant, and have other functional properties. There is a huge potential for coating materials in various applications. Some applications of the coating are in aircraft engines, automotive systems, boiler components, medical tools, power generation equipment, ships, and land-based and marine turbines. Nowadays, to study the behavior of coating, an experimental and computational method has been utilized. For a long time, the experimental method is already developed and studied for coating purposes. The computational technique is very new, very less utilized, and equally important to predict the properties and behavior of the coating materials. Several simulation methodologies like molecular dynamics (MD) simulation, quantum mechanics (QM) simulation, Monte Carlo (MC) simulation, and coarse-grained (CG) simulation have made tremendous progress in predicting various properties like interfacial energy, blend compatibilities, and surface characteristics [1–4]. These simulation techniques formulated using speed computing reduce the human efforts to get the materials' characteristics that were time-consuming and difficult to be synthesized experimentally. To use these simulation techniques, several researchers have successfully predicted the blend compatibility [5], adhesive and thin film characteristics [6–10], and thermodynamic properties [11, 12] at the molecular level. Several researchers and authors

P. Kumar (✉)
School of Nano Science and Technology, Indian Institute of Technology Kharagpur, West Bengal, India
e-mail: pankajbharti434@gmail.com

© The Author(s), under exclusive license to Springer Nature Singapore Pte Ltd. 2023   217
A. Verma et al. (eds.), *Coating Materials*, Materials Horizons: From Nature to Nanomaterials, https://doi.org/10.1007/978-981-99-3549-9_10

have successfully used various other models to simulate coating to predict the targeted properties of the materials.

2   Computational Techniques in Coating Materials

Computational techniques are used to theoretically construct the coating materials and to investigate the targeted properties and behavior. The computational technique can be classified into molecular dynamics (MD) simulation, quantum mechanics (QM) simulation, MC simulation, and coarse-grained (CG) simulation shown in Fig. 1. Recent studies of the computational methods for the coatings are described in the section below where the authors have applied computational techniques to support the experiments and develop coatings with improved functionalities and properties.

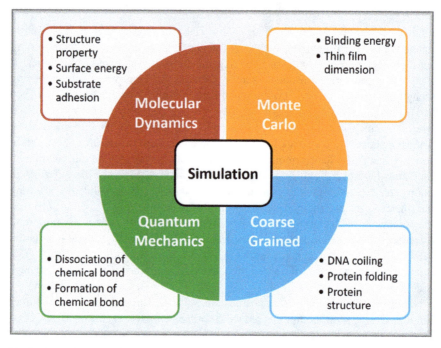

Fig. 1 Illustration of different simulation techniques with examples to investigate the coating materials

2.1 Molecular Dynamics (MD) Simulation

MD is a theoretical approach to studying the behavior of polymers and fillers in full atomic detail [13–15]. Recently, major improvements in simulation have been seen in speed, accuracy, and accessibility. The fundamental concept of an MD simulation is to numerically solve Newton's equations of motion for a system of interacting particles, where forces between the particles and their potential energies are frequently estimated using interatomic potentials or force fields from molecular mechanics. We can determine the impact of numerous molecular perturbations by contrasting simulations run under various settings. In the direction of coatings, several researchers have implemented this ability of MD simulation to predict water/oil contact angle, glass transition temperature, adhesion strength, transparency, etc. [13, 14, 16–27].

2.2 Quantum Mechanics (QM) Simulation

QM simulation is the most accurate and computationally expensive tool for simulation. It is fully reactive and can describe the electronic structure, polarization effects, bonding pattern charge transfer processes and atomic structure with all degrees of freedom and is widely used in physics, materials science, and biotechnology. QM simulations are confined to tiny system sizes of less than a thousand atoms and picoseconds in terms of time scales. This simulation technique creates clean realizations of specific systems, which makes it possible to precisely realize their properties and behaviors. The parameters of the system are precisely controlled over and broadly tunable which allows the changes of various parameters to be greatly disentangled.

2.3 Monte Carlo (MC) Simulation

MC simulation is an effective and productive tool to evaluate and analyze complex statistics using a model and to test the model to conclude the systems' behavior. A computer-based analytical tool called MC simulation uses statistical sampling methods to approximate a mathematical equation's or model's solution in a probabilistic manner. It uses random number sequences as inputs into a model which show results that are indications of the performance of the developed model. MC simulation technique is majorly applied in three problem classes: optimization, numerical integration, and generating draws from a probability distribution.

2.4   Coarse-Grained (CG) Simulation

A large class of computing methods is known as CG simulation. It is a successful tool to probe the length and time scales of systems. The main objective of CG simulation is to use course-grained (simplified) representation to simulate the behavior of complex systems. In multiscale applications, using a consistent approach to the development of compatible atomic-level force fields and CG is additionally significant.

3   Discussion

There are several classifications of the surface coating process, which are widely used in industries and laboratories. Several scientists and researchers used different coating processes to achieve the desired properties of the coating on a particular substrate or materials. The most common synthesis techniques are the cold spray process, chemical vapor deposition, dip coating, and spin coating. In the upcoming section, some authors have shown the effect of parameters on different coating processes on different substrates. Here, Joshi et al. [28] investigated how parameters affect coating during the cold spray process by MD simulation. In this study, it is shown that the bonding process cannot be adequately explained by finite element studies because during the cold spray process, bonding occurs at the molecular level. MD simulation method is a useful tool in such a situation. Additionally, it is employed to look at structure–property relations during the thermal spray process. They also have shown that stronger contact between the substrate and particles is produced by higher impact speeds [29]. Filali et al. [30] investigated the dip coating method numerically for Newtonian and Bingham fluids to analyze the free surface location under a variety of circumstances. The primary objective of the simulation is an assessment of the Finite Volume Method (FVM) combined with the Volume of Fluid (VOF) method incorporated in the commercial code (Fluent), while some data is also produced using the Finite Element Method (FEM) utilizing different commercial code (Polyflow). The numerical results for Newtonian and Bingham fluids were compared well with earlier simulations that were published in the literature. The effect of the yield stress for Bingham fluids and coating fluid characteristics in addition to surface withdrawal speed for Newtonian fluids are discussed. Javidi et al. [31] simulated a numerically controlled dip-coating method for depositing liquid films on a cylinder's substrate. For the deposited film thickness, the outcomes of the coating bath walls' proximity to the substrate are examined. In this paper, Carreau and power law models are performed to study the hydrodynamics of non-Newtonian liquid films. In a three-dimensional system, the free surface position is calculated by the volume of the fluid method, while the effects of surface tension, density, and viscosity are taken into consideration. Numerical simulation in an open-source CFD software package of OpenFOAM is used to analyzed the mass balance and momentum, alongside the constitutive equation for the dip coating process. Experimental data are used to validate numerical results. Additionally, taking into account the impacts of bath walls on the cylindrical substrate being removed from a coating bath, good agreement is established between the findings of numerical simulation and actual results for dip coating thickness. Kaushal et al. [32] investigated the gas phase behavior using the computational tool COMSOL in a specially designed chemical vapor deposition (CVD). They have mainly focused on the two major parts. The initial one is the precursor vaporization part and the final one is the coating part. Gas phase dynamics simulations in both zones of the CVD reactor were used to generate the optimized coating parameters. The overall outcome of heat flow, concentration profile, and fluid velocity revealed the ideal parameters for uniform coating in the CVD system. They have used non-corrosive and non-toxic silicon carbide precursors in CVD coating. At 900 °C, a uniform coating of SiC was produced on a zircaloy substrate.

4   Polymers

By using different polymers, superhydrophobic properties or the nature of the coating have been achieved. In this statement, the word 'nature' refers to the properties of the coating. Some of the authors optimized the polymer results with the help of a supercomputer to achieve the superhydrophobic nature of the coating. For steel pipeline protection, Al-Masoud et al. [33] have developed two novel nanocomposites based on polyaniline functionalized Zinc Oxide (ZnO) and Zinc Oxide–Silicon dioxide (ZnO–SiO$_2$) nanoparticles. The novelty of prepared films is that it has high conductivity and stability in acidic solution. MC simulation and molecular modeling using DFT demonstrated excellent agreement with experimental results. The binding energies of the compounds interface were showing a good match with the results. Sethi et al. [27] investigated the PVA-PDMS-based coating by employing MD simulation through Materials Studio software. All simulations were investigated using condensed-phase optimized molecular potentials for atomistic simulation studies (COMPASS) force field. In their study, the MD simulation method has been used to create a thin film structure or coating of PDMS with a PVA blend in different weight fractions. Figure 2 shows the images of the water droplet profiles through simulation and experiment [27]. The water contact angle estimated from the simulated results are showing good agreement with the experimental results.

MD simulations are adopted in industries to forecast the desired properties of coating before going for any product development and any material formulation. Similarly, Sethi et al. [18] investigated the impact of co-polymerizing PVAs with PDMS to develop an appropriate self-cleaning polymeric coating. The reported experimental data has been validated with the simulation data. The different wt.% of PVAc with PDMS-g-PVAc have been simulated, and cleaning properties have been evaluated through the estimation of substrate adhesion, surface energy, and transparency. The simulation observed reliability by showing good agreements of solubility parameter, simulated density, refractive index, and water contact angle.

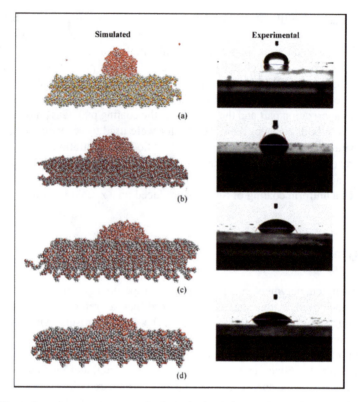

Fig. 2 Illustration of water contact angle through simulations and experimental technique of different percentages of hyd. PVAc. (Reprinted with permission from [27])

5 Steel

Ramezanzadeh et al. [34] showed the treatment of steel surfaces with Pr nanofilm. The results were showing increment adhesion bonds. The theoretical MD and QM and experimental methods were employed to study the corrosion protection, interfacial adhesion, and cathodic delamination rate of a coating. Structural characterization is used to analyze the morphology and chemical composition of the nanocoating. Furthermore, the increased coating adherence onto surface-treated steel was further supported by theoretical results from MD simulations of the metal/polymer interface carried out in the dry and wet environment. Theoretically, it was also analyzed that the primary force for the coating to adhere to untreated or treated steel substrates was interfacial electrostatic interactions. Mahidashti et al. [35] investigated the processing of steel substrate with cerium–lanthanum (Ce–La)-based conversion coatings. Experimental and computational studies were used to predict the impact of steel's surface treatment on the cathodic delamination, interfacial adhesion, and corrosion resistance of the epoxy coating. According to atomic force microscopy (AFM) measurements, the Ce–La and Ce–La-Post had the highest and lowest roughness levels, respectively. Ce–La-based conversion coating also produced superior outcomes in comparison to other samples. MD simulation results demonstrated that cross-linked epoxy resin adhered to the cerium oxide surface more firmly than it did to any of the three iron oxides in both dry and humid environments, further demonstrating the epoxy coating's enhanced corrosion–inhibition performance over surface-treated steel.

Hajimirza et al. [36] reported on the validation, fabrication, and computational design of a silicon-based multi-layer thin-film solar cell. It is mainly composed of amorphous silicon deposited on a layer of an aluminum back reflector and coated on top of transparent ITO conductive oxide. The entire framework was affixed on a glass substrate. To modify the geometry of the coating design, the authors have used numerical solvers of the electromagnetic equation along with constrained optimization techniques. The results predicted that photon absorptivity in the silicon-based thin film may be increased by up to 100% over the uncoated layer. Thin film deposition techniques were used to fabricate the proposed design. MC simulation was used to support the experimental results. Terasaka et al. [37] reported the sintering performance of yttria-stabilized zirconia (YSZ) coating for thermal barrier coatings (TBCs) using both experiment and simulation techniques. They have designed two different types of samples. The one with only a coating (free coating) and the other with a coating on the substrate (constrained coating). The sintering results in deformation and microstructure changes. The simulation using the finite element approach was used to conduct a calculation analysis. The simulation and experimental results were in good agreement. The experimental findings in terms of temperature dependency and time dependency may be replicated. The simulation results showed excellent agreement with the experimental results. Consequently, it is possible to predict the sintering behavior for the complex porous structures of TBCs by simulation research and experimental method. It could help in the creation of a system for coating delamination prediction. Wang et al. [38] explored the creation of an oxide cathode-stabilized, double-layer coating with a sulfide-based layer next to the thiophosphate electrolyte. Using MD simulation, they found numerous sulfide/halide couples that potentially outperform the known coating materials in ionic conductivity. Recently, several of the halides they mention have been determined to be novel candidates for electrolytes.

Kuthe et al. [39] investigated the possibility of structured, reinforced coatings to enhance wind turbine blade erosion prevention and stop surface deterioration over a longer period. The liquid impact of rain droplets on polymer coatings with internal structures is computed using a multiscale model. The local stress distribution in the structured coatings has been determined using the finite element model-based Coupled Eulerian–Lagrangian (CEL) approach combined with the sub-modeling approach. Computational experiments were used to investigate the stress concentration around the voids in the material and wave reflection on particle reinforcement in coatings. Xu et al. [40] tested a single particle to assess the performance of impact resistance of automobile coatings. The main pattern of coating-substrate failure in vehicle structures is impact-induced debonding. At the moment, efficiently and accurately simulating progressive debonding of automotive coating substrate interface under single-particle impact remains a difficult task. This work was mainly focused on creating a graphics processing unit (GPU)-based parallel computational framework to achieve the desired properties. By using the GPU parallel computing methodology, the created method's computational efficiency is significantly increased. Grohn et al. [41] simulated the coating process using computational fluid dynamics (CFD) coupled with the discrete element method (DEM) in a rotate rotor granulator. In DEM, the droplets are produced as a second particulate phase. The DEM model incorporates a liquid bridge model to account for the viscous forces and capillary during the wet contact of the particles into account. A model is created that takes into account both the growth of the particle and the drying of the liquid layer on the particles. When comparing the dry process simulation results to the simulations with liquid injection, it is clear that the liquid has a significant impact on particle dynamics. Pogrebnjak et al. [42] thoroughly investigated the phase changes of Al-doped NbN nanocomposite coatings, concentrating on the development of their microstructure and phase composition. To study the phase evolution of the Nb–Al–N system, a variety of structural characterization techniques and MD modeling are used. The roughness of the coating and the nanocomposite structures formed in the coatings decreased with increasing the Al concentration due to decreasing grain size.

Although we know that simulation techniques are a very useful tool for the decoration of coating and the study of the properties of the coating materials, however, some of the limitations of the computation have been shown in Table 1. These problems have been faced by researchers and scientists to develop the coating using computational techniques.

Table 1 Some limitations of the computational methods

Computational method	Limitations
MD simulation	• Results depend on the parameters of the selected interaction potential • It is computationally intensive and time-consuming, requiring supercomputers to study very large polymer systems [43–68] • It contains the redistribution of electrons approximately at best
QM calculation	• For sophisticated polymer sensing devices, it is too sluggish • The systems need too many approximations for simulation [69–72] • Studying very large polymer systems needs supercomputers, otherwise it is time-consuming and computationally intensive
Monte Carlo simulation	• It always provides us with statistical estimates of results instead of exact results • It is fairly complicated and can only be done with specialized software, which could be costly
Coarse-grained simulation	• A huge library with a broad spectrum of molecules is needed • It also needs a high-performance computer to execute the simulation model

6 Conclusion

Recent progress in computational techniques in various Coating materials has been discussed and the critical literature study has been presented with the impact of the parameters of the coatings. Several computational approaches and models have been developed and adopted to construct the coating on different substrate materials like steel and glass substrates. Different simulation techniques have been used as per the requirement of the model to desired properties of the coatings.

References

1. Es-haghi SS, Offenbach I, Debnath D, Weiss RA, Cakmak M (2017) Mechano-optical behavior of loosely crosslinked double-network hydrogels: modeling and real-time birefringence measurement during uniaxial extension. Polymer (Guildf) 115:239–245
2. Delgadillo-Velasco L et al (2018) Screening of commercial sorbents for the removal of phosphates from water and modeling by molecular simulation. J Mol Liq 262:443–450
3. Taheri MH, Mohammadpourfard M, Sadaghiani AK, Kosar A (2018) Wettability alterations and magnetic field effects on the nucleation of magnetic nanofluids: a molecular dynamics simulation. J Mol Liq 260:209–220
4. Saha S, Bhowmick AK (2017) Computer aided simulation of thermoplastic elastomer from poly (vinylidene fluoride)/hydrogenated nitrile rubber blend and its experimental verification. Polymer (Guildf) 112:402–413
5. Fu Y et al (2012) Molecular dynamics and mesoscopic dynamics simulations for prediction of miscibility in polypropylene/polyamide-11 blends. J Mol Struct 1012:113–118
6. Prathab B, Aminabhavi TM, Parthasarathi R, Manikandan P, Subramanian V (2006) Molecular modeling and atomistic simulation strategies to determine surface properties of perfluorinated homopolymers and their random copolymers. Polymer (Guildf) 47:6914–6924
7. Kumar N, Manik G (2016) Molecular dynamics simulations of polyvinyl acetate-perfluorooctane based anti-stain coatings. Polymer (Guildf) 100:194–205
8. Saha S, Bhowmick AK (2016) Computer simulation of thermoplastic elastomers from rubber-plastic blends and comparison with experiments. Polymer (Guildf) 103:233–242
9. Kasbe PS, Kumar N, Manik G (2017) A molecular simulation analysis of influence of ligno-sulphonate addition on properties of modified 2-ethyl hexyl acrylate/methyl methacrylate/acrylic acid based pressure sensitive adhesive. Int J Adhes Adhes 78:45–54
10. Yang JH, Bae YC (2018) Phase behaviors of polymer solutions using molecular simulation technique. 064902
11. Karavias F, Myers AL (1991) Isosteric heats of multicomponent adsorption: thermodynamics and computer simulations. Langmuir 7(12):3118–3126
12. Montmorillonite S (1995) Computer hydrates. 2734–2741
13. Agrawal G, Samal SK, Sethi SK, Manik G, Agrawal R (2019) Microgel/silica hybrid colloids: bioinspired synthesis and controlled release application. Polymer (Guildf) 178:121599
14. Gogoi R, Sethi SK, Manik G (2021) Surface functionalization and CNT coating induced improved interfacial interactions of carbon fiber with polypropylene matrix: a molecular dynamics study. Appl Surf Sci 539:148162
15. Shankar U, Oberoi D, Avasarala S, Ali S, Bandyopadhyay A (2019) Design and fabrication of a transparent, tough and UVC screening material as a substitute for glass substrate in display devices. J Mater Sci 54:6684–6698
16. Sethi SK, Manik G (2021) A combined theoretical and experimental investigation on the wettability of MWCNT filled PVAc-g-PDMS easy-clean coating. Prog Org Coat 151:106092

17. Sethi SK, Manik G (2018) Recent progress in super hydrophobic/hydrophilic self-cleaning surfaces for various industrial applications: a review. Polym Plast Technol Eng 57:1932–1952
18. Sethi SK, Shankar U, Manik G (2019) Fabrication and characterization of non-fluoro based transparent easy-clean coating formulations optimized from molecular dynamics simulation. Prog Org Coat 136:105306
19. Sethi S, Soni L, Manik G (2018) Blend compatibility studies using atomistic and mesoscale molecular dynamics simulations. In: 5th Annual international conference on materials science, metal & manufacturing—M3 2018. https://doi.org/10.5176/2251-1857_M318.35
20. Sethi SK, Gogoi R, Manik G (2021) Plastics in self-cleaning applications. Ref Mod Mater Sci Mater Eng. https://doi.org/10.1016/B978-0-12-820352-1.00113-9
21. Sethi SK, Kadian S, Manik G (2022) A review of recent progress in molecular dynamics and coarse-grain simulations assisted understanding of wettability. Arch Comp Methods Eng 2021(1):1–27
22. Sethi SK, Singh M, Manik G (2020) A multi-scale modeling and simulation study to investigate the effect of roughness of a surface on its self-cleaning performance. Mol Syst Des Eng 5:1277–1289. https://doi.org/10.1039/D0ME00068J
23. Sethi SK, Soni L, Manik G (2018) Component compatibility study of poly(dimethyl siloxane) with poly(vinyl acetate) of varying hydrolysis content: an atomistic and mesoscale simulation approach. J Mol Liq 272:73–83
24. Sethi SK, Manik G, Sahoo SK (2019) Fundamentals of superhydrophobic surfaces. In: Superhydrophobic polymer coatings. Elsevier, pp 3–29
25. Sethi SK, Gogoi R, Verma A, Manik G (2022) How can the geometry of a rough surface affect its wettability?–A coarse-grained simulation analysis. Prog Org Coat 172:107062
26. Maurya AK, Gogoi R, Sethi SK, Manik G (2021) A combined theoretical and experimental investigation of the valorization of mechanical and thermal properties of the fly ash-reinforced polypropylene hybrid composites. J Mater Sci 56:16976–16998
27. Sethi SK, Soni L, Shankar U, Chauhan RP, Manik G (2020) A molecular dynamics simulation study to investigate poly(vinyl acetate)-poly(dimethyl siloxane) based easy-clean coating: an insight into the surface behavior and substrate interaction. J Mol Struct 1202:127342
28. Santana A, Afonso PSLP, Zanin A, Wernke R (2018) Costing models for capacity optimization in Industry 4. 0: Trade-off between used capacity and operational efficiency. Procedia Manuf 26:190–197
29. Goel S, Haque N, Ratia V, Agrawal A, Stukowski A (2014) Atomistic investigation on the structure—property relationship during thermal spray nanoparticle impact. Comput Mater Sci 84:163–174
30. Filali A, Khezzar L, Mitsoulis E (2013) Some experiences with the numerical simulation of Newtonian and Bingham fluids in dip coating. Comput Fluids 82:110–121
31. Javidi M, Hrymak AN (2015) Numerical simulation of the dip-coating process with wall effects on the coating film thickness. J Coat Technol Res 12:843–853
32. Kaushal A, Prakash J, Dasgupta K, Chakravartty JK (2016) Simulation and experimental study of CVD process for low temperature nanocrystalline silicon carbide coating. Nucl Eng Des 303:122–131
33. Al-masoud MA, Khalaf MM, Mohamed IMA, Shalabi K, El-lateef HMA (2022) Computational, kinetic, and electrochemical studies of polyaniline functionalized ZnO and ZnO-SiO 2 nanoparticles as corrosion protection films on carbon steel in acidic sodium chloride solutions. J Ind Eng Chem 112:398–422
34. Ramezanzadeh M, Bahlakeh G, Sanaei Z, Ramezanzadeh B (2019) Interfacial adhesion and corrosion protection properties improvement of a polyester-melamine coating by deposition of a novel green praseodymium oxide nanofilm: a comprehensive experimental and computational study. J Ind Eng Chem 74:26–40
35. Mahidashti Z, Ramezanzadeh B, Bahlakeh G (2018) Screening the effect of chemical treatment of steel substrate by a composite cerium-lanthanum nanofilm on the adhesion and corrosion protection properties of a polyamide-cured epoxy coating; experimental and molecular dynamic simulations. Prog Org Coat 114:188–200

36. Hajimirza S, Howell JR (2014) Computational and experimental study of a multi-layer absorptivity enhanced thin film silicon solar cell. J Quant Spectrosc Radiat Transf 143:56–62
37. Terasaka S et al (2020) Experimental and computational study on sintering of ceramic coating layers with complex porous structures. J Am Ceram Soc 103:2035–2047
38. Wang C, Aoyagi K, Mueller T (2021) Computational design of double-layer cathode coatings in all-solid-state batteries. J Mater Chem A Mater 9:23206–23213
39. Kuthe N, Mahajan P, Ahmad S, Mishnaevsky L (2022) Engineered anti-erosion coating for wind turbine blade protection: computational analysis. Mater Today Commun 31:103362
40. Xu X, Zou C, Zang M, Chen S (2022) Development of a GPU parallel computational framework for impact debonding of coating–substrate interfaces. Thin-Walled Struct 175:109270
41. Grohn P, Lawall M, Oesau T, Heinrich S, Antonyuk S (2020) CFD-DEM simulation of a coating process in a fluidized bed rotor granulator. Processes 8
42. Pogrebnjak A et al (2017) Nanocomposite Nb-Al-N coatings: experimental and theoretical principles of phase transformations. J Alloys Compd 718:260–269
43. Verma A, Parashar A, Packirisamy M (2018) Atomistic modeling of graphene/hexagonal boron nitride polymer nanocomposites: a review. Wiley Interdiscipl Rev Comp Mol Sci 8(3):e1346
44. Verma A, Parashar A, Packirisamy M (2019) Effect of grain boundaries on the interfacial behaviour of graphene-polyethylene nanocomposite. Appl Surf Sci 470:1085–1092
45. Chaurasia A, Verma A, Parashar A, Mulik RS (2019) Experimental and computational studies to analyze the effect of h-BN nanosheets on mechanical behavior of h-BN/polyethylene nanocomposites. J Phys Chem C 123(32):20059–20070
46. Verma A, Kumar R, Parashar A (2019) Enhanced thermal transport across a bi-crystalline graphene–polymer interface: an atomistic approach. Phys Chem Chem Phys 21(11):6229–6237
47. Verma A, Parashar A (2020) Characterization of 2D nanomaterials for energy storage. In: Recent advances in theoretical, applied, computational and experimental mechanics. Springer, Singapore, pp 221–226
48. Kataria A, Verma A, Sanjay MR, Siengchin S (2022) Molecular modeling of 2D graphene grain boundaries: mechanical and fracture aspects. Mater Today Proc 52:2404–2408
49. Verma A, Parashar A, Singh SK, Jain N, Sanjay SM, Siengchin S (2020) Modeling and simulation in polymer coatings. In: Polymer coatings. CRC Press, pp 309–324
50. Verma A, Jain N, Sethi SK (2022) Modeling and simulation of graphene-based composites. In: Innovations in graphene-based polymer composites. Woodhead Publishing, pp 167–198
51. Chaturvedi S, Verma A, Singh SK, Ogata S (2022) EAM inter-atomic potential—its implication on nickel, copper, and aluminum (and their alloys). In: Force fields for atomistic-scale simulations: materials and applications. Springer, Singapore, pp 133–156
52. Verma A, Sharma S (2022) Atomistic simulations to study thermal effects and strain rate on mechanical and fracture properties of graphene like BC3. In: Forcefields for atomistic-scale simulations: materials and applications. Springer, Singapore, pp 237–252
53. Chaturvedi S, Verma A, Sethi SK, Ogata S (2022) Defect energy calculations of nickel, copper and aluminium (and their alloys): molecular dynamics approach. In: Forcefields for atomistic-scale simulations: materials and applications. Springer, Singapore, pp 157–186
54. Kataria A, Verma A, Sethi SK, Ogata S (2022) Introduction to interatomic potentials/forcefields. In: Forcefields for atomistic-scale simulations: materials and applications. Springer, Singapore, pp 21–49
55. Shankar U, Gogoi R, Sethi SK, Verma A (2022) Introduction to materials studio software for the atomistic-scale simulations. In: Forcefields for atomistic-scale simulations: materials and applications. Springer, Singapore, pp 299–313
56. Shankar U, Sethi SK, Verma A (2022) Forcefields and modeling of polymer coatings and nanocomposites. In: Forcefields for atomistic-scale simulations: materials and applications. Springer, Singapore, pp 81–98
57. Verma A, Ogata S (2022) Computational modelling of deformation and failure of bone at molecular scale. In: Forcefields for atomistic-scale simulations: materials and applications. Springer, Singapore, pp 253–268

58. Homer ER, Verma A, Britton D, Johnson OK, Thompson GB (2022) Simulated migration behavior of metastable Σ3 (11 8 5) incoherent twin grain boundaries. In: IOP conference series: materials science and engineering (vol. 1249, No. 1, p 012019). IOP Publishing

59. Verma A, Parashar A, van Duin AC (2022) Graphene-reinforced polymeric membranes for water desalination and gas separation/barrier applications. In: Innovations in graphene-based polymer composites. Woodhead Publishing, pp 133–165

60. Verma A, Parashar A (2017) The effect of STW defects on the mechanical properties and fracture toughness of pristine and hydrogenated graphene. Phys Chem Chem Phys 19(24):16023–16037

61. Verma A, Parashar A (2018) Molecular dynamics based simulations to study failure morphology of hydroxyl and epoxide functionalised graphene. Comput Mater Sci 143:15–26

62. Verma A, Parashar A (2018) Molecular dynamics based simulations to study the fracture strength of monolayer graphene oxide. Nanotechnology 29(11):115706

63. Verma A, Parashar A (2018) Structural and chemical insights into thermal transport for strained functionalised graphene: a molecular dynamics study. Mater Res Express 5(11):115605

64. Verma A, Parashar A, Packirisamy M (2018) Tailoring the failure morphology of 2D bicrystalline graphene oxide. J Appl Phys 124(1):015102

65. Verma A, Parashar A (2018) Reactive force field based atomistic simulations to study fracture toughness of bicrystalline graphene functionalised with oxide groups. Diam Relat Mater 88:193–203

66. Singla V, Verma A, Parashar A (2018) A molecular dynamics based study to estimate the point defects formation energies in graphene containing STW defects. Mater Res Express 6(1):015606

67. Verma A, Parashar A, Packirisamy M (2019) Role of chemical adatoms in fracture mechanics of graphene nanolayer. Mater Today Proc 11:920–924

68. Verma A, Zhang W, Van Duin AC (2021) ReaxFF reactive molecular dynamics simulations to study the interfacial dynamics between defective h-BN nanosheets and water nanodroplets. Phys Chem Chem Phys 23(18):10822–10834

69. Deji R, Verma A, Kaur N, Choudhary BC, Sharma RK (2022) Density functional theory study of carbon monoxide adsorption on transition metal doped armchair graphene nanoribbon. Mater Today Proc 54:771–776

70. Deji R, Verma A, Choudhary BC, Sharma RK (2022) New insights into no adsorption on alkali metal and transition metal doped graphene nanoribbon surface: a DFT approach. J Mol Graph Model 111:108109

71. Deji R, Jyoti R, Verma A, Choudhary BC, Sharma RK (2022) A theoretical study of HCN adsorption and width effect on co-doped armchair graphene nanoribbon. Comput Theor Chem 1209:113592

72. Deji R, Verma A, Kaur N, Choudhary BC, Sharma RK (2022) Adsorption chemistry of co-doped graphene nanoribbon and its derivatives towards carbon based gases for gas sensing applications: quantum DFT investigation. Mater Sci Semicond Process 146:106670

Chapter 11
Future and Challenges of Coating Materials

Md Mahamud Hasan Tusher⊙**, Alisan Imam, and Md. Shahidul Islam Shuvo**

1 Introduction

Recent industrial growth has made it necessary to design materials with specific properties that vary depending on the region for them to be acceptable for a variety of operating circumstances [1–3]. Each engineering part is initially exposed to thermal, friction, chemical, electrical, and electrochemical processes, each of which causes materials to deteriorate the portion while it is in use [4, 5]. The price of repairing or buying a new part for this system, as a result, is ultimately increased. The tribological and corrosion phenomena must be managed carefully in order to treat this deterioration [6, 7]. Therefore, using technological surfaces can aid in reducing the damage [8, 9]. The technology used in surface coating is used to improve the qualities of the substrate, such as corrosion resistance, wear performance, and hardness in order to effectively handle this issue [10–13].

Because surface coatings can be produced with tailored quantities, there are many perfect materials for a variety of applications. There have been more articles published on surface coatings in recent years as shown in Fig. 1.

Reviewing the history of coating technology is important before looking at the present state of the research and finding new advancements in surface coating materials because it emphasizes the achievements achieved since its conception and the growth of this industry.

It is crucial to mention that research on the characteristics of coatings has increased dramatically in recent years, with particular emphasis on corrosion resistance [15–17], mechanical behavior [18, 19] and wear behavior [20, 21]. This is because there are numerous methods for altering the characteristics of engineered materials' surfaces. Additionally, great progress has been made in the twenty-first century in

M. M. H. Tusher · A. Imam · Md. S. I. Shuvo (✉)
Department of Materials Science and Engineering, Rajshahi University of Engineering and Technology (RUET), Rajshahi 6204, Bangladesh
e-mail: shahidul1613021@gmail.com

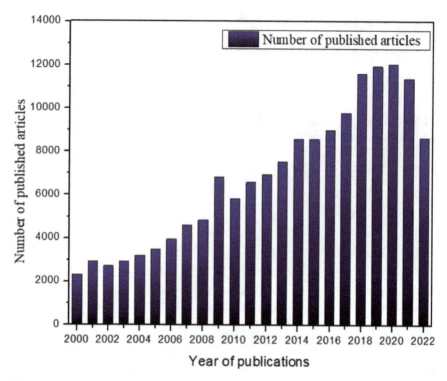

Fig. 1 Number of published documents on surface coatings, according to research website constellate [14]

the study of deposition materials for mono and multi-coatings [22–25]. Additionally, advancements in surface coating production methods have garnered a great deal of focus in the past ten years [26].

For example, metallic coatings are frequently used for surface coating in a variety of applications, such as the thermal management of sensors, electronics, nuclear power plants and concentrating photovoltaics. It has been demonstrated that both micro- and nano scale surfaces exhibit improved physical and mechanical performance by varying surface properties like area, wettability, and roughness. This chapter discusses the future and challenges of different types of surface coating that combine distinctive characteristics [27].

Monomers are repeating units of tiny molecules that make up polymers. Depending on how the monomers are bonded to one another, different polymers exhibit various properties. While some polymers, like bakelite and glass, are hard, others, including biopolymers, silk, and rubber, are soft. Numerous uses for polymers have been discovered, and both natural and synthetic polymers are widely used in contemporary culture.

Polymeric coatings are made of polymeric components. Solution casting procedures and dispersion are two common ways of applying polymeric coatings to various surfaces. Excellent adhesion and environmental protection are both offered by polymeric coatings. The way they are made prevents them from quickly peeling off the surface or deteriorating from moisture, heat, chemicals, or salt. They also stick effectively to the substrate. Additives, solvents, and thinners are only a few of the chemicals used in coating production. Coatings contain a variety of substances that serve a variety of purposes, such as functional qualities that are improved by additions [28].

A variety of industrial uses have led to the development and widespread use of ceramic coatings. Due to current technological advancements, ceramic coating has steadily evolved into the best substitute for any other coating. The coating lessens thermal deterioration, erosion, and corrosion.

Hydraulic turbine blades, gas turbine blades, boilers, and other structures have received composite coatings to increase their surface hardness, corrosion resistance, and wear resistance.

Since they can detect and react to environmental influences (such as temperature, light, pressure, etc.) and react in a predictable and reversible way, traditional smart coatings have undergone significant development and discussion recently. Advanced coatings can also be applied to materials to delay corrosion and prolong their usable lives as well as prevent rusting.

The study of organic coatings has advanced significantly in recent years, and new advances are opening up new research directions. New materials based on nanotechnology applications are dramatically changing the performance of organic coatings.

There are presently several industries where nanotechnology is commercially available. Since altering materials at the nanoscale can impart special thermal, chemical, surface, mechanical, and photophysical properties, nanotechnologies have been widely used in functional coatings to boost the functionalities necessary in urban built environments.

Coatings are now widely used in the marine, automotive, aerospace, aviation, membrane, food sector, and storage devices. They are additionally used in the encapsulation of electronic circuits, the oil and gas industry for corrosion control, the surface engineering industry for the production of multilayer coatings used in a variety of transportation and infrastructure applications, implantable devices, the textile industry for the creation of breathable clothing, and the production of materials for weatherproofing.

2 Metal Coating

Metal coatings are finishes that are put to metal in order to protect it and lessen deterioration. Due to exposure to the environment, a metal that is not protected may rust and corrode. An additional layer of defense is given by coating the metal [29]. It's more challenging to modify the qualities of metallic coatings than it is with polymer coatings. It is possible to build a micrometer- or nanometer-scale composite by combining a monomer, resin, filler, solvent, etc., with a phase encapsulating the inhibitor (for example, strontium chromate pigment) in a liquid state, giving the process a considerable deal of wiggle room. After being treated to a number of processes, the slurry eventually hardens at temperatures much below the polymer's Tg. In contrast, metallic coatings are often applied at high temperatures using a hot dip or spray procedure to provide a liquid metal. Such heating may negatively change the microstructure of the underlying alloy.

When the structure and features of coated materials are carefully examined, combined hardening may be implemented in reality when, for example, a coating provides greater wear- and heat resistance and a bulk-hardened base metal has a considerable margin of fracture resistance. This is achieved by utilizing all of the fundamental structure-controlling dislocation mechanisms, such as the formation of subgrains, polygons, cells, and granular microstructural barriers to provide bulk hardening; the separation of dispersed phases; the introduction of dissolved substitution and insertion atoms; and the increase in dislocation density to form unique surface properties. The finished composite will have excellent strength, durability, and dependability.

According to Fact. MR report, at a compound annual growth rate (CAGR) of 6.9% between 2022 and 2032, the worldwide market for metal coatings is expected to expand from an anticipated USD 15.3 Billion in 2022 to an estimated USD 30.8 Billion by 2032 [30].

2.1 Challenges

The lack of a defined system of testing procedures is the fundamental reason for the variations and incomparability of the results provided by studies of coated materials, as seen by the broad variety of experimental data from various laboratories. Guidelines for selecting and preparing samples, as well as the required number and geometry of specimens, as well as descriptions of recommended equipment and testing procedures, should be included in the newly established standards [31].

The development and widespread distribution of specialist devices for studying the properties of coatings and coated materials is, therefore, an undeniable requirement (porosity, bonding strength, internal stresses, thermal conductivity, etc.). These devices need to be reasonably priced, responsive to their environment, fully automated, and user-friendly.

3 Polymers Coating

Polymer coatings are finding more and more uses across an ever-expanding range of industries. Polymers provide a haven for their underlying hosts, whether in the form of a basic barrier coating or a complex nanotechnology-based composite [32].

Because they offer cheaper material and processing costs, as well as easier deposition procedures, but suffer from relatively poor hardness and decreased chemical and thermal durability [33]. Those coatings are primarily used to provide: (a) a surface resistant to all types of corrosion; (b) electrically insulating; (c) shock absorption; (d) non-sticking qualities; and (e) ornamentation [34]. Acrylic resins, polyurethanes, polyesters, alkyd resins, and epoxy resins are the most widely used polymers in the coatings industry [35]. The main areas of use of plastic coatings are biomedical [36], aeronautical [37], automobile [38], construction [39], food [40], agriculture [41], base metal production [42], chemical [43], and mining industries [44].

It has been hypothesized that micro- and nanoencapsulation active chemicals with edible polymer coatings may assist to regulate their release under specified circumstances, so protecting them from moisture, heat, or other harsh conditions and boosting their durability and viability. Nanolaminates may be applied to food by dipping it in a succession of solutions containing compounds that would be adsorbed to a meal's surface, or by spraying them directly onto the food. The functional compounds, such as antimicrobials, antibrowning agents, antioxidants, enzymes, flavorings, and colorants, in these nanolaminate coatings, might be created wholly from food-grade substances [45].

3.1 Challenges

Due to their softer nature, polymers may undergo plastic deformation at lower speeds than metals. In order to apply polymer coatings, standard coating methods, such as cold spray, need some modification. Low pressure cold spraying (LPCS) deposition of polyolefin powder has been explored by Xu et al., who use a cylindrical nozzle in place of the traditional convergent diverging nozzle and much lower velocities (up to 135 m s^{-1}) than those employed in the cold spray technique for metals.

4 Plastics Coatings

The usage of plastic coatings has skyrocketed in recent years as designers and engineers have realized their potential. Popularly used to prevent corrosion of various substrates, plastic-based coatings consist of a primer, an intermediate layer, and a topcoat made of different plastic compounds, which are applied sequentially to achieve a good corrosion-resistant property on the surfaces of metals, woods, ceramics, etc. [35].

The market for plastic coatings is divided into several submarkets based on the type of plastic (PVC, Polyurethane, Polyester, Fluoropolymer, Polyamide, Polyethylene, and Other Plastic Types), the form (Liquid and Powder), the end-user industry like building, aerospace, and defense, home appliance, automobile parts, electricals, and other end-user industries, and the geographical region like Europe, South America,

Asia–Pacific, North America, and Middle-East and Africa [46]. The building sector is one of the most prominent users of plastic coatings. Coatings consisting of polyvinyl chloride (PVC), for instance, plastisol, are often used for roofing [46]. The Institution of Civil Engineers (ICE) predicts that the building sector in China, India, and North America will boost the global construction market to a value of USD 8 trillion by 2030 [47]. According to Mordor Intelligence's report, during 2022 to 2027, the global plastic coatings market is predicted to expand at a CAGR of 5.2% and the fastest growing and the largest market will occur in the Asia Pacific region [46]. Over the projected period, Mordor Intelligence [46] analysts anticipate a rise in demand for plastic coatings in the Asia–Pacific region, particularly in China [46], as shown in Fig. 2.

According to the Allied Market Research report, with an expected compound annual growth rate (CAGR) of 5.1% between 2020 and 2027, the worldwide plastic coatings market is expected to expand from its 2019 valuation of $6.5 billion to $8.8 billion [48]. While epoxy resin was predicted to be the most in demand product, polyurethane was predicted to be the most lucrative. In terms of sales, the polyurethane category had a 25.3% share. This is due to its superior performance and ability to cure at lower baking temperatures, which make it a great solution for a variety of end-use sectors. By process, the dip coating sector accounted for 16.6% of the total revenue. This method is adaptable to situations including high-volume orders requiring quick delivery, various colors and finishes that may be manufactured rapidly and inexpensively, UV resistance, and others. According to their report, power coating is anticipated to be the most profitable market though spray coating is predicted to most demanded coating. In 2019, the automobile category accounted for 34.1% of sales by the end-use industry. This is attributable to the greater usage

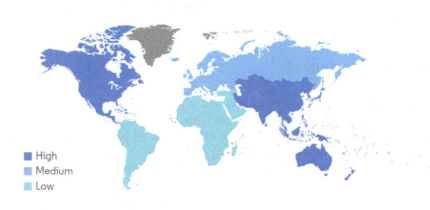

Source: Mordor Intelligence

Fig. 2 Plastic coatings market growth rate by region. (Photo is collected from mordor intelligence [46])

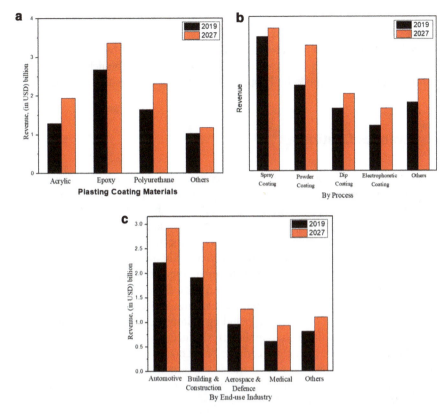

Fig. 3 Plastic coating materials **a** by type **b** by process **c** by end product industry. (Figures are recreated from collecting data from Allied Market Research [48])

of plastic coatings in automobile production and the growing concern about carbon emissions [48], which are shown in Fig. 3. Market research firm Vantage [49] Point estimates that by 2028, the global biopolymer coatings market would have grown from its 2021 worth of $1,062.4 billion to a staggering $2,440 billion [49].

4.1 Challenges

The increasing usage of various kinds of plastic materials is a significant challenge in the development of acceptable coating systems with improved performance and adherence without compromising the bulk qualities of the materials. Adhesion is difficult because many plastics absorb moisture from the air [50]. Baking and limiting the components' exposure to extreme temperatures and humidity before coating may help with this problem. Due to the temperature sensitivity of plastic materials, coatings with low curing temperatures and excellent storage durability are also highly demanded. Thus, it is necessary to use or invent novel cross-linking reactions [51]. It is preferable to design solvent free systems, such as appropriate aqueous or powder coating systems, because of solvent emission limits imposed by various clean air act requirements [52]. In this case, the substitution of hyper-branched binders for traditional binders may potentially open up novel paint composition options [51].

5 Ceramics Coatings

Ceramic coating has gradually become the best coating to replace any other coating as a result of today's technological advancement. It is the coating that reduces the thermal deterioration, erosion, and corrosion. It represents each of these outcomes in order to ensure the surface on which it was sprayed is safe. Though ceramic coating technology has applications throughout a broad spectrum. The coating's primary application is to safeguard structural materials and components. Its applications range from the construction industry to the automobile industry. To preserve the residual qualities, ceramic coating is generally applied to metal surfaces.

Because of its high anti-corrosion and abrasion resistance, ceramic coating is used extensively in the steel sector, including chimneys, bearings, ceramic pumps, and bricks. Due to qualities including shock resistance, corrosion resistance, hardness and wear resistance, the market for ceramic-based coatings is expected to grow until 2032. Some of the influential market trends for ceramic coatings include the main purpose of such coatings in turbine engines. As more cars are sold, so does the demand for ceramic coating, which is primarily used for aesthetic and protective purposes. Consumers are increasingly aware of vehicle safeguards against UV rays, discoloration, etching, and other factors that are expected to drive demand for ceramic coating [53].

Initially, the primary goal of ceramic coating development was high temperature applications to safeguard low-carbon steel for industrial uses. Coatings have become more useful in the automobiles, aircraft, and marine industries in recent decades. Its application in the automotive sector is due to its superior thermal, corrosion, and wear and tear resistance. As a result, ceramic coatings are a preferred choice for protecting automotive paint [54].

The ceramic-based coating market is anticipated to increase at a CAGR of 7.6 percent from 2022 to 2032 and reach a value of US$ 21.82 billion. The market has 10.49 billion dollars in revenue [53] (Table 1).

Table 1 Year based ceramic coating market value

Attributes	Details
Ceramic coating market CAGR (2022–2032)	7.6%
Value of the ceramic coatings market (2022)	US$ 10.49 Bn
Value of the ceramic coatings market (2032)	US$ 21.82 Bn

5.1   Challenges

The two key barriers to the adoption of ceramic-based protective coatings are significantly more expensive, and there are questions regarding their capacity to repeat features like stickiness, rigidity, structure, hardness and thickness.

Hard coatings have hyper-dense patterns, so there is less surface area available to transmit abrasion strain when a scratch happens. In times of pressure increases again, it concentrates in that tiny patch of the surface, causing a scratch to form. Such issues are most likely to slow the growth of the market for ceramic coatings within the expected span of time. Furthermore, a nonstick coating protects cookware from scratching or damage. Polytetrafluoroethylene (PTFE), as well as perfluorooctanoic acid, as well as perfluorooctanoic acid, are components of nonstick coatings (PFOA). IARC classified PFOA as "possibly human carcinogenic." As a result, the carcinogenic properties of PFOA are anticipated to limit the growth of the ceramic coating market [53].

Water spotting, etching, and coating premature failure are also some challenges that need to be overcome to establish the ceramic coating for future advanced applications [55].

6   Composite Coatings

A substrate, such as steel, concrete, or any other material, is covered with a sequence of protective layers known as composite coatings. A minimum of two ingredients must be combined to defend against substrate corrosion when employing composite coatings. Typically, these materials combine to create two or more layers of synthetic resins called polyurethane and epoxy [56].

The oil and gas sector's alluring growth prospects are anticipated to fuel the market for composite coatings in the near future. It is projected that technological advancements in procedures like pipe and downhole tubing systems will boost the market for composite coatings. Aircraft airframes are covered with composite coatings [57].

A prospective contender for the next-generation protecting system, composite coating is gaining popularity due to its unique protective qualities. ZnO, SiO2, TiO2, graphene, and CNT form the foundation of the lately well-liked composite nanocoating system. It is obvious that adding modified nanoparticles to a regular coating system might greatly enhance the coating's protective qualities, typically its corrosion-, UV-, and oxidation-resistance performance, and even its ability to self-clean. However, introducing nanoparticles in the right quantity is essential to maintaining their protective characteristics. To increase the corrosion resistance of a carbon-based composite coating system, either CNT or graphene modification should be used, or the coating system should be composited with other coating systems.

The scientific community and industries have utilized nanocoating, which has a broad range of real features, in a strategic way. The goal of furthering this field's protective capabilities should be to combine them with other functionality, such as self-cleaning, self-healing, UV resistance, antibacterial qualities, thermal resistance, and color-changing in a single nanocoating. The creation of a novel nanocoating method is also crucial, as are studies on the relationship between size minimization, optimization of the modification and amount, uniform dispersion of the nanoparticles, and the performance of nano-coatings. Also becoming more popular is the use of many pigments or organic nano-pigments in a single coating system to improve the coating's qualities.

6.1   Challenges

The structural integrity of the cladding and other components is severely compromised over an extended length of time by exposure to the elements. Delamination of the laminated material frequently causes this to begin. The substrate material or base coat no longer adheres to the adhesive. Debonding is the result of the layers of the composite separating.

Debonding eventually exposes fasteners, sheet ends, and other parts to the elements. The use of lower grade, less expensive coatings is the primary cause of premature delamination and debonding. Debonding also happens when the mechanical forces that keep the bond in place are disrupted. Usually, this happens as a result of an impact, repeated cyclic pressures, vibration, or an outside attack like UV rays [56].

7   Smart Surface Coatings

Traditional smart coatings have had a lot of development and discussion recently since they can sense and respond to an environmental influence (such as light, pressure, temperature, etc.) and react in a predictable and reversible way. Additionally, advanced coatings can be used to prevent rusting as well lengthen the useful life of materials in corrosive conditions. By lowering maintenance costs and monitoring intervals, they hence improve material efficiency. However, it is difficult to build a smart composite coating in applications where the outer surface is exposed to high temperatures because the coating and the substrate have sharp edges in their structural relationship.

Sharp edges concentrate tension, which can result in coating failure in smart conventional coatings. In order to avoid sharp edges, one solution to this issue is to use smart coatings with surface properties between the deposition medium and the substrate. Therefore, it is anticipated that one of the most significant future study fields for these coatings would be smart surface coatings. These coatings may help surface coatings spread to many applications and may even rule the industry. Aside from that, with the world moving toward intelligent technology. For biomedical issues like implants in the human body, smart surface coatings may be the answer. The objective is to offer the necessary elements as soon as contact is made and to deal with the challenges that are anticipated over time [58].

8 Organic Graded Coatings

Both scientifically and technologically, organic coatings have made impressive strides in recent years, and new developments are opening up new avenues for their study [59]. The advancement of multifunctional materials combining various properties, such as corrosion protection, aesthetic purposes, hydrophobic properties, and self-healing capability, is made possible by nanotechnology and surface science [60]. The performance of organic coatings is being significantly altered by new materials based on applications of nanotechnology (nanostructured polymeric matrixes, nano-pigments, and new surface pretreatments enhancing the physical and chemical stability of the interfaces) [61–63]. Modern organic coatings frequently need to support additional roles in addition to more conventional ones like aesthetic or corrosion protection, such as antibacterial, self-healing, and tribological properties [64].

Recent developments in experimental methods (electrochemical approaches, optical and scanning electron spectroscopy, chemical surface analysis, thermal analysis, etc.) used to study organic coatings offer a potent tool for the study and advancement of science in this field [65]. Environmental concerns are additional forces pushing innovation in the organic coatings sector. New materials are being developed in order to create new systems that combine high environmental sustainability with advanced performance, anticipating upcoming regulatory requirements. This lens is persuaded to view developments in organic coatings—the "skin of materials"—as one of the most fascinating and forward-thinking areas of innovation in the field of material science [66].

9 Nanostructure-Tailored Graded Coatings

Research on nanomaterials is going on in more than 30 different countries. The United States is the most active, with about a quarter of all publications being about nanotechnology. Japan, China, France, the United Kingdom, and Russia are next in line. 70% of all scientific papers on nanotechnology in the world come from just these countries. Nanotechnologies have been widely incorporated into functional coatings due to the fact that manipulating materials at the nanoscale can impart unique thermal, chemical, surface, mechanical, and photophysical properties that can be exploited to increase the functionalities required in urban built environments [67]. Coatings nanotechnology is now offered commercially in a variety of fields. These are: Self-cleaning coatings, water sheeting coatings, anti-corrosion

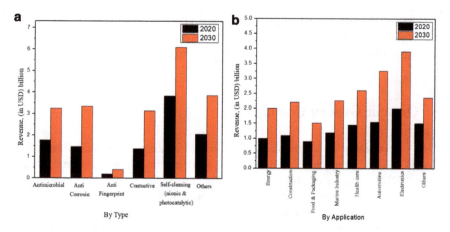

Fig. 4 Nano-coating materials **a** by type **b** by applications (Figures are recreated from collecting data from Allied Market Research [69])

coatings, ultraviolet (UV) light protection coatings, anti-graffiti coatings, insulative nano-coatings, water beading coatings, and depolluting coatings [68]. The use of nano-based coatings in that sector is increasing day by day. According to the Allied Market Research report, the worldwide market for nano-coatings was worth $10.7 billion in 2020. By 2030, the market is projected to be worth $20.1 billion, growing at a CAGR (compound annual growth rate) of 6.7% from 2021 to 2030 [69]. By type, the market share for self-cleaning (bionic and photocatalytic) systems is anticipated to continue growing over the next several years. The rise may be traced back to the rising popularity of self-cleaning nano-coatings in the construction, maritime, and residential (mostly bathroom) markets. By application, it is anticipated that the electronics industry would reap the greatest financial benefits from using nano-coatings. Nano-coatings are becoming more in demand as a result of growing consumer demand for electronic gadgets and their widespread usage in electronics and optical applications [69], are shown in Fig. 4. According to the Fortune Business Insights report, with a CAGR of 22.7% between 2020 and 2028, the worldwide nano-coatings market is expected to grow from a 2020 valuation of USD 7,825.3 mln to a 2028 valuation of USD 39,869.2 million [70].

9.1 Challenges

Electrical Performance: Testing for dielectric strength and insulation resistance at the nano-thickness application rate might be concerning. When compared to conventional coatings, this pace of application is sometimes insufficient [71].

Abrasion: Scratch resistance, as tested by Taber testing (abrasion), might be a serious concern at nano-thickness. This means that the boards may perform effectively during production and in the short term, but that they may fail in the long run in the demanding settings common in industrial and automotive sectors.

Exposure to Water vs. Vapor: The outstanding ability in repelling surface liquids is a consequence of the low surface energy of many nano-coatings. Resistance to vapors is anticipated to become considerably more challenging when applied at the nanoscale level. This is why it's often necessary to apply nano-coatings at thicknesses considerably closer than those typically used [71].

Solvent Protection: When nano-coatings are applied at nano-thicknesses, substantial concerns and possible hazards are raised about exposure to solvent and gas.

Dendritic Growth: There are worries regarding inadequate protection against nano-coatings, because of the very tiny application thicknesses. Dendritic development is mostly caused by nano-thick coatings, which might be an issue since water vapor can penetrate them and they can interact with active species on board (left over from flux residues).

In conclusion, several major issues arise when their nano-thicknesses are used, including abrasion and wear resistance, moisture and solvent vapor resistance, and others. Coatings from this family tend to be costly and may need major adjustments to the machinery used in their application.

10 Trending Materials

Before it arrives, the year 2023, which marks the start of a new decade and ushers in changes to the status quo, will have occurred:

- In that year, work is expected to be finished on the 11-mile underwater Fehmarn Belt Fixed Link that connects Germany and Denmark.
- A number of nations, including India, Malaysia, Chile, Trinidad & Tobago, Oman, Pakistan, and the Philippines, have vowed to achieve economic development by that time.
- By the start of the next decade, India claims, it will launch a manned space mission to Mars.
- Furthermore, according to the Russian Federal Space Agency, helium-3 will already be mined from the moon by that time.

However, what may be anticipated in the realm of coatings and finishing?

If you believe the industry experts consulted by Products Finishing magazine to forecast what would propel the plating, painting, powder coating, anodizing, electrocoating, and vacuum coating sectors at the turn of the next decade there are plenty of factors. For example.

According to Mark Main, product manager for BASF Automotive Coatings, the industry will put its attention on new technologies, innovations, products, and processes that help the client be more effective and ultimately more successful.

"By more efficient, we mean 'no coats' and 'partial coats' with noticeably less re-work for the end-user," adds Main." Our consumers request lower costs per component and less maintenance, as well as goods that are simpler to use and more ecologically friendly, along with the elimination of non-value-added tasks.

According to economists, the worldwide paint and coatings industry had a 2010 revenue of $99.5 billion and is projected to expand at a compound annual growth rate of 3.2 percent through 2015 to a 2010 revenue of $116.4 billion.

According to Brough Richey and Mary Burch, who wrote about the sector in the book Polymer Dispersions and Their Industrial Applications, "the most significant movement in the decorative and protective coatings markets has been to more environmentally acceptable coating materials." A significant outcome of this trend is the transition away from older, solvent-based technologies and toward more modern ones based on water-borne emulsion polymers, high solids coatings, and powder coatings [72].

Because of recent developments in numerous sectors, including logistics, construction, manufacturing, automotive, and others, along with commercial advancements, it is anticipated that Surface coatings made of popular materials including 2D materials, porous materials, and boron nitride nanotubes will become more popular. The development of porous coatings having a surface gradient in microstructure, porosity, and/or content aims to improve biological and mechanical properties. The benefits of surface porous coatings over uniform or homogeneous porous coatings and dense surface materials are numerous, which is driving demand for cutting-edge porous surface coatings [73].

For example, porous coatings used in biomedical applications, have been shown they can support the formation of artificial bone inside the body, while having poor mechanical performance [74].

Two-dimensional (2D) materials, which are simply described as crystalline substances made of single- or few-layer atoms, are another example of a trending substance that is projected to be used in the near future for putting surface coatings [75]. Hexagonal boron nitride and graphene are just two examples of 2D materials. Other examples include 2D metal carbides and nitrides (MXenes), transition metal dichalcogenides, 2D mono-elemental nanomaterials (Xenes) and phosphorene [73, 75].

11 Computational Coatings

Coating optimization is still mostly done by inspired trial and error methods. These coating technologies are costly, especially for complex machining. Experiments must be performed to determine whether the modified can withstand mechanical and thermal loads, substrate-coating connection, and materials adherence [76]. To develop new coating systems, experimental development should be combined with numerical simulations to investigate the influence of external loads on the coatings' stability, substrate-coating connection, and materials adherence, which depends on the tool's deformation state, stress distribution, and adhesion stress ratio in the surrounding material boundaries [77, 78]. These are computable using three-dimensional finite element simulation. This process is called computational coating [77]. Combining computational and experimental methods is a way to tackle issues that cannot be addressed by theoretical models alone. Computational tools may then utilize this refined model to analyze structure and characteristics, and optimize over a wide range of permutations, all thanks to the work of experimental tools that can enhance the model by factoring in fresh data. Besides the computational techniques have already been utilized to predict several other properties such as glass transition temperature, polymer-filler interaction, transparency, and polymer-substrate interactions [79–88].

Basically, computational coatings are a combination of computer science, applied mathematical and numerical methods, and different types of coatings engineering techniques that use computer-aided design (CAD) and computer-aided manufacturing (CAM) to put a coating on a surface, as shown in Fig. 5. By using this method, the coating's thickness, composition, and other attributes may be managed with pinpoint accuracy. It has been used in fields as diverse as automobile manufacturing, aeronautical engineering, and medicine [89]. This technique is used to generate coatings that are more durable, efficient, and cost-effective than conventional coatings.

Developing unique software to link experimental and computational tools is vital to improving the Factory of Future's Design Stage. Various research organizations and corporations produce these software technologies. Simpleware toolkit, 3D pixels, 6Ab initio, and Monte Carlo simulations are commonly used to examine the electronic, and mechanical structure, intermolecular forces, and different properties that determine the coating material's structure and dynamic characteristics [90]. These strategies help to understand coating material behavior. It allows coated tool users to determine the acceptability of various coatings for their purposes and manufacturers to design new coating systems using simulations. This reduces the cost of developing application-specific coatings [77].

The computational coating may be used to develop self-clean coatings [91–103], anti-corrosion and anti-fouling coatings, fire retardant or fire-resistive coatings, metal coatings, polymers coatings, plastics coatings, protective coatings, functional coatings, and decorative coatings, among others. In combination of computational techniques with conventional techniques, those coatings show unique, controllable properties.

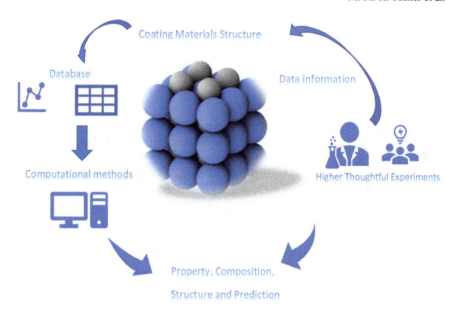

Fig. 5 The fundamental computational methods for coating materials structure modeling and property prediction

11.1 Limitations of Computational Coatings

The computational coating method is frequently used to produce coatings that outperform conventional coatings in terms of efficiency and durability. But there are also certain restrictions. These include, residual stress effects, equipment cost and quality control effects, gradient structure analysis, etc., which are discussed below.

11.1.1 Residual Stress Effects

In some cases, the significant distinction in thermal expansion between the coating layers and the substrate can cause thermal residual stresses that can cause computational coatings to fail early. The thermal expansion coefficient between the coating and the substrate can't be figured out using computational methods. Because there is a lot of plastic deformation, residual stresses cause the coating to change shape and warp. This means that the coating is more likely to wear away and break [104].

11.1.2 Equipment Cost and Quality Control Effects

Cost is one of the most important factors in the development and growth of every product. It also affects how products are made and what materials they are made of. Most decisions made by producers around the world have little to do with economics.

But it's important to find the right balance between coating time and quality by getting the highest spray rate in the shortest amount of time and a smoother gradient behavior in the coating properties. To do this, the computational technique is used to find the maximum spray rate with the shortest duration and the highest demand. It is also known that this means the need for more cost-effective equipment to make these coatings increases. So, more research should be done to find ways to lower the cost of equipment while keeping the right quality of coatings. This could help directly with the development of technology for these coatings.

11.1.3 Gradient Formation Analysis

Due to its strong reliance on the degree of gradation control and the amount of complexity that can be attained for the computational technique, gradient generation has a considerable impact on the qualities of the created coatings. Another issue with the computational coating process is that.

In computational technique, gradient control is a way to measure how much the qualities of the substrate change because of how the material is spread out during deposition along the desired direction. While the ability of the deposition process to make complex geometries in a number of directions or the possible gradient directions for the coatings made can be used to measure or describe the level of complexity of a coating, they are not the only ways [21].

11.1.4 Limitation in Commercial Application

Computational coating technology cannot be employed extensively in large-scale production because of its high equipment cost. Additionally, the complicated production system makes it more challenging for businesses to use the computational coating approach.

12 Conclusion

The coatings business has always contained segments that have served as catalysts for the adoption of high-tech advancements in a variety of industries, notably among early adopters such as sports, home appliances, and construction. Advanced engineering implementations necessitate the development of a new class of coating materials that protect surface components while also trying to meet the conflicting demands for the required characteristics to enhance the performance of several parts in biomedical, aviation, electronic parts, automotive, offshore, oil and gas pipelines, and electronics fields. The utilization of new sophisticated materials is transforming several mature sectors, including the coatings business. Coatings encounter increasing issues that necessitate more imaginative solutions, resulting in novel concepts and methodologies for coating formulation.

As a result, the coatings sector employing proven technology has witnessed the commercial deployment of numerous new sophisticated materials and process methods during the previous decade. Nano-structured deposition, chemical vapor deposition, and physical vapor deposition are all techniques used to create lighter, thinner, more sophisticated, and chemically complicated coatings for high-tech applications.

However, due to the simultaneous demand for opposing features in the same coating, surface coatings on numerous components have remained a tough problem for researchers.

These sophisticated coatings typically have greater cost factors to consider. But in the next years, it might offer superior performance advantages that are applicable to a wide range of applications. It will take some time before these new technologies and materials become commonplace in the formulation and manufacturing of coatings.

References

1. Saleh B et al (2020) 30 Years of functionally graded materials: an overview of manufacturing methods, applications and future challenges. Compos B Eng 201:108376. https://doi.org/10.1016/j.compositesb.2020.108376
2. Park S-M, Lim JH, Seong MR, Sohn D (2019) Efficient generator of random fiber distribution with diverse volume fractions by random fiber removal. Compos B Eng 167:302–316. https://doi.org/10.1016/j.compositesb.2018.12.042
3. Ghanavati R, Naffakh-Moosavy H, Moradi M (2021) Additive manufacturing of thin-walled SS316L-IN718 functionally graded materials by direct laser metal deposition. J Market Res 15:2673–2685. https://doi.org/10.1016/j.jmrt.2021.09.061
4. Sopronyi M et al (2021) Laser-assisted synthesis of carbon coatings with cobalt oxide nanoparticles embedded in gradient of composition and sizes. Surf Coat Technol 419:127301. https://doi.org/10.1016/j.surfcoat.2021.127301
5. Zhang Y, Fu T, Yu L, Shen F, Wang J, Cui K (2022) Improving oxidation resistance of TZM alloy by deposited Si–MoSi2 composite coating with high silicon concentration. Ceram Int 48(14):20895–20904. https://doi.org/10.1016/j.ceramint.2022.04.080
6. Hu X, Li F, Shi D, Xie Y, Li Z, Yin F (2020) A design of self-generated Ti–Al–Si gradient coatings on Ti–6Al–4V alloy based on silicon concentration gradient. J Alloys Compd 830:154670. https://doi.org/10.1016/j.jallcom.2020.154670
7. Jia M et al (2022) Microstructure and shear fracture behavior of Mo/AlN/Mo symmetrical compositionally graded materials. Mater Sci Eng A 834:142591. https://doi.org/10.1016/j.msea.2021.142591
8. Dumée LF, et al (2015) Graphene coatings make steel corrosion-resistant. Carbon N Y 87(C):395–408. https://doi.org/10.1016/j.carbon.2015.02.042
9. Tscheliessnig R et al (2012) Nano-coating protects biofunctional materials. Mater Today 15(9):394–404. https://doi.org/10.1016/S1369-7021(12)70166-9

10. Stathopoulos V et al (2016) Design of functionally graded multilayer thermal barrier coatings for gas turbine application. Surf Coat Technol 295:20–28. https://doi.org/10.1016/j.surfcoat.2015.11.054

11. Cui Y, Shen J, Geng K, Hu S (2021) Fabrication of FeCoCrNiMnAl0.5-FeCoCrNiMnAl gradient HEA coating by laser cladding technique. Surf Coat Technol 412:127077. https://doi.org/10.1016/j.surfcoat.2021.127077

12. Meschini S, Testoni R, Segantin S, Zucchetti M (2021) ARC reactor: a preliminary tritium environmental impact study. Fusion Eng Des 167:112340. https://doi.org/10.1016/j.fuseng des.2021.112340

13. Banthia S, Amid M, Sengupta S, Das S, Das K (2020) Reciprocating sliding wear of Cu, Cu-SiC functionally graded coating on electrical contact. J Mater Eng Perform 29(6):3930–3940. https://doi.org/10.1007/s11665-020-04878-8

14. Build a new dataset. https://constellate.org/builder/?unigrams=corrosion%2C+coating

15. Wu H et al (2021) Influence of spray trajectories on characteristics of cold-sprayed copper deposits. Surf Coat Technol 405:126703. https://doi.org/10.1016/j.surfcoat.2020.126703

16. McDonnell LP, et al (2021) Superposition of intra- and inter-layer excitons in twistronic MoSe2/WSe2 bilayers probed by resonant Raman scattering. 2d Mater 8(3):35009. https://doi.org/10.1088/2053-1583/abe778

17. Allahyarzadeh MH, Aliofkhazraei M, Sabour Rouhaghdam AR, Torabinejad V (2016) Functionally graded nickel–tungsten coating: electrodeposition, corrosion and wear behavior. Canadian Metall Q 55(3):303–311. https://doi.org/10.1080/00084433.2016.1190542

18. Torabi S, Valefi Z, Ehsani N (2022) The effect of the SiC content on the high duration erosion behavior of SiC/ZrB2– SiC/ZrB2 functionally gradient coating produced by shielding shrouded plasma spray method. Ceram Int 48(2):1699–1714. https://doi.org/10.1016/j.cer amint.2021.09.249

19. Wang Y, Liu Q, Zheng Q, Li T, Chong N, Bai Y (2021) Bonding and thermal-mechanical property of gradient nicocraly/ysz thermal barrier coatings with millimeter level thickness. Coatings 11(5). https://doi.org/10.3390/coatings11050600

20. Yi J, et al (2021) A novel Nb pre-diffusion method for fabricating wear-resistant NbN ceramic gradient coating. Vacuum 185:109993. https://doi.org/10.1016/j.vacuum.2020.109993

21. Szparaga Ł, Bartosik P, Gilewicz A, Mydłowska K, Ratajski J (2021) Optimisation of mechanical properties of gradient Zr–C coatings. Materials 14(2):1–21. https://doi.org/10.3390/ma14020296

22. Amado MM, Olaya JJ, Alfonso JE (2021) Mechanical behavior of YSZ coatings co-deposited with Al and Ag on AISI 316L via RF sputtering. Ceram Int 47(6):7497–7503. https://doi.org/10.1016/j.ceramint.2020.11.089

23. Peng C, Yao BH, Li JX, Niu JF (2016) Preparation and characterization of conjugated microspheres FeTCPP-SSA-TiO2. Mater Sci Forum 852:244–251. https://doi.org/10.4028/www.scientific.net/MSF.852.244

24. Vickers NJ (2017) Animal communication: when i'm calling you, will you answer too? Curr Biol 27(14):R713–R715. https://doi.org/10.1016/j.cub.2017.05.064

25. Pandey S et al (2021) Effect of fuel pressure, feed rate, and spray distance on cavitation erosion of Rodojet sprayed Al2O3+50%TiO2 coated AISI410 steel. Surf Coat Technol 410:126961. https://doi.org/10.1016/j.surfcoat.2021.126961

26. Kumar A, Patnaik PC, Chen K (2020) Damage assessment and fracture resistance of functionally graded advanced thermal barrier coating systems: experimental and analytical modeling approach. Coatings 10(5). https://doi.org/10.3390/COATINGS10050474

27. Khan SA, Sezer N, Koç M (2020) Design, synthesis, and characterization of hybrid micro-nano surface coatings for enhanced heat transfer applications. Int J Energy Res 44(15):12525–12534. https://doi.org/10.1002/er.5685

28. Sallis JGM (1976) Polymer Coatings. https://doi.org/10.1038/2201259a0

29. What Are Metal Coatings? https://toefco.com/what-are-metal-coatings/

30. Metal Coatings Market

31. Tushinsky L, Kovensky I, Plokhov A, Sindeyev V (2002) Coated metal. Engineering materials. https://doi.org/10.1007/978-3-662-06276-0
32. Smith JR, Lamprou DA (2014) Polymer coatings for biomedical applications: a review. Trans Inst Met Finish 92(1):9–19. https://doi.org/10.1179/0020296713Z.000000000157
33. Barroso G, Li Q, Bordia RK, Motz G (2019) Polymeric and ceramic silicon-based coatings-a review. J Mater Chem A Mater 7(5):1936–1963. https://doi.org/10.1039/c8ta09054h
34. Authors F (1963) Plastics coatings—their uses in industry. Anti-Corros Methods Mater 10(4):89–91. https://doi.org/10.1108/eb020056
35. Maiti T, et al (2021) Plastics in coating applications. In: Reference module in materials science and materials engineering. https://doi.org/10.1016/B978-0-12-820352-1.00176-0
36. Murata H, Chang B-J, Prucker O, Dahm M, Rühe J (2004) Polymeric coatings for biomedical devices. Surf Sci 570(1):111–118. https://doi.org/10.1016/j.susc.2004.06.185
37. Bedel V, Lonjon A, Dantras É, Bouquet M (2018) Influence of silver nanowires on thermal and electrical behaviors of a poly(epoxy) coating for aeronautical application. J Appl Polym Sci 135(47):46829. https://doi.org/10.1002/app.46829
38. Chandra AK, Kumar NR (2017) Polymer nanocomposites for automobile engineering applications BT–properties and applications of polymer nanocomposites: clay and carbon based polymer nanocomposites. In: Tripathy DK, Sahoo BP (eds). Berlin, Heidelberg: Springer Berlin Heidelberg, pp 139–172. https://doi.org/10.1007/978-3-662-53517-2_7
39. Ramos-Fernández JM, Guillem C, Lopez-Buendía A, Paulis M, Asua JM (2011) Synthesis of poly-(BA-co-MMA) latexes filled with SiO2 for coating in construction applications. Prog Org Coat 72(3):438–442. https://doi.org/10.1016/j.porgcoat.2011.05.017
40. Dewettinck K, Huyghebaert A (1999) Fluidized bed coating in food technology. Trends Food Sci Technol 10(4):163–168. https://doi.org/10.1016/S0924-2244(99)00041-2
41. Luangtana-anan M, Limmatvapirat S (2019) Shellac-based coating polymer for agricultural applications BT–polymers for agri-food applications. In: Gutiérrez TJ (ed). Springer International Publishing, Cham, pp 487–524. https://doi.org/10.1007/978-3-030-19416-1_24
42. Tüken T (2006) Zinc deposited polymer coatings for copper protection. Prog Org Coat 55(1):60–65. https://doi.org/10.1016/j.porgcoat.2005.11.008
43. Grate JW, Wise BM, Gallagher NB (2003) Classical least squares transformations of sensor array pattern vectors into vapor descriptors: simulation of arrays of polymer-coated surface acoustic wave sensors with mass-plus-volume transduction mechanisms. Anal Chim Acta 490(1–2):169–184. https://doi.org/10.1016/S0003-2670(03)00016-3
44. Lamminmäki T et al (2013) New silica coating pigment for inkjet papers from mining industry sidestreams. J Surf Eng Mater Adv Technol 03(03):224–234. https://doi.org/10.4236/jsemat.2013.33030
45. Shit SC, Shah PM (2014) Edible polymers: challenges and opportunities. J Polym 2014:1–13. https://doi.org/10.1155/2014/427259
46. Plastic coatings market - growth, trends, Covid-19 impact, and forecast (2022–2027). https://www.mordorintelligence.com/industry-reports/plastic-coatings-market
47. Robinson G (2016) Global construction market to grow $8 trillion by 2030: driven by China, US and India. 44:1–3
48. Amit Navale EP (2020) Plastic coatings market by type (polyurethane, acrylic, epoxy, and others), market process (dip coating, spray coating, powder coating, electrophoretic painting, and others), and end-use industry (automotive, aerospace & defense, building & construction). https://www.alliedmarketresearch.com/plastic-coatings-market-A07586
49. Biopolymer coatings market–global industry assessment & forecast (2021) https://www.vantagemarketresearch.com/industry-report/biopolymer-coatings-market-1684
50. Abadias G et al (2018) Review article: stress in thin films and coatings: current status, challenges, and prospects. J Vac Sci Technol A Vac Surf Films 36(2):020801. https://doi.org/10.1116/1.5011790
51. Paul S (2002) Painting of plastics: new challenges and possibilities. Surf Coat Int Part B Coat Int 85(2):79–86. https://doi.org/10.1007/BF02699746

52. Vidal D, Bertrand F (2006) Recent progress and challenges in the numerical modeling of coating structure development. In: 2006 TAPPI advanced coating fundamentals symposium, vol. 2006, pp. 228–250
53. Ceramic Coating Market Outlook (2022–2032) (2022) https://www.futuremarketinsights.com/reports/ceramic-coating-market
54. Why ceramic coatings are the future of the automotive industry https://gardgroup.com/blog/f/why-ceramic-coatings-are-the-future-of-the-automotive-industry?blogcategory=Exterior+Surfaces
55. Issues found with ceramic paint protection https://www.melbournemobiledetailing.com.au/ceramic-coating-the-negatives-issues-and-problems/
56. What are Composite Coatings? https://www.ddcoatings.co.uk/1167/what-are-composite-coatings
57. Dataintelo (2021) Composite coatings market research report. https://dataintelo.com/report/composite-coatings-market/
58. Fathi R et al (2022) Past and present of functionally graded coatings: advancements and future challenges. Appl Mater Today 26:101373. https://doi.org/10.1016/j.apmt.2022.101373
59. Polyester Resins. In Organic coatings. pp 141–150. https://doi.org/10.1002/9781119337201.ch10
60. Deflorian F (2020) Special issue: 'advances in organic coatings 2018.' Coatings 10(6):2018–2020. https://doi.org/10.3390/COATINGS10060555
61. Si Y, Guo Z (2015) Superhydrophobic nanocoatings: from materials to fabrications and to applications. Nanoscale 7(14):5922–5946. https://doi.org/10.1039/C4NR07554D
62. Figueira RB, Silva CJR, Pereira EV (2015) Organic–inorganic hybrid sol–gel coatings for metal corrosion protection: a review of recent progress. J Coat Technol Res 12(1):1–35. https://doi.org/10.1007/s11998-014-9595-6
63. Hornberger H, Virtanen S, Boccaccini AR (2012) Biomedical coatings on magnesium alloys–a review. Acta Biomater 8(7):2442–2455. https://doi.org/10.1016/j.actbio.2012.04.012
64. Montemor MF (2014) Functional and smart coatings for corrosion protection: a review of recent advances. Surf Coat Technol 258:17–37. https://doi.org/10.1016/j.surfcoat.2014.06.031
65. Huang VM, Wu SL, Orazem ME, Pébre N, Tribollet B, Vivier V (2011) Local electrochemical impedance spectroscopy: a review and some recent developments. Electrochim Acta 56(23):8048–8057. https://doi.org/10.1016/j.electacta.2011.03.018
66. Ataei S, Khorasani SN, Neisiany RE (2019) Biofriendly vegetable oil healing agents used for developing self-healing coatings: a review. Prog Org Coat 129:77–95. https://doi.org/10.1016/j.porgcoat.2019.01.012
67. Zhu Q et al (2022) Recent advances in nanotechnology-based functional coatings for the built environment. Mater Today Adv 15:100270. https://doi.org/10.1016/j.mtadv.2022.100270
68. Tator K (2017) Nanotechnology: the future of coatings-part 1. pp 1–4
69. Mamta EPS, Arpita K (2022) Nanocoatings market by type (anti-fingerprint, antimicrobial, self-cleaning (bionic & photocatalytic), anti-corrosion, conductive, and others) and application (electronics, energy, food & packaging, construction, marine, automotive, healthcare, others). https://www.alliedmarketresearch.com/nano-coatings-market
70. The global nanocoatings market size was USD 7,825.3 million in 2020 and is projected to reach USD 39,869.2 million by 2028, exhibiting a CAGR of 22.7% during the forecast period. (2021). FBI105023
71. Griffin D (2019) Limitations of nano-coatings for PCBs. https://blog.humiseal.com/6-limitations-of-nano-coatings-for-pcbs
72. 2020 vision: the future of coatings. (2011). https://www.pfonline.com/articles/2020-vision-the-future-of-coatings
73. Miao X, Sun D (2010) Graded/gradient porous biomaterials. Materials 3(1):26–47. https://doi.org/10.3390/ma3010026
74. Cui W, Qin G, Duan J, Wang H (2017) A graded nano-TiN coating on biomedical Ti alloy: low friction coefficient, good bonding and biocompatibility. Mater Sci Eng C 71:520–528. https://doi.org/10.1016/j.msec.2016.10.033

75. Xiong Z, Zhong L, Wang H, Li X (2021) Structural defects, mechanical behaviors and properties of two-dimensional materials. Materials 14(5):1–43. https://doi.org/10.3390/ma1405 1192

76. Kouitat Njiwa R, von Stebut J (1999) Boundary element numerical modelling as a surface engineering tool: application to very thin coatings. Surf Coat Technol. 116–119:573–579. https://doi.org/10.1016/S0257-8972(99)00230-3

77. Leopold J, Meisel M, Wohlgemuth R, Liebich J (2001) High performance computing of coating-substrate systems. Surf Coat Technol 142–144:916–922. https://doi.org/10.1016/S0257-8972(01)01247-6

78. Klocke F et al (1998) Improved cutting processes with adapted coating systems. CIRP Ann 47(1):65–68. https://doi.org/10.1016/S0007-8506(07)62786-3

79. Gogoi R, Sethi SK, Manik G (2021) Surface functionalization and CNT coating induced improved interfacial interactions of carbon fiber with polypropylene matrix: a molecular dynamics study. Appl Surf Sci 539. https://doi.org/10.1016/j.apsusc.2020.148162

80. Verma A, Jain N, Sethi SK (2022) Modeling and simulation of graphene-based composites. Innov Graph-Based Polym Comp 167–198. https://doi.org/10.1016/B978-0-12-823789-2.000 01-7

81. Gogoi R, Sethi SK, Manik G (2021) Surface functionalization and CNT coating induced improved interfacial interactions of carbon fiber with polypropylene matrix: a molecular dynamics study. Appl Surf Sci 539:148162. https://doi.org/10.1016/J.APSUSC.2020.148162

82. Kataria A, Verma A, Sethi SK, Ogata S (2022) Introduction to interatomic potentials/forcefields. Lect Notes Appl Comput Mech 99:21–49. https://doi.org/10.1007/978-981-19-3092-8_2/COVER

83. Maurya AK, Gogoi R, Sethi SK, Manik G (2021) A combined theoretical and experimental investigation of the valorization of mechanical and thermal properties of the fly ash-reinforced polypropylene hybrid composites. J Mater Sci 56(30):16976–16998. https://doi.org/10.1007/S10853-021-06383-2/FIGURES/10

84. Agrawal G, Samal SK, Sethi SK, Manik G, Agrawal R (2019) Microgel/silica hybrid colloids: bioinspired synthesis and controlled release application. Polymer (Guildf) 178. https://doi.org/10.1016/j.polymer.2019.121599

85. Dhiman A, Gupta A, Sethi SK, Manik G, Agrawal G (2022) Encapsulation of wax in complete silica microcapsules. J Mater Res 2022:1–14. https://doi.org/10.1557/S43578-022-00865-Y

86. Agrawal G, Samal SK, Sethi SK, Manik G, Agrawal R (2019) Microgel/silica hybrid colloids: bioinspired synthesis and controlled release application. Polymer (Guildf) 178:121599. https://www.sciencedirect.com/science/article/pii/S003238611930583X Last accessed June 24, 2019

87. Shankar U, Sethi SK, Singh BP, Kumar A, Manik G, Bandyopadhyay A (2021) Optically transparent and lightweight nanocomposite substrate of poly(methyl methacrylate-co-acrylonitrile)/MWCNT for optoelectronic applications: an experimental and theoretical insight. J Mater Sci 56(30):17040–17061. https://doi.org/10.1007/S10853-021-06390-3/FIGURES/17

88. Saini A et al (2021) Microdesigned nanocellulose-based flexible antibacterial aerogel architectures impregnated with bioactive Cinnamomum cassia. ACS Appl Mater Interfaces 13(4):4874–4885. https://doi.org/10.1021/acsami.0c20258

89. Singh V, Patra S, Murugan NA, Toncu DC, Tiwari A (2022) Recent trends in computational tools and data-driven modeling for advanced materials. Mater Adv 4069–4087. https://doi.org/10.1039/d2ma00067a

90. Shymchenko AV, Tereshchenko VV, Ryabov YA, Salkutsan SV, Borovkov AI (2017) Review of the computational approaches to advanced materials simulation in accordance with modern advanced manufacturing trends. Mater Phys Mech 32(3):328–352

91. Sethi SK, Manik G (2021) A combined theoretical and experimental investigation on the wettability of MWCNT filled PVAc-g-PDMS easy-clean coating. Prog Org Coat 151:106092. https://doi.org/10.1016/j.porgcoat.2020.106092

92. Sethi SK, Manik G (2018) Recent progress in super hydrophobic/hydrophilic self-cleaning surfaces for various industrial applications: a review. Polym Plast Technol Eng 57(18):1932–1952. https://doi.org/10.1080/03602559.2018.1447128

93. Sethi SK, Shankar U, Manik G (2019) Fabrication and characterization of non-fluoro based transparent easy-clean coating formulations optimized from molecular dynamics simulation. Prog Org Coat 136. https://doi.org/10.1016/j.porgcoat.2019.105306

94. Sethi SK, Gogoi R, Manik G (2021) Plastics in self-cleaning applications. In: Reference module in materials science and materials engineering. https://doi.org/10.1016/B978-0-12-820352-1.00113-9

95. Sethi SK, Kadian S, Manik G (2021) A review of recent progress in molecular dynamics and coarse-grain simulations assisted understanding of wettability. Arch Comp Methods Eng 1:1–27. https://doi.org/10.1007/S11831-021-09689-1

96. Sethi SK, Singh M, Manik G (2020) A multi-scale modeling and simulation study to investigate the effect of roughness of a surface on its self-cleaning performance. Mol Syst Des Eng 5(7):1277–1289. https://doi.org/10.1039/d0me00068j

97. Sethi SK, Singh M, Manik G (2020) A multi-scale modeling and simulation study to investigate the effect of roughness of a surface on its self-cleaning performance. Mol Syst Des Eng. https://doi.org/10.1039/D0ME00068J

98. Sethi SK, Soni L, Manik G (2018) Component compatibility study of poly(dimethyl siloxane) with poly(vinyl acetate) of varying hydrolysis content: an atomistic and mesoscale simulation approach. J Mol Liq 272:73–83. https://doi.org/10.1016/J.MOLLIQ.2018.09.048

99. Sethi SK, Shankar U, Manik G (2019) Fabrication and characterization of non-fluoro based transparent easy-clean coating formulations optimized from molecular dynamics simulation. Prog Org Coat 136:105306. https://doi.org/10.1016/j.porgcoat.2019.105306

100. Sethi SK, Manik G, Sahoo SK (2019) Fundamentals of superhydrophobic surfaces. In Super-hydrophobic polymer coatings. Elsevier, pp 3–29. https://linkinghub.elsevier.com/retrieve/pii/B9780128166710000011. Accessed Aug 29 Aug 2019

101. Sethi SK, Gogoi R, Verma A, Manik G (2022) How can the geometry of a rough surface affect its wettability?–a coarse-grained simulation analysis. Prog Org Coat 172:107062. https://doi.org/10.1016/J.PORGCOAT.2022.107062

102. Sethi SK, Soni L, Shankar U, Chauhan RP, Manik G (2020) A molecular dynamics simulation study to investigate poly(vinyl acetate)-poly(dimethyl siloxane) based easy-clean coating: an insight into the surface behavior and substrate interaction. J Mol Struct 1202:127342. https://doi.org/10.1016/j.molstruc.2019.127342

103. Sethi SK, Kadian S, Anubhav G, Chauhan RP, Manik G (2020) Fabrication and analysis of ZnO quantum dots based easy clean coating: a combined theoretical and experimental investigation. ChemistrySelect 5(29):8942–8950. https://doi.org/10.1002/slct.202001092

104. Montay G, Cherouat A, Nussair A, Lu J (2004) Residual stresses in coating technology. J Mater Sci Technol 20:81–84

Chapter 12
Emerging 2D Materials for Printing and Coating

Preetam Singh, Richa Mudgal, and Aditya Singh

1 Introduction

In 2004, the Nobel prize-winning discovery of graphene by Geim and Novoselov by scotch tape method has led to the foundation of a new research area in two-dimensional (2D) materials [1]. Later on, numerous 2D materials have been added to this vast family of nanomaterials, e.g., transition metal dichalcogenides (TMDs), MXenes, hexagonal boron nitride (h-BN), and black phosphorous (BP) [2–10]. 2D materials can be easily procured from their parent bulk crystals using scotch tape [11–14]. Numerous exotic properties of 2D materials, such as atomic thickness, mechanical flexibility, transparency, and electrical conductivity, have evidenced their scope in printing electronic devices [15–19]. The evolution of publications in 2D materials for printing applications is shown in Fig. 1.

There are various printing processes like inkjet, screen, roll-to-roll (R2R) gravure, and flexographic printing, primarily utilized for the patterning of ink onto stiff, flexible, and conformable surfaces. These routes facilitate the ultra-low-cost and large-scale manufacture of packaging materials, documents, magazines, newspapers, etc. Recently, incorporating 2D materials as an active pigment into the ink has attracted broad attention. This tactic takes the benefit of a well-established printing process to produce devices of 2D materials [20]. The 2D material printed device first came into the picture in 2012, when the liquid phase exfoliated (LPE) graphene

P. Singh · R. Mudgal
Department of Physics, Indian Institute of Technology Delhi, Hauz Khas, New Delhi 110016, India

P. Singh
Department of Physics, Rashtriya Kisan Post Graduate College, Shamli, Uttar Pradesh 247776, India

A. Singh (✉)
Department of Physics, Freie Universität Berlin, Arnimallee 14, 14195 Berlin, Germany
e-mail: adi.s@fu-berlin.de

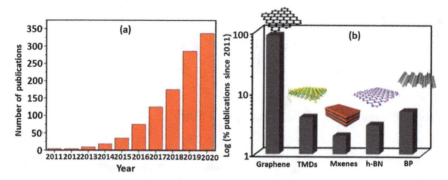

Fig. 1 Statistics of research publications collected from Web of Science: **a** Number of publications per year for printing with 2D materials, MXenes, h-BN, and BP; **b** relative % of publications since 2011 (adapted with permission from reference [28])

was inkjet-printed to fabricate the field effect transistor (FET) [21]. Since then, this field of research has observed remarkable evolvement in the fabrication of 2D printed devices. Other 2D materials, such as TMDs, black phosphorus, and h-BN, along with graphene, have been employed as active pigments [22–27]. The first-generation inks of 2D materials were not optimized, but new 2D functional materials ink formulations will emerge.

The feasibility of organic solvents, mixtures, and aqueous dispersion makes them suitable for ink formulation. Polymers and surfactants are primarily used in ink formulation as binders and additives [25]. The conditioning and tailoring of the formulation of the ink with 2D functional materials have permitted the fluidic properties, drying dynamics, and interaction with surfaces [29–31]. The printing of 2D materials has proven potential in electronics, optoelectronics, photonics, sensors, and energy storage devices [22–26]. The controlled patterning of 2D materials due to their layered nature allows us to fabricate fully printed heterostructures that avail the complementary properties of 2D materials for excellent device performance [23–25].

After the fabrication of the first inkjet-printed graphene transistor, there have been many advancements in the origination of functional ink and the application of printed devices of 2D materials. Here, in this chapter, we try to deliver an understanding of the printing and coating of 2D materials with a background of how graphene inks formulation and printing technologies could offer themselves to the printing of 2D materials. The chapter has been divided into different sections; in the first section, a concise outline of 2D materials and synthesis techniques have been discussed. The second section includes information about 2D material inks and printing, like ink composition, viscosity, ink processing, spreading and drying, inkjet printing, screen printing, gravure, and flexographic printing of 2D materials. The third section shows the applications of different printable 2D materials, and the chapter ends with a summary and future perspectives.

2   2D Materials

Graphene, the foundational member of the 2D materials family, contains a single layer of sp^2 hybridized carbon atoms arranged in a hexagonal lattice structure. It has many exciting properties, such as high mechanical strength, flexibility, high thermal conductivity, high carrier density, and optical transparency [15–19]. Graphene also shows extraordinary biomedical properties, such as antibacterial and high resistance towards a range of chemicals, like strong acids or bases, reductants, or oxidants [32, 33]. The primary commercial and cost-effective procedure for the manufacturing of graphene is its production from graphite through LPE, which permits the formation of a few layers (<10) of material [34]. Other methods, with partially changed chemical composition, can be achieved by oxidative interaction (graphene oxide, GO) and its partial structural restoration through chemical reduction (reduced graphene oxide, rGO) [35–37].

Apart from graphene, widely studied 2D materials for device applications are TMDs, MXenes, chemically synthesized h-BN, and BP [38–40]. TMDs are compounds of transition metal and group VI element (chalcogen), represented as MX_2, where M is transition metal (Mo, W, etc.), arranged in one plane, surrounded by two other planes of X (chalcogen atom) with covalent bonds. Depending upon the orientation of these three planes, the structure exists in tetragonal (T), hexagonal (H), and octahedral geometry, where the properties of TMDs vary from metallic to semiconductor [38]. TMDs have exotic properties of interest in nanotechnology and energy storage. They are robust and stable, absorb and emit light, and have high charge density and mobility. The first MoS_2 was exfoliated in 1986 [41], but the most intense research started after 2011 when the first MoS_2 layered transistor was demonstrated with an $I_{on}/I_{off} = 10^8$ [42]. Depending upon the number of layers of TMDs, the band gap can be easily tailored due to the quantum confinement, which makes them suitable for different device applications such as optoelectronics and photodetectors [43–46].

MXenes are another vast family of 2D materials [47], which consist of a single layer of transition metal carbides, nitrides, and carbonides with the chemical formula $M_{n+1}X_nT_x$, where M represents transition metals (Ti, Mo, Ta, etc.), X denotes carbon or nitrogen, and T_x stands for terminating surfaces (fluorine, oxygen, or hydroxyl) [48, 49]. Owing to chemical composition and surface terminating nature, MXenes have good hydrophilic nature and effects on the density of states near the Fermi level, and hence extraordinary electronic properties [50, 51]. In different applications, such as energy storage, biomedicine, photocatalysis, and other areas, MXenes are used as cathode material [52–58]. The synthesis of MXenes is mostly feasible by a top-down approach, i.e., etching of layered ternary carbides (MAX) phase with HCl, LiF, or HF [59].

After the first successful synthesis of BP by heating white phosphorus at high pressure in 1914, it took another 100 years to use BP as 2D material in transistors with a mobility of $1000\,cm^2V^{-1}S^{-1}$ and $I_{on}/I_{off} = 10^5$. BP is a rare allotrope of phosphorus and can be obtained as a 2D layer of phosphorus known as phosphorene. The direct band gap of BP makes it a potential contestant for optical devices, and it shows the wrinkled structure with high hole mobility and excellent mechanical, electrical, and optical properties [60]. The poor stability of BP due to chemical degradation and oxidation limits its applications in many electronic and optoelectronic devices. However, recently, it has been used for energy storage in batteries, capacitors and supercapacitor devices [61], chemical and biochemical applications [62], and sensors [63]. BP can be obtained from vapor phase growth, LPE, or red phosphorus [63–65].

h-BN is another vital member of the 2D materials family, which comprises B and N in a 2D plane forming a hexagonal structure with sp^2 hybridization between B and N [66]. h-BN shows a structure like graphite, but it is broadly used as a lubricant [67, 68], and it is a thermally conductive but electronic insulator, which makes it one of the most suitable materials for an encapsulator for other 2D materials [69]. It shows a nearly transparent nature with a wide band gap and is chemically and thermally stable. h-BN is utilized in lubrication, cosmetics, insulation, and other areas from an industrial point of view [68].

Besides these 2D materials, materials with 5 atom layers along the z-direction, like Bi_2Se_3 and Sb_2Te_3, can also be exfoliated layer by layer from their bulk counterparts [70]. Owing to their potential applications in electronics, optoelectronics, and thermoelectric power generation, these materials show high scientific interest [71, 72]. Recently, III-VI group layered compounds, such as $GeSe_2$, InSe, and $InSe_2$, have emerged as potential materials in the family of 2D materials for electronic and optoelectronic applications [73, 74].

Contemporarily, the importance of 2D materials is rising day by day (Fig. 1(a)) due to their extraordinary properties suitable for diverse device applications. The candidacy of 2D materials in printing is a growing aspect of research and evolving with different printed devices for different applications. The feasibility of 2D materials for ink formalism and printing will be discussed below.

3 Printing

Printing is a well-established process for functional device fabrication, which can be of different types named inkjet, R2R gravure, and flexographic printing [75]. Nowadays, it is a widely used technique to make the pattern of 2D materials on the desired substrate due to its capabilities of large-volume production and low-cost setup [20]. 2D materials have arisen as favorable materials due to their extraordinary properties, such as high strength, low weight, and different optical, electrical, and mechanical properties compared to conventional bulk materials [10, 76–81]. Various synthesis techniques are used to produce 2D materials, which are classified into bottom-up and top-down approaches. CVD is one of the most popular methods falling in the bottom-up approach category, which is commonly used to yield a large area of the thin film of 2D materials [82–88]. Despite producing a uniform thin film of 2D materials, CVD is often less preferred for production at the industrial level due to its demanding conditions, such as high temperature, various carrier gases, and precursors [89]. Another approach, i.e., the top-down approach, includes mechanical exfoliation [90] and solution processing methods [91].

Mechanical exfoliation of 2D materials on the desired substrate faces a major challenge due to the small size of exfoliated flakes [93–96]. In contrast, solution processing is a promising method of depositing 2D materials in terms of processing and cost-effectiveness [91]. In this method, monolayer or few-layer flakes of 2D materials are exfoliated in liquid, which is subsequently transferred over the desired substrate. Several techniques are available in the literature for processing this solution, which includes ion intercalation, ion exchange, and liquid phase exfoliation (LPE) [97, 98]. This solution is formulated as ink, which is later used to make a pattern by printing onto rigid, flexible, or conformal surfaces. Graphene was the first 2D material deposited using LPE in 2008 [16]. Later, this method was adopted for other 2D materials, such as MoS_2 and h-BN [99, 100]. As we discussed above, different types of printing differ by resolution, throughput, and scalability. For example, inkjet is a non-contact, high-resolution, and maskless patterning process and requires a low concentration of pigments in ink [75]. On the other hand, flexography is a high-throughput printing process requiring a higher concentration of pigments [75]. These things suggest that a better knowledge of the appropriate printing process and corresponding ink is required to achieve the desired pattern successfully. Here, we will discuss the details of the properties and processing of ink, followed by further information on different printing processes.

3.1 Ink Formulation

We discussed in the above section that ink properties are crucial to optimize before going to further processing. Ink formulation can be divided into the following three parts:

(a) **Composition**: Typical ink for modern graphics consists of pigments, binders, additives, and solvents. In conventional printing, pigments are used to give color to the soot; similarly, functional materials are used for the same purpose in electronic printing [20]. Sometimes, these are called active pigments, and they can be conducting, insulating, or semiconducting, depending upon the applications [101]. Ink, comprising of such active pigment, is used to deposit patterns over the substrate for device fabrication. To date, there are various examples of such ink formulation using nanoparticles and CNTs as active pigments [102, 103]. At present, the fabrication of functional devices using 2D materials such as active pigments is a field of great interest.

As clear from the name, binders are used to bind the pigment particles to each other and the substrate. They solidify after solvent evaporation and may contribute to the physical properties of the film, such as adhesion with substrate [104]. Surface tension and surface energy of ink are crucial parameters to decide the wetting behavior of thin film. These properties can be modified using an additive with < 5% wt. in ink solution [105]. Sometimes, they are also used to modify the pH of the solution. Solvents are water or organic solvent used to dilute the other elements of ink. The selection of solvent depends upon the printing process, substrate, and drying conditions. For example, volatile solvents (IPA, ethyl acetate, and n-propyl acetate) are required in flexographic printing [20], while comparatively less volatile solvents are used in screen printing [75].

(b) Viscosity: Viscosity is a critical parameter of liquid, which defines the resistance to flow. Inks of different viscosity are desired for different printing process; for example, low-viscosity inks are required for inkjet printing [106], while viscous one is needed for screen printing [75].

(c) Ink processing: It includes dissolved pigments in varnish. Varnish is defined as the mixture of binders, additives, and solvents, which governs the wetting properties of the printed film. In the raw form, pigments have lumps or aggregations, which is not acceptable to achieve a uniform film. Hence, dispersion is required to break pigments into primary particles. Various techniques are available in the literature to disperse pigments into varnish. These techniques include ball milling, agitation, and meshed screens [75, 107] and can be chosen depending on the viscosity of the varnish.

(d) Ink spreading and Drying: After ink processing, ink is spread over the substrate. The spreading of the ink can be understood by Young's equation [108]:

$$\gamma_{s-v} = \gamma_{s-i} + \gamma_{i-v}\cos\theta$$

Here, γ_{s-v}, γ_{s-i}, and γ_{i-v} are the interface tension between substrate-vacuum, substrate-ink, and ink-vacuum, respectively. θ is defined as the contact angle and decides the wetting of the substrate. A small value of θ gives a good wetting, while a larger value corresponds to poor wetting. As suggested by the literature, a suitable wetting (uniform film) can be accomplished if the surface tension of the ink is 7–10 mN-m^{-1} [109] lower than the substrate surface energy. Apart from the spreading, the ink's viscosity decides the thin film's morphology. Depending upon the viscosity, a printed film can be uniform, semicircular, or coffee ring-shaped [101].

3.2 Inkjet Printing

Principle: This is a high-resolution, maskless, and non-contact technique of patterning. This approach designs patterns electronically and controlled remotely [110]. As no hard mask is required for patterning, inkjet printing can design flexible patterns directly over the substrate. In this type of printing, ink jetting is followed by two mechanisms: continuous inkjet (CIJ) and drop-on-demand (DOD) inkjet [75]. In CIJ (Fig. 2(a)), charged droplets of ink are continuously jetted over the substrate and subsequently deflected selectively by deflection plates. However, CIJ is a speedy way of jetting ink, but the complexity of setup limits its application at a large scale.

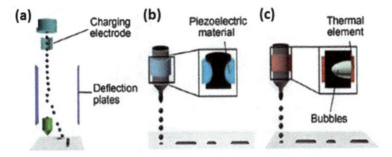

Fig. 2 **a** Schematic illustration of continuous inkjet printing; drop-on-demand inkjet printing with **b** piezoelectric **c** thermal head (adapted with permission from reference [92])

DOD is another mechanism and a good substitute for CIJ for ink jetting. In DOD, ink is sprayed over the substrate when demanded [75]. In this mechanism, the jetting process is controlled by the voltage pulse or temperature and is named as piezo-electric or thermal inkjet process. In the piezoelectric jetting process (Fig. 2(b)), the ink tank changes its shape by applying voltage. This results in pressure on ink and drops it out on the substrate. On the other hand, in thermal jetting (Fig. 2(c)), ink is heated up rapidly to generate bubbles, which push ink out from the tank as a droplet. Sometimes, secondary (satellite) droplets are also formed under a single voltage pulse; hence, achieving stable jetting is a significant challenge in the DOD mechanism [106].

The formation of the satellite drops mainly depends on the fluid properties. A characteristic parameter Z, named as inverse Ohnesorge (Oh) number, is defined to know the behavior of the fluid. The expression of Z is given as follows [111, 112]:

$$Z = \frac{1}{Oh} = \frac{\sqrt{(\gamma a \rho)}}{\eta}$$

where γ, ρ, η are the surface tension, density, and viscosity of the ink material, respectively, and 'a' is the diameter of the jetting nozzle. In the literature, a value of $Z > 14$ and $Z < 1$ results in droplet formation [21, 111, 113], while $Z = 3.5$ for stable jetting [111].

Preparation of 2D materials ink for inkjet printing: LPE dispersion is widely used as ink material for inkjet printing. Viscosity <2 mPa-s of solvents in LPE makes it appropriate for inkjet printing [21]. In LPE dispersion, the dispersed flakes have a lateral size of <200 nm, which also fulfill the conditions of inkjet ink; hence, no further ink formulation is required [21]. Torrisi et al. show the optical image of inkjet drops on plasma cleaned, pristine, and hexamethyldisilazane (HDMS)-treated substrate as shown in Figs. 3(b–d), respectively. However, the method faces challenges in terms of printing efficiency due to the low concentration of 2D materials. Another major issue in LPE dispersion is the high boiling point of ordinarily used solvents such as N-methylpyrrolidine (NMP) [16, 25, 114]. People have tried water and alcohol also

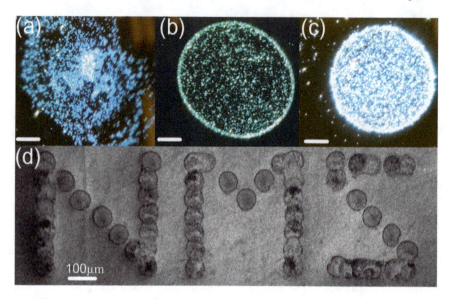

Fig. 3 Dark-field optical micrograph of inkjet-printed drops on **a** plasma-cleaned, **b** pristine, and **c** HMDS-treated substrate. The scale bar is 20 μm. **d** SEM micrograph of drops printed in a pattern (adapted with permission from reference [21])

as dispersing materials [25], but the high surface tension of these solvents is another concern. These low viscous solvents are subject to poor substrate wetting and may result in 'coffee ring'-shaped deposition [21, 25]. The use of polymeric binders in ink formulation has emerged as a potential solution to address above-mentioned challenges [115–117]. In this process, flakes are extracted from the dispersing solution via different methods, e.g., filtration and sedimentation, and redispersing in another solvent with polymer binders [75, 115]. Such solvents are highly concentrated in nature. These solvents are adapted for depositing various 2D materials, such as graphene and MoS_2 [118, 119]. After the printing process, the impurities of binders are removed from the printed film by high-temperature annealing to avoid the change in properties of 2D materials.

3.3 Screen Printing

Principle: This is a stencil-based process, where formulated ink is transferred over the substrate using a stencil screen. This screen can be made of a fine and porous mesh of fabric, silk, metal threads, or synthetic fiber [75, 120]. Mesh is stretched using some solid frame, and pores remain open in the imaging area to let the ink flow from there, and the rest are closed using photo-polymerized resin. In the process, ink is spread over the mesh and is squeezed to force the ink to let it fall from the pores. Simultaneously, the substrate is kept touching the screen to receive the ink.

The technique becomes essential when precisely overlapping layers of functional materials are required. Various semi-auto and fully automatic machines are available to avoid time consumption. Cylinder presses are also used for high-speed and energy-efficient R2R printing. In such a setup, presses and squeegees are stationary while the substrate, cylinder, and screen are moved simultaneously. In screen printing, the deposition and volume of ink are affected by thread count, thread diameter, the pressure of the squeegee, the angle of contact of the squeegee with the screen, and the speed of the blade [75, 104, 121].

Preparation of 2D materials ink for screen printing: As per the requirement of high viscosity ink in screen printing, polymer binders are used to disperse the pigments in solvents [104]. Firstly, Zhang et al. demonstrated the reduced graphene oxide (rGO) ink by dispersing it in terpineol and using ethyl cellulose as an ink binder [122].

This ink required annealing at a temperature of 400 °C to evaporate the binders. However, such a high temperature badly affects the film adhesion with the substrate. To date, a lot of work on screen ink formulation of solution-processed graphene has been done, and many are going on for incorporating h-BN and MoS_2 [123, 124]. Except for ethyl cellulose, other polymer binders, such as polyvinylpyrrolidone (PVP) and polyvinyl alcohol (PVA), are explored for screen ink formulation [31, 125]. K. Arapov et al. prepared the screen-printing ink by gelation of graphene dispersion.

3.4   Gravure and Flexographic Printing of 2D Materials

Gravure printing (GP) and flexographic printing (FXP) are two commonly used techniques in various packaging applications. Both processes are based on metal or ceramic rolls, which govern the deposition of ink over the substrate [75]. These rolls are called gravure cylinders in the case of GP and an anilox roller for FXP. For printing 2D materials, these techniques were rarely used to date due to their complex setup, the high cost of producing the gravure roller (in GP) or printing plates (in FXP), and a large amount of ink (>1L) required to stabilize the printing. GP is a kind of direct printing where ink is directly deposited over the substrate via a roller. Unlike GP, FXP is an indirect printing technique in which the transferred metered ink takes place from the anilox roller to the printing plate. This plate holds an image of a pattern before transferring it over the substrate [20, 75].

Principle of gravure printing (GP): The printing setup of GP comprises an ink trough, a metal gravure roller, a doctor blade, and an impression roller [75, 104, 126–128]. The ink trough is filled with fluid ink, where the gravure roller rotates. The doctor blade removes the excess amount of ink from the gravure roller. This results in transferring patterns over the metal roller in the form of cells, which are intended to get filled with ink when the roller rotates in the ink trough. This ink is transferred over the substrate when it passes between the gravure roller and impression roller, as schematically depicted in [104, 126]. The ink used in gravure printing has medium viscosity (100–1000 mPa s). Dense layers with wet thickness and resolution of 7 and 100 μm, respectively [120], can be achieved using this technique. A model for the ink transfer mechanism using GP was suggested by Nguyen et al. [128, 129]. The process of filling the gravure cell during the ink phase and being empty after successfully transferring ink to the substrate is governed by the forces F_{afs}, F_{afc}, and F_{cf}. These forces are named cohesive force (F_{cf}), adhesive force on cell (F_{afc}), and adhesive force on substrate (F_{afs}). The ink transfer will take place only if $F_{cf} < F_{afc}$ and $F_{cf} < F_{afs}$. If $F_{afs} > F_{afc}$, ink is transferred over the substrate without defects (pinhole and voids).

Principle of flexographic printing (FXP): FXP involves a bit more complex process of ink transferring compared to GP, where an image of a pattern is formed on a soft and flexible printing plate mounted on a plate cylinder [104, 130]. This pattern is subsequently transferred over the substrate. In the process, a screened anilox roller with engraved cells rolls through the ink trough, where ink is applied over its surface to fill the cells. Excess of ink scraps from the roller surface by a doctor blade. This evenly distributed ink, over an anilox roller, is transferred over a printing plate (mounted over a second cylinder) before depositing over the substrate [130]. In FXP, ink of medium viscosity (1000–2000 mPa s) is required [120]. A typical wet thickness of ~ 3 μm can be achieved using FXP, while resolution lies between 100 and 200 μm [120].

2D materials ink formulation for gravure and flexographic printing: Secor et al. first reported the GP of graphene in 2014 [32]. In the report, graphene ink was prepared by dispersing the graphene-ethyl cellulose (EC) powder in a mixture of ethanol and terpineol. In this method, the graphene-EC powder was achieved by solvent exfoliation of graphene, as mentioned in the reference [115]. The size of LPE graphene flakes was well suited for high-resolution gravure printing. Ink solvent and concentration were varied by controlling the load of graphene-EC to make it suitable for gravure printing. Baker et al. first reported flexographic printing of 2D materials [131]. In the report, graphene nanoplatelet/sodium carboxymethylcellulose (Na-CMC) ink was prepared in water/IPA solution. Here, the ratio of Na-CMC and graphene were controlled to suit it for flexographic printing.

4   Applications of Printable 2D Materials

There has been a lot of demonstration and fabrication of device architectures based on the solution of 2D materials over the last decade. The prime applications of printed 2D materials include conductive inks, electronics and optoelectronics, photonics, sensors, and energy storage. There are many other applications of printed 2D materials that are beyond this chapter's scope. These key applications are described below.

4.1   Conductive Inks

Printable technology holds great potential in logic circuits, sensors, optoelectronics, flexible displays, radio-frequency identification (RFID), and portable energy storage [20, 132]. Conductive inks are necessary for the printing of different cathodes and interconnect circuits. Traditional conducting inks for printing these electronic circuits were carbon- or metal-based. The conductivity of carbon did not suit some specific device applications. In contrast, metal-based conducting inks are costly and require curing and sintering processes [34]. Therefore, graphene-based conducting inks are promising for printing devices, as the conductivity of graphene-based inks is between carbon and metal-based inks [34]. The use of graphene-based ink has been demonstrated by the printing gate and source of a flexible transistor on Kapton tape [30]. The LPE dispersion has been started to investigate printed device applications. It has been shown that inkjet-printed graphene shows an electrical conductivity of ~100 Sm^{-1}, which is far below the metal conductivity (<1 × 10^5 Sm^{-1}), and even the carbon (~1000 Sm^{-1}) [133].

To overcome the above drawback, recent advancement suggests a large loading of graphene/binder. These inks show better, uniform printing of devices with a better inter-flake percolation and hence result in improved electrical conductivity. It could further be improved by adding nitrocellulose as a binder, but for such binder decomposition, high temperature (>200 °C) annealing is required [34]. This shows that the conducting inks are more suitable for printing on substrates that can withstand high temperatures, like SiO_2, glass, Kapton, etc. This annealing process reflects the success of graphene/binder conductive inks only after getting a sweet-temperature range, which could decompose the binder for improved percolation of flakes without damaging the substrate and printed device/structures.

To overcome the annealing issue, a rapid photonics lamp consisting of high-intensity xenon ions was utilized to decompose the binder after printing. It has been shown that this tactic can decompose ethyl cellulose without damaging printed graphene and polymeric substrate [134]. It is a strong method of printing robust electronic structures even after 1000 bending cycles. High electrical conductivity and cost-effectiveness have increased the expectations of researchers that printed graphene can be widely used in the field of printed electronic devices. However, the conductivity of graphene-based inks is still far below till date from the conductivity of metal-based inks and needs advancement. This may be further improved by controlled graphene doping to improve the inter-flake percolation [34, 134].

4.2   Printed Optoelectronics and Photonics

Though different exfoliated and CVD-grown optoelectronic and photonic devices have been demonstrated for different applications, their scalable device fabrication for real-world applications has been delimited owing to low yield and high production cost [135]. Therefore, efforts have been made toward printable optoelectronic and photonic materials.

The high mobility of graphene has motivated researchers towards graphene-printed devices. The first printed graphene device was demonstrated in 2012 [21], where the transistor channel was printed. The graphene-printed transistor (Fig. 4(a)) showed better carrier mobility, ~95 $cm^2V^{-1} s^{-1}$, in comparison to the printed organic transistors (usually, <1 $cm^2V^{-1} s^{-1}$) [136], but with compromised I_{on}/I_{off} (~10), but in the case of printed organic transistors, it was ~10^5. This is possibly due to semi-metallic nature of graphene [21].

There were only a few reports on graphene-printed transistors until inkjet-printed graphene heterojunction-based transistors have not come into the picture [37]. The photograph of the device, where the channels have been printed using an inkjet printer on the inkjet-printed h-BN dielectric substrate [37], has been shown in Fig. 4(b). The printed devices showed increased carrier mobility ~204 $cm^2V^{-1} s^{-1}$, high mechanical strength, and better durability [37]. Moreover, the typical I_{on}/I_{off} is found to be ~1.2, as shown in Fig. 4(c) [37].

Contrary to graphene, semiconducting TMDs have shown promise in electronic applications with better I_{on}/I_{off} [42, 137]. The inkjet-printed heterojunction transistors employ WSe_2 for channel printing rather than graphene, with an improved I_{on}/I_{off} ~ 600, which is far better than that of graphene-printed transistors.

Semiconducting TMDs and BP can also be used for photodetector circuits [38]. The first photodetector circuit was demonstrated in 2014 [22] by inkjet printing of graphene as electrodes and inkjet-printed MoS_2 as a photodetection channel. The photoresponsivity of this circuit was found to be <1 μA/W (at 532 nm). The

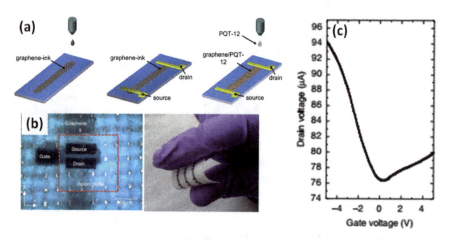

Fig. 4 **a** Schematic of the first printed graphene transistors (adapted with permission from reference [21]), **b** image of the printed graphene transistor on textile, and **c** associated transfer characteristics (adapted with permission from reference [37])

previous circuit junction, graphene/WSe$_2$/graphene, can also be used as a photodetector with an improved photoresponse <1 mA/W (at 514 nm) [25]. The photoresponsivity can further be improved by employing hybrid photodetectors, which have been constructed by the integration of inkjet-printed BP with graphene/Si Schottky junction [26]. Figure 5(a) shows the printed photodetector of graphene on Si, and the inset shows the schematic of this hybrid structure. As the graphene can be degraded due the oxidation, the perylene-C [26] is used as a capping layer.

Except for these optoelectronic properties, the nonlinear optical absorption and ultrafast carrier dynamics make the 2D materials important for a saturable absorber (SA) [26]. SAs have been demonstrated by inkjet printing of BP and can be used for ultrafast signal generation and many other applications [138]. Figure 5(b) shows the printed BP as SA, and due to the reactiveness of BP, the Perylene-C capping layer has been used. Figures 5(c and d) shows the ultrafast pulse generation signals from printed BP as SA [138]. Therefore, functional inks and printed devices show their importance in printing optoelectronic and photonic devices and electronic circuits.

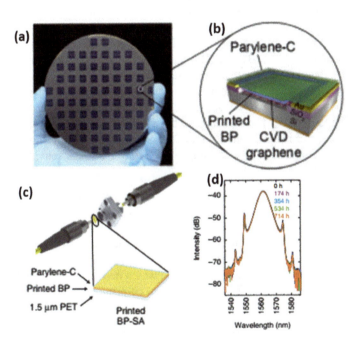

Fig. 5 **a** Inkjet-printed BP detector arrays across the Si wafer, **b** inset of device, showing a capping layer over BP to prevent exposure of BP into air, **c** schematic of inkjet-printed BP-SAs, and **d** associated spectrum (adapted with permission from reference [26])

4.3 Printed Sensors

Presently, many applications of sensors can be seen in electronics, biomedical, and aerospace. The printed 2D materials have many real applications in optoelectronics and photonics. The real-world application of 2D materials-based printed sensors is an emerging field of research due to their high carrier mobility, mechanical strength, and surface area [135]. When the 2D materials-based sensors are exposed to air, contact/non-contact with moisture or other elements can react and may convert it to signals, which can further be read out. The exfoliated and CVD-grown sensors have been demonstrated, but the fabrication process needs to emerge for real applications [139, 140].

Chemical sensors: Chemical sensors have many applications in the health sector, environment monitoring, industrial production, agriculture, and smart buildings. 2D materials have proved their importance in the field of sensors. MoS_2-based sensors can vary resistance when coming in contact with NH_3 and can sense it with a sensitivity of 5 ppm [141], while other materials, such as graphene and BP-based sensors, can be used for NO_2, NH_3, etc. [142]. In comparison to MoS_2, these materials show better sensitivity, like BP, better than 20 times with an improved response time, ~40 times. This suggests a vital role of BP in sensing applications but the highly reactive nature of BP limits its printable device applications. Better sensitivity and response time have been observed for 2D materials-based hybrid sensors. The $rGO-SO_3H$ decorated with the nanoparticle sensor has been shown in Fig. 6(a) [143]. The sensors have been printed using gravure printing using functional inks over metal contacts. Figure 6(b and c) shows a fast response for NO_2 and a linear response for humidity. Graphene-based sensors can sense humidity, as shown in Fig. 6(d) [33].

Temperature sensors: These sensors are extensively utilized in the field of research and industrial monitoring, diagnostics, thermal management, etc. [144, 145]. Research is growing in wearable temperature sensors that can monitor skin temperature and provide insight into the body and other physical activities [145]. Graphene and other 2D materials are potential candidates for printable temperature sensors. To date, there are limited sensor applications, which need to be improved by using printable 2D sensors. Figure 7(a) shows the inkjet-printed epidermal temperature sensor [146].

Biosensors: Technological advancements have made the sensors useful in many aspects of human life. Detecting enzymes, antibodies, proteins, and other biomolecules is an essential biological, technological, and pharmaceutical task [147]. This needs the miniaturization of sensing components and demands advancements in the underlying material technology and engineering [148]. 2D materials seem to be promising candidates for biosensing applications. The 2D materials-based sensing application is minimal, and primarily rGO-based sensors have been fabricated [149], possibly due to the properties of rGOs that can be decorated with any particle. Figure 7 (c) and (d) [28] shows the graphene-printed biosensors.

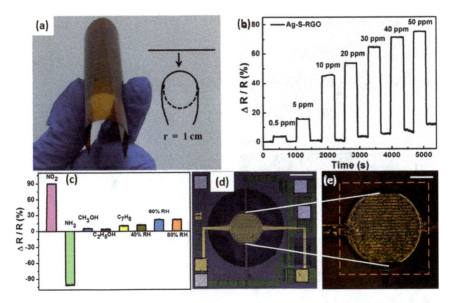

Fig. 6   a Photograph of inkjet-printed rGO-Ag gas sensor, **b** associated graph with respect to NO₂ concentration, **c** selective response to various gases and humidity levels (adapted with permission from reference [143]), **d** inkjet graphene/PVP polymer base composites for humidity sensing, and **e** enlarged view of a printed circuit (adapted with permission from reference [33])

Strain, touch, and pressure sensors: Strain sensors can be used in stretchable and skin-mountable wearables [150, 151]. The strain sensor entails a conductive pattern qualified for generating an electrical signal on the deformation or production of the strain [150]. 2D materials-based printed signals can be important materials in the field of strain sensors. The graphene-based inkjet-printed strain sensors have been demonstrated. The schematic and the printed sensors are depicted in Fig. 8(a and b) [152].

Pressure and touch sensors can detect pressure, contact, or proximity. These are other emerging technologies in the field of consumers, like smartphones and touch pads. Pressure sensors can be fabricated by sandwiching the sensor layer between two contacts. Although there have been a lot of device demonstrations at the commercial level, printed reports on touch or pressure sensors are less. The 2D materials in the field of pressure sensors can be important materials. Figure 8(c) shows the printed graphene thin film pressure sensing panel.

4.4   Printed Energy Storage

Energy storage technologies are used in everything from next-generation wearables and cars to industrial production and portable electronic gadgets [139]. The most

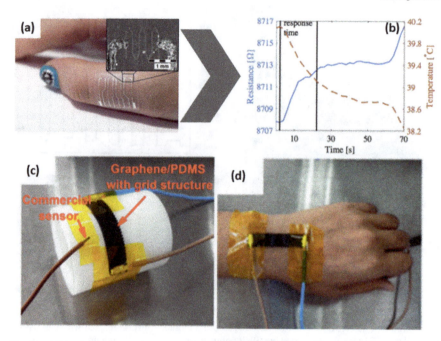

Fig. 7 **a** Photograph of printed graphene temperature sensors attached to the skin, **b** the associated response of change of temperature (adapted with permission from reference [146]), **c** experimental setup, and **d** simultaneous monitoring of change in the parameter of the body using printed graphene biosensors (adapted with permission from reference [28])

popular types of energy storage devices for various purposes are batteries and capacitors. Supercapacitors employ a twofold effect on electrodes to store energy, whereas batteries use chemical processes to accomplish so [140]. Optically conductive electrodes are required for efficient and dense energy storage. Due to their extensive surface area, 2D materials have become promising candidates for electrode function. Because of a reduction in electrode thickness and an increase in electrode-to-electrode contact, graphene-based electrodes can perform better than those now in use. It can also provide better performance due to high electrical conductivity. There have been many successful demonstrations of 2D materials (graphene [153–172], MoS_2, MXenes, etc.)-based energy storage devices. Inkjet printing has been proven to be one of the most efficient methods for the production of these energy storage devices. Many capacitors and micro capacitors have been inkjet-printed from 2D materials for electrodes.

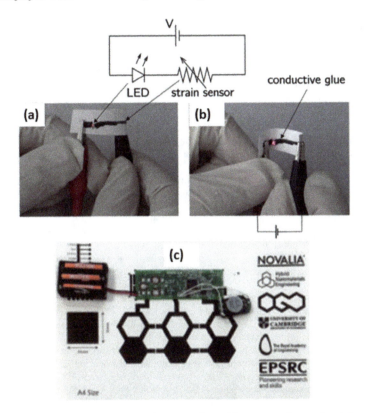

Fig. 8 Printed graphene strain sensors for **a** tensile and **b** compressive strains (adapted with permission from reference [152]), and **c** flexographic printed graphene-based printed capacitors

5 Summary and Future Perspective

We have outlined the current development and evolution of printing methods using 2D material-based inks in this chapter and covered their practical applications in electronics, optoelectronics, biomedicine, sensors, and energy storage. There have been many 2D materials investigated for various printing technologies, but there is still an opportunity for development and improvement. In fact, volumetric printing, two-photon polymerization printing, and selective laser sintering have not employed printing methods. Furthermore, there are numerous untapped potential uses for stereolithography and digital laser printing for producing 2D materials. The studied features of printed 2D materials have shown their significance in fields ranging from electronics to biomedicine, opening up new options to address problems with various 2D materials.

References

1. Ferrari C et al (2006) Raman spectrum of graphene and graphene layers. Phys Rev Lett 97:187401
2. Naumis GG, Barraza-Lopez S, Oliva-Leyva M, Terrones H (2017) Electronic and optical properties of strained graphene and other strained 2D materials: a review. Rep Prog Phys 80:096501
3. Huo C, Yan Z, Song X, Zeng H (2015) 2D Materials via liquid exfoliation: a review on fabrication and applications. Chinese Sci Bull 23:1994
4. Mas-Ballesté R, Gómez-Navarro C, Gómez-Herrero J, Zamora F (2011) 2D materials: to graphene and beyond. Nanoscale 3:20
5. Glavin NR, et al (2020) Emerging applications of elemental 2d materials. Adv Mater Wiley Online Library. https://doi.org/10.1002/adma.201904302
6. Iqbal A, et al (2020) 2D MXenes for electromagnetic shielding: a review. Adv Funct Mater Wiley Online Library. https://doi.org/10.1002/adfm.202000883
7. Tahir MB, Fatima N, Fatima U, Sagir M (2021) A review on the 2D black phosphorus materials for energy applications. Inorg Chem Commun 124:108242
8. Chhowalla M, Liu Z, Zhang H (2015) Two-dimensional transition metal dichalcogenide (TMD) nanosheets. Chem Soc Rev 44:2584
9. Bie C, Cheng B, Fan J, Ho W, Yu J (2021) Enhanced solar-to-chemical energy conversion of graphitic carbon nitride by two-dimensional cocatalysts. EnergyChem 3:100051
10. Aggarwal P, Singh A, Sorifi S, Sharma M, Singh R (2022) Characterization of 2D transition metal dichalcogenides. In: Das S (ed) 2D materials for electronics, sensors and devices: synthesis, characterization, fabrication and application. Elsevier
11. Zhang C, Tan J, Pan Y, Cai X, Zou X, Cheng H-M, Liu B (2020) Mass production of 2D materials by intermediate-assisted grinding exfoliation. Natl Sci Rev 7:324
12. Yao JD, Zheng ZQ, Yang GW (2019) Production of large-area 2D materials for high-performance photodetectors by pulsed-laser deposition. Prog Mater Sci 106:100573
13. Yang S, Zhang P, Nia AS, Feng X (2020) Emerging 2D materials produced via electrochemistry. Adv Mater 32:1907857
14. Chacham H et al (2020) Controlling the morphology of nanoflakes obtained by liquid-phase exfoliation: implications for the mass production of 2D materials. ACS Appl. Nano Mater. 3:12095
15. Park S, Ruoff RS (2009) Chemical methods for the production of graphenes. Nature Nanotech 4:4
16. Hernandez Y et al (2008) High-yield production of graphene by liquid-phase exfoliation of graphite. Nature Nanotech 3:9
17. Zhang D, Chi B, Li B, Gao Z, Du Y, Guo J, Wei J (2016) Fabrication of highly conductive graphene flexible circuits by 3D printing. Synth Met 217:79
18. Ye J, Craciun MF, Koshino M, Russo S, Inoue S, Yuan H, Shimotani H, Morpurgo AF, Iwasa Y (2011) Accessing the transport properties of graphene and its multilayers at high carrier density. Proc Natl Acad Sci 108:13002
19. Lee G-H et al (2013) High-strength chemical-vapor–deposited graphene and grain boundaries. Science 340:1073
20. Cui Z (2016) Printed electronics: materials, technologies and applications. Wiley/Higher Education Press, Hoboken, NJ : Solaris South Tower, Singapore
21. Torrisi F et al (2012) Inkjet-printed graphene electronics. ACS Nano 6:2992
22. Finn DJ, Lotya M, Cunningham G, Smith RJ, McCloskey D, Donegan JF, Coleman JN (2014) Inkjet deposition of liquid-exfoliated graphene and MoS2 nanosheets for printed device applications. J Mater Chem C 2:925
23. Withers F et al (2014) Heterostructures produced from nanosheet-based inks. Nano Lett 14:3987
24. Kelly G et al (2017) All-printed thin-film transistors from networks of liquid-exfoliated nanosheets. Science 356:69

25. McManus D et al (2017) Water-based and biocompatible 2D crystal inks for all-inkjet-printed heterostructures. Nature Nanotech 12:4
26. Hu G et al (2017) Black phosphorus ink formulation for inkjet printing of optoelectronics and photonics. Nat Commun 8:1
27. Bianchi V et al (2017) Terahertz saturable absorbers from liquid phase exfoliation of graphite. Nat Commun 8:1
28. Gómez J, Alegret N, Dominguez-Alfaro A, Vázquez Sulleiro M (2021) Recent advances on 2D materials towards 3D printing. Chemistry 3:1314
29. Xu Y, Schwab MG, Strudwick AJ, Hennig I, Feng X, Wu Z, Müllen K (2013) Screen-printable thin film supercapacitor device utilizing graphene/polyaniline inks. Adv Energy Mater 3:1035
30. Hyun WJ, Secor EB, Hersam MC, Frisbie CD, Francis LF (2015) High-resolution patterning of graphene by screen printing with a silicon stencil for highly flexible printed electronics. Adv Mater 27:109
31. Arapov K, Rubingh E, Abbel R, Laven J, de With G, Friedrich H (2016) Conductive screen printing inks by gelation of graphene dispersions. Adv Func Mater 26:586
32. Secor EB, Lim S, Zhang H, Frisbie CD, Francis LF, Hersam MC (2014) Gravure printing of graphene for large-area flexible electronics. Adv Mater 26:4533
33. Santra S, Hu G, Howe RCT, De Luca A, Ali SZ, Udrea F, Gardner JW, Ray SK, Guha PK, Hasan T (2015) CMOS integration of inkjet-printed graphene for humidity sensing. Sci Rep 5:1
34. Secor EB, Gao TZ, Islam AE, Rao R, Wallace SG, Zhu J, Putz KW, Maruyama B, Hersam MC (2017) Enhanced conductivity, adhesion, and environmental stability of printed graphene inks with nitrocellulose. Chem Mater 29:2332
35. Dodoo-Arhin D, Howe RCT, Hu G, Zhang Y, Hiralal P, Bello A, Amaratunga G, Hasan T (2016) Inkjet-printed graphene electrodes for dye-sensitized solar cells. Carbon 105:33
36. Novoselov KS, Geim AK, Morozov SV, Jiang D, Zhang Y, Dubonos SV, Grigorieva IV, Firsov AA (2004) Electric field effect in atomically thin carbon films. Science 306:666
37. Carey T, Cacovich S, Divitini G, Ren J, Mansouri A, Kim JM, Wang C, Ducati C, Sordan R, Torrisi F (2017) Fully inkjet-printed two-dimensional material field-effect heterojunctions for wearable and textile electronics. Nat Commun 8:1
38. Xu M, Liang T, Shi M, Chen H (2013) Graphene-like two-dimensional materials. Chem Rev 113:3766
39. Lipp K, Schwetz A, Hunold K (1989) Hexagonal boron nitride: fabrication, properties and applications. J Europ Ceram Soc 5:3
40. Li L, Yu Y, Ye GJ, Ge Q, Ou X, Wu H, Feng D, Chen XH, Zhang Y (2014) Black phosphorus field-effect transistors. Nature Nanotech 9:5
41. Joensen P, Frindt RF, Morrison SR (1986) Single-layer MoS2. Mater Res Bull 21:457
42. Radisavljevic B, Radenovic A, Brivio J, Giacometti V, Kis A (2011) Single-layer MoS2 transistors. Nat Nanotech 6:147
43. Chaves et al (2020) Bandgap engineering of two-dimensional semiconductor materials. Npj 2D Mater Appl 4:1
44. Singh A, Lynch J, Anantharaman SB, Hou J, Singh S, Kim G, Mohite AD, Singh R, Jariwala D (2022) Cavity-enhanced raman scattering from 2D hybrid perovskites. J Phys Chem C 126:11158
45. Moun M, Singh A, Tak BR, Singh R (2019) Study of the photoresponse behavior of a high barrier Pd/MoS2/Pd photodetector. J Phys D: Appl Phys 52:325102
46. Moun M, Singh A, Singh R (2018) Study of electrical behavior of metal-semiconductor contacts on exfoliated MoS2 flakes. Phys Status Solidi (a) 215:1800188
47. Naguib M, Kurtoglu M, Presser V, Lu J, Niu J, Heon M, Hultman L, Gogotsi Y, Barsoum MW (2011) Two-dimensional nanocrystals produced by exfoliation of Ti3AlC2. Adv Mater 23:4248
48. VahidMohammadi A, Rosen J, Gogotsi Y (2021) The world of two-dimensional carbides and nitrides (MXenes). Science 372:eabf1581

49. Jiang Q, Lei Y, Liang H, Xi K, Xia C, Alshareef HN (2020) Review of MXene electrochemical microsupercapacitors. Energy Storage Mater 27:78
50. Zhang C, et al. (2019) Additive-free MXene inks and direct printing of micro-supercapacitors. Nat Commun 10:1
51. Hart JL, Hantanasirisakul K, Lang AC, Anasori B, Pinto D, Pivak Y, van Omme JT, May SJ, Gogotsi Y, Taheri ML (2019) Control of MXenes' electronic properties through termination and intercalation. Nat Commun 10:1
52. Anasori B, Lukatskaya MR, Gogotsi Y (2017) 2D metal carbides and nitrides (MXenes) for energy storage. Nat Rev Mater 2:2
53. Bai Y, Liu C, Chen T, Li W, Zheng S, Pi Y, Luo Y, Pang H (2021) MXene-copper/cobalt hybrids via lewis acidic molten salts etching for high performance symmetric supercapacitors. Angew Chem Int Ed 60:25318
54. Huang K, Li Z, Lin J, Han G, Huang P (2018) Two-dimensional transition metal carbides and nitrides (MXenes) for biomedical applications. Chem Soc Rev 47:5109
55. Guo Z, Zhou J, Zhu L, Sun Z (2016) MXene: a promising photocatalyst for water splitting. J Mater Chem A 4:11446
56. Jun B-M, Kim S, Heo J, Park CM, Her N, Jang M, Huang Y, Han J, Yoon Y (2019) Review of MXenes as new nanomaterials for energy storage/delivery and selected environmental applications. Nano Res 12:471
57. Khazaei M, Mishra A, Venkataramanan NS, Singh AK, Yunoki S (2019) Recent advances in MXenes: from fundamentals to applications. Curr Opin Solid State Mater Sci 23:164
58. Sajid M (2021) MXenes: are they emerging materials for analytical chemistry applications?— a review. Anal Chim Acta 1143:267
59. Verger L, Natu V, Carey M, Barsoum MW (2019) MXenes: an introduction of their synthesis, select properties, and applications. Trends Chem 1:656
60. Gusmão R, Sofer Z, Pumera M (2017) Black phosphorus rediscovered: from bulk material to monolayers. Angew Chem Int Ed Engl 56:8052
61. Shen Z-K, Yuan Y-J, Pei L, Yu Z-T, Zou Z (2020) Black phosphorus photocatalysts for photocatalytic H2 generation: a review. Chem Eng J 386:123997
62. Li Q, Wu J-T, Liu Y, Qi X-M, Jin H-G, Yang C, Liu J, Li G-L, He Q-G (2021) Recent advances in black phosphorus-based electrochemical sensors: a review. Anal Chim Acta 1170:338480
63. Ferrara C, Vigo E, Albini B, Galinetto P, Milanese C, Tealdi C, Quartarone E, Passerini S, Mustarelli P (2019) Efficiency and quality issues in the production of black phosphorus by mechanochemical synthesis: a multi-technique approach. ACS Appl Energy Mater 2:2794
64. Antonatos N, Bouša D, Kovalska E, Sedmidubský D, Růžička K, Vrbka P, Veselý M, Hejtmánek J, Sofer Z (2020) Large-scale production of nanocrystalline black phosphorus ceramics. ACS Appl Mater Interfaces 12:7381
65. Brent JR, Savjani N, Lewis EA, Haigh SJ, Lewis DJ, O'Brien P (2014) Production of few-layer phosphorene by liquid exfoliation of black phosphorus. Chem Commun 50:13338
66. Ueda K, Tabata H, Kawai T (2001) Magnetic and electric properties of transition-metal-doped ZnO films. Appl Phys Lett 79:988
67. Liu L, Feng YP, Shen ZX (2003) Structural and electronic properties of H-BN. Phys Rev B 68:104102
68. Jiang X-F, Weng Q, Wang X-B, Li X, Zhang J, Golberg D, Bando Y (2015) Recent progress on fabrications and applications of boron nitride nanomaterials: a review. J Mater Sci Technol 31:589
69. Dean R et al (2010) Boron nitride substrates for high-quality graphene electronics. Nat Nanotech 5:10
70. Carroll E, Buckley D, Mogili NVV, McNulty D, Moreno MS, Glynn C, Collins G, Holmes JD, Razeeb KM, O'Dwyer C (2017) 2D nanosheet paint from solvent-exfoliated Bi2Te3 ink. Chem Mater 29:7390
71. Zheng W et al (2015) Patterning two-dimensional chalcogenide crystals of Bi2Se3 and In2Se3 and efficient photodetectors. Nat Commun 6:1

72. Das VD, Soundararajan N (1989) Thermoelectric power and electrical resistivity of crystalline antimony telluride (Sb2Te3) thin films: temperature and size effects. J Appl Phys 65:2332

73. Jacobs-Gedrim RB, Shanmugam M, Jain N, Durcan CA, Murphy MT, Murray TM, Matyi RJ, Moore RLI, Yu B (2014) Extraordinary photoresponse in two-dimensional In2Se3 nanosheets. ACS Nano 8:514

74. Bandurin et al (2017) High electron mobility quantum hall effect and anomalous optical response in atomically thin InSe. Nat Nanotech 12:3

75. Ng LWT, Hu G, Howe RC, Zhu X, Jones CG, Yang Z, Hasan T (2019) Printing of graphene and related 2D materials: technology, formulation and applications, 1st edn. Springer International Publishing : Imprint: Springer, Cham

76. Castellanos-Gomez A, Poot M, Steele GA, van der Zant HSJ, Agraït N, Rubio-Bollinger G (2012) Elastic properties of freely suspended MoS$_2$ nanosheets. Adv Mater 24:772

77. Li K, Wang W (2020) Effects of substrates on the optical properties of monolayer WS2. J Cryst Growth 540:125645

78. Le OK, Chihaia V, Pham-Ho M-P, Son DN (2020) Electronic and optical properties of monolayer MoS$_2$ under the influence of polyethyleneimine adsorption and pressure. RSC Adv 10:4201

79. Peng Q, De S (2013) Outstanding mechanical properties of monolayer MoS2 and its application in elastic energy storage. Phys Chem Chem Phys 15:19427

80. Sharma M, Aggarwal P, Singh A, Kaushik S, Singh R (2022) Flexible, transparent, and broadband trilayer photodetectors based on MoS2/WS2 nanostructures. ACS Appl Nano Mater 5:13637

81. Sharma M, Singh A, Aggarwal P, Singh R (2022) Large-area transfer of 2D TMDCs assisted by a water-soluble layer for potential device applications. ACS Omega 7:11731

82. Baek SH, Choi Y, Choi W (2015) Large-area growth of uniform single-layer MoS2 thin films by chemical vapor deposition. Nanoscale Res Lett 10:388

83. Okada M, Sawazaki T, Watanabe K, Taniguch T, Hibino H, Shinohara H, Kitaura R (2014) Direct chemical vapor deposition growth of WS$_2$ atomic layers on hexagonal boron nitride. ACS Nano 8:8273

84. Al-Hilfi SH, Derby B, Martin PA, Whitehead JC (2020) Chemical vapour deposition of graphene on copper-nickel alloys: the simulation of a thermodynamic and kinetic approach. Nanoscale 12:15283

85. Aggarwal P, Kaushik S, Bisht P, Sharma M, Singh A, Mehta BR, Singh R (2022) Centimeter-scale synthesis of monolayer WS2 using single-zone atmospheric-pressure chemical vapor deposition: a detailed study of parametric dependence, growth mechanism, and photodetector properties. Cryst Growth Design 22:3206

86. Singh A, Moun M, Sharma M, Barman A, Kumar Kapoor A, Singh R (2021) NaCl-assisted substrate dependent 2D planar nucleated growth of MoS2. Appl Surf Sci 538:148201

87. Singh A, Sharma M, Singh R (2021) NaCl-assisted CVD growth of large-area high-quality trilayer MoS2 and the role of the concentration boundary layer. Cryst Growth Design **21**:4940

88. Singh A, Sharma M, Singh R (2019) Effect of different precursors on CVD growth of molybdenum disulfide. J Alloys Comp 782:772

89. Ohring M (2002) Materials science of thin films: deposition and structure, 2nd edn. Academic Press, San Diego, CA

90. Gao E, Lin S-Z, Qin Z, Buehler MJ, Feng XQ, Xu Z (2018) Mechanical exfoliation of two-dimensional materials. J Mech Phys Solids 115:248

91. Bellani S, Bartolotta A, Agresti A, Calogero G, Grancini G, Di Carlo A, Kymakis E, Bonaccorso F (2021) Solution-processed two-dimensional materials for next-generation photovoltaics. Chem Soc Rev 50:11870

92. Yoo H, Kim C (2015) Experimental studies on formation, spreading and drying of inkjet drop of colloidal suspensions. Colloids Surf A 468:234

93. Konstantinov N et al (2021) Electrical read-out of light-induced spin transition in thin film spin crossover/graphene heterostructures. J. Mater. Chem. C 9:2712

94. López LEP, Moczko L, Wolff J, Singh A, Lorchat E, Romeo M, Taniguchi T, Watanabe K, Berciaud S (2021) Single- and narrow-line photoluminescence in a boron nitride-supported MoSe$_2$/graphene heterostructure. C R Phys 22:77
95. Singh A, Singh R (2020) γ-ray irradiation-induced chemical and structural changes in CVD monolayer MoS2. ECS J Solid State Sci Technol 9:093011
96. Sharma M, Singh A, Singh R (2020) Monolayer MoS2 transferred on arbitrary substrates for potential use in flexible electronics. ACS Appl Nano Mater. 3:4445
97. Eames C, Islam MS (2014) Ion Intercalation into two-dimensional transition-metal carbides: global screening for new high-capacity battery materials. J Am Chem Soc 136:16270
98. Shen J et al (2015) Liquid phase exfoliation of two-dimensional materials by directly probing and matching surface tension components. Nano Lett 15:5449
99. Sahoo D, Kumar B, Sinha J, Ghosh S, Roy SS, Kaviraj B (2020) Cost effective liquid phase exfoliation of MoS2 nanosheets and photocatalytic activity for wastewater treatment enforced by visible light. Sci Rep 10:10759
100. Zhang B, Wu Q, Yu H, Bulin C, Sun H, Li R, Ge X, Xing R (2017) High-efficient liquid exfoliation of boron nitride nanosheets using aqueous solution of alkanolamine. Nanoscale Res Lett 12:596
101. Suganuma K (2014) Introduction to printed electronics. Springer Science + Business Media, New York
102. Dang MC, Dang TMD, Fribourg-Blanc E (2014) Silver nanoparticles ink synthesis for conductive patterns fabrication using inkjet printing technology. Adv Nat Sci Nanosci Nanotechnol 6:015003
103. Menon H, Aiswarya R, Surendran KP (2017) Screen printable MWCNT inks for printed electronics. RSC Adv 7:44076
104. Leach RH (ed) (1993) The printing ink manual, 5th edn. Blueprint, London , New York
105. Flick W (1999) Printing ink and overprint varnish formulations, 2nd edn. Noyes Publications/ William Andrew Pub, Norwich, N.Y.
106. Hoath SD (ed) (2016) Fundamentals of inkjet printing: the science of inkjet and droplets. Wiley-VCH Verlag GmbH & Co. KGaA, Weinheim
107. Yang X, Li X-M, Kong Q-Q, Liu Z, Chen J-P, Jia H, Liu Y-Z, Xie L-J, Chen C-M (2020) One-pot ball-milling preparation of graphene/carbon black aqueous inks for highly conductive and flexible printed electronics. Sci China Mater 63:392
108. Fowkes M (ed) (1964) Contact angle, wettability, and adhesion, vol 43. American Chemical Society, Washington, D.C.
109. Robertson GL (2016) Food packaging. CRC Press
110. Hutchings M, Martin G (eds) (2013) Inkjet technology for digital fabrication. Wiley, Chichester, West Sussex, United Kingdom
111. Derby B (2010) Inkjet printing of functional and structural materials: fluid property requirements, feature stability, and resolution. Annu Rev Mater Res 40:395
112. Fromm JE (1984) Numerical calculation of the fluid dynamics of drop-on-demand jets. IBM J Res Dev 28:322
113. Jang D, Kim D, Moon J (2009) Influence of fluid physical properties on ink-jet printability. Langmuir 25:2629
114. Coleman JN et al (2011) Two-dimensional nanosheets produced by liquid exfoliation of layered materials. Science 331:568
115. Secor B, Prabhumirashi PL, Puntambekar K, Geier ML, Hersam MC (2013) Inkjet printing of high conductivity, flexible graphene patterns. J Phys Chem Lett 4:1347
116. Li J, Ye F, Vaziri S, Muhammed M, Lemme MC, Östling M (2013) Efficient inkjet printing of graphene. Adv Mater 25:3985
117. Michel M, Desai JA, Biswas C, Kaul AB (2016) Engineering chemically exfoliated dispersions of two-dimensional graphite and molybdenum disulphide for ink-jet printing. Nanotechnology 27:485602
118. Feng N, Meng R, Zu L, Feng Y, Peng C, Huang J, Liu G, Chen B, Yang J (2019) A polymer-direct-intercalation strategy for MoS2/carbon-derived heteroaerogels with ultrahigh pseudocapacitance. Nat Commun 10:1372

119. Song B, Wu F, Zhu Y, Hou Z, Moon K, Wong C-P (2018) Effect of polymer binders on graphene-based free-standing electrodes for supercapacitors. Electrochim Acta 267:213
120. Hu G, Kang J, Ng LWT, Zhu X, Howe RC, Jones CG, Hersam MC, Hasan T (2018) Functional Inks and printing of two-dimensional materials
121. Kang B, Lee WH, Cho K (2013) Recent advances in organic transistor printing processes. ACS Appl Mater Interfaces 5:2302
122. Zhang W, Li XD, Li HB, Chen S, Sun Z, Yin XJ, Huang SM (2011) Graphene-based counter electrode for dye-sensitized solar cells. Carbon 49:5382
123. Joseph M, Nagendra B, Bhoje Gowd E, Surendran KP (2016) Screen-printable electronic ink of ultrathin boron nitride nanosheets. ACS Omega 1:1220
124. Pavličková M, Lorencová L, Hatala M, Kováč M, Tkáč J, Gemeiner P (2022) Facile fabrication of screen-printed MoS2 electrodes for electrochemical sensing of dopamine. Sci Rep 12:11900
125. Xu, B. Xu, Y. Gu, Z. Xiong, J. Sun, and X. S. Zhao, *Graphene-Based Electrodes for Electrochemical Energy Storage*, Energy Environ. Sci. **6**, 1388 (2013).
126. Tracton AA, (ed) (2005) Coatings technology handbook, 3rd ed. Taylor & Francis, Boca Raton, FL
127. Lahti M, Leppävuori S, Lantto V (1999) Gravure-offset-printing technique for the fabrication of solid films. Appl Surf Sci 142:367
128. Nguyen D, Lee C, Shin K-H, Lee D (2015) An investigation of the ink-transfer mechanism during the printing phase of high-resolution roll-to-roll gravure printing. IEEE Trans Compon Packag Manufact Technol 5:1516
129. Nguyen D, Lee J, Kim CH, Shin K-H, Lee D (2013) An approach for controlling printed line-width in high resolution roll-to-roll gravure printing. J Micromech Microeng 23:095010
130. Streitberger H-J, Goldschmidt A (2018) BASF handbook basics of coating technology, 3rd edn. Vincentz, Hannover
131. Baker J, Deganello D, Gethin DT, Watson TM (2014) Flexographic printing of graphene nanoplatelet ink to replace platinum as counter electrode catalyst in flexible dye sensitised solar cell. Energy Mater **9**:86
132. Howe RC, Hu G, Yang Z, Hasan T (2015) Functional Inks of Graphene, Metal Dichalcogenides and Black Phosphorus for Photonics and (Opto) Electronics. In: Society of photo-optical instrumentation engineers (SPIE)
133. A Phillips C, Al-Ahmadi A, Potts S-J, Claypole T, Deganello D (2017) The effect of graphite and carbon black ratios on conductive ink performance. J Mater Sci 52:9520
134. Secor EB, Ahn BY, Gao TZ, Lewis JA, Hersam MC (2015) Rapid and versatile photonic annealing of graphene inks for flexible printed electronics. Adv Mater 27:6683
135. Nicolosi V, Chhowalla M, Kanatzidis MG, Strano MS, Coleman JN (2013) Liquid exfoliation of layered materials. Science 340:1226419
136. Sirringhaus H (2-14) 25th anniversary article: organic field-effect transistors: the path beyond amorphous silicon. Adv Mater 26:1319
137. Xia F, Wang H, Xiao D, Dubey M, Ramasubramaniam A (2014) Two-dimensional material nanophotonics. Nature Photon 8:12
138. Sun Z, Hasan T, Torrisi F, Popa D, Privitera G, Wang F, Bonaccorso F, Basko DM, Ferrari AC (2010) Graphene mode-locked ultrafast laser. ACS Nano 4:803
139. Schedin F, Geim AK, Morozov SV, Hill EW, Blake P, Katsnelson MI, Novoselov KS (2007) Detection of individual gas molecules adsorbed on graphene. Nature Mater 6:9
140. Mannoor MS, Tao H, Clayton JD, Sengupta A, Kaplan DL, Naik RR, Verma N, Omenetto FG, McAlpine MC (2012) Graphene-based wireless bacteria detection on tooth enamel. Nat Commun 3:1
141. Yao Y, Tolentino L, Yang Z, Song X, Zhang W, Chen Y, Wong C (2013) High-concentration aqueous dispersions of MoS2. Adv Func Mater 23:3577
142. Cho S-Y, Lee Y, Koh H-J, Jung H, Kim J-S, Yoo H-W, Kim J, Jung H-T (2016) Superior chemical sensing performance of black phosphorus: comparison with MoS2 and graphene. Adv Mater 28:7020

143. Huang L, Wang Z, Zhang J, Pu J, Lin Y, Xu S, Shen L, Chen Q, Shi W (2014) Fully printed, rapid-response sensors based on chemically modified graphene for detecting NO2 at room temperature. ACS Appl Mater Interfaces 6:7426

144. Kennedy W (1969) Thin film temperature sensor. Rev Sci Instrum 40:1169

145. Yeo W-H et al (2013) Multifunctional epidermal electronics printed directly onto the skin. Adv Mater 25:2773

146. Vuorinen T, Niittynen J, Kankkunen T, Kraft TM, Mäntysalo M (2016) Inkjet-printed graphene/PEDOT:PSS temperature sensors on a skin-conformable polyurethane substrate. Sci Rep 6:35289

147. Shavanova K, Bakakina Y, Burkova I, Shtepliuk I, Viter R, Ubelis A, Beni V, Starodub N, Yakimova R, Khranovskyy V (2016) Application of 2D non-graphene materials and 2d oxide nanostructures for biosensing technology. Sensors 16:223

148. Li J, Rossignol F, Macdonald J (2015) Inkjet printing for biosensor fabrication: combining chemistry and technology for advanced manufacturing. Lab Chip 15:2538

149. Zang D, Yan M, Ge S, Ge L, Yu J (2013) A disposable simultaneous electrochemical sensor array based on a molecularly imprinted film at a NH2-graphene modified screen-printed electrode for determination of psychotropic drugs. Analyst 138:2704

150. Amjadi M, Kyung K-U, Park I, Sitti M (2016) Stretchable, skin-mountable, and wearable strain sensors and their potential applications: a review. Adv Funct Mater 26:1678

151. Karim N, Afroj S, Tan S, He P, Fernando A, Carr C, Novoselov KS (2017) Scalable production of graphene-based wearable E-textiles. ACS Nano 11:12266

152. Casiraghi C, Macucci M, Parvez K, Worsley R, Shin Y, Bronte F, Borri C, Paggi M, Fiori G (2018) Inkjet printed 2D-crystal based strain gauges on paper. Carbon 129:462

153. Verma A, Parashar A, Packirisamy M (2018) Atomistic modeling of graphene/hexagonal boron nitride polymer nanocomposites: a review. Wiley Interdiscipl Rev Comp Mol Sci 8(3):e1346

154. Verma A, Parashar A, Packirisamy M (2019) Effect of grain boundaries on the interfacial behaviour of graphene-polyethylene nanocomposite. Appl Surf Sci 470:1085–1092

155. Verma A, Kumar R, Parashar A (2019) Enhanced thermal transport across a bi-crystalline graphene–polymer interface: an atomistic approach. Phys Chem Chem Phys 21(11):6229–6237

156. Kataria A, Verma A, Sanjay MR, Siengchin S (2022) Molecular modeling of 2D graphene grain boundaries: mechanical and fracture aspects. Mater Today Proc 52:2404–2408

157. Verma A, Jain N, Sethi SK (2022) Modeling and simulation of graphene-based composites. In: Innovations in graphene-based polymer composites. Woodhead Publishing, pp 167–198

158. Verma A, Sharma S (2022) Atomistic simulations to study thermal effects and strain rate on mechanical and fracture properties of graphene like BC3. In: Forcefields for atomistic-scale simulations: materials and applications. Springer, Singapore, pp 237–252

159. Verma A, Parashar A, van Duin AC (2022) Graphene-reinforced polymeric membranes for water desalination and gas separation/barrier applications. In: Innovations in graphene-based polymer composites. Woodhead Publishing, pp 133–165

160. Verma A, Parashar A (2017) The effect of STW defects on the mechanical properties and fracture toughness of pristine and hydrogenated graphene. Phys Chem Chem Phys 19(24):16023–16037

161. Verma A, Parashar A (2018) Molecular dynamics based simulations to study failure morphology of hydroxyl and epoxide functionalised graphene. Comput Mater Sci 143:15–26

162. Verma A, Parashar A (2018) Molecular dynamics based simulations to study the fracture strength of monolayer graphene oxide. Nanotechnology 29(11):115706

163. Verma A, Parashar A (2018) Structural and chemical insights into thermal transport for strained functionalised graphene: a molecular dynamics study. Mater Res Express 5(11):115605

164. Verma A, Parashar A, Packirisamy M (2018) Tailoring the failure morphology of 2D bicrystalline graphene oxide. J Appl Phys 124(1):015102

165. Verma A, Parashar A (2018) Reactive force field based atomistic simulations to study fracture toughness of bicrystalline graphene functionalised with oxide groups. Diam Relat Mater 88:193–203

166. Singla V, Verma A, Parashar A (2018) A molecular dynamics based study to estimate the point defects formation energies in graphene containing STW defects. Mater Res Express 6(1):015606

167. Verma A, Parashar A, Packirisamy M (2019) Role of chemical adatoms in fracture mechanics of graphene nanolayer. Mater Today Proc 11:920–924

168. Verma A, Zhang W, Van Duin AC (2021) ReaxFF reactive molecular dynamics simulations to study the interfacial dynamics between defective h-BN nanosheets and water nanodroplets. Phys Chem Chem Phys 23(18):10822–10834

169. Deji R, Verma A, Kaur N, Choudhary BC, Sharma RK (2022) Density functional theory study of carbon monoxide adsorption on transition metal doped armchair graphene nanoribbon. Mater Today Proc 54:771–776

170. Deji R, Verma A, Choudhary BC, Sharma RK (2022) New insights into NO adsorption on alkali metal and transition metal doped graphene nanoribbon surface: a DFT approach. J Mol Graph Model 111:108109

171. Deji R, Jyoti R, Verma A, Choudhary BC, Sharma RK (2022) A theoretical study of HCN adsorption and width effect on co-doped armchair graphene nanoribbon. Comput Theor Chem 1209:113592

172. Deji R, Verma A, Kaur N, Choudhary BC, Sharma RK (2022) Adsorption chemistry of co-doped graphene nanoribbon and its derivatives towards carbon based gases for gas sensing applications: quantum DFT investigation. Mater Sci Semicond Process 146:106670

Chapter 13
Coating Methods for Hydroxyapatite—A Bioceramic Material

Gagan Bansal, Rakesh Kumar Gautam, Joy Prakash Misra, and Abhilasha Mishra

1 Introduction

Currently, the design and development of bioimplants and its optimization have highly evolved the researcher to work in the multidisciplinary domain. Researchers from various engineering disciplines like biomedical engineering, mechanical engineering, chemical engineering, material science engineering, metallurgy engineering, ceramic engineering, etc., are collectively working to develop highly customized, biocompatible, bioactive, nontoxic, anti-inflammatory, anti-fungal, antibacterial, osteointegration, high thermal stability, non-cytotoxicity and high mechanical strength implants. Synthetic Implants are getting popular because of the advancement in the health sector [1].

Crystalline hydroxyapatite has a hexagonal structure with approximate lattice parameters as $a = b = 9.372$ Å, $c = 6.88$ Å, $\alpha = \beta = 90°$ and $\gamma = 120°$. The skeleton of the unit cell is a tetrahedral arrangement of phosphate ($PO4^{3-}$) that forms the HAP structure. Two of the oxygen are perpendicular to the c-axis, while the other two are in a horizontal plane [2]. Within the unit cell, phosphates are separated into two layers with corresponding heights of 1/4 and 3/4. The simple crystallite structure of HAP is shown in Fig. 1. Based on the application and requirements, biodegradable and non-biodegradable composites are being developed using hydroxyapatite (HAP) material. Basically, hydroxyapatite i.e., calcium hydroxyapatite ($Ca_{10}(PO_4)_6(OH)_2$), also chemically named as Pentacalcium hydroxide triphosphate ($Ca_5(PO_4)_3(OH)$) is the most adored stable inorganic bioceramic [3] material which has characteristics similar to human bone.

G. Bansal (✉) · R. K. Gautam · J. P. Misra
Indian Institute of Technology, Banaras Hindu University, Varanasi, India
e-mail: gaganbansal.rs.mec21@itbhu.ac.in

G. Bansal · A. Mishra
Graphic Era (deemed to be) University, Dehradun, India

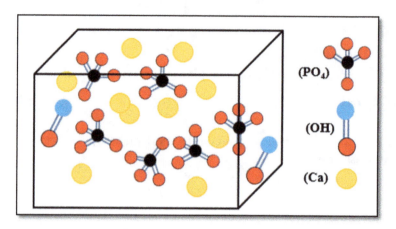

Fig. 1 Crystal structure of hydroxyapatite ($Ca_{10}(PO_4)_6(OH)_2$)

The motivation behind synthesizing and coating HAP with high commercial applications is its closeness to human bone. The inorganic phase of a natural bone consists of almost 65–70% HAP and 5–8% water. HAP has a molecular weight of 1004.6 g/mol. Also, the organic phase of human bone consists of protein fiber i.e., collagen, along with amino acid, which is responsible for the elastic resistance of the bone [3]. Several degradable and non-degradable biocomposites are being developed using hydroxyapatite as a matrix or the reinforcing agent.

Crystalline HaP has a high level of thermal stability with a Ca/P ratio of 1.67 [4, 5]. It enhances osteoconduction [6–9], Osteointegration [10, 11], bioactivity[12], etc. It has biocompatibility with several types of biocells like osteoblast, osteoclast, fibroblasts, periodontal ligament cells and macrophage [6, 13]. It is, therefore, the first choice of several researchers working in the field of artificial implants [14, 15]. Other than bioimplants, additional applications of HAP includes the pharmaceutical sector, water treatment, protein chromatography and fertilizer [16]. HAP can also be used for coating purposes on a large variety of substrates due to its effective adhesive nature and blending ability. High Strength and good surface morphology can be achieved by coating HAP-based composite with metallic substrates. Some resorbable polymers and inorganic compound-based HAP reinforcement composites are gaining popularity in biomedical applications. Apart from the coating of HAP and its composites, a small amount of doping in HAP also improves the characteristics and the desired outputs can be achieved. For example, coating of silver or copper doped HAP on any bioimpant improves the antibacterial nature of the coating material [17]. HAP delivers numerous applications in biomedical fields. Artificial limbs, joints, part replacements, prosthesis, soft tissue placements, resorbable surgery and drug delivery with HAP material is possible, and it's in-vivo and in-vitro characteristics are also verified [18].

Coating of HAP on any substrate is commercially acceptable if it possesses certain minimum specific properties as outlined by Food and Drug Administration guidelines

Table 1 Specifications required for HAP Coating (reprinted from reference [20] with permission from Elsevier)

Property	Specification
Coating thickness	As desired
Crystallinity	62% (minimum)
Phase purity	95% (minimum)
Ca/P	1.67–1.76
Density	2.98 g/cm^3
Heavy metals	Less than 50 ppm
Tensile strength	More than 50.8 MPa
Shear strength	More than 20 MPa
Abrasion	Not specific

as mentioned in Table 1. The thickness, crystallinity, phase purity, Ca/P ratio, density, presence of heavy metals, tensile strength, abrasion and shear strength properties and specifications are taken under the permissible limit for better adhesion. Yang et al. in 2004 [19] reported these properties desired for HAP coating, as mentioned in Table 1. It was again published in a review article on hydroxyapatite-based coating techniques by Asri et al. [20] in 2015.

2 Hydroxyapatite Based Coating

HAP-based coating on metallic implants provides biomimetic surface morphology, bioactivity and biocompatibility, along with strength and mechanical rigidity to the artificial implant. [21]. Metallic implant possesses outstanding mechanical qualities and adaptability within the physiological environment. Metallic biomaterials like 316L stainless steel, cobalt-based alloy, titanium and its alloys are extensively employed as medical implants [22, 23]. The inability of metallic implants to perfectly couple to the local tissue environment in the human body is now a serious problem in the field. The dissonance arises from the fact that metallic implants and human bone have different chemical compositions. All the governing parameters, like size, shape, strength, texture, morphology, compatibility, reactivity, etc., can be feasibly controlled and fabricated using numerous matrix-reinforcement combinations [24]. Coating otherwise bioinactive implants with hydroxyapatite, which has a bonelike structure and promotes osseointegration, allows for speedier healing and recovery. This coating has several applications outside of just orthopedic and dental implants, such as drug administration. The same mindset and upgradation have become possible for contact surface modification using advanced coating techniques. Focusing on hydroxyapatite, surface morphology characteristics and topography are also enhanced using HaP biocomposite coating on the metal, polymer and ceramic-based composites. Coating metallic biomaterials with HAP can increase their resistance to corrosion while also boosting their load-bearing capacity and

adherence to their substrate [25]. The nature of the substrate, its adhesive characteristics, compatibility with the coating material, durability and application are the
governing parameters of HAP coating. Coating methods include dip coating [26],
sputter coating [27], electrophoretic coating [28–30], pulsed laser deposition coating,
sol–gel coating, microarc oxidation [31] and thermal spraying [32].

Out of all the coating methods mentioned, the Sol–gel coating method provides
a minimum coating thickness (in nm). The advantage and disadvantages of several
coating methods are mentioned in Table 2.

Table 2 Coating methods adopted for hydroxyapatite coating [59]

Coating method	Coating thickness	Adhesion strength (MPa)	Advantages	Disadvantages
Sol gel method [50, 59]	100 nm–1 μm	25–35	Works at Low temperatures, can coat complex shapes, thin coating and cost-effective	It requires control atmosphere and some costly raw materials which are difficult to obtain
Pulsed laser deposition method [60]	Less than 5 μm	30–40	Uniform coating on simple and flat surfaces	Costly, time-consuming, produce amorphous coating, cannot coat complex geometries
Sputter coating [61]	0.02–1 μm	55–65	Uniform coating on simple and flat surfaces	It uses line of sight technique, costly, cannot coat complex geometries, produce amorphous coatings
Dip coating [26]	0.05–0.5 mm	28–32	Fast, low cost and can coat complex shapes	Requires high temperature for sintering, mismatch in thermal expansion
Electrophoretic deposition methods [28, 62]	0.1–2 mm	15–18	Fast process, uniform coating, can coat complex substrates	Requires high thermal temperature, difficult to produce defect-less coating
Thermal spray coating [63]	30–250 μm	25–35	Fast and high deposition rate	Line of sight coating, requires high temperature, it induces deposition, amorphous coating occurs due to rapid cooling
Biomimetic coating [10, 59, 60, 64, 65]	~1 mm	10	Fast Process, can be done on complex geometries	Low adhesion strength, surface treatment required, applicable for limited substrate

In order to modify the surface morphology, characteristics, hardness, biocompatibility, chemical insulation and degradation rate, hydroxyapatite coating is an effective solution [33]. As mentioned, coating methods like sol–gel method, pulsed laser deposition method, sputtering coating, electrophoretic deposition method, dip coating, thermal spray coating, and biomimetic coating method are highly appreciated for HAP and are being adopted based on the substrate, nature and state of coating material, adhesion strength requirement, application, coating thickness, etc. Every method has its own advantages and disadvantages. Galindo et al. [34] works on the coating of Nanosized hydroxyapatite on the biocompatible polymer matrix substrate to enhance the bioactivity of the polymer-based implant. The results revealed the improved rheological, thermal and mechanical properties of the final composite along with biocompatibility and bioactivity [34]. Maximum Bioceramics [35] have inert nature and require osteogenesis as it does not form any bonding with the bone [36]. Thus adding a fibrous hydroxyapatite coating on the surface enables the osteoconduction nature of the implant and the natural bone. Hydroxyapatite is a bioactive bioceramic with improved biocompatibility [15, 37–39]. The size, shape, composition and morphology of different graded hydroxyapatite regulate the coating characteristics. Pepla et al. [40] effectively reviewed the nanosized HaP and elaborated on the efficient adhesion, proliferation and mineralization of regenerative, restorative and preventive dentistry applications [40].

Coating of HaP on a metallic, polymeric, ceramic or hybrid substrate requires some standardized parameters as directed by Food and drug administration and ISO standards. These parameters include coating thickness (>1 µm), approx. 62% crystallinity, 95% purity, constant Ca/P ratio (between 1.67 – 1.76), weight density ($2980 \, \text{kg/m}^3$), tensile strength (more than 50 MPa), shear strength (more than 20 MPa) and permissible abrasion [41–43].

3 Thin Film Deposition Techniques

Coating and thin film deposition techniques [44] as described in Fig. 2 are the surface adhesion techniques that are used for a large variety of materials and substrates. Broad classifications of thin film deposition techniques are solution-based deposition technique, vapors-based deposition technique (VDT) and physical vapors deposition techniques. Further, solution-based deposition techniques are sub-classified as sol–gel coating, spin coating, dip coating, screen coating and combined coating technique. Vapor-based deposition technique includes metal–organic, low-pressure, atmospheric pressure and plasma-enhanced VDT. Also, physical vapor deposition techniques include thermal evaporation, pulsed laser deposition, molecular beam and sputtering deposition (DC and RF). DC sputtering includes direct current, whereas, for RF, the source of power is alternating current. Koch et al. [45] reported thin film coating of HAP and alpha-tricalcium phosphate using two step porosification by vapor deposition method and outlined its application in electronics, biomedical and nano-filtration membrane. The obtained coatings are approximately 35 nm in

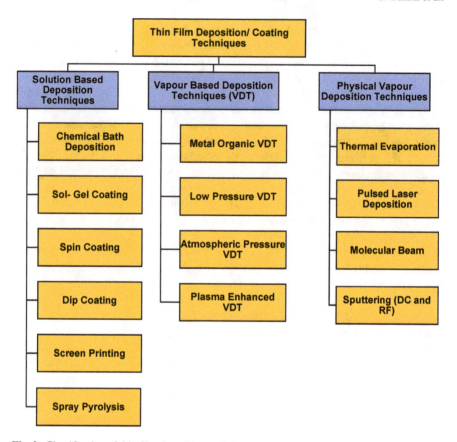

Fig. 2 Classification of thin film deposition techniques

thickness. Piccirillo et al. [46] also used the vapor deposition method for coating nano-sized HAP/TiO$_2$ composite using aerosol and reported excellent photocatalytic activity.

3.1 Deposition of HAP by Chemical Method

Using the aqueous solutions of calcium and phosphate salts, electrolysis produces their ions (calcium and phosphate ions). NaNO$_3$ and H$_2$O$_2$ are utilized to increase the electrolytic conductivity and decrease the hydrogen evolution, respectively, during the process [47, 48]. Several processes result in the synthesis of hydroxyapatite during electrochemical deposition. At the cathode, hydrogen is initially created as bubbles (Eq. (1)). The hydroxyl radicals then increase the pH near the cathode. In addition, the increase in hydroxide ion concentration causes a rise in the generation of hydrogen phosphate and phosphate as shown in Eqs. (2) and (3). Simultaneously,

HAP is produced by depositing calcium, phosphate groups, and hydroxyl radicals at the cathode, as depicted in Eq. (4). Also, other studies demonstrated the existence of carbonate ions (CO_3) alongside HAP. The carbonate ions, formed when carbon dioxide is dissolved in an aqueous electrolytic solution, can readily replace the phosphate group, generating carbonized hydroxyapatite (Eq. (5–7)).

$$2H_2O + 2e^- \rightarrow H_2 \uparrow + 2OH^- \tag{1}$$

$$H_2PO4^- + OH^- \rightarrow HPO_4^{2-} + H_2O \tag{2}$$

$$HPO_4^{2-} + OH^- \rightarrow PO_4^{3-} + H_2O \tag{3}$$

$$10Ca_2 + + 6PO_4^{3-} + 2OH^- \rightarrow Ca_{10}(PO_4)_6(OH)_2 \tag{4}$$

$$CO_2 + H_2O \rightarrow H_2CO_3 \tag{5}$$

$$H_2CO_3 + OH^- \rightarrow HCO^{3-} + H_2O \tag{6}$$

$$HCO^{3-} + OH^- \rightarrow CO_3^{2-} + H_2O \tag{7}$$

In addition, the mechanism of HAP deposition in the EP process is related to the synthesis of HAP powder utilizing the metathesis method 60 with calcium nitrate and di-ammonium hydrogen phosphate in a Ca/P ratio of 1.667%. (Eq. 8).

$$10Ca(NO_3)_2 + 6(NH_4)_2HPO_4 + 8NH_4OH \rightarrow Ca_{10}(PO_4)_6(OH)_2 \\ + 6H_2O + 20NH_4NO_3 \tag{8}$$

The produced HAP powder can be directly dissolved in any organic solvent (ethanol or isopropanol), the positively charged HAP particles migrate toward the cathode, and deposition takes place [49].

Hydroxyapatite coating on metallic, polymer or any biocompatible substrate requires certain commercial coating methods like sol–gel, pulsed deposition, sputter coating, dip coating, spin coating, thermal coating, microarc oxidation, electrophoretic deposition, biomimetic coating etc. These coating methods are discussed in detail below.

4 Commercial Techniques for HAP Coating on Metallic Implants

Coatings can be used to improve implant devices in terms of biocompatibility, reliability, and performance. Therefore, the majority of researchers have reported successful tests of HAP coating applied to different metallic implants concerning their biocompatibility and corrosion behavior. Biocompatibility and corrosion issues with metallic biomaterials have been addressed in recent years by employing many different HAP deposition techniques. Some examples of deposition methods include the sol–gel technique, pulsed laser, plasma spraying, electrochemical deposition, and high-velocity suspension plasma spraying. Some of them are discussed below.

4.1 Sol–Gel Coating Method

Jaafar et al. [50] used the titanium substrate and hydroxyapatite coating using the sol–gel coating method for bioimplant application. The elaborative review on Sol–Gel based coating method for uniform and homogeneous layering, its optimization and factors governing the coating quality like corrosion resistance, osteointegration with the substrate material, morphology, rheology, mechanical properties, bioactivity, etc. Micro coating of HaP on different substrate materials like stainless steel [51], titanium and titanium alloy [50, 52, 53], magnesium alloy [54], etc., provides desired surface modification for biomedical applications. Basically, for hydroxyapatite-based coating in different substrates, the aqueous sol containing the source of calcium as calcium nitrate tetrahydrate and the source of the phosphatic group as triammonium phosphate trihydrate is mostly used [52]. A repetitive dip and withdraw technique is applied for better adhesion. The advantages of using the sol–gel coating method are the low cost, low operating temperature, coating on complex substrates and thin coating. The limitation includes the controlled environment desired for coating and the post-processing in the form of annealing upto 600 °C as desired for enhanced adhesion.

Sol–Gel method along with the post and preprocessing, are used for synthesis as well as coating purposes. Tripathi et al. [55], while preparing a memristor with TiO_2 for electrical characterization, used the sol–gel process along with high temperature calcination for powder preparation and film coating. Figure 3 shows the simplest process diagram of converting the powdered precursors to Nanopowder and sol solution applicable for coating. A simple dissolution method is used to make the sol solution which is further used in spin coating, dip coating and a combination of both.

Figure 4 highlights the simple schematic representation of utilizing the precursor in forming aerogel, xerogel, powder or coating material. Super hydrophobicity [56], antibacterial surface morphology [57], good biocompatibility and Osteointegration

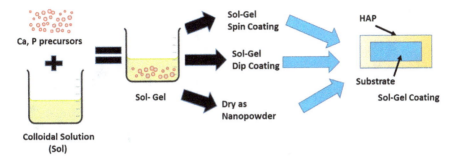

Fig. 3 Steps to convert precursors to nanopowder and coating material using sol–gel method [50]

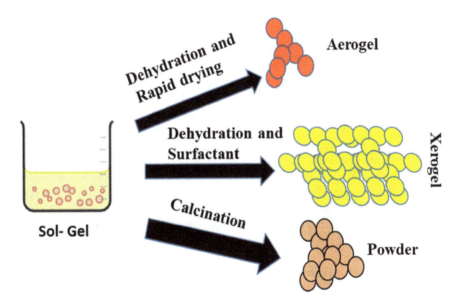

Fig. 4 Use of sol–gel method [55]

of substrate material can be improved using HAP coating over any metallic and non-metallic material. Heat treatment improves the characteristics of the final product obtained using the Sol–Gel Method. The sol–gel process is followed by spin coating or dip coating. Nanostructured coating materials can be achieved by the sol–gel method as stated and reported by Mishra et al. while coating solar panels [58]. Figure 5 shows the nanosized thickness of HAP coating on the commercially pure titanium substrate. Asri et al. [20], in the review article, highlighted the thickness of HAP coating using different coating methods and showed the high magnification image of HAP coating on commercially pure titanium substrate captured using scanning electron microscopy. The 100 nm scale reflects that the thickness of HAP coating is nanosized and uniform.

Fig. 5 SEM magnification
highlighting thickness of
HAP coating on
commercially pure titanium
from a fractured section
(Reprinted from [20] with
permission from Elsevier)

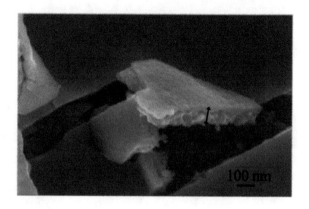

100 mm

4.2   Microarc Oxidation Coating

Microarc oxidation (MAO) or plasma electrolytic oxidation is an effective technique
for the surface treatment of metals like Ti, Al, Mg, Zr, Ta and their alloys. It is
a convenient and effective technique to produce porous, rough and hard ceramic
coatings on metals such as Ti, Mg, Al, Ta and Zr and their alloys. The MAO technique
relies on the anodic oxidation of metals in aqueous electrolyte solutions. Sparks can
be triggered by applying voltages higher than the dielectric breakdown voltage of
a thickening oxide film since this causes the creation of micro-discharge plasma
channels on the metal's surface oxide. Multiple simultaneous plasma discharges
cause intense pressure and heat to concentrate on the metal's surface. Qaid et al. [66]
performed the coating of eggshell-derived HAP on titanium substrates and achieved
good adhesion and corrosion resistance. Figure 6 shows the morphology of the cross-
section view of the MAO-based coating through FESEM. The figure shows clear
coating, epoxy and titanium substrate. The samples are coated with 0, 1, 1, 5 and
2% HAP coating on Ti6Al4V substrate, and the results reveal coating thicknesses
of 8.53, 10.65, 12.72 and 15.57 μm, respectively. This implies that MAO coating
method is used for thin coating applications.

4.3   Pulsed Laser Deposition Method

The pulsed laser deposition method (PLDM) requires the high intensity laser beam
focusing the vacuum chamber on striking a material to be coated. It requires preces-
sion and an expensive setup. A simple schematic diagram of PLDM is shown in Fig. 5.
It provides a uniform and homogeneous coating on flat surfaces with a high level
of accuracy and minimum thickness (less than 5 μm) [60]. PLDM provides a thin
and uniform coating of HAP on metallic substrates [67]. Dhinasekaran, D et al. [68]
used the bioactive glass along with HAP for coating and performed microstructural
and electrochemical characterization. A comparison of bioactivity on the titanium

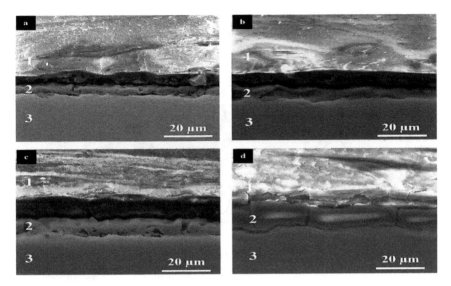

Fig. 6   FESEM image of cross-sectional view highlighting (1) Epoxy resin, (2) MAO coating region and (3) Ti6Al4V for the varying concentration of eggshell-based hydroxyapatite, i.e., **a** 0%, **b** 1%, **c** 1.5% and **d** 2% (reprinted figure from [66] with permission from Elsevier)

substrate concludes the Hemocompatibility and cytocompatibility of the HAP on the Ti Substrate. Dinda et al. [69] investigated the effect of heating during the PLDM for HAP coating and concluded the effective adherent characteristics of the coating. [36, 44]. Comparison between the PLDM and plasma spray coating techniques for HAP-based coating revealed the better output of PLDM in biomedical applications [71]. Figure 7 shows the PLDM Coating method used by Bosco et al. [72] using two pumps i.e., a Gas pump and a Vacuum pump. HAP coating using PLDM delivers food adhesion strength, uniformity, less material wastage and homogeneity. González-Estrada et al. [73] characterizes HAP coating by PLDM on 3d printed titanium alloy-based substrate. The research highlights that the incident energy of a pulsed laser is the governing parameter for morphology, strength and microhardness of the coating. Also, the small grain size of the incident laser improves the morphology.

4.4   Sputter Coating

Sputter coating (SC) as shown in Fig. 6, is a coating method to adhere the ultra-thin layer of conducting material onto a semiconductor or insulators as per the application desired. Safavi et al. [61] diagnosed the HAP-based composite with RF- magnetron multi-layered coating on HAP. Figure 8 shows the simple schematic diagram for the multilayer coating using sputtering. It uses inert gas, either Argon or Helium as an input gas. The precursor coating on the substrate is done using sputtering.

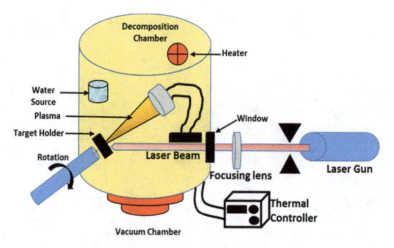

Fig. 7 Schematic diagram for PLDM coating method [72]

The setup involves the radio frequency generator along with a vacuum pump for effective targeting. The challenges and future scope are effectively elaborated in the manuscript. Mediaswanti et al. [74] investigated the effect of sputter-based coating of HAP on titanium and titanium alloys. The comparative analysis between two different coating methods i.e., the Electron Beam Evaporation coating method and the sputter coating method, was performed and characterized. The profilometer was used for measuring coating thickness. Apart from titanium and its alloys, the coating of HAP using the sputter coating technique and magnetron is performed on different substrates like polymer, wood, ceramics, metals, nonmetals, etc. [75]

4.5 Dip Coating

One of the easiest and most effective layering methods is the dip coating technique [26]. It is the process in which the substrate is symmetrically immersed into the tank containing the coating material. HAP in sol or reactive aqueous form can be coated on a substrate using dip coating. Li et al. [76] performed the coating of nanosized HAP in aqueous form and revealed the increase in the bone bond strength of the HAP coated implant as observed after four weeks of implantation. The limitation of the dip coating method is the requirement of high temperature for sintering which sometimes mismatches thermal expansion and leads to weaker adhesion characteristics. Titanium substrate, as optimized for bioimplant application, can also be coated for Osteointegration and better surface morphology by HAP nanoparticles coating using dip coating method [77, 78].

Figure 9 shows the schematic diagram of the stepwise dip coating method with sol–gel precursor [79, 80]. The process includes the repetitive dipping–withdrawing

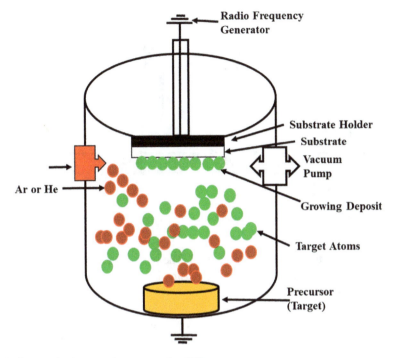

Fig. 8 Schematic diagram of sputter coating [61]

process for maintaining the thickness of the coating on a substrate. The process involves dipping, withdrawing and drying. The coating material with good cross-linking ability and adhesive characteristics are dip coated. Dip coating is also followed by the sol–gel preparation method as shown in Figs. 3 and 4 [80]. Here a precursor is a sol–gel that is to be dip coated on the substrate, followed by steering and drying. Modifying dip coating with sol–gel adhesion and spinning of quartz substrate is also designed and characterized [81]. The spinning of the substrate and dipping/withdrawing technique provides a thick adhesion of coating material. Based on the application and the layer thickness, the modification in dipping and withdrawing techniques are highlighted. The time interval between dipping a substrate into the coating material is also a controlling parameter.

4.6 Electrophoretic Deposition Method

The electrophoretic deposition coating method (EDM) is performed on the conducting substrate with the colloidal suspension material [28]. Modified HAP can be layered using EDM in In-Situ synthesis [6]. Asgari and Rajabi in 2021, experimentally performed the EDM of HAP on a different substrate to enhance the

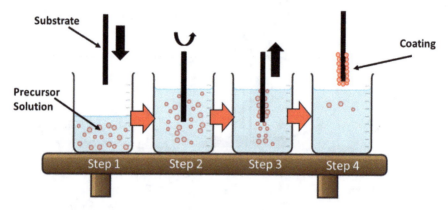

Fig. 9 Schematic diagram of dip coating method with sol–gel precursor [80]

mechanical strength of the final sample [82]. The biocompatible, low cytotoxic, and highly bioactive material can be generated using EDM coating, which modifies the surface morphology and improves adhesion between the substrate and coating, as well as the junction forming the biomedical implant during installation and fixation. Figure 10 shows the schematic diagram of the EDM coating of HAP-based bio nanocomposite material on titanium substrate. The 10 V supply with steel as cathode and titanium substrate as anode is electrophoretic in Isopropane and acetone solution.

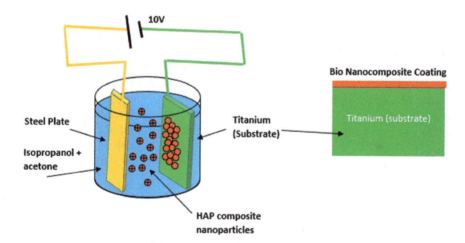

Fig. 10 Schematic diagram of EDM coating of HAP nanocomposite on titanium substrate [82]

4.7 Thermal Spray Coating

Melting and spraying of plasma to obtain fast and high quality layering on the sight of the coat is done using thermal spray coating. Amin et al. elaborated on the coating's methodology, limitations, physical and mechanical properties, etc., and the final sample was prepared using thermal spray coating [63]. It is a fast deposition method with approximately 30–250 μm coating thickness. It provides adhesion strength between 25 and 35 MPa. The limitation of the thermal spray coating is the line of sight, the requirement of high working temperature and the occurrence of the non-crystalline coating due to rapid cooling. Plasma spray is an industrial coating technique that can be uniaxial as well as multi-axial based on the application and requirement. The voltage regulation technique adopted by Ganvir et al. [83] for thermal spray coating of hydroxyapatite at dynamic voltage provides control over the deposition rate of the HAP on the substrate. Also, high adhesion between the substrate and coating leads to enhanced bond strength. The samples prepared are durable and of desired thickness. Safavi et al. [84] reviewed the bio-coating of HAP using electrical and thermal sources. The process, challenges and progress in the field of thermal spray coating of HAP are highlighted. Figure 11 shows the schematic diagram of the thermal spray coating method, where the spray plume from the precursor feeder is sprayed on the substrate using a spray device installed in the setup. It provides a denser implant coating. Currently, many researchers are working on thermal and plasm spray coating of HAP on different substrates and performing tribological, mechanical, biological and other characterizations. Hussain et al. [84] recently performed the mechanical and tribological characterization of seashell-derived HAP coating on titanium alloy using plasma spray coating. The results show good adhesion strength and wear resistance of HAP coating synthesized at 900 °C.

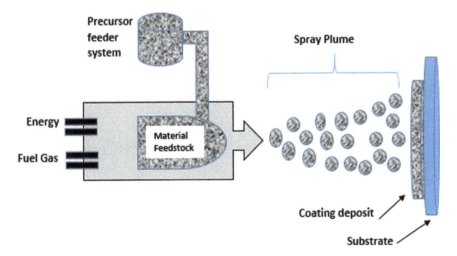

Fig. 11 Schematic diagram of thermal spray coating method [84]

Nucleation of HAP Coating   **Growth of HAP Coating**

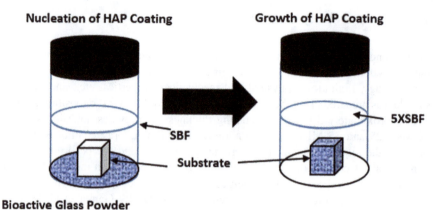

Bioactive Glass Powder

Fig. 12 Schematic diagram of biomimetic coating with bioactive glass powder using SBF

4.8 Biomimetic Coating

Koju et al. [85] performed the biomimetic coating of calcium phosphatic basic salt, mostly calcium hydroxyapatite on several different biocompatible substrates. The degradation of the coating material is analyzed using simulated body fluid (SBF). The biocompatible coating helps in using both degradable and non-degradable substrates in biomedical applications. The coating can be microwave-assisted and conventional. As observed, the thickness and adhesion characteristics also depend on the sintering and other post-processing processes. It is a slow deposition technique. The tabulated form of features of biomimetic coating on various polymers, metals, ceramics, composites and alloys by different researchers is highlighted by Koju et al. [85]. As shown in Fig. 12, the biomimetic coating method involves the bioactive glass powder for biomimetic coating on titanium alloy using SBF. It clearly highlights the nucleation and growth of HAP on a titanium alloy substrate with five times concentrated SBF. The thickness of the coating depends on the immersion time. The drawback of biomimetic coating is the non-uniform and agglomerated (non-crystalline) coating. Therefore, biomimetic coating is seldom used for HAP.

5 Advantages of HAP Coating on Biomedical Applications

- HAP provides an osteophilic surface that accelerates bonding between the implant and natural bone [86].
- HAP coatings are biocompatible, nontoxic and bioactive.
- HAP coating has effective osteointegration characteristics and matches the healing rate of natural bone tissue [87].

- HAP coating is cost-effective and can be synthesized through natural and artificial means.
- HAP coatings have good adhesion and surface morphology and can work on alloys, polymer, metallic, non-metallic and composite substrates.
- Plasma coating of HAP on prostheses offers improved wear resistance, low coefficient of friction, hardness, good antibacterial properties, and promotes osseointegration.

Several other potential coating materials are there [88–126], but that is out of the scope of this chapter.

6 Conclusions

Coating of HAP as a bioceramic material can be effectively done using several methods sol–gel, dip coating, spin coating, thermal spray, pulsed deposition, sputter, electrophoretic deposition, thermal, biomimetic, microarc oxidation and the combination of two or more coating methods. The method depends on the type of substrate, adhesion strength required, the thickness of the coating, etc. HAP coating as a bioceramic material deposition is highly appreciated in various biomedical applications because it possesses characteristics like biocompatibility, Osteointegration, osteoconductive, bioactivity, anticorrosion, nontoxic, antibacterial, anti-fungal and comparable mechanical and physical strength. Coating methods govern the adhesion strength, thickness, durability, compatibility, and physical and thermal characteristics of the coated material. The selection of the coating method depends on the type of substrate and the characteristics desired in the final product. For modified surface properties, coating of HAP, its blends or composites is highly recommended as per the results obtained by characterization by various researchers. Coating of HAP is applicable in polymers, metals, nonmetals, ceramics and composite substrates. Due to its multiple applications, it has a commendable future perspective in sustainable material development.

References

1. Kadambi P, Luniya P, Dhatrak P (2021) Current advancements in polymer/polymer matrix composites for dental implants: a systematic review. Mater Today Proc 46:740–745. https://doi.org/10.1016/j.matpr.2020.12.396
2. Shamray VF et al (2019) Study of the crystal structure of hydroxyapatite in plasma coating. Surf Coatings Technol 372:201–208. https://doi.org/10.1016/j.surfcoat.2019.05.037
3. Gomes DS, Santos AMC, Neves GA, Menezes RR (2019) A brief review on hydroxyapatite production and use in biomedicine. Ceramica 65(374):282–302. https://doi.org/10.1590/0366-69132019653742706
4. Valletregi M (2004) Calcium phosphates as substitution of bone tissues. Prog Solid State Chem 32(1–2):1–31. https://doi.org/10.1016/j.progsolidstchem.2004.07.001

5. Märten A, Fratzl P, Paris O, Zaslansky P (2010) On the mineral in collagen of human crown dentine. Biomaterials 31(20):5479–5490. https://doi.org/10.1016/j.biomaterials.2010.03.030
6. Hussain R et al (2016) In situ synthesis of mesoporous polyvinyl alcohol/hydroxyapatite composites for better biomedical coating adhesion. Appl Surf Sci 364:117–123. https://doi.org/10.1016/j.apsusc.2015.12.057
7. Kokubo T, Takadama H (2006) How useful is SBF in predicting in vivo bone bioactivity? Biomaterials 27(15):2907–2915. https://doi.org/10.1016/j.biomaterials.2006.01.017
8. Kim HW, Koh YH, Li LH, Lee S, Kim HE (2004) Hydroxyapatite coating on titanium substrate with titania buffer layer processed by sol-gel method. Biomaterials 25(13):2533–2538. https://doi.org/10.1016/j.biomaterials.2003.09.041
9. Gadow R, Killinger A, Stiegler N (2010) Hydroxyapatite coatings for biomedical applications deposited by different thermal spray techniques. Surf Coatings Technol 205(4):1157–1164. https://doi.org/10.1016/j.surfcoat.2010.03.059
10. Avci M, Yilmaz B, Tezcaner A, Evis Z (2017) Strontium doped hydroxyapatite biomimetic coatings on Ti6Al4V plates. Ceram Int 43(12):9431–9436. https://doi.org/10.1016/j.ceramint.2017.04.117
11. Xue T et al (2020) Surface modification techniques of titanium and its alloys to functionally optimize their biomedical properties: thematic review. Front Bioeng Biotechnol 8(November):1–19. https://doi.org/10.3389/fbioe.2020.603072
12. Szcześ A, Hołysz L, Chibowski E (2017) Synthesis of hydroxyapatite for biomedical applications. Adv Colloid Interface Sci 249(April):321–330. https://doi.org/10.1016/j.cis.2017.04.007
13. Xue K, Teng SH, Niu N, Wang P (2018) Biomimetic synthesis of novel polyvinyl alcohol/hydroxyapatite composite microspheres for biomedical applications. Mater Res Express 5(11). https://doi.org/10.1088/2053-1591/aadc7d
14. Pandey E, Srivastava K, Gupta S, Srivastava S, Mishra N (2016) Some biocompatible materials used in medical practices-a review. Int J Pharm Sci Res IJPSR 7(7):2748–2755. https://doi.org/10.13040/IJPSR.0975-8232.7(7).2748-55
15. Dorozhkin SV (2011) Calcium orthophosphates. Biomatter 1(2):121–164. https://doi.org/10.4161/biom.18790
16. Cao H, Zhang L, Zheng H, Wang Z (2010) Hydroxyapatite nanocrystals for biomedical applications. J Phys Chem C 114(43):18352–18357. https://doi.org/10.1021/jp106078b
17. Iconaru SL, Predoi D, Ciobanu CS, Motelica-Heino M, Guegan R, Bleotu C (2022) Development of silver doped hydroxyapatite thin films for biomedical applications. Coatings 12(3). https://doi.org/10.3390/coatings12030341
18. Li M et al (2018) An overview of graphene-based hydroxyapatite composites for orthopedic applications. Bioact Mater 3(1):1–18. https://doi.org/10.1016/j.bioactmat.2018.01.001
19. Yang B, Uchida M, Kim H-M, Zhang X, Kokubo T (2004) Preparation of bioactive titanium metal via anodic oxidation treatment. Biomaterials 25(6):1003–1010. https://doi.org/10.1016/S0142-9612(03)00626-4
20. Asri RIM, Harun WSW, Hassan MA, Ghani SAC, Buyong Z (2016) A review of hydroxyapatite-based coating techniques: sol-gel and electrochemical depositions on biocompatible metals. J Mech Behav Biomed Mater 57:95–108. https://doi.org/10.1016/j.jmbbm.2015.11.031
21. Bansal G, Jain A, Taluja R, Verma S (2018) Application of green composite material in sustainable architectural and automotive part development-a review (January):6–11
22. Müller V, Pagnier T, Tadier S, Gremillard L, Jobbagy M, Djurado E (2021) Design of advanced one-step hydroxyapatite coatings for biomedical applications using the electrostatic spray deposition. Appl Surf Sci 541. https://doi.org/10.1016/j.apsusc.2020.148462
23. Yilmaz B, Alshemary AZ, Evis Z (2019) Co-doped hydroxyapatites as potential materials for biomedical applications. Microchem J 144:443–453. https://doi.org/10.1016/j.microc.2018.10.007
24. Eliaz N, Metoki N (2017) Calcium phosphate bioceramics: a review of their history, structure, properties, coating technologies and biomedical applications. Materials (Basel) 10(4). https://doi.org/10.3390/ma10040334

25. Harun WSW, Asri RIM, Sulong AB, Ghani SAC, Ghazalli Z (2018) Hydroxyapatite-Based Coating on Biomedical Implant. Hydroxyapatite—Adv Compos Nanomater Biomed Appl Its Technol Facet (March 2019). https://doi.org/10.5772/intechopen.71063

26. Abrishamchian A, Hooshmand T, Mohammadi M, Najafi F (2013) Preparation and characterization of multi-walled carbon nanotube/hydroxyapatite nanocomposite film dip coated on Ti-6Al-4V by sol-gel method for biomedical applications: an in vitro study. Mater Sci Eng C 33(4):2002–2010. https://doi.org/10.1016/j.msec.2013.01.014

27. Wolke JGC, van Dijk K, Schaeken HG, de Groot K, Jansen JA (1994) Study of the surface characteristics of magnetron-sputter calcium phosphate coatings. J Biomed Mater Res 28(12):1477–1484. https://doi.org/10.1002/JBM.820281213

28. Nuswantoro NF et al (2021) Hydroxyapatite coating on titanium alloy TNTZ for increasing osseointegration and reducing inflammatory response in vivo on Rattus norvegicus Wistar rats. Ceram Int 47(11):16094–16100. https://doi.org/10.1016/j.ceramint.2021.02.184

29. Karimi N, Kharaziha M, Raeissi K (2019) Electrophoretic deposition of chitosan reinforced graphene oxide-hydroxyapatite on the anodized titanium to improve biological and electrochemical characteristics. Mater Sci Eng C 98:140–152. https://doi.org/10.1016/j.msec.2018.12.136

30. Wei M, Ruys AJ, Milthorpe BK, Sorrell CC (2005) Precipitation of hydroxyapatite nanoparticles: effects of precipitation method on electrophoretic deposition. J Mater Sci Mater Med 16(4):319–324. https://doi.org/10.1007/S10856-005-0630-0

31. Popkov D, Popkov AV, Gorbach EN, Kononovich NA, Tverdokhlebov SI, Shesterikov EV (2017) Bioactivity and osteointegration of hydroxyapatite-coated stainless steel and titanium wires used for intramedullary osteosynthesis. Strateg Trauma Limb Reconstr 12(2):107–113. https://doi.org/10.1007/s11751-017-0282-x

32. Sidane D et al (2015) Study of the mechanical behavior and corrosion resistance of hydroxyapatite sol-gel thin coatings on 316 L stainless steel pre-coated with titania film. Thin Solid Films 593:71–80. https://doi.org/10.1016/j.tsf.2015.09.037

33. Tao ZS et al (2016) A comparative study of zinc, magnesium, strontium-incorporated hydroxyapatite-coated titanium implants for osseointegration of osteopenic rats. Mater Sci Eng C 62:226–232. https://doi.org/10.1016/j.msec.2016.01.034

34. Galindo TGP, Chai Y, Tagaya M (2019) Hydroxyapatite nanoparticle coating on polymer for constructing effective biointeractive interfaces. J Nanomater 2019. https://doi.org/10.1155/2019/6495239

35. Kon E et al (2000) Autologous bone marrow stromal cells loaded onto porous hydroxyapatite ceramic accelerate bone repair in critical-size defects of sheep long bones. J Biomed Mater Res 49(3):328–337. https://doi.org/10.1002/(SICI)1097-4636(20000305)49:3%3c328::AID-JBM5%3e3.0.CO;2-Q

36. El Hassanin A, Quaremba G, Sammartino P, Adamo D, Miniello A, Marenzi G (2021) Effect of implant surface roughness and macro- and micro-structural composition on wear and metal particles released. Materials (Basel) 14(22):6800. https://doi.org/10.3390/ma14226800

37. Champion E (2013) Sintering of calcium phosphate bioceramics. Acta Biomater 9(4):5855–5875. https://doi.org/10.1016/j.actbio.2012.11.029

38. Balázsi C, Wéber F, Kövér Z, Horváth E, Németh C (2007) Preparation of calcium-phosphate bioceramics from natural resources. J Eur Ceram Soc 27(2–3):1601–1606. https://doi.org/10.1016/j.jeurceramsoc.2006.04.016

39. Dorozhkin SV (2016) Multiphasic calcium orthophosphate (CaPO4) bioceramics and their biomedical applications. Ceram Int 42(6):6529–6554. https://doi.org/10.1016/j.ceramint.2016.01.062

40. Pepla E, Besharat LK, Palaia G, Tenore G, Migliau G (2014) Nano-hydroxyapatite and its applications in preventive, restorative and regenerative dentistry: a review of literature. Ann Stomatol (Roma) 5(3):108–114. https://doi.org/10.1016/0304-4165(71)90042-0

41. Yang Y, Kim K, Ong J (2005) A review on calcium phosphate coatings produced using a sputtering process?an alternative to plasma spraying. Biomaterials 26(3):327–337. https://doi.org/10.1016/j.biomaterials.2004.02.029

42. Ellies LG, Nelson DGA, Featherstone JDB (1992) Crystallographic changes in calcium phosphates during plasma-spraying. Biomaterials 13(5):313–316. https://doi.org/10.1016/0142-9612(92)90055-S

43. Nimb L, Gotfredsen K, Steen Jensen J (1993) Mechanical failure of hydroxyapatite-coated titanium and cobalt-chromium-molybdenum alloy implants. An animal study. Acta Orthop Belg 59(4):333–338

44. Ukoba KO, Eloka-Eboka AC, Inambao FL (2018) Review of nanostructured NiO thin film deposition using the spray pyrolysis technique. Renew Sustain Energy Rev 82:2900–2915. https://doi.org/10.1016/j.rser.2017.10.041

45. Stuart BW, Murray JW, Grant DM (2018) Two step porosification of biomimetic thin-film hydroxyapatite/alpha-tri calcium phosphate coatings by pulsed electron beam irradiation. Sci Rep 8(1):14530. https://doi.org/10.1038/s41598-018-32612-x

46. Piccirillo C et al (2017) Aerosol assisted chemical vapour deposition of hydroxyapatite-embedded titanium dioxide composite thin films. J Photochem Photobiol A Chem 332:45–53. https://doi.org/10.1016/j.jphotochem.2016.08.010

47. Ofudje EA, Adeogun AI, Idowu MA, Kareem SO (2019) Synthesis and characterization of Zn-Doped hydroxyapatite: scaffold application, antibacterial and bioactivity studies. Heliyon 5(5):e01716. https://doi.org/10.1016/j.heliyon.2019.e01716

48. Manso M, Jiménez C, Morant C, Herrero P, Martínez-Duart J (2000) Electrodeposition of hydroxyapatite coatings in basic conditions. Biomaterials 21(17):1755–1761. https://doi.org/10.1016/S0142-9612(00)00061-2

49. Kumar P, Huo P, Zhang R, Liu B (2019) Antibacterial properties of graphene-based nanomaterials. Nanomaterials 9(5):737. https://doi.org/10.3390/nano9050737

50. Jaafar A, Hecker C, Árki P, Joseph Y (2020) Sol-gel derived hydroxyapatite coatings for titanium implants: a review. Bioengineering 7(4):1–23. https://doi.org/10.3390/bioengineering7040127

51. Liu DM, Yang Q, Troczynski T (2002) Sol-gel hydroxyapatite coatings on stainless steel substrates. Biomaterials 23(3):691–698. https://doi.org/10.1016/S0142-9612(01)00157-0

52. Stoch A et al. (2005) Sol-gel derived hydroxyapatite coatings on titanium and its alloy Ti6Al4V. J Mol Struct 744–747(SPEC. ISS):633–640. https://doi.org/10.1016/j.molstruc.2004.10.080

53. Kim HW, Kim HE, Knowles JC (2004) Fluor-hydroxyapatite sol-gel coating on titanium substrate for hard tissue implants. Biomaterials 25(17):3351–3358. https://doi.org/10.1016/j.biomaterials.2003.09.104

54. Wang X et al (2014) Fabrication and corrosion resistance of calcium phosphate glass-ceramic coated Mg alloy via a PEG assisted sol–gel method. Ceram Int 40(2):3389–3398. https://doi.org/10.1016/j.ceramint.2013.09.093

55. Tripathi SK, Kaur R, Kaur H, Rani M, Kaur J, Kaur H (2015) Fabrication and electrical characterization of memristor with TiO2 as an active layer, p 110027. https://doi.org/10.1063/1.4915472

56. Nguyen-Tri P et al (2019) Recent progress in the preparation, properties and applications of superhydrophobic nano-based coatings and surfaces: a review. Prog Org Coatings 132:235–256. https://doi.org/10.1016/j.porgcoat.2019.03.042

57. Wilcock CJ et al (2017) Preparation and antibacterial properties of silver-doped nanoscale hydroxyapatite pastes for bone repair and augmentation. J Biomed Nanotechnol 13(9):1168–1176. https://doi.org/10.1166/jbn.2017.2387

58. Mishra A, Bhatt N, Bajpai AK (2019) Nanostructured superhydrophobic coatings for solar panel applications. In: Nanomaterials-based coatings, Elsevier, pp 397–424

59. Choudhury P, Agrawal DC (2012) Hydroxyapatite (HA) coatings for biomaterials. Woodhead Publishing Limited

60. Duta L, Popescu AC (2019) Current status on pulsed laser deposition of coatings from animal-origin calcium phosphate sources. Coatings 9(5). https://doi.org/10.3390/coatings9050335

61. Safavi MS, Surmeneva MA, Surmenev RA, Khalil-Allafi J (2021) RF-magnetron sputter deposited hydroxyapatite-based composite & multilayer coatings: a systematic review

from mechanical, corrosion, and biological points of view. Ceram Int 47(3):3031–3053. https://doi.org/10.1016/j.ceramint.2020.09.274

62. Li M et al (2014) Graphene oxide/hydroxyapatite composite coatings fabricated by electrophoretic nanotechnology for biological applications. Carbon N. Y. 67:185–197. https://doi.org/10.1016/j.carbon.2013.09.080

63. Amin S, Panchal H, Professor A (2016) A review on thermal spray coating processes. Int J Curr Trends Eng Res Sci J Impact Factor 2(4):556–563. http://www.ijcter.com

64. Baino F, Yamaguchi S (2020) The use of simulated body fluid (SBF) for assessing materials bioactivity in the context of tissue engineering: review and challenges. Biomimetics 5(4):1–19. https://doi.org/10.3390/biomimetics5040057

65. Al-Sanabani JS, Madfa AA, Al-Sanabani FA (2013) Application of calcium phosphate materials in dentistry. Int J Biomater 2013. https://doi.org/10.1155/2013/876132

66. Qaid TH et al (2019) Micro-arc oxidation of bioceramic coatings containing eggshell-derived hydroxyapatite on titanium substrate. Ceram Int 45(15):18371–18381. https://doi.org/10.1016/j.ceramint.2019.06.052

67. Gomes GC, Borghi FF, Ospina RO, López EO, Borges FO, Mello A (2017) Nd:YAG (532 nm) pulsed laser deposition produces crystalline hydroxyapatite thin coatings at room temperature. Surf Coatings Technol 329:174–183. https://doi.org/10.1016/j.surfcoat.2017.09.008

68. Dhinasekaran D et al (2021) Pulsed laser deposition of nanostructured bioactive glass and hydroxyapatite coatings: microstructural and electrochemical characterization. Mater Sci Eng C 130:112459. https://doi.org/10.1016/j.msec.2021.112459

69. Dinda GP, Shin J, Mazumder J (2009) Pulsed laser deposition of hydroxyapatite thin films on Ti-6Al-4V: effect of heat treatment on structure and properties. Acta Biomater 5(5):1821–1830. https://doi.org/10.1016/j.actbio.2009.01.027

70. Fathi AM, Ahmed MK, Afifi M, Menazea AA, Uskoković V (2021) Taking hydroxyapatite-coated titanium implants two steps forward: surface modification using graphene mesolayers and a hydroxyapatite-reinforced polymeric scaffold. ACS Biomater Sci Eng 7(1):360–372. https://doi.org/10.1021/acsbiomaterials.0c01105

71. García-Sanz FJ, Mayor MB, Arias JL, Pou J, León B, Pérez-Amor M (1997) Hydroxyapatite coatings: a comparative study between plasma-spray and pulsed laser deposition techniques. J Mater Sci Mater Med 8(12):861–865. https://doi.org/10.1023/a:1018549720873

72. Bosco R, Van Den Beucken J, Leeuwenburgh S, Jansen J (2012) Surface engineering for bone implants: a trend from passive to active surfaces. Coatings 2(3):95–119. https://doi.org/10.3390/coatings2030095

73. González-Estrada OA, Pertuz Comas AD, Ospina R (2022) Characterization of hydroxyapatite coatings produced by pulsed-laser deposition on additive manufacturing Ti6Al4V ELI. Thin Solid Films 763:139592. https://doi.org/10.1016/j.tsf.2022.139592

74. Mediaswanti K, Wen C, Ivanova EP, Berndt CC, Wang J (2013) Sputtered hydroxyapatite nanocoatings on novel titanium alloys for biomedical applications. In: Titanium alloys—advances in properties control, InTech

75. Lee K, Shin G (2015) RF magnetron sputtering coating of hydroxyapatite on alkali solution treated titanate nanorods. Arch Metall Mater 60(2):1319–1322. https://doi.org/10.1515/amm-2015-0122

76. Li T, Lee J, Kobayashi T, Aoki H (1996) Hydroxyapatite coating by dipping method, and bone bonding strength. J Mater Sci Mater Med 7(6):355–357. https://doi.org/10.1007/BF00154548

77. Baptista R et al (2016) Characterization of titanium-hydroxyapatite biocomposites processed by dip coating. Bull Mater Sci 39(1):263–272. https://doi.org/10.1007/s12034-015-1122-6

78. Mavis B, Taş AC (2004) Dip coating of calcium hydroxyapatite on Ti-6Al-4V substrates. J Am Ceram Soc 83(4):989–991. https://doi.org/10.1111/j.1151-2916.2000.tb01314.x

79. Sims L, Egelhaaf HJ, Hauch JA, Kogler FR, Steim R (2012) Plastic solar cells. Compr Renew Energy 1:439–480. https://doi.org/10.1016/B978-0-08-087872-0.00120-7

80. Nwanna EC, Imoisili PE, Jen T-C (2020) Fabrication and synthesis of SnOX thin films: a review. Int J Adv Manuf Technol 111(9–10):2809–2831. https://doi.org/10.1007/s00170-020-06223-8

81. Malakauskaite-Petruleviciene M, Stankeviciute Z, Beganskiene A, Kareiva A (2014) Sol-gel synthesis of calcium hydroxyapatite thin films on quartz substrate using dip-coating and spin-coating techniques. J Sol-Gel Sci Technol 71(3):437–446. https://doi.org/10.1007/s10971-014-3394-5

82. Asgari N, Rajabi M (2021) Enhancement of mechanical properties of hydroxyapatite coating prepared by electrophoretic deposition method. Int J Appl Ceram Technol 18(1):147–153. https://doi.org/10.1111/ijac.13638

83. Meng X, Kwon T-Y, Kim K-H (2008) Hydroxyapatite coating by electrophoretic deposition at dynamic voltage. Dent Mater J 27(5):666–671. https://doi.org/10.4012/dmj.27.666

84. Safavi MS, Walsh FC, Surmeneva MA, Surmenev RA, Khalil-Allafi J (2021) Electrodeposited hydroxyapatite-based biocoatings: recent progress and future challenges. Coatings 11(1):110. https://doi.org/10.3390/coatings11010110

85. Koju N, Sikder P, Ren Y, Zhou H, Bhaduri SB (2017) Biomimetic coating technology for orthopedic implants. Curr Opin Chem Eng 15:49–55. https://doi.org/10.1016/j.coche.2016.11.005

86. Shah NJ, Hong J, Hyder MN, Hammond PT (2012) Osteophilic multilayer coatings for accelerated bone tissue growth. Adv Mater 24(11):1445–1450. https://doi.org/10.1002/adma.201104475

87. Cox SC, Thornby JA, Gibbons GJ, Williams MA, Mallick KK (2015) 3D printing of porous hydroxyapatite scaffolds intended for use in bone tissue engineering applications. Mater Sci Eng C 47:237–247. https://doi.org/10.1016/j.msec.2014.11.024

88. Bharath KN, Madhu P, Gowda TY, Verma A, Sanjay MR, Siengchin S (2021) Mechanical and chemical properties evaluation of sheep wool fiber–reinforced vinylester and polyester composites. Mater Perform Characterization 10(1):99–109

89. Singh K, Jain N, Verma A, Singh VK, Chauhan S (2020) Functionalized graphite–reinforced cross-linked poly (vinyl alcohol) nanocomposites for vibration isolator application: morphology, mechanical, and thermal assessment. Mater Perform Characterization 9(1):215–230

90. Sethi SK, Gogoi R, Verma A, Manik G (2022) How can the geometry of a rough surface affect its wettability?-a coarse-grained simulation analysis. Prog Org Coat 172:107062

91. Verma A, Budiyal L, Sanjay MR, Siengchin S (2019) Processing and characterization analysis of pyrolyzed oil rubber (from waste tires)-epoxy polymer blend composite for lightweight structures and coatings applications. Polym Eng Sci 59(10):2041–2051

92. Verma A, Joshi K, Gaur A, Singh VK (2018) Starch-jute fiber hybrid biocomposite modified with an epoxy resin coating: fabrication and experimental characterization. J Mech Behav Mater 27(5–6):1–16

93. Verma A, Baurai K, Sanjay MR, Siengchin S (2020) Mechanical, microstructural, and thermal characterization insights of pyrolyzed carbon black from waste tires reinforced epoxy nanocomposites for coating application. Polym Compos 41(1):338–349

94. Rastogi S, Verma A, Singh VK (2020) Experimental response of nonwoven waste cellulose fabric–reinforced epoxy composites for high toughness and coating applications. Mater Perform Characterization 9(1):151–172

95. Verma A, Parashar A, Singh SK, Jain N, Sanjay SM, Siengchin S (2020) Modeling and simulation in polymer coatings. In: Polymer coatings. CRC Press, pp 309–324

96. Verma A, Jain N, Rastogi S, Dogra V, Sanjay SM, Siengchin S, Mansour R (2020) Mechanism, anti-corrosion protection and components of anti-corrosion polymer coatings. In: Polymer coatings. CRC Press, pp 53–66

97. Shankar U, Sethi SK, Verma A (2022) Forcefields and modeling of polymer coatings and nanocomposites. In: Forcefields for atomistic-scale simulations: materials and applications. Springer, Singapore, pp 81–98

98. Verma A, Parashar A, Packirisamy M (2018) Atomistic modeling of graphene/hexagonal boron nitride polymer nanocomposites: a review. Wiley Interdisc Rev: Comput Mol Sci 8(3):e1346

99. Verma A, Kumar R, Parashar A (2019) Enhanced thermal transport across a bi-crystalline graphene–polymer interface: an atomistic approach. Phys Chem Chem Phys 21(11):6229–6237

100. Verma A, Singh C, Singh VK, Jain N (2019) Fabrication and characterization of chitosan-coated sisal fiber–phytagel modified soy protein-based green composite. J Compos Mater 53(18):2481–2504

101. Verma A, Parashar A, Packirisamy M (2019) Effect of grain boundaries on the interfacial behaviour of graphene-polyethylene nanocomposite. Appl Surf Sci 470:1085–1092

102. Kataria A, Verma A, Sanjay MR, Siengchin S (2022) Molecular modeling of 2D graphene grain boundaries: mechanical and fracture aspects. Mater Today: Proc 52:2404–2408

103. Verma A, Jain N, Sethi SK (2022) Modeling and simulation of graphene-based composites. In: Innovations in graphene-based polymer composites. Woodhead Publishing, pp 167–198

104. Verma A, Sharma S (2022) Atomistic simulations to study thermal effects and strain rate on mechanical and fracture properties of graphene like BC3. In: Forcefields for atomistic-scale simulations: materials and applications. Springer, Singapore, pp 237–252

105. Verma A, Parashar A, van Duin AC (2022) Graphene-reinforced polymeric membranes for water desalination and gas separation/barrier applications. In: Innovations in graphene-based polymer composites. Woodhead Publishing, pp 133–165

106. Verma A, Parashar A (2017) The effect of STW defects on the mechanical properties and fracture toughness of pristine and hydrogenated graphene. Phys Chem Chem Phys 19(24):16023–16037

107. Verma A, Parashar A (2018) Molecular dynamics based simulations to study failure morphology of hydroxyl and epoxide functionalised graphene. Comput Mater Sci 143:15–26

108. Verma A, Parashar A (2018) Molecular dynamics based simulations to study the fracture strength of monolayer graphene oxide. Nanotechnology 29(11):115706

109. Verma A, Parashar A (2018) Structural and chemical insights into thermal transport for strained functionalised graphene: a molecular dynamics study. Mater Res Express 5(11):115605

110. Verma A, Parashar A, Packirisamy M (2018) Tailoring the failure morphology of 2D bicrystalline graphene oxide. J Appl Phys 124(1):015102

111. Verma A, Parashar A (2018) Reactive force field based atomistic simulations to study fracture toughness of bicrystalline graphene functionalised with oxide groups. Diam Relat Mater 88:193–203

112. Singla V, Verma A, Parashar A (2018) A molecular dynamics based study to estimate the point defects formation energies in graphene containing STW defects. Mater Res Express 6(1):015606

113. Verma A, Parashar A, Packirisamy M (2019) Role of chemical adatoms in fracture mechanics of graphene nanolayer. Mater Today: Proc 11:920–924

114. Verma A, Zhang W, Van Duin AC (2021) ReaxFF reactive molecular dynamics simulations to study the interfacial dynamics between defective h-BN nanosheets and water nanodroplets. Phys Chem Chem Phys 23(18):10822–10834

115. Deji R, Verma A, Kaur N, Choudhary BC, Sharma RK (2022) Density functional theory study of carbon monoxide adsorption on transition metal doped armchair graphene nanoribbon. Mater Today: Proc 54:771–776

116. Deji R, Verma A, Choudhary BC, Sharma RK (2022) New insights into NO adsorption on alkali metal and transition metal doped graphene nanoribbon surface: a DFT approach. J Mol Graph Model 111:108109

117. Deji R, Jyoti R, Verma A, Choudhary BC, Sharma RK (2022) A theoretical study of HCN adsorption and width effect on co-doped armchair graphene nanoribbon. Comput Theor Chem 1209:113592

118. Deji R, Verma A, Kaur N, Choudhary BC, Sharma RK (2022) Adsorption chemistry of co-doped graphene nanoribbon and its derivatives towards carbon based gases for gas sensing applications: quantum DFT investigation. Mater Sci Semicond Process 146:106670

119. Singh A, Lynch J, Anantharaman SB, Hou J, Singh S, Kim G, Mohite AD, Singh R, Jariwala D (2022) Cavity-enhanced raman scattering from 2d hybrid perovskites. J Phys Chem C 126:11158

120. Moun M, Singh A, Tak BR, Singh R (2019) Study of the photoresponse behavior of a high barrier Pd/MoS2/Pd photodetector. J Phys D: Appl Phys 52:325102
121. Moun M, Singh A, Singh R (2018) Study of electrical behavior of metal-semiconductor contacts on exfoliated MoS2 flakes. Physica Status Solidi (a) 215:1800188
122. Aggarwal P, Kaushik S, Bisht P, Sharma M, Singh A, Mehta BR, Singh R (2022) Centimeter-scale synthesis of monolayer WS2 using single-zone atmospheric-pressure chemical vapor deposition: a detailed study of parametric dependence. Growth Mech, Photodetector Properties, Crystal Growth Des 22:3206
123. Singh A, Moun M, Sharma M, Barman A, Kumar Kapoor A, Singh R (2021) NaCl-assisted substrate dependent 2D planar nucleated growth of MoS2. Appl Surf Sci 538:148201
124. Singh A, Sharma M, Singh R (2021) NaCl-assisted CVD growth of large-area high-quality trilayer MoS2 and the role of the concentration boundary layer. Cryst Growth Des 21:4940
125. Singh A, Moun M, Singh R (2019) Effect of different precursors on CVD growth of molybdenum disulfide. J Alloy Compd 782:772
126. Arpitha GR, Jain N, Verma A, Madhusudhan M (2022) Corncob bio-waste and boron nitride particles reinforced epoxy-based composites for lightweight applications: fabrication and characterization. Biomass Convers Biorefinery, 1–8. https://doi.org/10.1007/s13399-022-037 17-1

Chapter 14
Superhydrophobic Coating: Stability and Potential Applications

Rajeev Gupta

1 Introduction

The superhydrophobic coating is mainly attributed by their water-repellent nature followed by high static water contact angle and low contact angle hysteresis, and it can be obtained by taking reduced energy surface material with high surface micro or nano features [1]. These specific properties of surface make the water droplet more spherical and prevent it to spread over the surface or will reduce the interaction between water molecule and substrate molecule. This interaction will decide the wettability of the liquid in the surface [2]. It is determined by making the balance of the surface energy at the interface involving liquid, air, and solid resources or wettability could be the propensity of a liquid to remain in touch with the solid surrounded by another fluid (liquid or gas). A balance of forces among cohesive and adhesive forces determines the degree of wetting (wettability) [3]. The composition of the chemicals and the shape structures of solid surfaces directs the wettability. This low wettability or reduced interaction will change the properties of original surface and will be very helpful in many applications like anti-corrosion [4], anti-wearing, self-cleaning [5–7], anti-bacterial, anti-fungal, solar reflective [8], anti-wax, etc. [9]. In nature, there are a large number of biological surfaces like plants, animals and insects exhibiting superhydrophobicity. The lotus plant (*Nelumbo nucifera*) is considered to be one of the good examples of natural SHS, which are characterized by CA $\theta_w >$ 150° very low adhesion towards water and self-maintaining properties [10]. Although it develops in mucky water, its leaves always appear perfectly clean because, during the night, when the dew drops come to the leaves, the droplet becomes spherical and rolling off down, picking the dirt, dust contaminations away from the surface [11]. Many plants, insects and animals that flourish in nature have their specific natural and

R. Gupta (✉)
Department of Physics, School of Engineering, University of Petroleum and Energy Studies,
Dehradun, Uttarakhand 248007, India
e-mail: rajeevguptaupes@gmail.com

© The Author(s), under exclusive license to Springer Nature Singapore Pte Ltd. 2023
A. Verma et al. (eds.), *Coating Materials*, Materials Horizons: From Nature
to Nanomaterials, https://doi.org/10.1007/978-981-99-3549-9_14

exceptional surface, physical and chemical properties. For example, superhydrophobicity of salvinia leaves, rose petals, and lotus leaves; hydrophilicity and hydrophobicity of carapace of Namib beetles; drag reduction property of shark scales; and
water repellency of legs of water striders are unique chemical, structural, or physical properties that help them adapt to environment (Fig. 1) [12]. In 1973, German
botanist, Wilhelm Barthlott with his squad at the University of Bonn, did a lot of
work on it to understand the self-cleaning process in lotus leaves. With the help of
SEM, they observed numbers of Papillose-based cells of epidermal in micrometer
order enclosed by the nm order natural epicuticle wax tubules over it [13]. Sizes
connecting the two groves are not as much of the minimum size of the droplets of
water and because of this reason, the water droplet is not able to get inside between
the groves and this makes the water droplet more spherical [14]. Particles of dust
sit only at the tip of the wax crystals that result on a very smaller surface area for
contact with the surface of the plant. The cohesive forces amongst the molecules of
water are greater compared to the adhesive force between lotus leaves surface and
dust particles because of this reason [15]. If water falls onto a lotus leaf surface, the
interaction of the surface tension and the low attraction forces within the surface and
the water results in the production of spherical water drop which only sits at the wax
structure's tip. The self-cleaning of dust and dirt from the lotus leaves is called the
lotus effect. The coexistence of two scales of roughness contributes significantly to
the quality of the superhydrophobicity and is derived with CB state [16].

2 Rudiments of Super Hydrophobicity

To understand the superhydrophobicity, there are few terms like wettability, contact
angle, contact angle hysteresis, sliding angle etc., which are very important to understand and without it, it is not possible to know hydrophobicity and how it can be used
for industrial applications [17]. Wettability for a solid surface is defined as its ability
to become wet by the liquid and it can be determined by making the balance of the
surface energy at the interface involving liquid, air, and solid resources. A balance of
forces among cohesive and adhesive forces determines the degree of wetting (wettability) [18]. In general, the wettability between phases of the liquid and solid is
expressed by the following Young–Laplace's equation $\gamma_{SV} = \gamma_{LV}\cos\theta + \gamma_{SL}$

Where, γ_{SA}, γ_{SL}, and γ_{LA} are the interfacial energy within the solid–air, solid–
liquid and air–liquid state, and these interfacial energies decide whether the water
droplet over a solid substrate will convert into the water film or into a spherical
droplet. The contact angle will help us to know the degree of wetting, i.e. wettability
is poor when the CA is larger than 90° and wettability will be good if the CA is
less than 90°. The value of CA is also dependent on various factors like roughness
possessed by a surface, CA hysteresis, sliding angle, gravity, temperatures, functional
group present in the surface, impurities, porosity, physico–chemical properties such
as chemical reactions between liquid alloy and solid oxide, oxygen partial pressure,
crystal structure of oxide phase [14]. Its value is obtained from the tangent of the

Fig. 1 Various micro- and nanoscale surfaces of living organisms in nature. The surfaces of the living organisms have complex and unique structural, chemical, and physical properties, serving as design principles for the development of engineering platforms and devices [Reproduced with permission [16] Copyright 2017, Elsevier]

drop of liquid at the contact point between the phases of liquid, solid and gas having the solid phase plane in the three-phase limits or the boundary. The water on any solid surface has three possibilities to show their behavior of water on solid surface. First is the water immediately wet on the surface and it will be achieved when $0° < \theta_Y < 90°$, i.e. the value of θ_Y lies between $0°$ and $90°$ then the surface is said to be hydrophilic. If $\theta_Y = 0$, i.e. water turned up into flat water film, i.e. the contact area between solid and liquid will be maximum then it shows the completely wetting tendency. When water drop makes the CA $90° < \theta_Y < 120°$ then the surface is said to be the hydrophobic. If the CA is more than $120°$ then the surface is said to be the superhydrophobic. For low-surface materials, the CA can reach upto $120°$ and can be further increased by creating an exclusive type of topography through micro/nano order surface roughness. The three original theoretical models used to describe the wetting principles are Young's Model (Young's) (Fig. 2), Wenzel Model, and CB Model with an objective of expressing an association between CA [CA], roughness, and the surface energy.

Neglecting the effect of gravity, the exact shape of the liquid drops on ideally flat surface can be found out by the Young equation $\gamma_{SV} = \gamma_{LV} \cos\theta_Y + \gamma_{SL}$ or $\cos\theta_Y = \frac{\gamma_{SV} - \gamma_{SL}}{\gamma_{LV}}$. For the industrial applications, the Young's equation is not applicable because

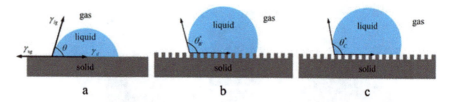

Fig. 2 Wetting models of **a** Yang's theory, **b** Wenzel theory, and **c** Cassie–Baxter theory [19]

this equation is valid for smooth solid surface not for uneven or rough surfaces. For overcome their limitation, the Wenzel considers the influence of roughness of a surface to modify the Young's equation. Wenzel's model explains the wetting, where there is a contact of the liquid drop with the peaks and valleys present in the rough surface [27] with the following equation $cos\theta_W = rcos\theta_Y = r\frac{\gamma_{SV}-\gamma_{SL}}{\gamma_{LV}}$. Here, r is defined as roughness factor, and it can be called as a ratio of the actual or the true area of the solid surface to its projected area [20] $r = \frac{Actual\ Surface\ Area}{Planar\ Area\ Beneath\ Droplet}$. Here, roughness factor is a non-dimensional parameter and is always greater than 1 for a surface which is rough. After the revolutionary effort of Wenzel on the wettability of rough surfaces, Cassie-Baxter [11] investigated the wettability for the porous planes. They stated that if the surface is more roughened or may be the order of nanometer where the entire surface is not wetted by the liquid. According to Cassie − Baxter, $cos\theta_{CB} = f_1cos\theta_1 + f_2cos\theta_2$. Here, θ_1 and θ_2 represent the CAs of water–solid and water–air, respectively, and θ_{CB} is the apparent contact angle of the composite. f_1, f_2 represent the area fractions occupied by solids and air, respectively, and $f_1 + f_2 = 1$. The above equation can be written as $cos\theta_{CB} = f_scos\theta_Y + (1 - f_s)cos\theta_{air}$. For $\theta_{air} = 180^0$, $cos\theta_{CB} = f_scos\theta_Y + (1 - f_s)(-1) = cos\theta_{CB} = f_s(cos\theta_Y + 1) - 1$. This model usually described water-repellent surfaces like the surface of Lotus leaf. Other than CA, contact angle hysteresis is also very significant to characterizing the surface wettability. If θ_A is defined as a proceeding CA and θ_R is the receding CA then the contact angle hysteresis as $\theta_H = \theta_A - \theta_R$. In practice, contact angle hysteresis is caused by chemical heterogeneity, drop size, surface roughness, molecular orientation, contamination and deformation, and liquid molecular conveyance [21].

3 Applications of Superhydrophobic Coatings

Self-Cleaning

In our daily life, many surfaces eventually get contaminated due to accumulation of dirt, dust, or through air pollution, which require a lot of time, energy, and money to clean these surfaces. Due to high contact angle (>150°) and low contact angle hysteresis, nanostructured superhydrophobic coating, thus, water adhesion is less than 9.6 mJ m^{-2} and is relatively weak [22]. This weak adhesion reduces the friction between the drops and the surface, thus allowing the drops to easily move along the surface, and ultimately keeping the dirt, dust, and contamination away automatically, i.e. clean the surface regularly. Self-cleaning surfaces have a wide range of applications in numerous fields such as textile, automobile, optical, marine, and aerospace industries, windows, and solar modules [23]. The forces/interactions applied between particles and smooth/nanotextured surfaces are dominating and are responsible for self-cleaning phenomena are double-layer interactions, van der Waals interactions, capillary forces, and lastly the line tension [24]. The self-cleaning mechanism, i.e. removal/detachment of dust particle from surface is illustrated in Fig. 3 [25].

Many scientists work to develop the superhydrophobic coating to remove the dust/ contamination from the surface. If the substrate surface is transparent, the superhydrophobic coating must be transparent. Transparency and superhydrophobicity are two conflicting properties [26, 27]. Indeed, it is possible to have both superhydrophobic and transparent properties by playing on the size of the nanostructures, as the decrease in the transparency is due to the light scattering inside the surface roughness [28]. The types, size, concentration, and methods are the deciding parameters of nano surface roughness. Nowadays, scientists are preferring fluorine-free chemical to make the coating environmental friendly. Nurul et al. [29] fabricated fluorine free, environmentally friendly, robust, transparent and stable superhydrophobic coatings. The used octadecyl trichlorosilane (OTS) modified TiO_2 to develop the required

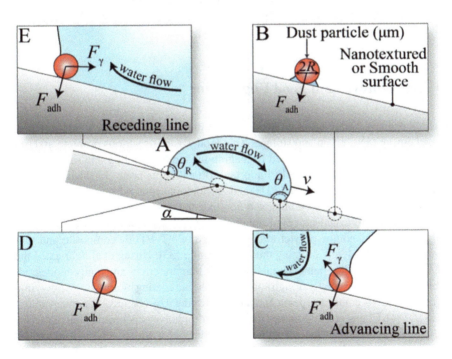

Fig. 3 The self-cleaning mechanism, i.e. removal/detachment of dust particle from surface [Reproduced with permission [25] Copyright 2019, ACS Publications]

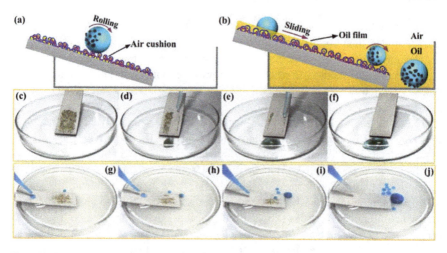

Fig. 4 **a–b** Schematic illustration of the self-cleaning processes for the EP + PDMS@SiO2-coated Mg alloy surface in air **c–f** and **g–j** oil-contamination [Reproduced with permission [30] Copyright 2019, Elsevier]

roughness, which gives water contact angle of $158 \pm 2°$ and sliding angle of $4 \pm 1°$ and showed an excellent self-cleaning ability against dirt particles after washed with water. The developed coating shows good performance after UV irradiation, chemical immersion, and physical abrasion (Fig. 4).

4 Agricultural and Food Engineering

Due to external and internal unfavorable conditions, food packaging gets contaminated during distribution and storage. This food waste is a serious problem and can be easily eliminated by improved packaging having anti-fouling and self-cleaning capabilities, which are critically important to tackle this issue. In this regard, the antibacterial [30], antimicrobial [31], and antifouling superhydrophobicity surfaces prepared from ecofriendly, edible, and nontoxic materials [32].

Yoon et al. fabricated superhydrophobic layers by annealing stainless steel plates with a carbon nanotube-polytetrafluoroethylene (CNT–PTFE) composite and titanium dioxide (TiO_2) by utilizing the spray coating [33]. Huang et al. fabricated the lignin-coated cellulose nanocrystals (L-CNC) that are environmentally biodegradable and friendly, to create a rough superhydrophobic coating by spraying a polyvinyl alcohol/ L-CNC composite paint, and then modifying it by chemical vapor deposition (CVD) [34]. Cheng et al. fabricated the superhydrophobic paper packaging platforms inspired by lotus leaf surface structures via coating paper with functionalized varnish modified with PDMS silicone oil and R812S silica nanoparticles [35].

5 Anti-biofouling and Anti-bacterial

The effects of bio-fouling cost marine, transportation, and other overall endeavors billions of dollars reliably. The biofilm can change the morphology of the base and lead to disintegration of metals [36]. The counter fouling covering is to prevent the headway of biofilms with and without biocide [37]. This sort of covering can be placed into two classes, for example, self-cleaning covering (SPC) and fluorine-based coating (FRC). Toughness, great grip, erosion restraint, perfection, simple application, quick drying, minimal effort, and simple accessibility are the properties of a decent enemy of fouling covering. As of late, superhydrophobic covering has developed reasonable for the counter fouling covering as the fluorine-based covering neglects to forestall colonization of biofilm, however, restrains the attachment of fouling under practically all unique conditions and expansion of poisonous segments like Pb+ can harm biological system [38]. The requirement for anti-bacterial layers recently has drawn an incredible attention among the scientists because of its huge potential application in different fields, particularly in open zones where the bacterial hazard is high [39]. Hence, the advancement of controlled discharge techniques is essential to control the organic action of the coating. Most normal strategy here is to utilize Ag and its subordinates [40]. Starting late, a procedure to design multifunctional silk has in like manner been made. The silk was secured a superhydrophobic layer using gigantic water repellent blends and silica nanoparticles. This got a multifunctional material having hostile to microbial properties with superhydrophobic and superoleophobic properties against water and oil [41, 42].

6 Anti-icing

Coatings with anti-icing performance possess hydrophobicity and low ice adhesion strength, which delay ice formation and make ice removal easier. Anti-icing is notable and complex. The comprehension of these marvels requires a multi-disciplinary recognition as they relate with: (i) interactions between the water and the solid surface, (ii) ice nucleation components, and (iii) ice adhesion parameters. Superhydrophobic coatings show low interfacial energy and diminish the ice attachment. The latter is essential to expel the ice by outside mechanical power without any problem [39, 43].

7 Fabrics and Textiles

Engineering of superhydrophobic textile surfaces has gained significant scientific and industrial interest for its potential applications in outdoor wear and protective textiles because of brilliant water-repellency [44]. The super water repellency property along with low tilt angle of water droplets in designed textures grants self-cleaning and anti-moisture properties to woven materials. A major challenge remains in preparing superhydrophobic textiles that withstand the impact of high hydrostatic pressure. Scientists are working in this direction to improve the quality and durability of the coated textiles. Chao et al. [45] coated textiles with PDMS/SiO$_2$ mixture followed by hydrophobized with tetraethoxysilane and cetyltrimethoxysilane by an in situ Stöber reaction, which shows high static water contact angle of 162.5° with a low sliding angle <10°. This coating shows high resistance to hydrostatic pressure up to 38.6 kPa, which is high water pressure resistance and is promising for a number of potential applications, such as outdoor never-wet tents (Fig. 5).

8 Stability of Superhydrophobic Coating

To use superhydrophobic coating for industrial application, its stability like chemical, mechanical, thermal stability matters. If coating is not having these stabilities, then coating cannot be durable. The bonding between coating molecules and substrate molecule played an important role in these stabilities. Many scientists are working in this direction and some of the reports are here. Enhancing structural stability contributes to resist the external force to maintain original facial structures or improve the mechanical stability which is a key issue in application of superhydrophobic surface. Many scientist reported mechanical stability evaluated through manners linear abrasion, tape-peeling, knife scratching, finger pressing, bending, etc. However, mechanical stability through many approaches has been published, there is still lack of standardization to make comparison of different surfaces.

A superhydrophobic coating may degrade its chemical stability due to setback in surface roughness or due to breaking down of surface layer upon prologue to a destructive domain [46]. Solvent stability is a noteworthy piece of chemical stability for ensuring application in various fields. The outside of a material is reliably disposed to ambush by different solvents depending upon its application. The covering material should show soundness toward solvents. The synthesized superhydrophobic fabric texture with improved non-wettability by a one-pot process comprising of nanoparticles of silica and a fluorinated alkyl silane [24]. A study on the solvent take-up limit of the texture as well as its solvent resistance when exposed to various natural solvents was conducted.

Deterioration of superhydrophobicity due to thermal instability of materials and thermal effect makes them unsuitable for application under elevated temperatures [47]. To check durability of the nanostructured surfaces, which are superhydrophobic in nature, plasma polymerization of the surface was trailed by coating utilizing hydrophobic diamond like carbon (DLC) [48, 49]. The WCA experienced a continuous abatement from 90 to 60° while the superhydrophobicity was kept up even at temperature bigger than 150° for strengthening temperatures somewhere in the range of 25–300 °C.

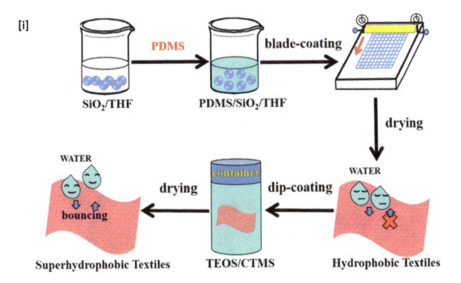

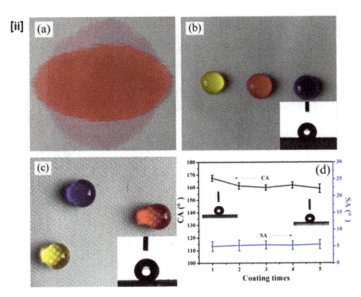

Fig. 5 [i] Illustration of the fabrication of superhydrophobic textiles with water pressure resistance. [ii] Dyed water droplet on **a** pristine PET textile; **b** PDMS/SiO2@PET textile after blade-coating four times; **c** S-PDMS/SiO2@PET textile. **d** variation in CA and SA of S-PDMS/SiO2@PET textiles with the coating times [Reproduced with permission [45] Copyright 2017, Elsevier]

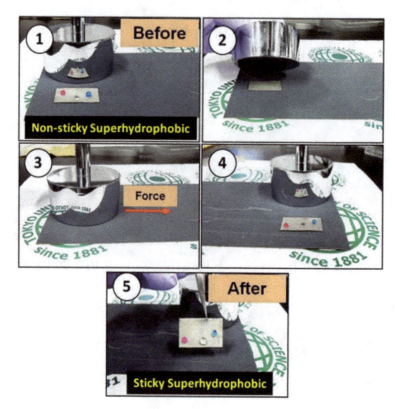

Fig. 6 Sandpaper abrasion test on the superhydrophobic M-430-SS/8h surface

Kim et al. made thermal stability of superhydrophobic surfaces was affirmed by warming it from 100 and 300 °C at a time frame for 1 h. The water CA changed here and there at 150° from 100 to 300°C [50]. The bend appeared uncovers the temperature security of the surface up to 300 °C. Rasouli et al. [51] watched the thermal stability of the surfaces of non-porous silica thin films up to 300 °C.

9 Conclusion

In this review, we presented an overview of the overall properties of the natural super-hydrophobic lotus leaf surface. This is our source of inspiration for unique strategies to create and manipulate of multifunctional engineering platforms having various industrial applications. As stability is one of the critical issues, so, the mechanical, chemical, and thermal study of superhydrophobic coating was also discussed.

References

1. Gupta KR, Kumar P, Yadav V, et al. Challenges and opportunities in fabrication of transparent superhydrophobic surfaces
2. Hooda A, Goyat MS, Pandey JK, Kumar A, Gupta R (2020) A review on fundamentals, constraints and fabrication techniques of superhydrophobic coatings. Prog Org Coatings 142:105557. https://doi.org/10.1016/j.porgcoat.2020.105557
3. Satyarathi J, Kumar V, Kango S, Verma R, Sharma N, Gupta R (2022) Comparative study of fabricated superhydrophobic surfaces via LST with carbon soot, chemical etching and auto-oxidation. Surf Topogr Metrol Prop 10(4):045017. https://doi.org/10.1088/2051-672X/ac9c76
4. Sharma V, Goyat MS, Hooda A, et al. (2020) Recent progress in nano-oxides and CNTs based corrosion resistant superhydrophobic coatings: a critical review. Prog Org Coatings 140. doi:https://doi.org/10.1016/j.porgcoat.2019.105512
5. Mahato S, Gupta A, Justin J, et al. Development of environment friendly superhydrophobic polystyrene/SiO_2 coatings via sol-gel route, 3–8
6. Hooda A, Goyat MS, Kumar A et al (2018) A facile approach to develop modified nano-silica embedded polystyrene based transparent superhydrophobic coating. Mater Lett 233:340–343. https://doi.org/10.1016/j.matlet.2018.09.043
7. Hooda A, Goyat MS, Kumar A, Gupta R (2018) A facile approach to develop modified nano-silica embedded polystyrene based transparent superhydrophobic coating. Mater Lett 233:340–343. https://doi.org/10.1016/j.matlet.2018.09.043
8. Saini RK, Saini DK, Gupta R, et al. Effects of dust on the performance of solar panels—a review update from 2015–2020. Energy Environ. Published online June 14, 2022. doi:https://doi.org/10.1177/0958305X221105267/ASSET/IMAGES/LARGE/10.1177_0958305X221105267-FIG12.JPEG
9. Gupta R, Verma R, Kango S et al (2023) A critical review on recent progress, open challenges, and applications of corrosion-resistant superhydrophobic coating. Mater Today Commun 34:105201. https://doi.org/10.1016/j.mtcomm.2022.105201
10. Barthlott W, Neinhuis C (1997) Purity of the sacred lotus, or escape from contamination in biological surfaces. Planta 202(1):1–8. https://doi.org/10.1007/s004250050096
11. Deo D, Singh SP, Mohanty S, Guhathakurata S, Pal D, Mallik S (2022) Biomimicking of phyto-based super-hydrophobic surfaces towards prospective applications: a review. J Mater Sci 57(19):8569–8596. https://doi.org/10.1007/S10853-022-07172-1
12. Zhang Y, Zhang Z, Yang J, Yue Y, Zhang H (2022) A review of recent advances in superhydrophobic surfaces and their applications in drag reduction and heat transfer. Nanomaterials 12(1). https://doi.org/10.3390/nano12010044
13. Mengnan Q, Jinmei H, Junyan Z (2010) Superhydrophobicity, Learn from the lotus leaf. Biomimetics learn from nat. Published online, 325–343
14. Sharma V, Sharma V, Goyat MS et al (2020) Recent progress in nano-oxides and CNTs based corrosion resistant superhydrophobic coatings: a critical review. Prog Org Coatings 140:105512. https://doi.org/10.1016/j.porgcoat.2019.105512
15. Gu G, Dang H, Zhang Z, Wu Z (2006) Fabrication and characterization of transparent super-hydrophobic thin films based on silica nanoparticles. Appl Phys A 83(1):131–132. https://doi.org/10.1007/s00339-005-3473-0
16. Kim W, Kim D, Park S, Lee D, Hyun H, Kim J (2018) Engineering lotus leaf-inspired micro- and nanostructures for the manipulation of functional engineering platforms. J Ind Eng Chem 61:39–52. https://doi.org/10.1016/j.jiec.2017.11.045
17. Dewan R, Kumar N, Yadav B et al (2014) Modification in surface chemistry of the polyetrafluoroethylene through chemical graft copolymerization for potential oil contamination control. Part Sci Technol 32(2):158–163. https://doi.org/10.1080/02726351.2013.850459
18. Barranco A, Borras A, Gonzalez-Elipe AR, Palmero A (2016) Perspectives on oblique angle deposition of thin films: from fundamentals to devices. Prog Mater Sci 76:59–153. https://doi.org/10.1016/j.pmatsci.2015.06.003

19. Han S, Yang R, Li C, Yang L (2019) The wettability and numerical model of different silicon microstructural surfaces. Appl Sci 9(3):566. https://doi.org/10.3390/app9030566
20. Bahgat Radwan A, Abdullah AM, Alnuaimi NA (2018) Recent advances in corrosion resistant superhydrophobic coatings. Corros Rev 36(2):127–153.https://doi.org/10.1515/corrrev-2017-0012
21. Zhang X, Wang L, Levänen E (2013) Superhydrophobic surfaces for the reduction of bacterial adhesion. RSC Adv 3(30). https://doi.org/10.1039/c3ra40497h
22. Chen W, Wang W, Luong DX et al (2022) Robust superhydrophobic surfaces via the sand-in method. ACS Appl Mater Interfaces 14(30):35053–35063. https://doi.org/10.1021/ACSAMI.2C05076/SUPPL_FILE/AM2C05076_SI_005.AVI
23. Somasundaram S, Kumaravel V (2019) Application of nanoparticles for self-cleaning surfaces. Springer, Cham, pp 471–498. https://doi.org/10.1007/978-3-030-04474-9_11
24. Parvate S, Dixit P, Chattopadhyay S (2020) Superhydrophobic surfaces: insights from theory and experiment. J Phys Chem B 124(8):1323–1360. https://doi.org/10.1021/ACS.JPCB.9B08567/ASSET/IMAGES/LARGE/JP9B08567_0025.JPEG
25. Heckenthaler T, Sadhujan S, Morgenstern Y, Natarajan P, Bashouti M, Kaufman Y (2019) Self-cleaning mechanism: why nanotexture and hydrophobicity matter. Langmuir 35(48):15526–15534. https://doi.org/10.1021/ACS.LANGMUIR.9B01874/SUPPL_FILE/LA9B01874_SI_001.PDF
26. Gupta R, Hooda A, Goyat MS, et al. (2018) Synthesis of polystyrene / ZnO super hydrophobic coatings with high transparency super-hydrophobic surfaces reference 13(March):248007
27. Hooda A, Goyat MS, Gupta R, Prateek M, Agrawal M, Biswas A (2017) Synthesis of nano-textured polystyrene/ZnO coatings with excellent transparency and superhydrophobicity. Mater Chem Phys 193:447–452. https://doi.org/10.1016/j.matchemphys.2017.03.011
28. Darmanin T, Guittard F (2015) Superhydrophobic and superoleophobic properties in nature. Mater Today 18(5):273–285. https://doi.org/10.1016/j.mattod.2015.01.001
29. Pratiwi N, Zulhadjri, Arief S, Admi, Wellia DV (2020) Self-cleaning material based on superhydrophobic coatings through an environmentally friendly sol–gel method. J Sol-Gel Sci Technol 96(3):669–678. https://doi.org/10.1007/S10971-020-05389-7/METRICS
30. Gupta R (2020) Materials for diagnosis, prevention and control of covid-19 pandemic. Biotechnol Kiosk 2(9):5–18. https://doi.org/10.37756/bk.20.2.9.1
31. Krishna Kudapa V, Mittal A, Agrawal I, K. Gupta T, Gupta R (2022) Role of graphene and graphene derived materials to fight with COVID-19. In: Biotechnology to Combat COVID-19. IntechOpen. https://doi.org/10.5772/intechopen.96284
32. Ruzi M, Celik N, Onses MS (2022) Superhydrophobic coatings for food packaging applications: a review. Food Packag Shelf Life 32:100823. https://doi.org/10.1016/j.fpsl.2022.100823
33. Yoon SH, Rungraeng N, Song W, Jun S (2014) Superhydrophobic and superhydrophilic nanocomposite coatings for preventing Escherichia coli K-12 adhesion on food contact surface. J Food Eng 131:135–141. https://doi.org/10.1016/j.jfoodeng.2014.01.031
34. Huang J, Wang S, Lyu S, Fu F (2018) Preparation of a robust cellulose nanocrystal superhydrophobic coating for self-cleaning and oil-water separation only by spraying. Ind Crops Prod 122:438–447. https://doi.org/10.1016/j.indcrop.2018.06.015
35. Chang KC, Lu HI, Peng CW et al (2013) Nanocasting technique to prepare lotus-leaf-like superhydrophobic electroactive polyimide as advanced anticorrosive coatings. ACS Appl Mater Interfaces 5(4):1460–1467. https://doi.org/10.1021/AM3029377/ASSET/IMAGES/MEDIUM/AM-2012-029377_0011.GIF
36. Li Y, Ning C (2019) Latest research progress of marine microbiological corrosion and bio-fouling, and new approaches of marine anti-corrosion and anti-fouling. Bioact Mater 4:189–195. https://doi.org/10.1016/j.bioactmat.2019.04.003
37. Roy R, Tiwari M, Donelli G, Tiwari V (2018) Strategies for combating bacterial biofilms: a focus on anti-biofilm agents and their mechanisms of action. Virulence 9(1):522–554. https://doi.org/10.1080/21505594.2017.1313372

38. Bae S, Lee YJ, Kim MK, Kwak Y, Choi CH, Kim DG (2022) Antifouling effects of super-hydrophobic coating on sessile marine invertebrates. Int J Environ Res Public Health 19(13). https://doi.org/10.3390/IJERPH19137973/S1

39. Wang L, Hu C, Shao L (2017) The antimicrobial activity of nanoparticles: present situation and prospects for the future. Int J Nanomedicine 12:1227–1249. https://doi.org/10.2147/IJN.S121956

40. Lawrencia D, Wong SK, Low DYS et al (2021) Controlled release fertilizers: a review on coating materials and mechanism of release. Plants 10(2):238. https://doi.org/10.3390/plants10020238

41. Ielo I, Giacobello F, Castellano A, Sfameni S, Rando G, Plutino MR (2021) Development of antibacterial and antifouling innovative and eco-sustainable sol-gel based materials: from marine areas protection to healthcare applications. Gels 8(1):26. https://doi.org/10.3390/gels8010026

42. Arora S, Yadav V, Kumar P, Gupta R, Kumar D (2013) Polymer based antimicrobial coatings as potential biomaterial: a review. Int J Pharm Sci Rev Res 23(2):279–290

43. Bhushan B (2016) Roughness-induced superliquiphilic/phobic surfaces: lessons from nature, 23–33. https://doi.org/10.1007/978-3-319-28284-8_2

44. Mao T, Xiao R, Liu P, et al. (2022) Facile fabrication of durable superhydrophobic fabrics by silicon polyurethane membrane for oil/water separation. Chin J Chem Eng. Published online May 21, 2022. https://doi.org/10.1016/j.cjche.2022.05.003

45. Xue CH, Li M, Guo XJ, Li X, An QF, Jia ST (2017) Fabrication of superhydrophobic textiles with high water pressure resistance. Surf Coatings Technol 310:134–142. https://doi.org/10.1016/j.surfcoat.2016.12.049

46. Zhang C, Liang F, Zhang W et al (2020) Constructing mechanochemical durable and self-healing superhydrophobic surfaces. ACS Omega 5(2):986–994. https://doi.org/10.1021/acsomega.9b03912

47. He H, Guo Z (2021) Superhydrophobic materials used for anti-icing theory, application, and development. iScience 24(11):103357. https://doi.org/10.1016/j.isci.2021.103357

48. Jiang S, Zhou S, Du B, Luo R (2021) Preparation of the temperature-responsive superhy-drophobic paper with high stability. ACS Omega 6(24):16016–16028. https://doi.org/10.1021/ACSOMEGA.1C01861/SUPPL_FILE/AO1C01861_SI_001.PDF

49. Vidal K, Gómez E, Goitandia AM, Angulo-Ibáñez A, Aranzabe E (2019) The synthesis of a superhydrophobic and thermal stable silica coating via sol-gel process. Coatings 9(10):627. https://doi.org/10.3390/coatings9100627

50. Kim H, Nam K, Lee DY (2020) Fabrication of robust superhydrophobic surfaces with dual-curing siloxane resin and controlled dispersion of nanoparticles. Polymers (Basel) 12(6):1420. https://doi.org/10.3390/polym12061420

51. Rasouli S, Rezaei N, Hamedi H, Zendehboudi S, Duan X (2021) Superhydrophobic and super-oleophilic membranes for oil-water separation application: a comprehensive review. Mater Des 204:109599. https://doi.org/10.1016/j.matdes.2021.109599

Chapter 15
Study of New Generation Thermal Barrier Coatings for High-Temperature Applications

Sumit Choudhary and Vidit Gaur

1 Introduction

The application of TBCs with nickel based superalloys results in an efficient class of low thermal conductive ceramic coatings. Successful deposition of these coatings raises the working temperature of hot section components (stator/rotor) in both the land-based and aircraft gas turbine engines, improving their performance, efficiency, and longevity [1]. The typical system of TBCs is designed with three layers: a creep-resistant, high tensile strength Ni-based superalloy component/substrate layer, and a bond coat (BC) which is resistant to oxidation. The top coat (TC) of a standard TBC system mostly contains yttria-stabilized zirconia (YSZ) layer [2]. In order to increase the endurance of the substrates, TBCs are applied. This lowers the temperature of the substrate due to which the driving force for the creep and thermal fatigue failure of substrates reduced significantly. The thermal barrier and resistance to corrosion of the substrate are major functions of the ceramic top coat. However, the layer of bond coat protects the substrate from high-temperature oxidation, corrosion, and promotes adhesion between the metal substrate and ceramic TC. Although zirconia has a higher CTE compared to other ceramic materials ($10-12 \times 10^{-6}$ °C), it is less than that of metal substrates ($17-21 \times 10^{-6}$ °C). Consequently, an overlay bond coat of MCrAlY or a NiPtAl diffusion is employed to reduce the CTE mismatch between coatings and the substrate. The M could be Ni, Co, NiCo, CoNi, or Fe depending on the type of substrate (superalloy) and the working environment conditions [3].

The strength of the adhesion, temperature mismatch at the BC and TC interface, and applied thermal load profiles are only a few of the variables that affect how long TBCs last [4]. Since each of these parameters has a unique impact on the induced internal stresses in the various constituents during service circumstances, the system will fail in a different way depending on how it reacts [5]. Because of this, it is

S. Choudhary (✉) · V. Gaur
Indian Institute of Technology Roorkee, Roorkee, Uttarakhand 247667, India
e-mail: s_choudhary@me.iitr.ac.in

© The Author(s), under exclusive license to Springer Nature Singapore Pte Ltd. 2023   317
A. Verma et al. (eds.), *Coating Materials*, Materials Horizons: From Nature to Nanomaterials, https://doi.org/10.1007/978-981-99-3549-9_15

necessary to have a thorough grasp of the effects of the contributing components and how they interact in order to create advanced TBCs. By changing their microstructure and architecture, traditional TBCs' attributes can be adjusted. This can be done by shedding light on the failure mechanisms that lead to these failures in service.

Due to its sintered state at 1300 °C, YSZ loses its damage tolerance due to phase stability, rendering it incompatible with the jet or gas turbine engines of next generation (with service temperatures exceeding 1400 °C). In addition, lower thermal conductivity of top coat is necessary to raise the working temperature of gas turbine and/or jet engines [6]. Consequently, it is crucial to create new TBC ceramic materials with decreased thermal conductivity and enhanced stability at high temperature. The TBC materials need to have no phase transitions during thermal cycling, high melting point, low sintering rate, good corrosion resistance, good damage tolerance, and thermal expansion match with the metallic substrate in order to function as a thermal barrier in addition to meeting other performance criteria.

This study reviews the potential deterioration mechanisms that the TBC system may encounter during its service life. Also covered were the most recent studies on modifying the chemical and structural makeup of YSZ to better fit specific desired qualities. The final step was an evaluation of innovative TBC materials and their properties.

2   Structure of Thermal Barrier Coatings

The thermal barrier coating system mostly contains three layers of coatings over the substrate. The substrate part is the metallic component, to protect that from high-temperature exposure, oxidation, and to maintain the mechanical properties during the service life the thermal barrier coating system is designed. In this system, the three layers generally exist over the substrate: (i) Bond coat (BC), (ii) Top coat, and (iii) Thermally grown oxides (TGO) layer as shown in Fig. 1.

The layer of bond coat MCrAlY(M =) first deposits on the metallic substrate having the CTE higher than the top coat but lower than the substrate, and it is customized using various mixtures of metal alloys to tailor the CTE. The bond coat

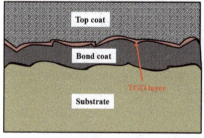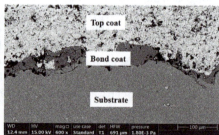

Fig. 1 Schematic of thermal barrier coating system structure with FE-SEM Image

deposited thickness typically exists in the range of 30 to 120 μm. The bond coat has properties to enhance the adhesion between the ceramic top coat and metallic substrate by managing the stresses generated due to thermal mismatch between them. It also provides resistance to oxidation during the thermal cycles, i.e., heating and cooling. The protection of substrates from hot corrosion can be improved significantly by using the suitable bond coat because the top coat ceramics have a porous structure and cannot completely protect the substrate alone as YSZ top coats are transparent to oxygen [7]. Hence, an oxides layer of around 2–10 μm forms at the border of the top coat and bond coat due to the reaction occurring between the metal elements existing in the bond coat with the oxygen passing through the top coat. This oxide layer formation is named the thermally grown oxide (TGO) layer which is rich in alumina (Al_2O_3). During the service period at high temperature, the thickness of this TGO layer gradually increases and that creates the critical failure mechanism of the TBCs. However, if the thickness growth rate is slowed down and it adheres well to the bond coat, the high-temperature oxidation protection of the substrate will improve [8, 9]. To control the failure that occurs due to TGO layer formation, many researchers have studied different materials for the top coat. There are some important parameters and materials that they have suggested which are discussed in the next section.

3   Selection of Top Coat Material

The top coat should have some key characteristics such as high-temperature oxidation resistance, excellent erosion resistance, high sintering resistance, very high melting point, resistance to thermal shocks, phase stability up to high-temperature limit, good fracture toughness, and relatively high coefficient of thermal expansion (CTE) and low thermal conductivity. Moreover, they should provide good mechanical integrity and high thermodynamic stability. Till now, no single material is available to satisfy all the requirements. However, Zirconia (ZrO_2) ceramic accomplishes most of the desired requirements due to which it becomes the prominent material for TBC applications. However, the ZrO_2 is susceptible to polymorphic phase transformation due to the quenching effect generated during the plasma spraying [10].

The nature of pure ZrO_2 is polymorphic. But the phase transformation occurs from monoclinic to tetragonal at ~1160 °C, and tetragonal to cubic at ~2350 °C, respectively. As a result, around 4% volume change during cooling occurs due to tetragonal to monoclinic transformation. This volume change creates out-of-plane stresses that lead to the failure or delamination of the thermal barrier coatings [11]. However, the volume change at higher temperatures can be prevented by adding 3–8 weight percent phase-stabilizing agents (CeO_2, Y_2O_3, MgO, CaO, Al_2O_3, and CeO_2) to the zirconia. Among these stabilizing agents, Y_2O_3 is having the closest vapor pressure and ionic radius to the zirconia due to which it has better stability compared to other stabilizing agents and hence suggested as a better stabilizer. At present, yttria-stabilized zirconia (YSZ) is the most preferred commercial top coat ceramic material because it has reduced thermal conductivity, good stability against chemicals, and relatively higher thermal expansion coefficient and fracture toughness [12]. The melting point of the top coat materials (>2000 °C) is typically very high. Therefore, to fabricate them over a metallic substrate, a special fabrication process was used. In the next section, the fabrication process is explained.

4   Fabrication of Thermal Barrier Coatings

The methods that are frequently used for the fabrication of TBCs include electron beam-physical vapor deposition (EB-PVD), plasma spray-physical vapor deposition (PS-PVD) [13], and atmospheric plasma spraying (APS) [8, 14]. However, APS is the more widely used coating technology due to its simplicity, comparatively less cost, and excellent spraying efficiency [15–17].

The APS technique is broadly used to overlay the TBC over hot-section engine components such as nozzles, combustors, shrouds, and ignitors. The TBC deposited by APS typically has a thickness in the variation of 250–500 μm. In APS, the plasma flume is created between the cathode and anode (nozzle itself), and the electric arc is generated by the ionization of mixed gases containing argon and hydrogen [18]. After creating the plasma flume, the first step is feeding (gm/min) the powder into the plasma flume and stabilizing the powder feed rate as per the required layer deposition thickness. The complete or partial melting of any ceramic or metallic particle can be easily achieved in the APS because it can easily operate/adjust in the temperature range of 1600–13,000 °C. The pressurized argon and hydrogen (as a fuel) gases are streamed continuously to form plasma and generate high temperatures to melt the feedstock and provide a high velocity in the range of 80–300 m/sec [19]. The velocity of the molten splat is a function of the powder particle size and the plasma flume characteristics. The detailed schematic of the atmospheric plasma spray coating is represented in Fig. 2. to understand the process in a better way.

The highly accelerated molten particles impinge on the metallic substrate and rapidly solidify in the form of flattened particles known as splats. These successfully impinged particles are interlocked over each other due to peaks and valleys existing on the rough surface and create a strong bonding between each other and form the coating by layer-by-layer deposition. Vertical cracks called intra-splat cracks develop within distinct splats as a result of shrinkage during splat cooling restricted by the depositing substrate surface. In addition, the bonded contact between the splats cracks results in inter-lamellar pores. Because it increases strain tolerance while lowering heat conductivity, the integrated porosity existing between the splats and network crack-like voids is preferred for sprayed coatings [20]. The 10–15% porosity volume develops in the APS which can be controlled by the APS/process parameters.

The APS technique is frequently used to prepare TBCs with lamellar structure and low thermal conductivity in turbine airfoils and engine combustion chambers due to its operational stability, increased economic feasibility, and deposition rate. In recent years, some researchers had tried to optimize the process parameters through numerical studies of the in-flight particles by considering their temperature and velocity to

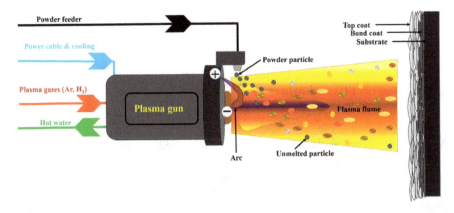

Fig. 2 Schematic of the complete process of atmospheric plasma spray TBC deposition over substrate

control the coating properties [21]. However, the typical ceramics used in the TBC system have some limitations which are tried to elaborate in the next section.

5   Limitations of Conventional TBCs

The YSZ ceramic top coat in TBCs has been extensively used for the last four decades; even at present, commercially many options are not available in the market. However, an increase in operating temperature transforms the phases of the YSZ along with the sintering which may lead to an increase in thermal conductivity and the formation of cracks. When the operating temperature is above 1200 C, the phase transformation starts occurring from toughened tetragonal prime to tetragonal and Cubic phases. In 9 h of operation at 1300 C, almost 60% of toughened tetragonal prime transforms into tetragonal and cubic phases [22]. The tetragonal phase changes to a monoclinic structure when YSZ cools down to room temperature and that causes the ~4% volume change; Young's modulus and thermal conductivity also increase due to the sintering effect. These factors would accelerate the failure of the thermal barrier coatings [23]. Thus, the next generation high-performance engine development demands the development of new top coat materials that can fullfill all the properties of the thermal barrier coatings. Additionally, the coatings must be flexible enough to generate a well-controlled microstructure with a stoichiometric composition employing a variety of production techniques. Depending on the preparation process, the morphology, homogeneity, and characteristics of the coatings would change [24]. With a single material, none of these requirements could be met. Recently, a number of ceramic compounds, typically rare-earth oxide ceramic, have been proposed as novel top coat materials suitable for TBCs. Particularly doped aluminates, fluorite, zirconium, perovskites, and pyrochlores are some of these novel ceramic materials. The TBCs mostly failed through different modes due to thermal cycles executed during the operation of the gas turbine engines.

6 Modes of TBCS Failures

The applications of TBCs are highly associated with their long-lasting stability. Mostly, TBCs are either failed due to thermal cyclic load or a combination of thermal and mechanical cyclic load.

The TBC system has layers of different materials having different CTE due to which a large thermal gradient is developed across the layers during the thermal cyclic load that generates very high stresses within the layers or at their interfaces. The stresses are generated along the layers and transverse to them, which creates vertical and horizontal cracks. These cracks propagate as the thermal cycles increase and finally the spallation of the coating occurs.

Predominantly during the cooling duration, vertical fissures may develop over the coating's surface as a result of high tensile stress and a temperature difference across the coating thickness caused by transitory thermal shock on the top coat outer surface. Due to the tensile strains, there is a chance that the surface fracture will move toward the interface. When the crack's tip contacts the interface, it may deflect along the interface or pierce the next layer. In reality, as interface fractures spread and merge, the coatings spall. Therefore, surface cracking and induced interfacial delamination are the main causes of failure in TBCs. Delaying the development of interface fractures might increase the coating's service durability [25].

There is a chance that additional cracks might occur from weak spots in addition to the expansion of main fractures, resulting in the appearance of several cracks in the coatings at once. For instance, the surface fracture that developed in the top coat material under loaded strain may have been able to deflect and expand along the TC/BC contact. New vertical fractures that form in the bond coat may also cause some cracks to appear at the boundary of the bond coat and substrate. The competitive propagation of these two interface fractures continues as the applied stress is increased, until one of them extends the critical length. Hence, the common fracture modes in the TBC system, apart from the separation of the top coat and bond coat from an interface, are spallation from the substrate and bond coat interface or from both interfaces [26, 27].

The evolution of the fracture characteristics during the failure process and the impact of surface crack density on the emergence of failure modes need to be clarified. As is well known, high temperatures are used in the thermal spraying process to fabricate coatings. Therefore, as the coating cools down to ambient temperature, preliminary thermal strains and residual stresses may be created due to the thermal expansion mismatch between the coating's various layers. The residual stresses have a considerable impact on the driving force at the crack's tip and the development of fracture modes. Therefore, to develope a better understanding of how residual

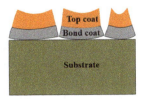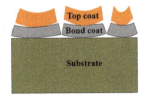

Fig. 3 schematic representation of the different modes of TBC failure: **a** Mode A, **b** Mode B, and **c** Mode C

stresses impact the failure modes in TBCs having vertical fracture cracks shows three different modes i.e., Mode A, Mode B, and Mode C [28, 29].

In conclusion, the thermal barrier coatings failed as a 3-D fracture phenomenon, despite the fact that their characterization may be done using the 2-D model of plane strain, as shown in Fig. 3. In Fig. 3, the typical failure mechanisms (in three modes) of A, B, and C are depicted. When TC and BC come together, mode A causes a vertical surface crack to deflect at their interface, and it then proceeds to spread until TC spalls off. In mode B, the propagation of surface fractures at the boundary of BC and the substrate without separating at the interface of TC and BC occurs, followed by the penetration of cracks existing on the surface into BC. However, in failure mode C the delamination occurs at both the interface of TC/ BC, and BC/substrate simultaneously. Since the deflection or penetration of surface cracks occurs at the interface of TC and BC, it appears that the investigation of modes A and B is quite straightforward. However, mode C's new fracture initiation, propagation, and branching are complicated processes [30]. It should be highlight here that typically bond coats like MCrAlY and metallic alloy substrates having ductile properties due to which the cracks do not frequently develop in them. So, with brittle bond coats TBCs, the modes B and C typically occur.

In researching the dependability and durability of TBCs, numerical analysis and modeling are very crucial. Finite element modeling (FEM) is particularly good in calculating the thermal insulation provided and the mechanism of fracture failure of TBCs. The failure issues of the TBCs in actual service conditions, the residual stress, and the thermal insulation qualities can all be modeled and calculated [31].

7   Thermal Fatigue Failure

One of the weaknesses of brittle ceramics is their poor thermal fatigue (shock) resistance. Thermal conductivity, fracture toughness, Poisson's ratio, CTE, and elastic modulus are some factors that typically influence the resistance of the TBC system against thermal fatigue. The thermal stresses are generated in the TBC-coated components due to temperature differences that occur at the surface and center of the component during heating and cooling. During the heating cycle, the top coat surface has a higher temperature compared to the metallic substrate temperature due to which the tensile stresses are generated in the substrate/center while compressive stresses are induced in the top coat. However, during the cooling cycle compressive stresses are induced in the metallic substrate/center and tensile stresses are generated in the top coat. Hence, the cracks are induced in the top coat during the cooling cycle. Ceramic materials have a high melting point, which makes them appropriate for components used at high temperatures. However, TBCs have an issue of reliability during service operating conditions. Material characteristics alter as a result of extreme temperature variations. For example, the repeated start and stop operations of gas turbines activate phenomena including thermal expansion, high-temperature friction, and the sintering effect, which continuously change the magnitude and directions of the internal stresses in turbine blades [32]. Throughout the operation, the cracks are continuously propagating and closing, and the elastic modulus is varying that significantly impacts the lifespan of the TBC. The investigation of their behavior during service became difficult due to frequent thermal cycles. Hence, the thermal endurance of the coatings during operational life is highly dependent on the resistance to the thermal shocks [33]. Thermal fatigue resistance of the coatings for extremely high-temperature applications can be improved by either small changes in the coatings' material composition or substrate composition to reduce the thermal mismatch between them. In some studies, the nanoparticles such as carbon nanotubes (CNTs), graphene nano platelets (GNPs), and mixing of two different ceramic powders are used to tailor the CTE so that significant improvement in the thermal shock resistance of the TBC system can be achieved [34]. Singh et al. [35] studied the hybrid composite coatings of the GNP/CNT-reinforced lanthanum-cerate coatings that could be a potential thermal barrier coating which can sustain the 1721 thermal cycles at 1800 °C without any major spallation failure. The schematic of the thermal shock setup along with temperature distribution at the front and backside of the TBC system in a detailed way is shown in Fig. 4.

8   New Ceramics (Pyrochlore Oxides) Materials in TBCs

The temperature abilities and demands of the recent generation of nickel-based or cobalt-based superalloys have reached their melting temperature, due to which it becomes difficult for them to continue operating at high temperatures. Consequently, research into novel materials that could be employed as TBCs is necessary. In recent years, many oxides are investigated for the TBC system such as fluorite oxides, pyrochlore oxides, perovskite oxides, and Lanthanum hexaaluminate. This chapter is limited to the pyrochlore oxides.

Due to the demand and development of more efficient and powerful gas turbine engines the gas inlet temperature has been reached to 1500–1600 °C. Therefore, several rare-earth oxides are complex pyrochlore and are coming in the list of potential materials for the applications of TBCs. These complex pyrochlore have the typical formula of $A_2B_2O_7$ [36]. Here, A and B represent the 3+ and 4+ cations, respectively. 'A' includes elements from La to Lu and 'B' includes Zr, Ti, and Hf, respectively.

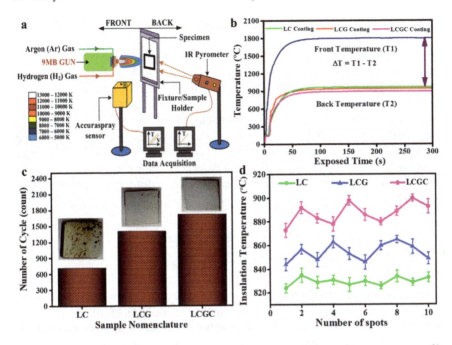

Fig. 4 a Schematic shows the high-temperature thermal shock setup, **b** temperature profile collected from the top coat and substrate of the specimen through pyrometer for LC, LCG, and LCGC coatings; **c** thermal shocks were executed at ~1800 °C on different coatings (inset: after-shocks digital image); **d** temperature variation on substrate during thermal shocks given to LC, LCG, and LCGC coated substrates. **LC: La₂Ce₂O₇, LCG: La₂Ce₂O + 1 wt.% GNP, LCGC: La₂Ce₂O + 1 wt.% GNP + 0.5 wt.% CNT** Ref. [35]

High melting point, relatively high CTE, and low thermal conductivity are some of the properties which promise for the suitability of these oxides for TBC applications. These materials are gaining attention because the pyrochlore and fluorite structures have a close relationship. When the eight fluorite unit cells are combined, they form a unit cell of the pyrochlore and each unit cell of this on average contains one vacancy of oxygen.

The pyrochlore structures can be stabilized by doping with the larger ions. It has also been done to dope oxide into the pyrochlore compounds. In this method, cation A is partially replaced by other cations, such as $A_{1-x}M_xB_2O_7$, where M is a rare-earth cation or another cation that ranges from 0 to 0.5. In contrast to un-doped lanthanum zirconate $(La_2Zr_2O_7)$, rare-earth-oxide-substituted pyrochlores like $(La, Yb)_2Zr_2O_7$, $(La, Gd)_2Zr_2O_7$, and $(La, Gd, Yb)_2Zr_2O_7$ demonstrated decreased thermal conductivity. The thermal conductivity is reduced by around 30% when Gd and Yb are co-doped. The findings show that doping, and particularly co-doping, of pyrochlore oxides can significantly lower their thermal conductivity [37].

Zirconate pyrochlores are an intriguing class of materials since several of them have decreased thermal conductivities. Because the cations in the crystal are in set locations, they also have good thermal stability. Co-doping can reduce these oxides' heat conductivity and boost their sintering resistance.

Li et al. [38] have tried to improve the sintering resistance of the pyrochlore oxide-based TBC coating by producing rare-earth zirconate high entropy single phase having the general formula $(5RE_{1/5})_2Zr_2O_7$. The rare-earth oxides and ZrO_2 are reacted through a solid-state process. The comparative densities of the pyrochlore oxides were varied in the range 70–80 percent after sintering at 1500 °C, which was thought to be the result of the slow diffusion in the high entropy material. The thermal conductivities of the produced oxides were lower than 1.0 Wm^{-1} K^{-1} in the range of 300 to 1200 °C temperature. Rather than these new ceramic materials some, most important oxide materials also exist for the TBC system that are mentioned in detail in the below sections.

9 Other Oxide Materials for TBCs

In recent years, for TBC applications certain oxides were investigated such as $SrAl_{12}O_{19}$ [39], $YaAlO_3$ (YAP) [40], Mg_2SiO_4 [41], $Y_3Al_5O_{12}$ [40], and $Y_4Al_2O_9$ (YAM) [42], and still exploring new oxides. Between room temperature to melting point, i.e., 1913 °C, yttrium aluminate perovskite (YAP) doesn't exhibit any phase transformation and has an average value of CTE [43]. Although $Y_3Al_5O_{12}$ has outstanding mechanical properties and a very low oxygen diffusivity, its somewhat high thermal conductivity is the key factor that prevents it from being used in hot section component applications. It has been noted that YAM is thermally stable almost up to 1350 °C. It exhibits a mild CTE in the 27–1200 °C temperature range. The thermal conductivity is significantly lower than that of YSZ's. Furthermore, $Y_4Al_2O_9$ has a lower density than YSZ and has a higher resistance to sintering [42]. Mg_2SiO_4 oxide has the stability of phase up to 1300 °C and the thermal conductivity inferior compared to YSZ, its CTE is mild, and its mechanical behavior is almost similar to the YSZ. Moreover, it has good resistance against thermal shocks and sintering [41]. The main properties of these oxide materials are precise in Table 1.

After investigating new oxide material of the TBC, some challenges still exist and researchers at the global level are trying to encounter them.

10 Challenges and Upcoming Insights

The main function of thermal barrier coatings is to act as the barrier to thermal damage, but they must also meet some strict performance parameters required such as phase transformation stability at least within the service conditions, high melting point, good corrosion resistance at high temperature, excellent sintering resistance, good damage tolerance, and thermal mismatch should be as low as possible with the

Table 1 New oxide properties used in TBCs

Oxides	Fracture toughness (MPa. m$^{1/2}$)	Elastic modulus (MPa)	Thermal conductivity (W/m K) at 1000 C	CTE (10^{-6}/ K) 30–1000 C	T$_{melting}$ (K)
SrAl$_{12}$O$_{19}$	–	–	1.36 [39]	7.52 [39]	–
YAlO$_3$	–	318 [40]	1.61 [40]	–	2186 [44]
Mg$_2$SiO$_4$	2.8 [41]	185 ± 10 [41]	1.68	8.6–11.3 [41]	2446 [45]
Y$_3$Al$_5$O$_{12}$	–	290 [40]	1.59 [40]	–	2213 [41]
Y$_4$Al$_2$O$_9$	3.36 [46]	191 [47]	1.13 [47]	7.37 [46]	2273 [44]
La$_2$Zr$_2$O$_7$	1.3 ± 0.3 [34]	175 [48]	1.56 [49]	9.1 [49]	2573 [49]
Gd$_2$Zr$_2$O$_7$				10.4 [50]	

metallic components. The second major challenge is accurately assessing TBC material performance in service. The operating conditions for TBC are extremely harsh due to the combination of high temperatures, quick temperature changes, sudden temperature gradients, added corrosion, mechanical loads, and high pressure of hot gases. Additionally, the majority of research now underway is restricted to recognizing one or a few oxides with reduced thermal conductivity. As a result, many multi-oxides with intricate crystal structures have not been studied for the applications of TBCs, and it is still unclear what causes the structural and chemical bond characteristics that underlie low thermal conductivities.

To further develop our understanding of material's performance at high temperatures, precise measuring procedures for characteristics like fracture toughness and hardness are required. A TBC lifetime prediction model under complex service loads also needs precise temperature readings at all the interfaces and surfaces. To control the process in a better way, online sensors are also preferred to boost reliability.

The phonon scattering mechanics, which are linked to the inherent properties of materials including atomic bonding, architecture, and structure, limit the coating materials' thermal conductivity. It is problematic to further reduce thermal conductivity at a specific layer thickness. Recent studies aim to modify ceramics, composition, microstructure, and density to obtain the necessary qualities. Additionally, more investigation is needed to understand the physical mechanisms underlying the thermal protection capabilities of nano-structured oxides and how they might be used to raise insulation temperature for usage in real-world applications and to advance scientific understanding.

It is expensive and time-consuming to investigate the wide range of chemical compositions connected to all refractory oxides in order to generate compositions that show promise. Experiments can go more quickly forward with the help of corresponding atomistic level models and expand our understanding of crystal structures. In future, the additively manufactured Nickel-based superalloy can be used for complex components of the hot section which can be protected through plasma spray coatings of noble ceramic materials upto very high operating temperatures [51, 52]. Moreover, the bond coat of the TBC system can be removed from the TBCs system to reduce the bulkiness of the components and to improve the overall efficiency of the engines [53].

11   Conclusions

TBCs have shown to be an important technique for lowering the exterior temperature of metallic components since they are crucial for the protection of turbines and engines against thermal socks, because they operate in a hostile and high-temperature condition that might shorten their lives. Among these issues include sintering, corrosion, thermal fatigue, transitions in phases, erosion, and damage from foreign objects. The TBC system itself is dynamic when in use. The components of the TBC system continuously changes in microstructure, crystalline phases, and structure during the course of their service life. At various points during their service lives, these adjustments cause changes in the mechanical and physical properties of TBCs.

For contemporary TBC applications, there is a rising need for the creation of novel ceramics having low thermal conductivities. Low thermal conductivity causes phonons to scatter as a result of their interactions with lattice flaws in actual crystal formations. Atoms with different masses, grain boundaries, vacancies, dislocations, and other phonons are a few examples of defects. The new ceramic materials must have large mean atomic masses, a lower Young's modulus, low thermal conductivity, and a high density of point defects with low distribution. There are a number of intriguing ceramics in the next generation of state-of-the-art ceramic TBCs with significantly lower thermal conductivity, including $LaMgAl_{11}O_{19}$, $Gd_2Zr_2O_7$, $La_2Ce_2O_7$, $SrZrO_3$, and $La_2Zr_2O_7$ ceramics. Additionally, nanostructured materials have been shown to have reduced thermal conductivities due to the intensified scattering of grain boundaries.

Acknowledgements The author feels gratitude toward the Mechanical & Industrial Engineering Department IIT Roorkee, Ministry of Education, Govt. of India, for awarding the Prime Minister's Research Fellowship (Grant no. PM-31-22-672-414) and his Supervisor (Dr. Vidit Gaur) for supporting during the writing of this chapter.

References

1. Pakseresht AH, Javadi AH, Nejati M, Shirvanimoghaddam K, Ghasali E, Teimouri R (2014) Statistical analysis and multiobjective optimization of process parameters in plasma spraying of partially stabilized zirconia. Int J Adv Manuf Technol 75(5–8):739–753. https://doi.org/10.1007/S00170-014-6169-9/METRICS

2. Pakseresht AH, Rahimipour MR, Alizadeh M, Hadavi SMM, Shahbazkhan A (2016) Concept of advanced thermal barrier functional coatings in high temperature engineering components 396–419. https://doi.org/10.4018/978-1-5225-0066-7.CH015

3. Jam A, Derakhshandeh SMR, Rajaei H, Pakseresht AH (2017) Evaluation of microstructure and electrochemical behavior of dual-layer NiCrAlY/mullite plasma sprayed coating on high silicon cast iron alloy. Ceram Int 43(16):14146–14155. https://doi.org/10.1016/J.CERAMINT.2017.07.155

4. Pakseresht AH, Saremi M, Omidvar H, Alizadeh M (2019) Micro-structural study and wear resistance of thermal barrier coating reinforced by alumina whisker. Surf Coatings Technol 366:338–348. https://doi.org/10.1016/J.SURFCOAT.2019.03.059

5. Chen LB (2012) Yttria-stabilized zirconia thermal barrier coatings—a review 13(5):535–544. https://doi.org/10.1142/S0218625X06008670

6. Clarke DR, Oechsner M, Padture NP (2012) Thermal-barrier coatings for more efficient gas-turbine engines. MRS Bull 37(10):891–898. https://doi.org/10.1557/MRS.2012.232/FIGURES/8

7. Karaoglanli AC, Ogawa K, Turk A, Ozdemir I (2013) Thermal shock and cycling behavior of thermal barrier coatings (TBCs) used in gas turbines. Prog Gas Turbine Perform. https://doi.org/10.5772/54412

8. Dong H, Yang GJ, Li CX, Luo XT, Li CJ (2014) Effect of TGO thickness on thermal cyclic lifetime and failure mode of plasma-sprayed TBCs. J Am Ceram Soc 97(4):1226–1232. https://doi.org/10.1111/JACE.12868

9. Abuchenari A et al (2020) A review on development and application of self-healing thermal barrier composite coatings. J Compos Compd 2(4):147–154. https://doi.org/10.29252/JCC.2.3.6

10. Zhang K, Van Le Q (2020) Bioactive glass coated zirconia for dental implants: a review. J Compos Compd 2(2):10–17. https://doi.org/10.29252/JCC.2.1.2

11. Rahimi S, SharifianJazi F, Esmaeilkhanian A, Moradi M, Safi Samghabadi AH (2020) Effect of SiO$_2$ content on Y-TZP/Al$_2$O$_3$ ceramic-nanocomposite properties as potential dental applications. Ceram Int 46(8):10910–10916. https://doi.org/10.1016/J.CERAMINT.2020.01.105

12. Gao L, Guo H, Wei L, Li C, Gong S, Xu H (2015) Microstructure and mechanical properties of yttria stabilized zirconia coatings prepared by plasma spray physical vapor deposition. Ceram Int 41(7):8305–8311. https://doi.org/10.1016/J.CERAMINT.2015.02.141

13. Liu MJ et al (2019) Transport and deposition behaviors of vapor coating materials in plasma spray-physical vapor deposition. Appl Surf Sci 486:80–92. https://doi.org/10.1016/J.APSUSC.2019.04.224

14. Pakseresht AH, Ghasali E, Nejati M, Shirvanimoghaddam K, Javadi AH, Teimouri R (2015) Development empirical-intelligent relationship between plasma spray parameters and coating performance of Yttria-Stabilized Zirconia. Int J Adv Manuf Technol 76(5–8):1031–1045. https://doi.org/10.1007/S00170-014-6212-X/METRICS

15. Nejati M, Rahimipour MR, Mobasherpour I (2014) Evaluation of hot corrosion behavior of CSZ, CSZ/micro Al$_2$O$_3$ and CSZ/nano Al$_2$O$_3$ plasma sprayed thermal barrier coatings. Ceram Int 40(3):4579–4590. https://doi.org/10.1016/J.CERAMINT.2013.08.135

16. Pandey KK, Singh RK, Rahman OA, Choudhary S, Verma R, Keshri AK (2020) Insulator-conductor transition in carbon nanotube and graphene nanoplatelates reinforced plasma sprayed alumina single splat: Experimental evidence by conductive atomic force microscopy. Ceram Int 46(15):24557–24563. https://doi.org/10.1016/J.CERAMINT.2020.06.243

17. Pandey KK et al (2021) Microstructural and mechanical properties of plasma sprayed boron nitride nanotubes reinforced alumina coating. Ceram Int 47(7):9194–9202. https://doi.org/10.1016/J.CERAMINT.2020.12.045

18. Das D, Saini AK, Pathak MK, Kumar S (2012) Application of functionally graded materials as thermal insulator in high temperature engineering components. IJREAS 2(2). Accessed 05 Feb 2023. http://www.euroasiapub.org

19. Wolfe D, Singh J (1998) Functionally gradient ceramic/metallic coatings for gas turbine components by high-energy beams for high-temperature applications. J Mater Sci 33(14):3677–3692. https://doi.org/10.1023/A:1004675900887/METRICS

20. Xu H, Guo H, Gong S (2008) Thermal barrier coatings. Dev High-Temper Corros Prot Mater:476–491. https://doi.org/10.1533/9781845694258.2.476

21. Batra RC, Taetragool U (2020) Numerical techniques to find optimal input parameters for achieving mean particles' temperature and axial velocity in atmospheric plasma spray process. Sci. Rep 10(1):1–11. https://doi.org/10.1038/s41598-020-78424-w

22. Ren X, Pan W (2014) Mechanical properties of high-temperature-degraded yttria-stabilized zirconia. Acta Mater 69:397–406. https://doi.org/10.1016/J.ACTAMAT.2014.01.017

23. Li GR, Wang LS, Yang GJ (2019) Achieving self-enhanced thermal barrier performance through a novel hybrid-layered coating design. Mater Des 167:107647. https://doi.org/10.1016/J.MATDES.2019.107647

24. Bazli L, Siavashi M, Shiravi A (2019) A review of carbon nanotube/TiO_2 composite prepared via sol-gel method. J Compos Compd 1(1):1–9. https://doi.org/10.29252/JCC.1.1.1

25. Fan XL, Xu R, Zhang WX, Wang TJ (2012) Effect of periodic surface cracks on the interfacial fracture of thermal barrier coating system. Appl Surf Sci 258(24):9816–9823. https://doi.org/10.1016/J.APSUSC.2012.06.036

26. Qian L, Zhu S, Kagawa Y, Kubo T (2003) Tensile damage evolution behavior in plasma-sprayed thermal barrier coating system. Surf Coatings Technol 173(2–3):178–184. https://doi.org/10.1016/S0257-8972(03)00429-8

27. Li XN, Liang LH, Xie JJ, Chen L, Wei YG (2014) Thickness-dependent fracture characteristics of ceramic coatings bonded on the alloy substrates. Surf Coatings Technol 258:1039–1047. https://doi.org/10.1016/J.SURFCOAT.2014.07.031

28. Zhou B, Kokini K (2004) Effect of preexisting surface cracks on the interfacial thermal fracture of thermal barrier coatings: an experimental study. Surf Coatings Technol 187(1):17–25. https://doi.org/10.1016/J.SURFCOAT.2004.01.028

29. Guo HB, Vaßen R, Stöver D (2004) Atmospheric plasma sprayed thick thermal barrier coatings with high segmentation crack density. Surf Coatings Technol 186(3):353–363. https://doi.org/10.1016/J.SURFCOAT.2004.01.002

30. Li B, Fan X, Okada H, Wang T (2018) Mechanisms governing the failure modes of dense vertically cracked thermal barrier coatings. Eng Fract Mech 189:451–480. https://doi.org/10.1016/J.ENGFRACMECH.2017.11.037

31. Wang L et al (2016) Modeling of thermal properties and failure of thermal barrier coatings with the use of finite element methods: a review. J Eur Ceram Soc 36(6):1313–1331. https://doi.org/10.1016/J.JEURCERAMSOC.2015.12.038

32. Nejati M, Rahimipour MR, Mobasherpour I, Pakseresht AH (2015) Microstructural analysis and thermal shock behavior of plasma sprayed ceria-stabilized zirconia thermal barrier coatings with micro and nano Al_2O_3 as a third layer. Surf Coatings Technol 282:129–138. https://doi.org/10.1016/J.SURFCOAT.2015.10.030

33. Li M, Sun X, Hu W, Guan H (2007) Thermal shock behavior of EB-PVD thermal barrier coatings. Surf Coatings Technol 201(16–17):7387–7391. https://doi.org/10.1016/J.SURFCOAT.2007.02.003

34. Islam A et al (2019) Exceptionally high fracture toughness of carbon nanotube reinforced plasma sprayed lanthanum zirconate coatings. J Alloys Compd 777:1133–1144. https://doi.org/10.1016/J.JALLCOM.2018.11.125

35. Singh P et al (2022) Plasma sprayed graphene/carbon nanotube reinforced lanthanum-cerate hybrid composite coating. Ceram Int. https://doi.org/10.1016/J.CERAMINT.2022.11.313

36. Chen X, Kou CC, Zhang SL, Li CX, Yang GJ, Li CJ (2020) Effects of powder structure and size on Gd_2O_3 preferential vaporization during plasma spraying of $Gd_2Zr_2O_7$. J Therm Spray Technol 29(1–2):105–114. https://doi.org/10.1007/S11666-019-00944-3

37. Shrirao PN, Pawar AN, Borade AB (2011) An overview on thermal barrier coating(TBC) materials and its effect on engine performance and emission. researchgate.net. Accessed 05

Feb 2023. https://www.researchgate.net/profile/Mondher-Zidi/publication/282726174_Eff ect_of_the_resin_type_on_the_acoustic_activity_and_the_mechanical_behavior_of_e_gla sspolymer_resin_55_filament_wound_pipes_under_axial_loading/links/5b3215cc4585150 d23d4a256/Effect-of-the-resin-type-on-the-acoustic-activity-and-the-mechanical-behavior-of-e-glass-polymer-resin-55-filament-wound-pipes-under-axial-loading.pdf#page=184

38. Li F, Zhou L, Liu JX, Liang Y, Zhang GJ (2019) High-entropy pyrochlores with low thermal conductivity for thermal barrier coating materials. J Adv Ceram 8(4):576–582. https://doi.org/ 10.1007/S40145-019-0342-4/METRICS

39. Zhou X et al (2020) Thermophysical properties and cyclic lifetime of plasma sprayed $SrAl_{12}O_{19}$ for thermal barrier coating applications. J Am Ceram Soc 103(10):5599–5611. https://doi.org/ 10.1111/JACE.17319

40. Zhan X, Li Z, Liu B, Wang J, Zhou Y, Hu Z (2012) Theoretical prediction of elastic stiffness and minimum lattice thermal conductivity of $Y_3Al_5O_{12}$, $YAlO_3$ and $Y_4Al_2O_9$. J Am Ceram Soc 95(4):1429–1434. https://doi.org/10.1111/J.1551-2916.2012.05118.X

41. Chen S et al (2019) Mg_2SiO_4 as a novel thermal barrier coating material for gas turbine applications. J Eur Ceram Soc 39(7):2397–2408. https://doi.org/10.1016/J.JEURCERAMSOC. 2019.02.016

42. Zhou X, Xu Z, Fan X, Zhao S, Cao X, He L (2014) $Y_4Al_2O_9$ ceramics as a novel thermal barrier coating material for high-temperature applications. Mater Lett 134:146–148. https:// doi.org/10.1016/J.MATLET.2014.07.027

43. Turcer LR, Krause AR, Garces HF, Zhang L, Padture NP (2018) Environmental-barrier coating ceramics for resistance against attack by molten calcia-magnesia-aluminosilicate (CMAS) glass: part I, $YAlO_3$ and γ-$Y_2Si_2O_7$. J Eur Ceram Soc 38(11):3905–3913. https://doi.org/ 10.1016/J.JEURCERAMSOC.2018.03.021

44. Lemański K, Michalska M, Ptak M, Małecka M, Szysiak A (2020) Surface modification using silver nanoparticles for $Y_4Al_2O_9$: Nd–synthesis and their selected studies. J Mol Struct 1202:127363. https://doi.org/10.1016/J.MOLSTRUC.2019.127363

45. Cojocaru CV, Lamarre JM, Legoux JG, Marple BR (2013) Atmospheric plasma sprayed forsterite (Mg_2SiO_4) coatings: an investigation of the processing-microstructure-performance relationship. J Therm Spray Technol 22(2–3):145–151. https://doi.org/10.1007/S11666-012-9856-9/FIGURES/8

46. Zhou Y, Lu X, Xiang H, Feng Z (2015) Preparation, mechanical, and thermal properties of a promising thermal barrier material: $Y_4Al_2O_9$. J Adv Ceram 4(2):94–102. https://doi.org/10. 1007/S40145-015-0141-5/METRICS

47. Zhou Y, Xiang H, Lu X, Feng Z, Li Z (2015) Theoretical prediction on mechanical and thermal properties of a promising thermal barrier material: $Y_4Al_2O_9$. J Adv Ceram 4(2):83–93. https:/ /doi.org/10.1007/S40145-015-0140-6/METRICS

48. Ma W, Mack D, Malzbender J, Vaßen R, Stöver D (2008) Yb_2O_3 and Gd_2O_3 doped strontium zirconate for thermal barrier coatings. J Eur Ceram Soc 28(16):3071–3081. https://doi.org/10. 1016/J.JEURCERAMSOC.2008.05.013

49. Lecomte-Beckers J, Schubert F, Ennis P (1998) Materials for advanced power engineering 1998. Accessed 05 Feb 2023. https://juser.fz-juelich.de/record/811813

50. Lehmann H, Pitzer D, Pracht G, Vassen R, Stöver D (2003) Thermal conductivity and thermal expansion coefficients of the lanthanum rare-earth-element zirconate system. J Am Ceram Soc 86(8):1338–1344. https://doi.org/10.1111/J.1151-2916.2003.TB03473.X

51. Choudhary S, Pandey A, Gaur V (2023) Role of microstructural phases in enhanced mechanical properties of additively manufactured IN718 alloy. Mater Sci Eng A 862:144484. https://doi. org/10.1016/J.MSEA.2022.144484

52. Choudhary S, Gaur V (2022) Study of mechanical properties and applications of aluminium based composites manufactured using laser based additive techniques:261–300. https://doi. org/10.1007/978-981-16-7377-1_12

53. Choudhary S et al (2021) Plasma sprayed Lanthanum zirconate coating over additively manu-factured carbon nanotube reinforced Ni-based composite: unique performance of thermal barrier coating system without bondcoat. Appl Surf Sci 550. https://doi.org/10.1016/J.APS USC.2021.149397

Chapter 16
A Review on Nickel Composite Coatings Deposited by Jet Electrodeposition

Jhalak and Dishant Beniwal

1 Introduction

Ni and Ni–Co-based coatings are used in numerous applications due to their unique chemical and mechanical properties. Ni–Co coatings exhibit excellent magnetic, anti-wear, and anti-corrosion properties which can be further improved through the incorporation of ceramic particles in the alloy matrix. These coatings can be deposited using various techniques such as electroless-deposition and electrodeposition.

The electrodeposition technique has gained considerable attention due to its simple and economical nature along with the flexibility to deposit a variety of materials including metals, ceramics, and composites. It provides micro- and nanostructured coatings with excellent mechanical, tribological, and anti-corrosion properties. The inherent properties of the matrix and the reinforced phase govern coating properties [1–4]. In addition to this, the final properties of the coatings also depend on nucleation and growth kinetics. Researchers have extensively investigated Ni and Ni–Co composite coatings developed with electrodeposition using different current control techniques viz. direct current (DC), pulsed current (PC), and pulsed reversed current (PRC). Most of the studies focused on the effect of bath composition, pH, temperature, stirring rate, additives, and current density. Researchers have reported significant improvement in mechanical, tribological, and corrosion-resistant properties of coatings through optimization of these parameters. However, electrodeposition suffers from disadvantages in that it cannot provide high current densities and can cause corrosion of the cathode [5].

Recently, the jet electrodeposition method has been gaining attention as it can provide high deposition rates, ranging from ten to hundred times more than those obtained through commonly used deposition processes. The higher deposition rate

Jhalak (✉) · D. Beniwal
Department of Metallurgical and Materials Engineering, Indian Institute of Technology Ropar, Rupnagar, Punjab 140001, India
e-mail: jhalak.20mmz0010@iitrpr.ac.in

leads to lower immersion time, which in turn significantly reduces the chances of electrode corrosion. Studies have shown that combining jet electrodeposition with pulse electrodeposition provides better coating properties in terms of improved uniformity, higher corrosion and wear resistance, lower internal stresses, and porosity than direct current methods [6–8]. Jet electrodeposition has also been shown as an efficient process to obtain refined grains and studies have been conducted to evaluate the characteristics of nanocrystalline coatings obtained using jet electrodeposition in combination with pulse-, ultrasonic-, and magnetic-assisted methods [5, 7, 9]. Figure 1 shows the timeline of various developments related to Ni-based composite coatings deposited using jet electrodeposition.

This review summarizes

(a) the effect of various parameters such as bath composition (Sect. 4.1.1), external forces (Sect. 4.1.1), current density (Sect. 4.1.2), jet speed (Sect. 4.1.3), and temperature (Sect. 4.1.3) on the coating composition, microstructure, and surface morphology.
(b) the effect of adding ceramic particles (SiC, TiN, AlN, and BN) and graphene oxide (GO) on the microhardness (Sect. 4.2), tribological properties (Sect. 4.2), and corrosion properties (Sect. 4.3) of the coatings.
(c) the possible applications (Sect. 5) and future research scope (Sect. 5) for jet-electrodeposited Ni-based composite coatings.

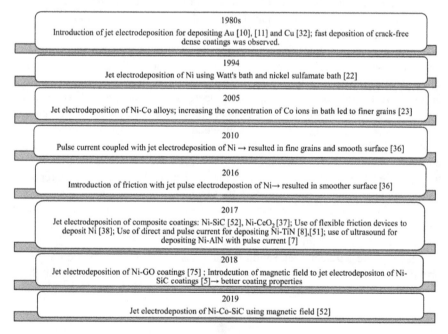

Fig. 1 Timeline of various developments related to Ni-based composite coatings deposited using jet electrodeposition

2 Electrodeposition Process

The electrodeposition process, also known as electroplating, utilizes electric current to deposit metallic coatings. The anode and cathode are immersed in an electrolytic bath containing ionic salts, and current is passed across them leading to anode dissolution that replenishes metal ions for deposition. These ions reduce on the cathode to deposit a layer over the surface. When ceramic particles, suspended in an electrolytic bath, are deposited along with metals or alloys, the process is known as electrophoretic deposition. Metal ions are adsorbed on the surface of these particles which facilitates the incorporation of particles on the substrate surface along with metallic deposits [10–15].

Electrodeposition can be carried out using direct current (DC), pulsed current (PC), or pulsed reversed current (PRC) techniques. In the DC method, the constant value of current is applied without any interruption. In the PC method, positive current pulses are applied for a duration (T_{on}) and each pulse is separated by a zero current time duration (T_{off}) [16]. Pulsed process parameters are evaluated based on average current density, pulse frequency, and duty cycle given by the equations below.

Pulse frequency,

$$f = \frac{1}{T_{on} + T_{off}} \tag{2.1}$$

Duty cycle,

$$d = \left(\frac{T_{on}}{T_{on} + T_{off}} \right) \times 100\% \tag{2.2}$$

Average current density,

$$I_{avg} = \left(\frac{T_{on}}{T_{on} + T_{off}} \right) I_{peak} \tag{2.3}$$

where I_{peak} is the peak current density.

In the PRC method, each positive current pulse is followed by a negative current pulse. It has been reported that PC and PRC methods provide better coating properties including less porosity, and improved mechanical-, wear-, and corrosion-resistant properties than those obtained using the DC method [17]. Switching off the current in the PC method and change of polarity in the PRC method, both lead to discharging of the diffusion layer formed at the electrode surface, thereby increasing the ionic diffusion toward the cathode surface resulting in improved properties [18].

3   Jet Electrodeposition Process

The jet electrodeposition process is based on the electrochemical phenomenon in which a jet of electrolyte is transferred to the cathode surface. This process utilizes an electric field established due to the flow of current along with the flow stream. The deposition occurs on the cathode surface where the stream strikes directly. It can produce metallic as well as composite coatings with ceramic particle incorporation. The deposition mechanism of composite coatings has been explained through Guglielmi's [19] two-step model: (1) Adsorption of metal ions on the particles' surface, and (2) entrapment of particles along with metal ions on the cathode surface. This process is very effective for the components where localized plating with a high deposition rate is required. It provides a large mass transfer on the cathode surface and enables high deposition rates as current densities of the order of hundreds of A/ dm^2 may be applied [5, 20, 21]. However, this process may result in uneven coating surfaces during the crystal growth as the metal ions start to deposit at active sites such as microscopic defects on the cathode surface. Non-uniform growth, commonly known as edge effect, and particle agglomeration occur in these regions [22]. Figure 2 shows the schematic of the jet electrodeposition process.

Nowadays, research is being carried out to prevent particle agglomeration and improve coating properties by integrating the jet electrodeposition process with various external forces such as pulse, ultrasonic waves, magnetic field, laser, rubbing, and mechanical grinding. These are classified as non-contact and contact forces based on their contact with the substrate [23–27]. Contact force fields like friction-assisted electrodeposition provide compact and uniform coatings with refined grains. However, these may cause abrasion of coatings if the parameters are not optimized [28].

3.1   Advantages and Challenges of Jet Electrodeposition

Deposition of the coating occurs by reduction of cations present near the cathode. During electrodeposition, a diffusion layer exists between the cathode and bulk solution containing metal ions and other suspended particles. When the diffusion layer is thick, diffusion of cations from the bulk solution to the cathode surface becomes slow and cations deplete from the cathode surface. This leads to hydrogen ion reduction at the cathode surface resulting in poor coating properties. Being a high-speed process, jet electrodeposition increases the flow velocity of electrolytes, which results in reduced thickness of the diffusion layer [29]. Continuous vigorous stirring and high flow rates of electrolyte enhance mass transfer and provide high coating rates [30].

For the electrodeposition of composite coatings, the transportation rate of particles, which depends on the electric field gradient also known as electrophoresis,

Fig. 2 Schematic of jet
electrodeposition process

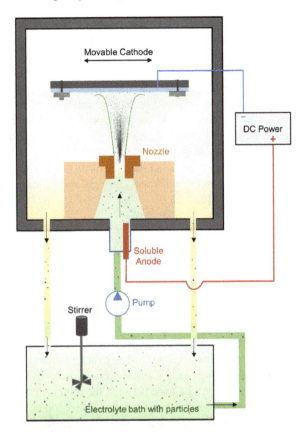

concentration gradient, and velocity gradient are an important factor. Rate of deposition of coating thickness is given by the Hamker equation [4, 31], which shows electrophoretic mobility of particles increases the deposition rate of thickness. Hamker law is given by Eq. 3.1:

$$\frac{dx}{dt} = \varepsilon \mu \rho C_w \frac{dE}{dx} \tag{3.1}$$

where x is coating thickness, t is time, ε is efficiency factor, μ is electrophoretic mobility of charged species, dE/dx is potential field gradient, and C_w is concentration of particles in the electrolyte.

Jet electrodeposition provides high-speed selective plating of small components. Electronic components like electric connectors and sliding contacts require high conductivity and high corrosion and wear resistance. Localized and fast coating processes for mass production are important for such applications, and jet electrodeposition is an economical and high-productivity process [20]. Proper control of process parameters such as flow rate, temperature, and current density provides stable and controlled deposition.

Another advantage of jet electrodeposition is the minimal corrosion of the cathode or substrate as compared to the traditional electrodeposition process. In the traditional electrodeposition process, anode and cathode are immersed in an electrolytic bath, whereas, in the jet electrodeposition process, the cathode lies outside the bath, which minimizes the chances of substrate corrosion.

As the jet electrodeposition process utilizes a high-speed jet, it becomes difficult to control input operating parameters. A model with a theoretical approach and quantitative analysis is important for designing experiments to have precise control over the various parameters. Researchers have developed mathematical models to optimize various factors, but the models do not consider all aspects such as the effect of change in coating thickness during deposition and current values higher than limiting current [22].

3.2   Friction-Assisted Jet Electrodeposition

Friction-assisted electrodeposition involves creating friction on the surface using flexible devices such as bio-bristles, natural fibers, polymers, and rigid devices like glass balls, ceramic balls, hard particles, and friction blocks. When crystal growth occurs on the coating surface, friction helps in mechanical leveling of coatings by removing the excess growth. Moreover, it enhances local flow for homogeneous coating morphology [32]. Crystal growth increases with increase in current density, thus there must be a balance between the removal rate due to friction and deposition rate due to the increased current density to obtain improved coating surface [33]. Liu et al. [34] suggested that flexible friction techniques do not provide continuous pressure on the surface; hence, it is necessary to investigate ways of producing controlled friction to apply continuous and uniform pressure throughout the cathode surface [35] as it can improve the deposition efficiency and coating quality. The position of the friction medium (or device), scanning speed, and amount of pressure play an important role in transforming the coating properties [35, 36].

3.3   Pulse-Assisted Jet Electrodeposition

A modulated power unit is used along with the jet nozzle through which the plating electrolyte is sprayed on the cathode surface. The pulse power unit used is either constant current or constant voltage type [37]. Metal ions are reduced and deposited on the cathode surface during T_{on} time. When the current is switched off, metal ions are replenished in the bath during T_{off} time. Very short T_{on} time results in very high peak current density that increases overpotential. This reduces hydrogen evolution on the cathode surface and improves coating quality [38]. If the limiting current density exceeds, then burnt coatings are produced [29]. The use of pulse current significantly improves limiting current density. Pulse current along with jet electrodeposition allows the use of high current densities of above 100 A/dm^2. This increases the nucleation rate and facilitates grain refinement increasing hardness, reduces microstructural defects, and improves the surface quality of the deposits [39].

3.4   Ultrasonic-Assisted Jet Electrodeposition

In ultrasonic-assisted jet electrodeposition, an ultrasonic transmitter is used to generate continuous or pulsed ultrasonic waves [40]. The presence of ultrasonic waves increases particle entrapment, restricts agglomerates formation, and enhances their homogenized distribution in the metal matrix [41–44]. Pulsed waves provide a high deposition rate and better surface uniformity with increasing pulse-on-time than that obtained using continuous waves. The waves produce a thermal effect by increasing the temperature of the electrolyte and also create impact forces on the coatings. Thermal effect increases deposition, but the impact peels off the surface resulting in reduced coating thickness. When pulsed waves are used, the thermal effect dominates during pulse-on-time. However, when continuous waves are used, the impact becomes dominant, leading to reduced deposition thickness. Studies have reported reduced particle incorporation in coatings due to the use of ultrasonic waves [44]. It depends on the type of waves that are being used, as the use of continuous waves imparts impact for longer durations, which may abrade the adsorbed particles from the coated surface. Excellent coating properties can be achieved with the use of optimized pulse-on-time for ultrasonic vibrations [40].

3.5   Magnetic-Assisted Jet Electrodeposition

In magnetic-assisted jet electrodeposition, a magnetic field is used along with an electric field and the metal ions experience a combined action of both. Application of magnetic field restricts vertical adsorption of ions and crystal growth that occurs due to edge effect. Hence, it facilitates uniform growth of crystals producing flat coatings. It also reduces concentration polarization at the interface [24, 45, 46]. The principle of magneto-hydrodynamics allows ion deflection, which causes microagitation in the bath inhibiting the formation of particle agglomerates. This enables more metal ions to adsorb particles' surface. Due to this, more particles can incorporate uniformly along with metal ions in growing coatings [5, 8]. Magnetic fields affect the properties of water by elongating the bond between H$^+$ and OH$^-$ ions due to the paramagnetic nature of hydrogen and the diamagnetic nature of oxygen. This helps in maintaining pH on the cathode surface [47]. The combined action of the magnetic field and high-speed jet alleviates hydrogen bubbles from the cathode surface [8]. Recent studies showing the effect of magnetic field on jet electrodeposition for depositing nickel composite coatings are discussed in Sect. 4.2.

4 Effect of Parameters on the Properties of Coatings Obtained by Jet Electrodeposition

This section reviews the effect of process parameters on coating composition, microstructure, surface morphology, microhardness, tribological, and corrosion properties. Selection of bath composition, current density, plating temperature, pH, jet speed, and other controls such as the application of pulse, ultrasound, and magnetic field play an important role in achieving the best coating properties. Optimum values for various parameters have been investigated for depositing Ni and Ni-based composite coatings.

4.1 Coating Composition, Microstructure, and Surface Morphology

The effect of various parameters on coating composition, grain size, surface appearance, and roughness has been discussed.

4.1.1 Effect of Electrolyte Composition and External Forces

Electrolyte composition is an important parameter that affects anode dissolution, development of internal stresses in the coating, and other coating properties. Sulfate–chloride bath or Watt's bath is commonly used for plating purposes. The presence of chloride ions prevents the passivation of the anode and promotes its dissolution, but an excess amount of chloride ions may induce internal stresses in the deposits [48]. It has been reported that the addition of saccharine to the plating solution reduces internal stresses in the coating and refines grain structure [21].

Qiao et al. [21] investigated the effect of Ni and Co ion concentration in a bath on the composition of the deposited coating. It was observed that Co/Ni ratio in the deposits was higher than that in the bath due to an anomalous Co deposition. At around 3 wt.% of Co ions in the bath, Co content in the coating was approximately 26 wt.% which kept on increasing with increasing Co ions in the bath up to around 75 wt.% Co content in coating at 25 wt.% Co ions in bath. It was suggested that preferential reduction of Co ions occurs resulting in the depletion of these ions from the diffusion layer at the cathode. When the concentration of Co ions increases in the bath, it replenishes the ions in the layer resulting in an increased Co content in the deposits. Addition of saccharine to the bath results in a decreased Co content in deposits as it prohibits Co deposition and facilitates Ni deposition [21].

Reduction in the grain size with an increasing Co/Ni ions ratio in the electrolytic bath was observed for Ni–Co deposits. The increased cobalt content in deposits leads to lattice strains and lattice distortion due to the atomic size difference between the two atoms. In addition to this, when Co is added to fcc Ni, the crystal structure transforms to fcc + hcp crystal structure. These factors lead to recrystallization resulting in grain refinement [21]. A significant reduction in grain size has been reported with the addition of saccharine to the bath.

Researchers [5, 9] have studied the effect of nanoparticle incorporation in Ni matrix. For Ni–SiC coatings, it was observed that SiC content in the coating is maximum at an optimum concentration of 3 g/L in the electrolytic bath for both traditional and magnetic jet electrodeposition. However, the maximum particle incorporation achieved with magnetic jet electrodeposition is approximately 2.5 times more as compared to traditional jet electrodeposition [5]. At very high concentrations of SiC in bath, agglomeration of SiC occurs which hinders the uniform incorporation of particles in the matrix resulting in bulges or protrusions on the coated surface. The use of magnetic field increases the convection which reduces agglomeration and facilitates particle incorporation. Particle incorporation affects the grain size since a larger number of uniformly dispersed particles provide more sites for nucleation. For TiN incorporation in Ni matrix using jet electrodeposition, an optimal bath load of 5 g/L resulted in a minimum average grain size value of 44.5 nm [49].

Jiang et al. [9] have investigated the effect of nano SiC incorporation in Ni–Co matrix deposited using traditional jet electrodeposition (TJE) and magnetic-assisted electrodeposition (MJE). SiC incorporation in Ni–Co matrix resulted in grain refinement with a smooth surface. While the addition of nanoparticles reduces grain size by promoting nucleation, their increased concentration in bath may lead to agglomeration and the formation of nodules in deposits. More protrusions or spherical nodules were observed as SiC concentration in bath increased from 4 g/L to 6 g/L during traditional jet electrodeposition as reported in the literature [9]. Similar results were reported for nano SiC incorporation in the Ni matrix [5]. It was suggested that when defects such as protrusions or nodules are formed on the growth interface, metal ions are preferentially adsorbed on these defects. This shields the adjacent pits resulting in an uneven coating surface [5]. Jiang et al. [9] reported lesser nodules or protrusions at increased SiC concentrations for coatings obtained using magnetic-assisted jet electrodeposition. The distribution of nanoparticles under magnetic electrodeposition was more uniform for the same amount of SiC content in bath. In the case of magnetic-assisted jet electrodeposition, the diffusion layer thickness is reduced by magneto-hydrodynamics leading to an increased ionic concentration gradient which increases the coating deposition. This promotes nanoparticle entrapment along with metallic ions [46, 50]. Additionally, high jet speed facilitates convection, which inhibits particle agglomeration in the bath and results in uniform dispersion of particles in the coating [51]. Although particle agglomeration was observed at SiC concentration of 6 g/L for both TJE and MJE processes, it was lower in the case of MJE [9]. Particle agglomeration due to excess SiC leads to poor adhesion of the coating. It was suggested that the grain refinement due to SiC incorporation resulted in reduced surface roughness [52]. A similar roughness pattern was also observed for Ni–TiN coatings deposited using jet pulse electrodeposition [8].

The incorporation of SiC nanoparticles increases Co content in coatings as Co ions are adsorbed preferentially on ceramic particles' surface [9, 53]. Moreover, the SiC incorporation is more uniform in the case of magnetic jet electrodeposition, which results in more Co content in deposits as compared to traditional jet electrodeposition.

Ma et al. [7] investigated Ni–AlN coatings and observed similar results as Ni–SiC prepared using traditional jet electrodeposition. The formation of coarse grains for pure Ni coating deposited using jet pulse electrodeposition was reported. However, after AlN addition, a smoother structure with refined grains was observed. Moreover, coatings deposited using jet pulse electrodeposition in the presence of ultrasound exhibited the smoothest structure with more refined grains as compared to the other two deposits. In the presence of ultrasound, incorporation of nanoparticles increases. Application of ultrasound increases the kinetic energy, which results in increased and uniform dispersion of particles with less chances of agglomeration [7]. An increase in Al content of the coating from 14.9 to 21.4 atomic percent was reported with the use of ultrasound along with jet pulse electrodeposition.

Cui et al. [54] explained the effect of kinetic energy on the microstructure of Ni–SiC deposits using nozzles with varying diameters of 4 m, 6 mm, and 8 mm. It was observed that the coatings obtained using 8 mm diameter had finer grains with smoother and more uniform structures than the coatings obtained using smaller diameter nozzles. A larger amount of particle deposition was obtained using an 8 mm diameter nozzle due to higher jet force and higher kinetic energy. High kinetic energy promotes dispersion and uniform distribution of particles throughout the matrix which provides more nucleation and inhibits grain growth resulting in smaller grains [54–57]. Average grain sizes of 674 nm and 98 nm respectively were reported for Ni and SiC in the coatings achieved using a 4 mm diameter nozzle, whereas 344 nm and 75 nm respectively for coatings achieved using an 8 mm diameter nozzle [54].

The introduction of friction on the surface by the application of pressure promotes coating uniformity and also reduces grain size. The mechanical force of friction eliminates hydrogen from the surface and reduces the formation of bubbles, pinholes, pits, and other defects. An optimum value of force provides good-quality deposits with lower roughness and refined grain size for the deposition of nanocrystalline nickel coatings [36]. The grain size was reduced from 15.9 mm without force to 12.8 mm with a force of 2.6 N [36].

4.1.2 Effect of Current Density

Current density is another factor that dictates coating composition and microstructure. In the case of Ni–Co deposition, studies have reported limited Co deposition in coatings at high current densities. Nickel deposition is activation-controlled while Co deposition is diffusion-controlled. At higher current densities, cathodic overpotential becomes high which results in more activation at the electrode surface. This behavior promotes Ni deposition on the cathode [21, 58].

Qiao et al. [21] observed a reduction in grain size of Ni–Co deposits from about 15 nm to 4 nm with increasing current density in between 159 and 477 A/dm². It was reported that increased current density leads to an increase in cathode overpotential which affects grain size. Grain size is related to cathode overpotential as shown in

Eq. 4.1 [20]:

$$P = B exp\left(\frac{-b}{\eta_k^2}\right) \tag{4.1}$$

where P is probability of nucleation, η_k is overpotential, and B and b are constants. This equation clearly depicts that an increase in the value of overpotential can increase nucleation. Therefore, grain refinement occurs at higher values of current density.

When high current densities are applied, the coating surface may become uneven due to the rapid growth of some grains. Every substrate has some microscopic defects such as cracks and protrusions, and rapid deposition of metal occurs at the edges and tips of these defects. Coarse grains form at these regions which shield the adjacent pits, thereby resulting in a reduced deposition rate and smaller grain size in depressed regions. This results in uneven coating growth as the coarser grains grow vertically [59–61]. Liu et al. [34] confirmed that this uneven growth can be controlled by using a controllable friction technique along with jet electrodeposition. Even at increased current densities, flat nanocrystalline Ni coatings can be obtained as the application of friction removes any excess growth by mechanical action. In addition, the use of a rolling nozzle in this mechanism stirs the electrolyte and reduces the thickness of the diffusion layer, thereby improving the overall deposition rate [34, 62]. Liu et al. [34] observed flat and smoother coated surfaces using friction-assisted jet electrode-position as compared to uneven and rough surfaces obtained using traditional jet electrodeposition.

4.1.3 Effect of Temperature and Jet Speed

A reduction in Co deposition has been observed with increasing bath temperature [21]. While the deposition of Co is a diffusion-controlled process, Ni deposition is activation-controlled. Therefore, an increase in temperature provides the energy required to overcome the activation barrier and promotes Ni deposition over Co deposition. In most of the studies, the temperature for carrying out electrodeposition lies approximately from 40 to 70 °C.

Increased Co deposition has been reported with an increase in jet speed. The high speed of jet reduces the diffusion layer thickness allowing more diffusion of Co ions toward the cathode surface, resulting in an increased Co content in coatings [21].

An increase in jet speed reduces the thickness of the diffusion layer which is related to limiting current density as given by the equation [20]:

$$i_L = nFD\frac{C^0}{\delta} \tag{4.2}$$

where i_L is limiting current density, F is Faraday's constant, D is diffusivity of ions, and δ is diffusion layer thickness. Equation 4.2 indicates that a decrease in diffusion layer thickness would increase the limiting current density, which will result in an increase in cathodic overpotential. This increases the probability of nucleation and results in a reduced crystallite size as depicted by Eq. 4.1. A grain size reduction from about 18.5 to 14.5 nm was reported upon increasing jet speed from 4.23 to 8.5 m/s [21].

The jet speed also affects particle incorporation in metal deposits. For TiN addition to Ni matrix deposited using pulse jet electrodeposition at speeds of 1, 3, and 5 m/s, maximum particle incorporation and minimum grain size were observed at the speed of 3 m/s [8]. Increased jet speed accelerates the electrochemical reactions and increases the rate of metal ions deposition; this also leads to an increase in the incorporation of ceramic particles, which have metal ions adsorbed on their surface. However, when the jet speed is too high, ceramic particles dislodge due to greater force which prevents particle incorporation in the metal matrix. Studies show similar results for high stirring speeds of electrolytic baths for ZnO, SiC, Al_2O_3, and Si_3N_4 incorporation in Ni and Ni–Co coatings deposited using a conventional electrodeposition process [53, 63–71]. Tables 1 and 2 summarize values of particle content in coating and coating morphology respectively for some jet-electrodeposited Ni-based composite coatings.

4.2 Microhardness and Tribological Properties

Effect of magnetic field, SiC, and TiN: Cui et al. [54] have reported the effect of nozzle size on microhardness of jet-electrodeposited coatings. For Ni–SiC coatings, the maximum nanohardness values of 33.4 GPa and 23.7 GPa were obtained using 8 mm and 4 mm diameter nozzles, respectively. This was attributed to the reduced crystallite size, maximum SiC incorporation, and uniform dispersion of SiC nanoparticles due to high kinetic energy imparted by larger nozzle diameter.

Jiang et al. [5] reported maximum microhardness values of 581.4 HV and 626.71 HV for Ni–SiC coatings deposited using TJE and MJE, respectively. These values coincided with the maximum incorporation and uniform distribution of SiC nanoparticles in the matrix with the use of magnetic field. The increase in microhardness values is attributed to dispersion strengthening and smaller grain size due to more nucleation sites provided by SiC nanoparticles. A smaller crystallite size increases the grain boundary area which acts as a hindrance to the dislocation movement, thereby increasing the strength of coatings [73–75]. For the Ni–TiN coatings developed using JPE, maximum hardness values were obtained for an optimal speed of 3 m/s [8]. Xia et al. [49] have reported an increased nanohardness and uniform surface for jet-electrodeposited Ni–TiN. The maximum nanohardness of 34.5 GPa was obtained at an optimal TiN concentration of 5 g/L in bath, whereas nanohardness of 25.8 GPa was obtained at 3 g/L TiN in bath [49].

Effect of AlN and ultrasound: Ma et al. [7] observed an increase in coating microhardness after incorporation of AlN nanoparticles in the Ni matrix using jet pulse electrodeposition. Incorporation of AlN particles increased the hardness from 431.1

Table 1 Particle content and coating thickness for jet-electrodeposited Ni-based composited coatings

Coating	Technique	Particle concentration in bath (g/L)	Particle incorporation in coating	Deposition time (minutes)	Coating thickness μm	References
Ni–GO (graphene oxide)	JE	0.5	27 wt.% C	5 20 150	3 12 350	[72]
Ni–AlN	JPE	6	14.9 at.% Al	80	56	[7]
Ni–AlN	JPE + US	6	21.4 at.% Al	80	56	[7]
Ni–TiN	JE	3 5 8	–	60	200	[49]
Ni–TiN	JPE (jet speed) 1 m/s 3 m/s 5 m/s	7 7 7	8.6 at% TiN 19.8 at% TiN ~15at% TiN	45 45 45	–	[8]
Ni–SiC	JE	1 3 5	2.13 wt.%	30	–	[5]
Ni–SiC	JE + MF	1 3 5	3.27 wt.%	30	–	[5]
Ni–Co–SiC	JE	0 2 4	–	30	–	[9]
Ni–Co–SiC	JE + MF	4	–	30	–	[9]

HV to 665 HV, and the use of ultrasound further enhanced the hardness of Ni–AlN coatings to 767.9 HV. Deep scratches with wear debris were visible on the surface of pure nickel coating indicating grinding wear. These scratches were significantly reduced after AlN addition. Coatings deposited using ultrasound demonstrated the lowest coefficient of friction and few small-sized scratches on the worn surface. A reduction in coefficient of friction was also observed upon incorporation of AlN in Ni coatings. The improved wear properties were attributed to a reduced crystallite size and more amount of uniformly dispersed AlN nanoparticles in coatings [7]. The use of ultrasound increases mass transfer and restricts particle agglomeration resulting in more deposition [76].

Effect of BN particles: Kang et al. [77] investigated the wear characteristics of jet-electrodeposited Ni–Co–P coatings reinforced with hexagonal BN nanoparticles. BN(h) nanoparticles are soft, with a structure similar to hexagonal graphite structure, and provide easy sliding. Their addition reduces the surface roughness and coefficient of friction. A reduction in coefficient of friction from 0.514 to 0.305 due to incorporation of BN(h) nanoparticles was observed. Also, the microhardness of Ni–Co–P

Table 2 Coating morphology after nanoparticles incorporation

Coating matrix	Surface morphology	Coating after particle incorporation	Surface morphology after particle incorporation	Matrix grain size (nm)	References
Ni	Smooth and compact	Ni–GO	Rough with uniformly distributed GO agglomerates	0 g/L: 49.2 0.5 g/L: 15.1 1 g/L: 13.2	[72]
Ni	Coarse grains	Ni–AlN (AlN—25 nm)	Refined grains, smooth surface	–	[7]
Ni	Coarse grains	Ni–AlN using US	More refined grains compared to those obtained without US	–	[7]
Ni	–	Ni–TiN (TiN—20 nm)	Best coating (uniform and refined structure) at 5 g/L Agglomeration at higher concentrations of TiN	3 g/L: 96.4 5 g/L: 44.5 8 g/L: 67.8	[49]
Ni	Bulges, microcracks	Ni–SiC	Few pores and microcracks in Ni coating under magnetic field Addition of SiC eliminates cracks and bulges, smooth surface	Without magnetic field: 0 g/L: 13.5 3 g/L: 12.7 With magnetic field: 3 g/L: 11.6	[5]
Ni–Co	Cellular protrusions	Ni–Co–SiC	Refined grains Reduced cellular protrusions More uniform coatings using magnetic field	–	[9]

coatings, measured as 700.8 HV, reduced to 681.5 HV after BN(h) incorporation. The reduced microhardness was attributed to the formation of large crystals and the softness of nanoparticles. However, heat treatment at 400 °C improved the microhardness and wear resistance of the coatings. During heat treatment, the formation of hard Ni_3P intermetallics leads to an increase in the hardness of coatings [78]. The coefficient of friction values for heat-treated Ni–Co–P and Ni–Co–P–BN(h) coatings were reported as 0.758 and 0.724 respectively, and the increment was attributed to the crystallization of coatings [77].

Effect of Graphene oxide: Ji et al. [72] observed increased hardness values after graphene oxide (GO) incorporation into the Ni coatings. Despite some agglomeration, the hardness increased due to a uniform distribution of GO throughout the coating matrix. Nanohardness values of 7.92 GPa, 8.52 GPa, and 8.70 GPa were reported for Ni, Ni-0.5GO, and Ni-1.0GO coatings, respectively. Grain refinement of the Ni matrix was observed after GO addition, resulting in increased strength as the GO sheets induce a pinning effect that prevents grain boundary sliding [79, 80]. In addition to this, GO sheets have high mechanical strength that may lead to the enhancement of the overall mechanical properties of composites. However, an optimal concentration of GO is important for achieving the best coating properties as very high concentrations may lead to increased agglomeration and other structural defects [81, 82]. Table 3 lists the microhardness and surface roughness of various jet-electrodeposited composite coatings. Xia et al. [49] suggested that hardness plays an important role in determining the wear-resistant properties of a coating which is supported by the fact that in most of the studies, coatings with increased hardness due to dispersion strengthening exhibited excellent tribological properties.

4.3 Corrosion Properties

Coatings with small grain size and smooth structure provide high corrosion resistance as observed by Cui et al. [54] since rough surfaces with coarse grains retain corrosive solutions for a longer duration than smoother and uniform surfaces.

Effect of nozzle diameter: A high impedance value with an average current density of $66 \ \mu A/dm^2$ was reported for Ni–SiC coatings obtained using 8 mm diameter nozzles indicating its high corrosion resistance in 3.5 wt. % NaCl. Poor corrosion resistance, with an average current density of $173 \ \mu A/dm^2$, was observed for coating deposited using a 4 mm diameter nozzle. This has been attributed to the higher kinetic energy generated with larger nozzle size which leads to a more uniform coating deposition.

Effect of SiC and magnetic field: Jiang et al. [9] studied the effect of SiC nanoparticle incorporation and magnetic field on the corrosion resistance of Ni–Co deposits. Incorporation of SiC in Ni–Co matrix reduced the corrosion current density by about 4.8 times indicating an improved corrosion resistance in 3.5 wt % NaCl. Moreover, the introduction of the magnetic field during jet electrodeposition of Ni–Co–SiC further improved corrosion resistance. Coatings deposited using magnetic-assisted jet electrodeposition with 4 g/L SiC content in bath demonstrated maximum impedance value indicating the highest corrosion resistance. The uniform dispersion of ceramic particles is responsible for this as it reduces the roughness and effective surface area for the retention of corrosive electrolytes [83, 84]. Along with that, reduction in

Table 3 Microhardness and surface roughness for various jet-electrodeposited composite coatings

Coating	Technique	Particle concentration in bath (g/L)	Microhardness (HV)/ Nanohardness (GPa)	Surface roughness (nm)	Coefficient of friction	References
Ni–GO	JE	0 0.5 1	7.92 GPa 8.52 GPa 8.70 GPa	6 12 28	–	[72]
Ni–AlN	JPE	0 6	431.1 HV ~650 HV	–	1.43 ~0.9	[7]
Ni–AlN	JPE + US	6	767.9 HV	–	0.52	[7]
Ni–TiN	JE	3 5 8	25.8 GPa 34.5 GPa ~31 GPa	–	–	[49]
Ni–TiN	JPE (jet speed) 1 m/s 3 m/s 5 m/s	7 7 7	789.5 HV 876.2 HV 849.9 HV	95.431 30.091 58.454	–	[8]
Ni–SiC	JE	3	584.1 HV	–	–	[5]
Ni–SiC	JE + MF	3	626.71 HV	–	–	[5]
Ni–Co–SiC	JE	0 2 4	–	1.960 1.114 0.751	–	[9]
Ni–Co–SiC	JE + MF	4	–	0.539	–	[9]

defects such as pits, microcracks, and pores due to nanoparticle addition restricts the penetration of corrosive electrolyte into the coating, thereby providing better corrosion-resistant properties [5, 52, 85, 86]. Jiang et al. [5] observed a similar trend where the corrosion current density was reduced by about 2.5 times after SiC incorporation. Also, a reduction of almost 3 times was obtained in the corrosion current density of pure Ni coatings upon the use of a magnetic field during jet electrodeposition. A similar increase in corrosion resistance has also been observed with the incorporation of TiN in Ni coatings, wherein a uniform, smooth, and compact surface was obtained using pulse jet electrodeposition which exhibited maximum corrosion resistance [8, 49].

Effect of AlN and ultrasound: Ma et al. [7] reported a reduction in corrosion current density of Ni–AlN coatings after AlN nanoparticle addition. The incorporation of AlN particles decreased the corrosion current density to 3.273×10^{-2} μA/cm^2 as compared to 8.879×10^{-2} μA/cm^2 for pure Ni coating, and the use of ultrasound during jet electrodeposition further decreased the corrosion density of Ni–AlN coating to 6.363×10^{-3} μA/cm^2. This is due to the formation of a smooth and nanostructured surface in Ni–AlN coatings which resist the penetration of corrosive fluid in the coating resulting in high corrosion resistance [7].

Effect of graphene oxide: It has been reported that the corrosion resistance of Ni coatings increased with the addition of an optimal amount of graphene oxide to the matrix. Corrosion current density values of 14.4 $\mu A/cm^2$, 5.7 $\mu A/cm^2$, and 20.3 $\mu A/cm^2$ were reported for Ni, Ni-0.5GO, and Ni-1.0GO, respectively. Ni-0.5GO coatings exhibited maximum corrosion resistance. On increasing the graphene oxide concentration to 1 g/L in bath, agglomeration was observed which increased the coating roughness. This promotes the retention of corrosive solution at the coating surface and increases the penetration through the coatings [72].

The corrosion current densities for various Ni-based jet-electrodeposited composite coatings have been compiled in Table 4.

Table 4 Corrosion current densities for various nickel-based composite coatings

Coating	Technique	Particle concentration in bath (g/L)	Corrosion current density (I_{corr}) ($\mu A/cm^2$) in NaCl	References
Ni–GO	JE	0 0.5 1	14.4 5.7 20.3	[72]
Ni–AlN	JPE	0 6	8.879×10^{-2} 3.272×10^{-2}	[7]
Ni–AlN	JPE + US	6	6.363×10^{-3}	[7]
Ni–TiN	JE	3 5 8	1.66 1.08 1.37	[49]
Ni–TiN	JPE (jet speed) 1 m/s 3 m/s 5 m/s	7 7 7	1.34 1.02 1.26	[8]
Ni–SiC	JE	0 1 3 5	31.39 18.57 12.67 16.86	[5]
Ni–SiC	JE + MF	0 1 3 5	11.06 5.640 4.320 6.66	[5]
Ni–Co–SiC	JE	0 2 4	23.08 15.90 13.97	[9]
Ni–Co–SiC	JE + MF	4	4.768	[9]

5 Conclusions and Future Scope

Jet electrodeposition provides a high deposition rate as it can be operated at very high current densities. Being a non-immersion electrochemical process, it also reduces the chances of corrosion of substrate, as long exposure times are not required. The recent research trends observed in deposition of Ni-based composite coatings using jet electrodeposition include

1. Focus on the effect of parameters such as jet speed, bath temperature, current density, and composition on coating properties.
2. Use of particles such as SiC, TiN, AlN, BN, and graphene oxide to improve hardness, tribological, and corrosion properties of coatings.
3. Improved coating surface quality by combining ultrasound, friction, and magnetic field along with jet electrodeposition.
4. Application of pulse current for fast deposition, grain refinement, and increased microhardness.
5. Fabrication of coatings with less microstructural defects and smoother surfaces for better corrosion properties.
6. Flow field simulation for optimization of operating parameters.

Precise control of current density, jet speed, and other process parameters helps in achieving desired coating properties. Along with a high-speed jet, the presence of ultrasonic waves, magnetic field, and pulsed current in a moderated amount leads to (a) homogeneous particle incorporation in the matrix, (b) fine crystallite size, and (c) uniform growth of the grains resulting in a smoother coating with better corrosion resistance.

Pulse electrodeposition facilitates the use of high current values to achieve faster speed for mass production. Higher current values promote nucleation resulting in refined grains with an increased hardness of coatings. Ultrasonic agitation increases particle entrapment, restricts agglomeration, and enhances their homogenized distribution in the metal matrix. Both pulse and continuous ultrasonic waves can be used during the coating process. Pulsed waves provide a high deposition rate and better surface uniformity than that obtained using continuous waves.

Use of friction on the cathode surface during deposition results in a smooth surface with reduced pits and voids. When grains grow non-uniformly, mechanical action polishes the surface and reduces uneven growth. Various types of friction devices can be used for mild abrasion of the surface.

Application of a magnetic field induces magneto-hydrodynamic action that causes microagitation in the electrolytic bath, thereby increasing particle incorporation. Magnetic field also reduces the vertical growth of grains and reduces the formation of hydrogen bubbles at the cathode surface that results in flat and smooth surfaces.

Uniform dispersion of ceramic particles such as SiC, AlN, and TiN in the matrix provides dispersion strengthening and grain refinement, which increases the mechanical strength and hardness of coatings. However, the incorporation of hexagonal BN nanoparticles reduced microhardness as they result in a reduced coefficient of friction due to the sliding nature of BN(h) nanoparticles. The introduction of graphene oxide in large concentrations increases agglomeration, but at optimal concentration, it improves microhardness and coating properties due to the inherent properties of graphene oxide. It can be concluded that the inherent properties of particles and their optimal concentration in electrolytes would result in desired coating properties.

Jet electrodeposition has gained attention in the last two decades, and comparatively fewer studies are available on Ni-based jet-electrodeposited composite coatings. Future studies in this area may explore

1. The individual and hybrid effect of incorporating different ceramic particles such as WC, ZrO_2, and TiO_2 in Ni/Ni–Co-based coatings using jet electrodeposition.
2. Development of multi-layer structured coatings with jet electrodeposition that may exhibit improved corrosion and wear properties.
3. Thermal stability of Ni/Ni–Co-based particle-reinforced jet-electrodeposited coatings and effect of external forces and different particles on thermal stability.
4. Formulation of empirical and semi-empirical models that can adequately capture the effect of various parameters on coating parameters and thus serve as a preliminary tool for the selection of parameters.
5. Implementation of machine learning techniques such as artificial neural networks, decision trees, genetic algorithms, and generative models to guide the development of novel Ni/Ni–Co coatings with optimal parameters, as done extensively for various materials science applications in recent years [87–93].
6. Development of computational models [94–101], including mathematical models, molecular dynamics simulations, and finite-element simulations, to simulate various aspects of the jet electrodeposition process.

Since jet electrodeposition leads to improved mechanical and chemical properties, these can be used in numerous applications and can replace traditional coatings. Figure 3 shows possible applications of jet-electrodeposited Ni-based coatings in future.

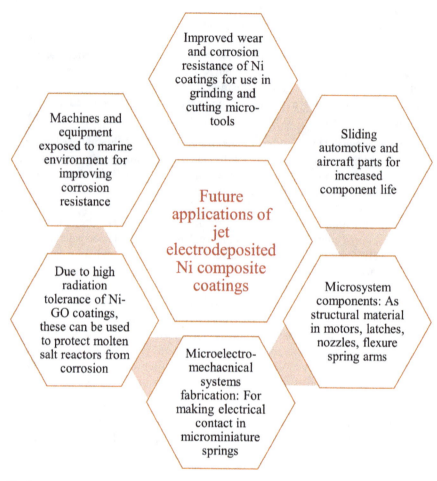

Fig. 3 Possible future applications of jet-electrodeposited Ni-based coatings

References

1. Koch CC (ed) (2007) Nanostructured materials: processing, properties, and applications, 2nd edn, vol 612. William Andrew Pub, Norwich, NY
2. Schario M (2007) Troubleshooting decorative nickel plating solutions (Part I of III installments): any experimentation involving nickel concentration must take into account several variables, namely the temperature, agitation, and the nickel-chloride mix. Met Finish 105(4):34–36
3. Boccaccini AR, Zhitomirsky I (2002) Application of electrophoretic and electrolytic deposition techniques in ceramics processing. Curr Opin Solid State Mater Sci 6(3):251–260
4. Besra L, Liu M (2007) A review on fundamentals and applications of electrophoretic deposition (EPD). Prog Mater Sci 52(1):1–61
5. Jiang W, Shen L, Qiu M, Wang X, Fan M, Tian Z (2018) Preparation of Ni-SiC composite coatings by magnetic field-enhanced jet electrodeposition. J Alloys Compd 762:115–124. https://doi.org/10.1016/j.jallcom.2018.05.097

6. Bahrololoom ME, Sani R (2005) The influence of pulse plating parameters on the hardness and wear resistance of nickel–alumina composite coatings. Surf Coat Technol 192(2):154–163. https://doi.org/10.1016/j.surfcoat.2004.09.023

7. Ma C, Yu W, Jiang M, Cui W, Xia F (2018) Jet pulse electrodeposition and characterization of Ni–AlN nanocoatings in presence of ultrasound. Ceram Int 44(5):5163–5170. https://doi.org/10.1016/j.ceramint.2017.12.121

8. Xia F, Jia W, Jiang M, Cui W, Wang J (2017) Microstructure and corrosion properties of Ni–TiN nanocoatings prepared by jet pulse electrodeposition. Ceram Int 43(17):14623–14628. https://doi.org/10.1016/j.ceramint.2017.07.117

9. Jiang W, Shen L, Xu M, Wang Z, Tian Z (2019) Mechanical properties and corrosion resistance of Ni-Co-SiC composite coatings by magnetic field-induced jet electrodeposition. J Alloys Compd 791:847–855. https://doi.org/10.1016/j.jallcom.2019.03.391

10. Bercot P, Pena-Munoz E, Pagetti J (2002) Electrolytic composite Ni–PTFE coatings: an adaptation of Guglielmi's model for the phenomena of incorporation. Surf Coat Technol 157(2–3):282–289

11. Narayan R, Narayana BH (1981) Electrodeposited Chromium-Graphite Composite Coatings. J Electrochem Soc 128(8):1704. https://doi.org/10.1149/1.2127714

12. Wang S-C, Wei W-CJ (2003) Kinetics of electroplating process of nano-sized ceramic particle/Ni composite. Mater Chem Phys 78(3):574–580

13. Chang Y-C, Chang Y-Y, Lin C-I (1998) Process aspects of the electrolytic codeposition of molybdenum disulfide with nickel. Electrochim Acta 43(3–4):315–324

14. Liu H, Chen W (2005) Electrodeposited Ni–Al composite coatings with high Al content by sediment co-deposition. Surf Coat Technol 191(2–3):341–350

15. Di Bari GA (2000) Electrodeposition of nickel. Mod Electroplat 5:79–114

16. Borkar T, Harimkar SP (2011) Effect of electrodeposition conditions and reinforcement content on microstructure and tribological properties of nickel composite coatings. Surf Coat Technol 205(17–18):4124–4134

17. Balasubramanian A, Srikumar DS, Raja G, Saravanan G, Mohan S (2009) Effect of pulse parameter on pulsed electrodeposition of copper on stainless steel. Surf Eng 25(5):389–392

18. Chandrasekar MS, Pushpavanam M (2008) Pulse and pulse reverse plating—Conceptual, advantages and applications. Electrochim Acta 53(8):3313–3322

19. Guglielmi N (1972) Kinetics of the deposition of inert particles from electrolytic baths. J Electrochem Soc 119(8):1009–1012

20. Karakus C, Chin D-T (1994) Metal distribution in jet plating. J Electrochem Soc 141(3):691. https://doi.org/10.1149/1.2054793

21. Qiao G et al (2005) High-speed jet electrodeposition and microstructure of nanocrystalline Ni–Co alloys. Electrochim Acta 51(1):85–92. https://doi.org/10.1016/j.electacta.2005.03.050

22. Rajput MS, Pandey PM, Jha S (2015) Micromanufacturing by selective jet electrodeposition process. Int J Adv Manuf Technol 76(1):61–67. https://doi.org/10.1007/s00170-013-5470-3

23. Tudela I, Zhang Y, Pal M, Kerr I, Mason TJ, Cobley AJ (2015) Ultrasound-assisted electrodeposition of nickel: effect of ultrasonic power on the characteristics of thin coatings. Surf Coat Technol 264:49–59. https://doi.org/10.1016/j.surfcoat.2015.01.020

24. Aaboubi O, Msellak K (2017) Magnetic field effects on the electrodeposition of CoNiMo alloys. Appl Surf Sci 396:375–383. https://doi.org/10.1016/j.apsusc.2016.10.164

25. Liu T, Shao G, Ji M (2014) Electrodeposition of Ni(OH)$_2$/Ni/graphene composites under supergravity field for supercapacitor application. Mater Lett 122:273–276. https://doi.org/10.1016/j.matlet.2014.02.035

26. Ban C, Shao X, Ma J, Chen H (2012) Effect of mechanical attrition on microstructure and property of electroplated Ni coating. Trans Nonferrous Met Soc China 22(8):1989–1994. https://doi.org/10.1016/S1003-6326(11)61418-0

27. Lü B, Hu Z, Wang X, Xu B (2015) Thermal stability of electrodeposited nanocrystalline nickel assisted by flexible friction. Trans Nonferrous Met Soc China 25(10):3297–3304. https://doi.org/10.1016/S1003-6326(15)63967-X

28. Biao LV, Zhenfeng HU, Xiaohe W, Binshi XU (2014) Effect of relative moving speed on microstructure of flexible friction assisted electrodeposited Ni coating. Chin J Mater Res 28(4):255–261

29. De Vogelaere M, Sommer V, Springborn H, Michelsen-Mohammadein U (2001) High-speed plating for electronic applications. Electrochim Acta 47(1):109–116. https://doi.org/10.1016/S0013-4686(01)00555-2

30. Alkire RC, Chen T-J (1982) High-speed selective electroplating with single circular jets. J Electrochem Soc 129(11):2424. https://doi.org/10.1149/1.2123560

31. Corni I, Ryan MP, Boccaccini AR (2008) Electrophoretic deposition: from traditional ceramics to nanotechnology. J Eur Ceram Soc 28(7):1353–1367. https://doi.org/10.1016/j.jeurceram soc.2007.12.011

32. Walsh FC, Wang S, Zhou N (2020) The electrodeposition of composite coatings: Diversity, applications and challenges. Curr Opin Electrochem 20:8–19. https://doi.org/10.1016/j.coe lec.2020.01.011

33. Jiang F (2016) Effect of cathodic current density on performance of tungsten coatings on molybdenum prepared by electrodeposition in molten salt. Appl Surf Sci 363:389–394. https://doi.org/10.1016/j.apsusc.2015.12.040

34. Liu X, Shen L, Qiu M, Tian Z, Wang Y, Zhao K (2016) Jet electrodeposition of nanocrystalline nickel assisted by controllable friction. Surf Coat Technol 305:231–240. https://doi.org/10.1016/j.surfcoat.2016.08.043

35. Wang C, Shen L, Qiu M, Tian Z, Jiang W (2017) Characterizations of Ni-CeO2 nanocomposite coating by interlaced jet electrodeposition. J Alloys Compd 727:269–277. https://doi.org/10.1016/j.jallcom.2017.08.105

36. Zhuo W, Shen L, Qiu M, Tian Z, Jiang W (2018) Effects of flexible friction on the properties of nanocrystalline nickel prepared by jet electrodeposition. Surf Coat Technol 333:87–95. https://doi.org/10.1016/j.surfcoat.2017.10.058

37. Tian Z, Wang D, Wang G, Shen L, Liu Z, Huang Y (2010) Microstructure and properties of nanocrystalline nickel coatings prepared by pulse jet electrodeposition. Trans Nonferrous Met Soc China 20(6):1037–1042. https://doi.org/10.1016/S1003-6326(09)60254-5

38. Jiang S, Pan Y, Tang T, Zhou Y (2007) Preparation of nanocrystalline nickel coating by pulse jet-electrodeposition. Mater Prot WUHAN 40(3):49

39. Zhao K, Shen L, Qiu M, Tian Z, Wei J (2017) Preparation and properties of nanocomposite coatings by pulsed current-jet electrodeposition. Int J Electrochem Sci:8578–8590. https://doi.org/10.20964/2017.09.04

40. Zheng X, Wang M, Song H, Wu D, Liu X, Tan J (2017) Effect of ultrasonic power and pulse-on time on the particle content and mechanical property of Co-Cr3C2 composite coatings by jet electrodeposition. Surf Coat Technol 325:181–189. https://doi.org/10.1016/j.surfcoat.2017.06.062

41. Zanella C, Lekka M, Rossi S, Deflorian F (2011) Study of the influence of sonication during the electrodeposition of nickel matrix nanocomposite coatings on the protective properties. Corros Rev 29(5–6):253–260. https://doi.org/10.1515/CORRREV.2011.005

42. García-Lecina E, García-Urrutia I, Díez JA, Morgiel J, Indyka P (2012) A comparative study of the effect of mechanical and ultrasound agitation on the properties of electrodeposited Ni/Al2O3 nanocomposite coatings. Surf Coat Technol 206(11):2998–3005. https://doi.org/10.1016/j.surfcoat.2011.12.037

43. Dietrich D, Scharf I, Nickel D, Shi L, Grund T, Lampke T (2011) Ultrasound technique as a tool for high-rate incorporation of Al2O3 in NiCo layers. J Solid State Electrochem 15(5):1041–1048. https://doi.org/10.1007/s10008-011-1348-1

44. Rezrazi M, Doche ML, Berçot P, Hihn JY (2005) Au–PTFE composite coatings elaborated under ultrasonic stirring. Surf Coat Technol 192(1):124–130. https://doi.org/10.1016/j.surfcoat.2004.04.067

45. Wang C et al (2008) Effects of parallel magnetic field on electrocodeposition behavior of Ni/nanoparticle composite electroplating. Appl Surf Sci 254(18):5649–5654. https://doi.org/10.1016/j.apsusc.2008.03.072

46. Wang C et al (2009) Effect of magnetic field on electroplating Ni/nano-Al_2O_3 composite coating. J Electroanal Chem 630(1):42–48. https://doi.org/10.1016/j.jelechem.2009.02.018
47. Pang X-F, Deng B, Tang B (2012) Influences of magnetic field on macroscopic properties of water. Mod Phys Lett B 26(11):1250069
48. Myung N, Nobe K (2001) Electrodeposited iron group thin-film alloys: structure-property relationships. J Electrochem Soc 148(3):C136–C144
49. Xia FF, Jia WC, Ma CY, Yang R, Wang Y, Potts M (2018) Synthesis and characterization of Ni-doped TiN thin films deposited by jet electrodeposition. Appl Surf Sci 434:228–233. https://doi.org/10.1016/j.apsusc.2017.10.203
50. Marikkannu KR, Kala KS, Kalaignan GP, Vasudevan T (2008) Electroplating of nickel from acetate based bath–Hull Cell studies. Trans. IMF 86(3):172–176
51. Stuyven B et al (2007) Magnetic field assisted nanoparticle dispersion. Chem Commun (1). https://doi.org/10.1039/B816171B
52. Bakhit B, Akbari A, Nasirpouri F, Hosseini MG (2014) Corrosion resistance of Ni–Co alloy and Ni–Co/SiC nanocomposite coatings electrodeposited by sediment codeposition technique. Appl Surf Sci 307:351–359. https://doi.org/10.1016/j.apsusc.2014.04.037
53. Bahadormanesh B, Dolati A (2010) The kinetics of Ni–Co/SiC composite coatings electrodeposition. J Alloys Compd 504(2):514–518. https://doi.org/10.1016/j.jallcom.2010.05.154
54. Cui W, Wang K, Xia F, Wang P (2018) Simulation and characterization of Ni–doped SiC nanocoatings prepared by jet electrodeposition. Ceram Int 44(5):5500–5505. https://doi.org/10.1016/j.ceramint.2017.12.189
55. Nikitin D, Sivkov A (2015) On the impact of the plasma jet energy on the product of plasma-dynamic synthesis in the Si-C system. IOP Conf Ser Mater Sci Eng 93:012039. https://doi.org/10.1088/1757-899X/93/1/012039
56. Xia F, Wu M, Wang F, Jia Z, Wang A (2009) Nanocomposite Ni–TiN coatings prepared by ultrasonic electrodeposition. Curr Appl Phys 9(1):44–47. https://doi.org/10.1016/j.cap.2007.11.014
57. Calderón JA, Henao JE, Gómez MA (2014) Erosion–corrosion resistance of Ni composite coatings with embedded SiC nanoparticles. Electrochim Acta 124:190–198. https://doi.org/10.1016/j.electacta.2013.08.185
58. Burzyńska L, Rudnik E (2000) The influence of electrolysis parameters on the composition and morphology of Co–Ni alloys. Hydrometallurgy 54(2):133–149. https://doi.org/10.1016/S0304-386X(99)00060-2
59. Guo J et al (2016) Effects of glycine and current density on the mechanism of electrodeposition, composition and properties of Ni–Mn films prepared in ionic liquid. Appl Surf Sci 365:31–37. https://doi.org/10.1016/j.apsusc.2015.12.248
60. Li Y, Jiang H, Huang W, Tian H (2008) Effects of peak current density on the mechanical properties of nanocrystalline Ni–Co alloys produced by pulse electrodeposition. Appl Surf Sci 254(21):6865–6869. https://doi.org/10.1016/j.apsusc.2008.04.087
61. Wasekar NP, Haridoss P, Seshadri SK, Sundararajan G (2016) Influence of mode of electrodeposition, current density and saccharin on the microstructure and hardness of electrodeposited nanocrystalline nickel coatings. Surf Coat Technol 291:130–140. https://doi.org/10.1016/j.surfcoat.2016.02.024
62. Ning Z, He Y, Gao W (2008) Mechanical attrition enhanced Ni electroplating. Surf Coat Technol 202(10):2139–2146. https://doi.org/10.1016/j.surfcoat.2007.08.062
63. Shi L, Sun C, Gao P, Zhou F, Liu W (2006) Mechanical properties and wear and corrosion resistance of electrodeposited Ni–Co/SiC nanocomposite coating. Appl Surf Sci 252(10):3591–3599. https://doi.org/10.1016/j.apsusc.2005.05.035
64. Imanian Ghazanlou S, Farhood A, Ahmadiyeh S, Ziyaei E, Rasooli A, Saman H (2019) Characterization of pulse and direct current methods for electrodeposition of Ni-Co composite coatings reinforced with nano and micro ZnO particles. Metall Mater Trans A 50:1922–1935. https://doi.org/10.1007/s11661-019-05118-y
65. Ghazanlou SI, Farhood AHS, Hosouli S, Ahmadiyeh S, Rasooli A (2017) Pulse frequency and duty cycle effects on the electrodeposited Ni–Co reinforced with micro and nano-sized

ZnO. J Mater Sci Mater Electron 28(20):15537–15551. https://doi.org/10.1007/s10854-017-7442-0

66. Srivastava M, Grips VKW, Rajam KS (2007) Electrochemical deposition and tribological behaviour of Ni and Ni–Co metal matrix composites with SiC nano-particles. Appl Surf Sci 253(8):3814–3824. https://doi.org/10.1016/j.apsusc.2006.08.022

67. Wu G, Li N, Zhou D, Mitsuo K (2004) Electrodeposited Co–Ni–Al2O3 composite coatings. Surf Coat Technol 176(2):157–164. https://doi.org/10.1016/S0257-8972(03)00739-4

68. Vaezi MR, Sadrnezhaad SK, Nikzad L (2008) Electrodeposition of Ni–SiC nano-composite coatings and evaluation of wear and corrosion resistance and electroplating characteristics. Colloids Surf Physicochem Eng Asp 315(1):176–182. https://doi.org/10.1016/j.colsurfa.2007.07.027

69. Yao Y, Yao S, Zhang L, Wang H (2007) Electrodeposition and mechanical and corrosion resistance properties of Ni–W/SiC nanocomposite coatings. Mater Lett 61(1):67–70. https://doi.org/10.1016/j.matlet.2006.04.007

70. Juneghani MA, Farzam M, Zohdirad H (2013) Wear and corrosion resistance and electroplating characteristics of electrodeposited Cr–SiC nano-composite coatings. Trans Nonferrous Met Soc China 23(7):1993–2001. https://doi.org/10.1016/S1003-6326(13)62688-6

71. Özkan S, Hapçı G, Orhan G, Kazmanlı K (2013) Electrodeposited Ni/SiC nanocomposite coatings and evaluation of wear and corrosion properties. Surf Coat Technol 232:734–741. https://doi.org/10.1016/j.surfcoat.2013.06.089

72. Ji L, Chen F, Huang H, Sun X, Yan Y, Tang X (2018) Preparation of nickel–graphene composites by jet electrodeposition and the influence of graphene oxide concentration on the morphologies and properties. Surf Coat Technol 351:212–219. https://doi.org/10.1016/j.surfcoat.2018.07.083

73. Bakhit B, Akbari A (2013) Synthesis and characterization of Ni–Co/SiC nanocomposite coatings using sediment co-deposition technique. J Alloys Compd 560:92–104. https://doi.org/10.1016/j.jallcom.2013.01.122

74. Ke LU et al (1994) The hall-petch relation in nanocrystalline materials. Chin J Mater Res 8(5):385–391

75. Roland T, Retraint D, Lu K, Lu J (2006) Fatigue life improvement through surface nanostructuring of stainless steel by means of surface mechanical attrition treatment. Scr Mater 54(11):1949–1954. https://doi.org/10.1016/j.scriptamat.2006.01.049

76. Wu M, Jia W, Lv P (2017) Electrodepositing Ni-TiN nanocomposite layers with applying action of ultrasonic waves. Procedia Eng. 174:717–723. https://doi.org/10.1016/j.proeng.2017.01.211

77. Kang M, Zhang Y, Li HZ (2018) Study on the performances of Ni-Co-P/BN(h)nanocomposite coatings made by jet electrodeposition. Procedia CIRP 68:221–226. https://doi.org/10.1016/j.procir.2017.12.052

78. Yuan Q-L, Cao J-J, Li S-M (2010) Effect of heat treatment temperature on microstructure and properties of Ni-Co-P alloy plating. Cailiao Rechuli Xuebao Trans Mater Heat Treat 31:73–76

79. Allahyarzadeh MH, Aliofkhazraei M, Rezvanian AR, Torabinejad V, Sabour Rouhaghdam AR (2016) Ni-W electrodeposited coatings: characterization, properties and applications. Surf Coat Technol 307:978–1010. https://doi.org/10.1016/j.surfcoat.2016.09.052

80. Allahyarzadeh MH, Aliofkhazraei M, Rouhaghdam ARS, Torabinejad V (2016) Structure and wettability of pulsed electrodeposited Ni-W-Cu-(α-alumina) nanocomposite. Surf Coat Technol 307:525–533. https://doi.org/10.1016/j.surfcoat.2016.09.036

81. Borkar T, Harimkar S (2011) Microstructure and wear behaviour of pulse electrodeposited Ni–CNT composite coatings. Surf Eng 27(7):524–530. https://doi.org/10.1179/1743294410Y.0000000001

82. Zheng Q, Li Z, Geng Y, Wang S, Kim JK (2010) Molecular dynamics study of the effect of chemical functionalization on the elastic properties of graphene sheets. https://www.ingentaconnect.com/content/asp/jnn/2010/00000010/00000011/art00016. Accessed 16 May 2020

83. Danaie M, Asmussen RM, Jakupi P, Shoesmith DW, Botton GA (2013) The role of aluminum distribution on the local corrosion resistance of the microstructure in a sand-cast AM50 alloy. Corros Sci 77:151–163. https://doi.org/10.1016/j.corsci.2013.07.038

84. Suter T, Böhni H (2001) Microelectrodes for corrosion studies in microsystems. Electrochim Acta 47(1):191–199. https://doi.org/10.1016/S0013-4686(01)00551-5

85. Yang X, Li Q, Hu J, Zhong X, Zhang S (2010) The electrochemical corrosion behavior of sealed Ni–TiO 2 composite coating for sintered NdFeB magnet. J Appl Electrochem 40(1):39–47. https://doi.org/10.1007/s10800-009-9961-8

86. Bahadormanesh B, Dolati A, Ahmadi MR (2011) Electrodeposition and characterization of Ni–Co/SiC nanocomposite coatings. J Alloys Compd 509(39):9406–9412. https://doi.org/10.1016/j.jallcom.2011.07.054

87. G. L. W. Hart, T. Mueller, C. Toher, and S. Curtarolo, "Machine learning for alloys," *Nat. Rev. Mater.*, vol. 6, no. 8, Art. no. 8, Aug. 2021, doi: https://doi.org/10.1038/s41578-021-00340-w.

88. K. T. Butler, D. W. Davies, H. Cartwright, O. Isayev, and A. Walsh, "Machine learning for molecular and materials science," *Nature*, vol. 559, no. 7715, Art. no. 7715, Jul. 2018, doi: https://doi.org/10.1038/s41586-018-0337-2.

89. R. Ramprasad, R. Batra, G. Pilania, A. Mannodi-Kanakkithodi, and C. Kim, "Machine learning in materials informatics: recent applications and prospects," *Npj Comput. Mater.*, vol. 3, no. 1, Art. no. 1, Dec. 2017, doi: https://doi.org/10.1038/s41524-017-0056-5.

90. D. Beniwal, P. Singh, S. Gupta, M. J. Kramer, D. D. Johnson, and P. K. Ray, "Distilling physical origins of hardness in multi-principal element alloys directly from ensemble neural network models," *Npj Comput. Mater.*, vol. 8, no. 1, Art. no. 1, Jul. 2022, doi: https://doi.org/10.1038/s41524-022-00842-3.

91. Beniwal D, Ray PK (2021) Learning phase selection and assemblages in High-Entropy Alloys through a stochastic ensemble-averaging model. Comput Mater Sci 197:110647. https://doi.org/10.1016/j.commatsci.2021.110647

92. D. Beniwal and P. K. Ray, "FCC vs. BCC phase selection in high-entropy alloys via simplified and interpretable reduction of machine learning models," *Materialia*, p. 101632, Nov. 2022. https://doi.org/10.1016/j.mtla.2022.101632.

93. DBeniwal, Jhalak, and P. K. Ray, "Data-Driven Phase Selection, Property Prediction and Force-Field Development in Multi-Principal Element Alloys," in *Forcefields for Atomistic-Scale Simulations: Materials and Applications*, A. Verma, S. Mavinkere Rangappa, S. Ogata, and S. Siengchin, Eds. Singapore: Springer Nature, 2022, pp. 315–347. doi: https://doi.org/10.1007/978-981-19-3092-8_16.

94. Brant AM, Sundaram M (2017) A fundamental study of nano electrodeposition using a combined molecular dynamics and quantum mechanical electron force field approach. Proc Manuf 10:253–264. https://doi.org/10.1016/j.promfg.2017.07.054

95. Chaturvedi S, Verma A, Singh SK, Ogata S (2022) EAM inter-atomic potential—its implication on nickel, copper, and aluminum (and their alloys). In: Verma A, Mavinkere Rangappa S, Ogata S, Siengchin S (Eds) Forcefields for atomistic-scale simulations: materials and applications. Springer Nature, Singapore, pp 133–156. https://doi.org/10.1007/978-981-19-3092-8_7

96. Chaturvedi S, Verma A, Sethi SK, Ogata S (2022) Defect energy calculations of nickel, copper and aluminium (and their alloys): molecular dynamics approach. In: Verma A, Mavinkere Rangappa S, Ogata S, Siengchin S (eds) Forcefields for atomistic-scale simulations: materials and applications. Springer Nature, Singapore, pp 157–186. https://doi.org/10.1007/978-981-19-3092-8_8

97. Homer ER, Verma A, Britton D, Johnson OK, Thompson GB (2022) Simulated migration behavior of metastable Σ3 (11 8 5) incoherent twin grain boundaries. IOP Conf Ser Mater Sci Eng 1249(1):012019. https://doi.org/10.1088/1757-899X/1249/1/012019

98. Guymon CG, Harb JN, Rowley RL, Wheeler DR (2008) MPSA effects on copper electrode-position investigated by molecular dynamics simulations. J Chem Phys 128(4):044717. https://doi.org/10.1063/1.2824928

99. Kaneko Y, Mikami T, Hiwatari Y, Ohara K (2005) Computer simulation of electrodeposition: hybrid of molecular dynamics and Monte Carlo. Mol Simul 31(6–7):429–433. https://doi.org/10.1080/08927020412331332758

100. Belov I, Zanella C, Edström C, Leisner P (2016) Finite element modeling of silver electrodeposition for evaluation of thickness distribution on complex geometries. Mater Des 90:693–703. https://doi.org/10.1016/j.matdes.2015.11.005
101. Brant AM, Sundaram MM, Kamaraj AB (2015) Finite element simulation of localized electrochemical deposition for maskless electrochemical additive manufacturing. J Manuf Sci Eng 137(1). https://doi.org/10.1115/1.4028198

Chapter 17
Microstructure, Microhardness and Tribological Properties of Electrodeposited Ni–Co Based Particle Reinforced Composite Coatings

Jhalak, Dishant Beniwal, and Rajnish Garg

1 Introduction

Any interaction between a material and its environment, whether mechanical, chemical, thermal or electrochemical, starts at the surface. These interactions often result in damage to the material surface due to corrosion and wear, and thus necessitate some form of surface modification. Different surface modification techniques, such as surface coatings, mechanical treatment and heat treatment, are commonly employed to enhance surface properties. Coatings can act as a barrier between the substrate and the environment, thereby providing protection against corrosion, wear and thermal effects. While many factors dictate the properties of coatings, a major control parameter is the deposition process. The selection of an appropriate coating technique can be critical for obtaining the desired properties that can prevent or delay damage to the component, thereby extending its life [1]. As depicted in Fig. 1, wear and corrosion resistant coatings are used in a wide range of industries to protect surfaces from degradation and to enhance the overall performance of components. Over the decades, many different coating techniques have been developed such as thin

Rajnish Garg, at the time of preparation of manuscript, author (Rajnish Garg) was associated with University of Petroleum and Energy Studies. At the time of submission, author is not associated with any institution.

Jhalak (✉) · D. Beniwal
Department of Metallurgical and Materials Engineering, Indian Institute of Technology Ropar, Rupnagar, Punjab 140001, India
e-mail: jhalak.20mmz0010@iitrpr.ac.in

R. Garg
Department of Mechanical Engineering, University of Petroleum and Energy Studies, Dehradun, Uttarakhand 248007, India

Fig. 1 Various industrial applications of wear and corrosion resistant coatings [2]

film deposition, sputtering deposition, thermal spraying, ion implantation, electroless plating and electroplating (also known as electrodeposition). Nowadays, many hybrid techniques such as plasma enhanced chemical vapor deposition (PECVD) with magnetron sputtering and polymer flash evaporation with physical vapor deposition have also been used for coating deposition [2]. Some of the common techniques have been discussed here in brief.

1.1 Thin Film Deposition

Chemical vapor deposition and physical vapor deposition are the most commonly used techniques for fabricating thin films.

a. Chemical Vapor Deposition (CVD)

CVD can be used to deposit both metallic and non-metallic coatings and is widely used in many applications such as electronics, optical, opto-electronics, and wear and corrosion resistant coatings especially for tools and high strength fibres. CVD process utilizes chemical vapors that condense and deposit on the substrate surface.

It can be used to deposit thin films as well as coatings with thickness greater than 1 mm [3]. CVD has many advantages as it allows deposition of refractory materials below their melting point and provides control over the grain size and orientation of coating which improves its bonding with the substrate. However, CVD is carried out at temperatures greater than 600 °C and thus can only be used for substrates that are thermally stable at these operating temperatures. Also, certain precursors used in the CVD process have been classified as hazardous as their reactions produce toxic byproducts.

b. **Physical Vapor Deposition (PVD)**

In the PVD process, first the atoms are generated using an ionized gas. These atoms, which eject from a vaporization source, are bombarded on the solid substrate resulting in deposition. PVD can be used to deposit a variety of materials but it's application is limited by the fact that it can only produce coatings with thickness ranging from 100 nm to 100 μm [3].

1.2 Sputtering

Sputtering involves ejection of atoms from the surface of a material due to the momentum transfer when the surface is bombarded by high energy particles or ions. These sputtered atoms are then deposited on the substrate surface. Sputtering offers various advantages in that it is environment friendly and can deposit a large variety of films such as metals, ceramics, alloys and inorganic compounds at low temperatures. But there are certain disadvantages such as high operating costs associated with maintaining a vacuum environment, restriction on the form of source material which should be a sheet or tube and a lower energy efficiency since most of the energy manifests as thermal energy which ends up heating the source material [4, 5].

1.3 Thermal Spraying

Thermal spraying employs a torch or spray-gun fed with the material to be coated in powder, rod or wire form. The material is heated inside the gun up to its melting point and a stream of molten material is sprayed on to the substrate. Multiple passes of coating are done to achieve the required thickness. A variety of materials, including metals, ceramics and polymers, that can melt without decomposition can be coated using thermal spraying.

1.4 Ion Implantation

In ion implantation, an ionized beam of atoms is bombarded on the surface of material to deposit a coating. Ion implantation creates an effect similar to surface alloying and is extensively used in semiconductor industries. A high level of adhesion is obtained through ion implantation which mitigates the chances of delamination, thereby improving the fatigue, oxidation, corrosive and tribological properties of metals, alloys and polymeric substrates [6]. Since it can be carried out at ambient temperatures, ion implantation changes surface properties without affecting the bulk properties of the substrate. However, ion implantation is an expensive and complex process that requires intensive training for depositing coatings [7].

1.5 Diffusion Coating

Diffusion coating is a solid-state diffusion process in which the substrate is heated while in contact with the powder to be coated. The temperature and exposure time are controlled to determine the thickness and structure of the deposited coating. Diffusion coating technique is generally used for depositing aluminium, chromium and silicon-based coatings on superalloys, steels and refractory metals. These coatings are widely used in gas turbine components.

1.6 Electro-less Plating

Electro-less plating involves deposition of coatings on the substrate without using any electric current. Metal cations are reduced with the help of a reducing agent and are deposited on the substrate surface. Electro-less deposition provides uniform coatings with excellent corrosion and wear resistant properties [8]. But it is expensive and has slower deposition rates as compared to the electroplating process.

1.7 Electroplating or Electrodeposition

Electrodeposition is an electrochemical process of depositing a single or multi-layered coating using an electroplating bath. Ions present in the bath are deposited on the electrode surface due to electrochemical reactions taking place at the electrode/electrolyte interface [9]. Table 1 summarizes the different popular coating processes along with their respective advantages and limitations.

Table 1 Various coating processes

Coating type	Method	Benefit	Challenges/limitation	References
Thin film deposition	PVD	High purity films of a variety of materials can be deposited	Coating thickness limited from 100 nm to 100 μm Coating properties may not meet requirements	[3]
	CVD	Both thin and thick metallic, non-metallic and intermetallic films can be deposited	Some precursors are hazardous as the reactions produce toxic solid byproducts; operating temperature above 600 °C	
Sputtering	Ionized beam is used to pull out atoms from a source that are then deposited on substrate	Low temperature deposition of refractory materials is possible	Sheet or tube form of source material is required; expensive operation as vacuum is required; low energy efficiency as the material heats up	[4, 5]
Thermal spraying	Stream of molten metal is sprayed against the substrate	Metallic, ceramic, polymeric coating can be deposited	Can't be used for parts or areas where the torch/gun can't reach	[3]
Ion implantation	Ionized beam of atoms is bombarded on the surface of material	High adhesion; can be carried out at room temperature so does not affect bulk properties	Limited thickness; expensive; complex to use	[6, 7]
Diffusion coatings	A solid-state diffusion process; substrate is heated while in contact with the powder to be coated	Provides excellent properties for high temperature applications	High operating temperatures	[3]
Electro-less plating	Metal cations are reduced with the help of reducing agent; no use of electric current	External current is not required; excellent uniformity of deposited coatings	Slow deposition rate; expensive as compared to electroplating	[8]

(continued)

Table 1 (continued)

Coating type	Method	Benefit	Challenges/limitation	References
Electroplating	Electric current is supplied to reduce ions to be deposited	Economical process; varying deposition rates can be achieved by controlling current	Pores and cracks may develop if parameters are not controlled properly	[9, 10]

Out of the various techniques available, electrodeposition is the most economical and widely used technique for developing thin as well as thick nanostructured coatings.

Electrodeposition has been used to deposit a variety of metals, alloys, ceramics and composites on a conductive substrate. Technologically it is a simple process capable of producing coatings with good chemical stability even on surfaces with complex geometries. But electrodeposition being a simple process, requires an optimal control over key parameters such as bath composition, pH, current density and agitation to achieve a good coating [10].

Electrodeposited coatings are used extensively as protective coatings in various industries such as electronics, automotive, musical instruments, medical, agricultural, marine, and aircraft industries. Pure nickel electrodeposited coatings are commonly used for decorative and engineering purposes due to their appearance and protective properties. The properties of these coatings can be varied by controlling bath composition and other operating parameters. Engineering applications of Ni plating include enhanced corrosion & wear resistance, buildup of worn parts, magnetic properties modification and surface preparation for organic coatings [11] but the most common application of Ni plating is as an undercoat for chromium coatings. Depositing thick chromium coatings is difficult due to their susceptibility to cracking as internal stresses build up. Moreover, chromium coatings require a low current density leading to a slower deposition rate. Thus, when thick chromium coatings are required, a Ni undercoat is applied initially and a very thin chromium layer is deposited over it [12]. It is reported that Ni has a lower coefficient of friction as compared to Cr, Zn and Cu [13]. Addition of Co to nickel can further improve tribological properties of the coatings, and thus Ni–Co based coatings when used in combination with hard particles impart high hardness, strength and wear resistance. Electrodeposited Ni–Co coatings offer better wear and corrosion properties than pure Ni coatings and can replace hard Cr coatings which involves hazardous use of materials [14].

Chromium plating involves use of hazardous substances that include hexavalent chromium. Although Ni and Co compounds are also classified as hazardous, these are comparatively less harmful and toxic than Cr compounds [15]. While Ni and Co coatings are expensive due to high cost of raw materials, they offer many advantages: (1) these coatings can act as dry lubricants, (2) higher hardness and tensile strength, (3) good corrosion and oxidation resistance, (4) good magnetic properties, (5) high surface area achieved through formation of three dimensional morphologies [16–18].

The major challenges involved in obtaining Ni–Co based coatings are: (1) getting desired properties (2) managing environmental hazards. These can be reduced by selecting proper bath composition and by optimizing coating parameters.

The aim of this review is to draw attention on:

(i)   Effect of using various baths for Ni–Co electrodeposition.
(ii) Effect of various process parameters on Ni–Co electrodeposition.
(iii) Incorporation of ceramic particles in Ni–Co matrix.
(iv) Tribological properties of Ni–Co particle reinforced composite coatings.

2   Impact of Ni–Co Electrodeposition on Environment and Health

For many tribological applications, chromium coatings are used due to their unique wear resistant and other mechanical properties [19]. However, the baths used for chrome-plating include acids which release toxic hexavalent chromium ions. In recent decades, Ni–Co based coatings were extensively studied as an alternative to hard chromium coatings. Electroplating of Ni and Co is done using baths that contain Ni, Co and other ions. It produces hazardous waste that effects environment and human health since Ni and Co ions are toxic and may cause skin irritation and other health problems [20]. However, these are less harmful than chromate ions used for electroplating of hard chromium coatings. Releasing of toxic and non-biodegradable ceramic particles, inclusions and surfactants into the environment can also pose a threat, but there are some non-toxic organic surfactants that are used for electrodeposition. Also, waste water treatment must be done to eliminate metal ions and other toxic species [21, 22].

3   Electrodeposition Process

The setup for electrodeposition of Ni–Co coatings consists of: (a) pure nickel as anode, (b) the substrate to be coated as cathode, and (c) an electrolytic bath that enables ionic flow between the electrodes. For creating particle-reinforced coatings, suitable ceramic particles are suspended in the electrolytic bath. Current is supplied through an external power supply connected between anode and cathode. Schematic in Fig. 2 shows the setup for electrodeposition process.

3.1   Electrodeposition of Nickel

The electrodeposition process requires two electrodes (anode and cathode) to be immersed in a conductive bath of Ni salts and current is passed across these electrodes

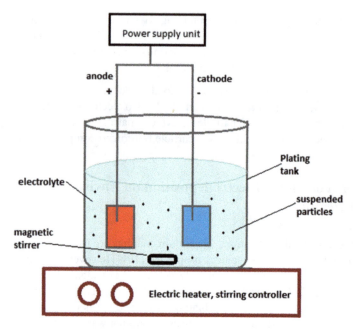

Fig. 2 Schematic showing electrodeposition process

leading to the dissolution of anode (pure Ni) and deposition of Ni at the cathode (substrate). The dissolution and deposition of Ni respectively at anode and cathode is accompanied by the reduction of hydrogen ions at cathode according to following reactions [11]:

(i) Deposition of Ni at cathode

$$Ni^{2+} + 2e^- \rightarrow Ni \tag{1}$$

(ii) Reduction of hydrogen ions at cathode

$$H^+ + e^- \rightarrow H \tag{2}$$

Hydrogen ions, present in water, are discharged at the cathode surface as hydrogen bubbles. This hydrogen evolution accounts for a reduced cathode efficiency for Ni deposition and also leads to formation of porous deposits at the cathode surface. Hydrogen evolution may reduce the cathode efficiency to as low as 92% depending on nature of the electrolyte bath. Under normal conditions, the hydroxyl ions are not discharged from water and thus the anodic dissolution efficiency is maintained at 100%. But, certain conditions, such as a high pH, can lead to discharge of hydroxyl ions resulting in oxygen evolution which passivates the Ni anode, thereby ceasing the anodic dissolution process [6].

3.2 Electro-codeposition Mechanism of Ni–Co

During electrodeposition, cobalt can be deposited along with nickel using soluble Co anodes or Co salts in the bath. Over the years, various models have been developed to explain the mechanism of Ni–Co codeposition. According to Maltosz model, metal deposition is a two-step process: (1) cation reduction as monovalent ion and adsorption at cathode surface, and (2) reduction of the ion to metallic form [23]. This mechanism has been validated by many researchers using electrochemical impedance spectroscopy (EIS) and it was observed that Co deposits preferentially over Ni and thus inhibits Ni deposition [24, 25]. This phenomenon has been further explained through the formation of a Ni–Co intermediate [26]. Following reactions depict the various steps involved in the process:

$$Ni^{2+} + e^- \rightarrow Ni^+ \tag{3}$$

$$Ni^+ + e^- \rightarrow Ni \tag{4}$$

$$Co^{2+} + e^- \rightarrow Co^+ \tag{5}$$

$$Co^{2+} + e^- \rightarrow Co \tag{6}$$

$$Ni^{2+} + Co^{2+} + e^- \rightarrow NiCo^{3+} \tag{7}$$

$$Ni^{2+} + Co^+ + e^- \rightarrow Ni^{2+} + Co \tag{8}$$

$$Ni^{2+} + Co^+ + e^- \rightarrow Ni + Co^{2+} \tag{9}$$

First, an intermediate of Ni–Co is formed and subsequently adsorbed on the substrate surface. This acts as a catalyst for preferential Co deposition. The complete surface is covered by Ni–Co intermediate, thus blocking the sites for further adsorption of Ni and Co ions. This results in reduction of Ni ions to form Ni deposits [3, 15, 20]

Another model suggests the reduction of H^+ ions and formation of metal hydroxyl ions. These ions are adsorbed on the substrate surface and form a coating after reduction [27, 28]. Following reactions are involved in this process:

$$2H^+ + 2e^- \rightarrow H_2 \tag{10}$$

$$2H_2O + 2e^- \rightarrow H_2 + 2OH^- \tag{11}$$

$$M^{2+} + OH^- \rightarrow M(OH)^+_{ads} + OH^- \tag{12}$$

$$M(OH)^+ + 2e^- \rightarrow M + OH^- \tag{13}$$

Here, M indicates a metal atom (Ni or Co).

3.3 Electrodeposition of Ni–Co Coatings Reinforced with Ceramic Particles

Electrodeposition of particles embedded in a metal matrix involves deposition of charged ceramic particles, dispersed in an electrolytic solution, on the substrate with an opposite charge. This is also known as 'electrophoretic' deposition and can be easily modified for specific applications such as deposition of complex compounds and ceramic laminates. This process offers an advantage that it can be used for deposition on different shapes through minor changes in design and positioning of electrodes. It needs less time and a simple apparatus [29–31]. Various applications of this process include production of anti-oxidant and wear resistant ceramic coatings, functional films for fuel cells, bioactive coatings for medical implants and multi-layered composite coatings [24, 26, 32–34].

Deposition of particles along with the metallic deposits has been explained by Guglielmi [35] model as shown in Fig. 3. According to this model, particle deposition on the cathode takes place in two steps: (1) the first step is adsorption of particles on metallic ions present in the solution which then diffuse towards cathode and attach loosely to the cathodic surface; (2) the second step involves encapsulation and incorporation of these particles as the reduction of ions takes place at the cathode [36–40].

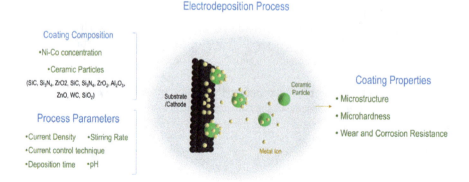

Fig. 3 Schematic showing overview of electrodeposition process and Guglielmi's model for particles incorporation

4 Various Baths for Electrodeposition of Nickel Based Coatings

Various baths are used to deposit nickel coating on a substrate and different modifications can be done by adding salts to codeposit other metals along with nickel. Some common baths used to deposit nickel have been discussed in this section.

4.1 Sulfate and Chloride Bath

Sulfate and chloride baths are economical and widely used for many industrial applications. The sulfate bath, also known as Watt's bath, has nickel sulfate as the main source of nickel ion while chloride baths contain nickel chloride as source of nickel. Nickel chloride is added to sulfate baths in moderate quantities. Addition of nickel chloride provides anode dissolution, increased conductivity of electrolyte and uniformity of coating thickness [6, 7, 41, 42]. It also provides harder coatings however, it may increase the corrosivity of the solution and may induce internal stresses in the coatings if used in excessive amount. Higher values of current efficiencies are obtained in chloride containing baths due to lower limiting current for H^+ ions in chloride baths [43, 44]. To deposit Co along with Ni, cobalt sulfate is added to the bath as a source of Co ions. Sulfate baths are economical and simple to use with low risk of equipment corrosion. Coatings deposited using sulfate baths have low internal stresses than chloride baths [45]. Deposits of sulfate bath have tendency to form strong passive film that enhances corrosion resistance of coatings [46]. As local pH may increase during the coating process, boric acid is used as buffering agent in these baths [26].

4.2 Sulfamate Bath

Sulfamate baths utilize sulfamate salts to deposit coatings. Nickel sulfamate and cobalt sulfamate are used as the source of Ni ions and Co ions respectively. As compared to sulfate baths, sulfamate baths provide higher rates of deposition with lower internal stresses in the deposited coatings. The deposition rates are high as high current densities are obtained with these baths due to their high solubility in water [47]. However, these baths are highly sensitive to the presence of impurities and are more expensive due to the high cost of chemicals [45]. During the deposition of iron-group metals (Fe, Ni and Co), the bath pH increases locally in areas adjacent to the cathode surface. This may lead to precipitation of hydroxides on the cathode surface, thereby blocking the surface sites to produce non-uniform deposition [48]. Usage of boric acid as a buffering agent helps in preventing this problem by regulating the pH values [49].

4.3 Fluoroborate Bath

Fluoroborate baths contain nickel fluoroborate as the source of ions. Nickel ions are replenished at a faster rate in these baths due to a higher nickel dissolution rate as compared to the sulfate, chloride and sulfamate baths. These baths inhibit a local change in pH near the cathode surface that may result in better coating properties. Additives, such as sulfur, that may induce internal stresses in the coating are not used in fluoroborate baths. While the deposition rates are high, these baths suffer from high cost and a corrosive nature [50].

4.4 Acetate Bath

Acetate baths use nickel acetate and cobalt acetate as the Ni and Co ion source respectively. These baths have low toxicity and are more ecofriendly [51]. Buffer solution is not required for acetate baths and often Cl^- ions are added to the bath to increase the dissolution rate of nickel, [52]. However, very few studies have been carried out for coatings deposition using these baths.

4.5 Ionic Baths

Ionic baths used for electrodeposition consist of ionic liquids. Deep Eutectic Systems (DES) have been used in the past to deposit Cr, Ni, Co, Zn and Cu. Ionic baths are ecofriendly but expensive [53]. Most common DES to deposit Ni and Ni–Co coatings is choline chloride/ethylene glycol [54]. Since the boiling temperature of ethylene glycol is higher than water, this bath can be used at higher temperatures than aqueous baths. In addition, these baths have higher stability due to certain properties of ethylene glycol such as high viscosity, high density and high surface wettability. Another advantage of ionic baths is that their use can effectively rule out any hydrogen evolution, thereby leading to an almost negligible hydrogen content in the deposited coating [55, 56].

4.6 Additives Used in Baths

Additives such as boric acid, citric acid, ascorbic acid and saccharin are added in electrolytic bath to get coatings with desired properties. Functions of some common additives are discussed below.

4.6.1 Boric Acid

In aqueous baths, hydrogen evolution can occur on cathode surface resulting in reduced current efficiency and increased pH value in the vicinity of cathode. This affects metal reduction and coating deposition on the cathode surface. It has been reported that the addition of boric acid in electrolyte prevents any increase in pH level near the substrate[57, 58]. In addition, it also improves the coating appearance. At lower concentrations, the coatings may appear frosty or burnt and cracked, but a higher concentration of boric acid can improve the coating appearance and also reduces its brittleness [59, 60].

4.6.2 Citric Acid

Citric acid is added to sulfate baths while depositing Ni and Fe. It acts as a complexing agent for Ni^{2+}, Fe^{2+} and Fe^{3+} ions and increases the Fe content in coating. Addition of citric acid to baths significantly reduces the current efficiency resulting in a local increase in the pH level [61, 62]. This change in pH leads to the formation of a hydroxide layer on cathode surface. Despite this, the addition of citric acid does lead to an increase in the number of nucleation sites which increases nucleation rate at the cathodic surface[63].

4.6.3 Saccharin

Saccharin is added to electrolytic baths to reduce grain size and internal stresses in the coating. It has been reported to form complex compounds at the cathode surface which reduce the diffusion rate of adsorbed Ni ions and thus helps in increasing nucleation sites [64]. Saccharin affects the polarization behavior of the cathode by increasing overpotential values which in turn increases the number of nucleation sites, thereby leading to a reduced grain size [65].

5 Electrodeposition Process Parameters

This section reviews various electrodeposition process parameters that affect the coating properties. These include bath composition, current density, pH, bath load, stirring rate, temperature and deposition time.

Bath composition: Bath composition includes the concentration of various salts and particles along with other additives that are present in the electrolytic bath. Besides all the other parameters, bath composition plays an important role in determining the composition, quality and properties of the coating. Various baths and additives were discussed in Sect. 4. Addition of surfactants to bath affects zeta potential of particles, which affect particle incorporation in coating [66].

Current density: Current density affects the surface roughness, deposition rate, coating thickness and particle incorporation in the coating [38, 67–69]. High current density increases the generation rate of metal ions and electrons. These metal ions are then redistributed uniformly over substrate giving a smooth surface finish to the coating [69]. High current density increases the deposition rate at the cathode surface and increases the coating thickness that can be obtained for a given process duration. Current density also affects the particle incorporation in coating matrix. Increase in TiO_2 incorporation in Ni matrix has been reported with an increase in the current density at slow agitation [24]. However, incorporation of diamond and Cr particles has been found to decrease at higher current density [70, 71].

pH and zeta potential: Under normal operating conditions, pH of the bath slightly increases as the electrodeposition progresses. Nickel carbonate can be added to the bath for reducing pH to the required level. This results in the formation of metal hydroxides due to the presence of metallic impurities such as iron, aluminium and silicon. However, a precise control of pH is critical since the lowering of pH is accompanied by a decrease in the Ni ions concentration that leads to improper coating deposition at the cathode.

Particle incorporation in coating depends on process parameters as well as on electrolyte parameters such as its composition, conductivity, viscosity and zeta potential of particles. The potential difference between the electrolyte and particle surface is zeta potential. It affects particle agglomeration in the electrolyte, thereby affecting the incorporation in the coating. High and low zeta potential values in stable range enhances particles repulsion and restricts agglomeration, whereas, at low zeta potential values, attractive forces dominate over repulsive forces and particles agglomerate [72, 73]. Electrolyte pH affects zeta potential values. At lower pH values, zeta potential is positive and when pH increases by addition of alkali, the zeta potential starts decreasing after a certain pH value. After further increment in pH value, zeta potential tends to zero which is known as isoelectric point. On further increasing pH, zeta potential moves towards negative side and a stable range is achieved again at higher pH of electrolyte [74]. Ghazanlou et al. [75] reported pH values of less than 5.5 and 6.5 for micro and nano SiO_2 particles for attaining a positive zeta potential. At pH values of above 5.5 and 6.5 micro and nano SiO_2 particles respectively attained negative values of potential. Optimum pH value of 4.3 for microparticles and 4.6 nanoparticles was reported at which maximum particles incorporation occurred. Lower pH values to obtain higher zeta potential are required for larger particles as they need more force to be incorporated on cathode surface along with cations [66].

Chang et al. [76] reported increase in zeta potential value of Al_2O_3 particles after addition of $Ce_2(SO_4)_3$ in bath. Moderate concentration of $Ce_2(SO_4)_3$ enhanced Al_2O_3 incorporation in Ni–Co coatings.

Bath load: Bath load represents the concentration of particles in the electrolytic bath. Bath load affects the final concentration of particles obtained in the deposited coating. Though the relationship is not linear, an optimum value of bath load has to be decided. Guglielmi's model of adsorption explains how the particle incorporation changes with bath load. At low particle concentration in bath, there is a limited supply of particles for deposition which results in a low particle concentration in the coating.

On increasing particle loading in bath, the particle concentration in coating also increases. However, at very high bath loads, the incorporation of particles in deposited coating is not proportional to their concentration in the bath. In such conditions, particle settling becomes significant [77] and particle agglomeration results in poor adsorption at the cathode surface [15, 37, 78].

Stirring speed: Agitation of bath is done at varying speeds to keep the particles and ions suspended uniformly in the bath by preventing them from settling at the bottom. As agitation increases, the surface becomes rough initially due to the formation of primary irregularities. The side faces of these irregularities act as nucleation sites for secondary irregularities because the agitation leads to an increased supply of ions and an easy discharge of metal at such sites. Thus, agitation results in the production of a smoother coating with increased microhardness. It also decreases the residual stresses developed in the coating [69]. In general, an increased agitation increases the codeposition rate of particles as larger number of particles arrive at the cathode surface. However, if the agitation is too high, the particles residence time at the electrode surface becomes too small, thereby lowering the deposition rate of particles in the growing metallic film [79].

Temperature: High bath temperature increases the diffusivity of ions in electrolytic bath thus allowing the ions to migrate long distances towards the substrate. Increase in temperature leads to a higher growth rate resulting in a larger grain size and an irregular coating surface [69].

Deposition time: Usually, an increase in deposition time increases the coating thickness. However, as the thickness increases, cracks may develop in the coating due to internal stresses and the coating may spall off.

6   Current Control in Electrodeposition Process

Electrodeposition process can be carried out using different current-control methods: direct current (DC), pulsed current (PC) or pulsed reversed current (PRC) techniques. These methods are explained in this section.

6.1   Direct Current (DC) Method

In DC method, constant current is applied to the system without any interruption throughout the coating process. This is the oldest and conventional method that has been successfully used for depositing various metal coatings. It is a simple and economical process as compared to PC and PRC techniques. However, it may result in a slower deposition rate and lead to coating defects such as porosity and poor adhesion. On the other hand, PC and PRC techniques provide improved coating properties such as high wear and corrosion resistant properties [80]. Figure 4 shows

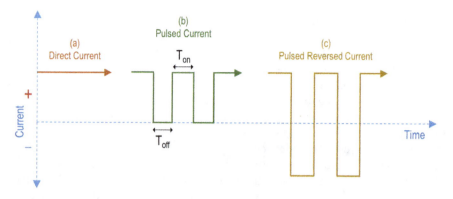

Fig. 4 Current control in **a** DC method **b** PC method **c** PRC method

the schematic of current supply during the three current control techniques (DC, PC and PRC).

6.2 Pulsed Current (PC) Method

In PC method, the current oscillates between a positive value and zero [81]. Positive current is applied for a time known as pulse on time (T_{on}) and zero current is maintained for a time known as pulse-off time (T_{off}). A negatively charged layer grows on the cathode surface during the T_{on} and the electrical double layer inhibits the diffusion of cations towards the cathode surface. When current flow is stopped during T_{off}, this layer discharges thus allowing the diffusion of metal ions towards the cathode surface [82]. The coating quality and parameters are a function of peak current density (I_{peak}), pulse-on time and pulse-off time. These parameters are evaluated in the terms of pulse frequency, duty cycle and average current density, which are calculated using the following relations [83]:

Pulse frequency,

$$f = \frac{1}{T_{on} + T_{off}} \tag{14}$$

Duty Cycle,

$$d = \left(\frac{T_{on}}{T_{on} + T_{off}} \right) \times 100\% \tag{15}$$

Average current density,

$$I_{avg} = \left(\frac{T_{on}}{T_{on} + T_{off}} \right) I_{peak} \tag{16}$$

In case of nickel coatings, PC electrodeposited coatings have been reported to have better hardness as compared to the DC deposited coatings [84].

6.3 Pulsed Reversed Current (PRC) Method

In PRC method, the current is applied in short pulses wherein the polarity is reversed in each subsequent pulse. Periodic change of current from cathodic to anodic polarization provides a better ionic distribution leading to improved coating properties [85]. There are very few studies available using PRC technique for depositing Ni–Co based coatings.

7 Microstructure, Morphology and Properties of Ni–Co Based Reinforced Coatings

Ni–Co coatings are very popular because of their excellent mechanical properties, wear and corrosion resistance, thermal stability and magnetic properties [86–92]. Figure 5 shows key properties of Ni–Co alloy coatings reinforced with various ceramic particles. The properties of composite coatings depend on: (i) electrodeposition process parameters, (ii) properties of matrix material and dispersed phase, (iii) fraction, shape, size and orientation of dispersed phase and (iv) resultant microstructure of the coating. Optimum selection and control of these parameters is essential to achieve coatings with desired properties.

This section reviews the effect of ceramic particles addition on coating properties and the effect of various parameters on particle incorporation. Table 2 summarizes the key observations for electrodeposited Ni–Co coatings reinforced with various particles.

7.1 Effect of Process Parameters on Particle Incorporation, Morphology and Microstructure of Coatings

Particle incorporation in the metal matrix depends on bath parameters such as bath composition, bath load, current density as well as on the particle characteristics such as composition, crystallographic phase, size, shape and density. It was reported that under same deposition conditions, the amount of TiO_2 particles incorporated in Ni matrix was three times higher as compared to Al_2O_3 particles [107]. Researchers have

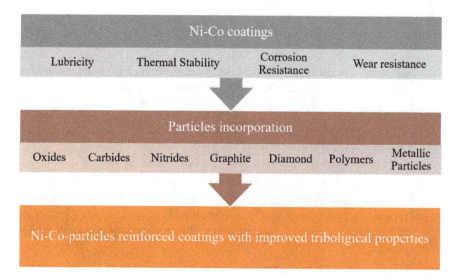

Fig. 5 Effect of reinforcement on properties of Ni–Co alloy coatings

observed an increase in the amount of particle incorporation in Ni-Al$_2$O$_3$, Ni-SiC and CuAl$_2$O$_3$ coatings with an increase in the particle size. However, some studies have also reported negligible effect of particle size on particle incorporation for Ni-Al$_2$O$_3$ and SnNi-SiC coatings [108].

With Co addition to Ni matrix, deposition of particles increases with an increase in Co content in bath as well as in coating. Zamani et al. [105] and Srivastava et al. [102] observed an increase in Co content in the coating with increase in Co ions concentration in bath. It has also been observed that SiC particles incorporation increases to 4.4 wt. % from 0.3 wt. % when Co content in bath was increased to 28 g/L from 1 g/L [102]. Similar observations were made by Bhakit and Akbari [98]. Absorptivity of Co ions on particles surface is better than that for Ni ions. Therefore, more amount of Co is adsorbed and a large positive charge is accumulated on particles' surface leading to more attraction of particles towards the cathode. Hence, more amount of particles are co-deposited on cathode surface in the presence of Co ions [47, 60, 92].

Particle incorporation in the coating increases on increasing bath load or particle concentration in the electrolyte. However, after optimal concentration, the particle incorporation decreases. Shi et. al [16] have reported a sixfold increase in SiC incorporation in Ni–Co matrix as the bath load was increased from 1 g/L to 5 g/L. On further increasing the particle concentration in bath, SiC incorporation reduced. It was suggested that at higher concentrations, particles settle down and agglomeration occurs, thereby reducing their codeposition in coatings. Studies [16, 95] have reported an increase in Si$_3$N$_4$ codeposition in Ni–B coatings on increasing the particle concentration in bath up to an optimal value beyond which a reduction was observed on further increase in bath load [109].

Table 2 Key observations and suggested mechanisms for electrodeposited Ni–Co- based particle-reinforced coatings

Coating	Technique	Observations	Suggested mechanism/insights	References
Ni–Co-micro ZnO; Ni–Co-nano ZnO	Direct current and Pulsed current electrodeposition	Improved microhardness, wear and corrosion resistance at optimum values of parameters (stirring rate, average current density, ZnO concentration in bath) Better properties for nano ZnO coatings with PC method Maximum microhardness was achieved at optimum values of pulse frequency and duty cycle	At low stirring rates, particles may agglomerate due to their limited mobility towards cathode Very high stirring rate causes particles to bounce back due to high kinetic energy/low current density leads to less reduction of cations High current density causes faster reduction of cations but the particles are left behind Higher particle concentration in coating leads to better microhardness, wear and corrosion resistance	[93, 94]
Ni–W/Si₃N₄	Pulsed electrodeposition with Nickel precursors and additives	Improved hardness, wear and corrosion properties at optimal concentration of 15 g/L Si₃N₄ PC method leads to higher incorporation of Si₃N₄ in coating than DC method	When concentration in bath is very high, particle agglomeration occurs In PC method, replenishment of Si₃N₄ near cathode during T_{off} leads to higher incorporation Higher particle concentration provides more nucleation sites and inhibits grain growth leading to reduced grain size, increased microhardness and better wear properties	[95]
Ni–Co-nano SiO₂	Pulsed electrodeposition	Maximum particle concentration, grain size reduction and microhardness at optimal current density of 6A/dm²	Nano particles decrease grain boundary mobility and prevent grain growth Nano particles provide sites for heterogeneous nucleation	[96]
N–WC; Ni–Co–WC	Direct current electrodeposition using watt's bath	WC incorporation reduced the grain size Maximum microhardness at 8 g/L WC concentration	Increased nucleation resulted in smaller grains and increased microhardness Overall grain boundary area increases, which hinders dislocation movement	[97]

(continued)

Table 2 (continued)

Coating	Technique	Observations	Suggested mechanism/insights	References
Ni–Co–SiC	Pulsed electrodeposition	Maximum particle incorporation at optimal current density of 3 A/dm^2 and optimal bath load of 20 g/L Optimal level of Co leads to increased SiC deposition and maximum microhardness	Particle agglomeration occurs at high bath load leading to less particle incorporation At higher Co content, reduction in microhardness due to formation of ($\alpha + \epsilon$) two phase matrix and reduced SiC incorporation	[98]
Ni–Co and Ni–Co–SiC		Maximum deposition of SiC at optimal stirring rate Co deposition decreased with increasing current density Presence of SiC increased Co content in coating	Particles swept away from cathode at very high stirring rates Deposition of Co is diffusion controlled due to low concentration in bath and that of nickel is kinetically controlled—higher current densities facilitates deposition of kinetically controlled species Adsorbability of Co ions on SiC is more than Ni ions	[99]
Ni–Co–SiC	Electrodeposition using Watt's bath	Microhardness increased by 1.5 times after SiC addition Maximum microhardness and wear resistance observed at optimal current density of 50 mA/dm^2 Enhanced corrosion resistance after SiC incorporation	Addition of SiC particles provides dispersion strengthening, restricts grain growth, thus leading to fine grains and increased microhardness At low current densities, less energy is available to entrap particles into metal ions Increasing current density enhances mobility and diffusion of SiC along with metal ions At very high current densities, preferential reduction of metal ions leads to less SiC incorporation SiC incorporation reduces porosity of coatings and provides better corrosion resistance	[100]
Ni–Co–Al$_2$O$_3$	DC and PRC electrodeposition	Reduction in wear loss on increasing particle concentration Higher wear resistance of coatings in PRC than DC	Particles provide dispersion strengthening and pin down grain growth and dislocation movement Particle incorporation higher in PRC method as compared to DC	[101]

(continued)

Table 2 (continued)

Coating	Technique	Observations	Suggested mechanism/insights	References
Ni–Co–Si$_3$N$_4$	Direct current Electrodeposition using sulfamate bath	Optimum level of Co addition increased Si$_3$N$_4$ incorporation in coating At 41 wt.% Co in coating, hardness improved by 1.5 times and wear properties also enhanced Particle incorporation increased at higher Co concentrations without change in microhardness	Co has better wettability than Ni that enhances particle incorporation At higher concentrations of Co, structure changes to hcp	[102]
Ni–SiC, and Ni–Co–SiC (micro and nano SiC were used)	DC electrodeposition using Sulfamate bath	Higher particle incorporation and smaller grain size for nano SiC as compared to micro SiC Increased microhardness and wear properties Increase in particle incorporation after Co addition due to higher wetting ability of Co	Micro SiC particles favored incorporation of Co while nano SiC particles resulted in more particle incorporation Smaller particle size leads to larger active surface area for adsorption of cations, thus facilitating higher concentration of particles in coating	[103]
Ni–Co–Si$_3$N$_4$	DC electrodeposition	Structure changed from fibrous to refined particulate type after Si$_3$N$_4$ addition Increased microhardness and reduced coefficient of friction and wear loss	Change in morphology due to agglomeration of nanoparticles up to some extent Increased microhardness due to reduced grain size and dispersion strengthening Formation of hydroxylated silicon acts as lubricant	[104]
Ni–Co	Direct current electrodeposition using watt's bath	Microhardness increased with Co content up to an optimal concentration	Microhardness increased due to decrease in grain size and solid solution strengthening	[105]

(continued)

Table 2 (continued)

Coating	Technique	Observations	Suggested mechanism/insights	References
Ni–Co	Direct current electrodeposition using Sulfamate bath	High microhardness and better wear and corrosion properties at optimal Co concentrations	Addition of optimum level of Co reduced grain size At high Co concentrations morphology changed from spherical clusters to branched acicular structures Corrosion properties were better due to fcc structure of Ni at low Co concentrations than hcp structure of Co at high concentrations	[89]
Ni–Co	Direct current	Increased microhardness on increasing deposition time keeping current density and other parameters constant	Microhardness was improved due to increased Ni content in Ni–Co coating with increasing time	[106]

By regulating stirring rates, particle agglomeration and transportation to cathode surface can be controlled. Stirring homogenizes the electrolyte, enhances mass transfer and eliminates concentration gradient within the electrolyte. It removes hydrogen bubbles from cathode surface thereby increasing the cathode efficiency and improves coating properties [98]. Ghazanlou et al. [93] have observed an optimal stirring rate of 200 rpm for Ni–Co-nano ZnO coating using PC technique, 300 rpm for the same coating with DC technique and 400 rpm for Ni–Co-micro ZnO coating using DC technique. Bahadormanesh et al. [99] have reported 480 rpm as the optimal stirring rate for achieving maximum incorporation of SiC particles in Ni–Co. Various studies have also reported similar trends of optimal stirring rate for Al_2O_3 and SiC incorporation in Ni and Ni–Co matrix [16, 110–115]. Stirring of electrolyte inhibits clustering of particles and thus enhances their transport to cathode surface. It has been suggested that at very high stirring rates, kinetic energy of these particles becomes very high which enables the particles to bounce back after striking the cathode surface, thereby decreasing the deposition rate of particles. Bakhit et al. [98] have reported a similar trend for Ni–Co-SiC coatings.

High current density increases the deposition rate of metal on cathode surface by inducing faster reduction rate of metal ions. However, this is possible only if the deposition is kinetically controlled. If the concentration of ions in electrolyte is very high, the deposition is kinetically controlled, but if the ionic concentration is low, then the metal deposition is diffusion controlled. Bahodarmanesh et al. [99] observed a reduction in Co deposition in Ni–Co-SiC deposits with increasing current density. This was attributed to a very low Co content in the bath. However, after particle incorporation, Co content in the deposited coating was found to increase.

Bakhit and Akbari [98] observed an increase in particle deposition on increasing current density up to 3 A/dm^2 followed by a reduction in co-deposition at current densities beyond 3 A/dm^2. Similar trends were observed in case of Ni–Co-SiC coating by Shi et al. [16], for Co–Ni-Al_2O_3 deposits by Wu et al. [111], for ZnNi-SiC deposits by Muller et al. [116] and for Al_2O_3-Cu-Sn deposits by Wang et al. [78]. Shi et al. reported a current density of 20 mA/cm^2 as the optimal value for achieving maximum particle incorporation. It was suggested that at very high current densities, the deposition of metals is very rapid and leads to fewer particles getting incorporated in the coating since the particles and metals ions are unable to move synergistically [9]. Authors suggested that an optimal current density value is required for achieving a proper coating. Nanoparticles can get completely surrounded by metal cations forming clusters or clouds. At very low current densities, the electrical charge developed is less and thus the force of attraction between the positive clouds of ZnO surrounded by metal cations and the cathode is very weak. This reduces the amount of deposition and particle incorporation. On the other hand, at very high current densities, reduction of metal cations on the cathode surface becomes very fast and metal ions move towards cathode without effectively entrapping the ceramic particles along with them [96]. Hence, the amount of particle incorporation in the coating is decreased.

Many researchers have also studied effect of current control on particle deposition. It was found that the incorporation of particles is more in PC technique than that in

DC technique for Ni–Co-ZnO coatings. In PC method, current is applied during 'ON' state and removed during 'OFF' state. The maximum current is applied during 'ON' state, which is similar to direct current method, and Ni and Co ions are reduced with limited ZnO particle incorporation during this state. During 'OFF' time, adsorption of Ni and Co ions occurs on ZnO particles making the ZnO incorporation dominant [94].

In an electrodeposition process, current density affects the particle deposition and grain size of coatings. It was observed that increase in current density up to an optimal value leads to more nucleation which results in a smaller grain size. However, at very high current densities, the grain size increases due to change in surface energy and growth mechanisms [117]. Ghazanlou et al. [96] observed approximately 1.5 times reduction in average grain size of Ni–Co-SiO$_2$ coating as the current density was increased from 2 A/dm^2 to 6 A/dm^2 whereas the grain size again increased with a further increase in the current density. Addition of particles to the coating reduces the grain size by pinning down the grain boundary movement and thereby restricting the grain growth[118]. SiC incorporation to Zn coating resulted in smaller grains by hindering the Zn grain growth [82]. Particle incorporation increases the number of nucleation sites which results in more number of grains with smaller average grain size [115, 119]. It was also reported that the amount of deposition of nano ZnO particles in Ni–Co coating was twice as much compared to the deposition of micro ZnO particles. Also, more force is required for the incorporation of larger particles and thus use of higher stirring rates is more suitable [93].

It was found that at high current densities and high duty cycles, discharge rate increases. In such conditions, the surface observed for Ni–Co-SiO$_2$ coatings was disordered and rough with fine globules. This was attributed to an increased precipitation rate of coatings, which results in fine grains. At low current densities and low duty cycles, the discharge rate is less and the surface observed in such conditions was ordered, smooth and gross globular [96]. Ghazanlou et al. [93] have reported a rough and disordered surface of Ni–Co-ZnO coating when micro ZnO particles were used, whereas a smooth and ordered surface was observed with nano sized ZnO particles. Nano sized particles provide more nucleation sites as compared to micro sized particles and thus lead to fine grain size and smoother surface finish.

The type of current control i.e. direct current, pulsed current or pulsed reversed current significantly changes the coating composition and properties. For Ni–Co coatings reinforced with micro and nano ZnO deposited, a crystallite size of 20 nm was reported for PC deposition of nano ZnO particles. Whereas, use of DC technique resulted in a crystallite size of 25 nm for Ni–Co nano ZnO coating and 30 nm for Ni–Co micro ZnO coating. The surface obtained for PC deposited Ni–Co-nano ZnO coating was smooth with fine grains with some globules. However, with the DC technique, the surface observed was rough and disordered with a high density of globules. Discharge rate is low in DC method leading to a lower solidification or nucleation rate. This results in a lesser number of grains with larger average size. Among these three iterations, the surface was roughest and most disordered in case of micro ZnO embedded coatings obtained with DC technique [93, 94]. For Ni–Co-SiO$_2$ coatings also, studies have reported a smoother and more refined grain structure using pulsed current technique as compared to that obtained using direct current [96].

While studying the effect of particle concentration in electrolytic bath, Lee et al. [120] have observed a breakage of coating beyond an optimum level of bath load. It has been suggested that the incorporation of a large number of particles generates stresses between the substrate and the coating and nickel being a weak bonding medium results in cracks and coating spallation. Addition of other elements such as Co, P and ceramic particles such as Si$_3$N$_4$, BN, WC, SiO$_2$ and diamond has been shown to affect the coating properties. Many researchers have studied coating properties after Si$_3$N$_4$ addition to Ni and Ni–Co matrix due to its excellent properties like strength, hardness and high temperature stability up to 1200 °C [95, 121].

At low concentrations of Co in Ni coatings, the coating structure observed was fcc since Ni has fcc crystal structure. Coating structure changed to combination of fcc + hcp on increasing the Co content and became only hcp at Co content above 70 wt.%. For Ni–Co coatings and Ni–Co-Si$_3$N$_4$ with low particle content, Shi et al. [104] observed the crystallization of a needle like structure. After addition of particles, the morphology changed to finer grains that are more particulate like. This indicated a uniform distribution of Si$_3$N$_4$ in Ni–Co matrix leading to grain refinement and an increased hardness of the coatings [47, 104]. At Si$_3$N$_4$ concentration of 3.4 wt.%, the microstructure was fibrous which changed to granular on increasing the particle content to 6.1 wt.% in coating. The granular structure indicates agglomeration of particles up to some extent. Researchers have observed similar behavior for Ni-SiC coatings [68, 122]. Bakhit et al. [98] reported a nodular structure for Ni–Co coatings both with and without SiC particles incorporation. However, the crystallite size obtained after 8.1 vol.% SiC particles addition was smaller. The 8.1 vol.% SiC represents the maximum incorporation of SiC that was obtained at optimal bath load and current density. A uniform distribution of Si in Ni–Co matrix was reported.

Krishnaveni et al. [95] observed that incorporation of Si$_3$N$_4$ particles in Ni–B matrix reduced the luster and increased the surface roughness of Ni–B coatings. The protruded ceramic particles made the coating surface uneven with an average roughness of 1.7 µm[95]. It was observed that there were cracks in Ni–B coating due to internal stresses whereas incorporation of Si$_3$N$_4$ resulted in a homogeneous structure with uniformly distributed particles throughout the coating [95]. A decrease in the grain size of coatings was also observed after addition of Si$_3$N$_4$ particles.

7.2 Effect of Various Parameters on Microhardness and Tribological Properties

The microhardness of Ni coatings increases on addition of Co due to solid solution strengthening. Addition of particles to the Ni–Co matrix further increases the hardness due to dispersion strengthening. At very high concentrations of Co (above 50 wt.%), reduced microhardness of the coating was observed by Zamani et al. [105], and it was attributed to hcp crystal structure of Co [26, 38].

Li et al. [106] have studied the effect of time and current density on microhardness values of Ni–Co coatings. It was observed that increasing current density increases microhardness due to change in coating composition and crystal structure. Authors reported an increase in microhardness with increased deposition time while keeping current density and other parameters constant. Ni content of about 70 wt % and 75 wt % was reported when deposition was carried out for 10 and 30 min respectively. This infers that coating composition and crystallinity can be controlled by controlling and optimizing deposition time of the coatings.

Microhardness of electrodeposited Ni, Ni–B and Ni–B–Si_3N_4 coatings ranges between 250–350 HV, 609 ± 15 and 640 ± 16 respectively. It is clear that addition of Si_3N_4 particles has increased microhardness values of Ni–B coatings. Heat treatment of Ni–B–Si_3N_4 coatings resulted in rapid increase in microhardness value up to an optimal temperature and then reduced on further increasing the temperature [39]. Krishnaveni et al. observed marginal improvement in corrosion properties after addition of Si_3N_4 particles to Ni–B coatings [109]. Li et al. [95] reported that at an optimal concentration of Si_3N_4 in bath, coatings exhibit maximum wear resistance and microhardness. Recently, Li et al. [106] have reported 15 g/L as optimum bath load for maximum nano ZrO_2 deposition and hardness in Ni–Co matrix. Agglomeration of nanoparticles was reported at bath loads higher than 15 g/L which is evident from Fig. 13d. These agglomerates inhibit dispersion strengthening and leads to reduced microhardness.

Srivastava et al. [89] studied electrodeposited Ni–Co coatings using sulfamate baths and reported an increase in the microhardness on addition of Co up to an optimal concentration, after which the microhardness dropped down with further Co addition. On addition of cobalt, a decrease in crystallite size and change in crystal structure of coating from fcc to mixed fcc + hcp was reported. However, when Co concentration is increased to more than 50 wt.% in coating, microhardness value is decreased [89]. Authors also studied the effect of Co addition on Si_3N_4 particle incorporation in Ni coating. Increase in particle incorporation after addition of an optimal amount of Co to the coating was reported. Maximum microhardness value was observed for 41 wt.% Co in coating. It was due to dispersion strengthening provided by incorporated particles, solid solution strengthening by Co addition and reduced crystallite size due to higher nucleation rates [102]. Effect of Al_2O_3, SiC and Si_3N_4 particles in Ni coating was studied by Srivastava et al. [123] and it was reported that coatings reinforced with Si_3N_4 particles exhibited maximum microhardness and best wear resistance among the three coatings. Yang et al. [124] have also reported an increased microhardness of Ni–Co coatings after Si_3N_4 reinforcement due to dispersion strengthening effect.

It has been found that the microhardness of Ni–Co-SiO_2 coatings is less at very low and very high values of current densities This is because of larger grain size and less particle incorporation in the coating at these current densities. As the force of attraction between positively charged ion clouds and cathode is very less, entrapped particles between them may dislodge due to bath agitation. It leads to less particle incorporation which reduces the microhardness. At optimal current values, micro-hardness increased by up to 1.2 times. On further increment in current density, a rapid reduction in microhardness was observed due to less particles incorporation [96]. Similar observations were made for Ni–Co-ZnO coatings. Ghazanlou et al. [93] have reported an optimal current density of 4 A/dm^2 for maximum deposition of nano ZnO particles in Ni–Co coating using PC method which results in a microhardness value of 400 HV and optimal current density of 3 A/dm^2 for DC method with a resultant microhardness value of 300 HV. Increased hardness was attributed to the dispersion strengthening provided by incorporation of more particles obtained with PC method.

Sknar et al. [125] studied the effect of current density on hardness of Ni–Co coat-ings obtained using two different baths viz. methansulfonate electrolyte and sulfate electrolyte. For both the baths, microhardness of the coating was increased with increase in the current density. However, the coatings developed using methansul-fonate bath have higher hardness and lower internal stresses. It was suggested that methansulfonate electrolyte affects the magnetic properties and Co content in the coatings, thereby affecting the internal stresses.

Ghazanlou et al. [93] observed maximum microhardness values for Ni–Co-nano ZnO coatings at optimal stirring rates of 200 rpm using PC technique and 300 rpm using DC technique, and 400 rpm for Ni–Co-micro ZnO coating using DC technique. Bakhit et al. [98] reported similar trend with maximum microhardness of 615 HV at 350 rpm for Ni–Co-SiC coatings.

It was observed that wear properties of Ni–Co-ZnO coatings with ZnO nanopar-ticles were better than those obtained with ZnO microparticles. It was reported that the friction coefficient values of coatings reinforced with nanoparticles were much lower than those reinforced with microparticles. Coatings with microparticles had a rougher surface leading to a higher coefficient of friction. A uniform dispersion of nanoparticles limits the plastic deformation of coatings providing better wear resis-tance. In addition, ZnO particles act as self-lubricating agents which detach easily from coating when the coated surface is rubbed against another surface. In case of nanoparticle-reinforced coating, large number of detached particles spread uniformly over the surface to provide lubrication. As a result, an increase in coefficient of fric-tion was observed initially which gradually became stable. This lubricating effect was attributed to be responsible for better wear properties of coatings with nanoparticles as compared to microparticles. It has been reported that Ni–Co coating embedded with ceramic particles exhibit better microhardness, wear and corrosion properties than Ni–Co alloy coatings [94]. Allahyarzadeh et al. [126] have reported an increased wear and corrosion resistance for Ni-W coatings after incorporation of alumina parti-cles. It was observed that uniform dispersion of nano-sized ceramic particles in metallic deposits reduces chemical activity and improves structural stability, chem-ical stability and mechanical properties such as hardness and strength [16, 124]. It was reported that ceramic particles being hard, restrict penetration in the coatings and protect it against abrasion and wear [127].

Chang et al. [101] have observed reduced wear loss of Ni–Co coatings after Al$_2$O$_3$ addition using DC and PRC techniques. In both methods, it was observed that the wear loss of coatings decreased at higher particle concentrations. However, coatings obtained using PRC technique showed comparatively higher wear resistance. Studies [101] reveal adhesive wear in case of DC coatings and abrasive wear in PRC coatings. It was suggested that pinning of grain boundaries and restricted dislocation movement to be responsible for the reduced wear and plastic deformation of coating.

Shi et al. [104] reported reduction in the coefficient of friction and wear loss on addition of Si_3N_4 particles. It was suggested that formation of hydroxylated silicon oxide occurs during sliding that acts as a solid lubricant and is responsible for reducing the shear strength of sliding surfaces. Severe wear with very high plastic deformation, adhesion and scuffing along with wear debris was reported for Ni–Co coatings which reduced significantly after Si_3N_4 addition. For Ni–Co coatings with particle concentration of 3.7%, less scuffing was observed as compared to those with 6.1% Si_3N_4. It was suggested that higher particle content in the coating leads to dislodging of more particles, thereby creating more wear debris and resulting in increased wear.

7.3 Corrosion Properties

Addition of ceramic particles to metal matrix increases corrosion resistance of coatings. Ghazanlou et al. [94] studied the effect of micro and nano-sized particles on corrosion resistance of Ni–Co-ZnO coatings. It was found that the coatings with a higher concentration of embedded particles exhibited better corrosion resistance. It was explained that the ceramic particles incorporated in metal matrix act as micro-imperfections that hinder the formation of corrosion defects. Being smaller, nanoparticles are distributed homogeneously, thereby proving a finer gran size as compared to microparticles. Since corrosion starts from the surface pits and imperfections, uniformly dispersed nanoparticles hinder the formation of pits and facilitate passivation process. Grain boundaries and particles act as sites for heterogeneous nucleation and fine grains provide more grain boundary area for nucleation and growth of passive layer. Hence, an increased resistance against corrosion is imparted by addition of nanoparticles. Yang et al. [124] suggested that incorporation of inert SiC particles in Ni–Co coating decreases the effective surface area available for cathodic reduction in presence of corrosive species. This leads to reduction in anodic dissolution thereby increasing corrosion resistance.

8   Conclusion and Future Scope

Electrodeposition of Ni–Co based coatings is done using metal salts dissolved in an electrolytic bath which can be of various types. The most common, simple and economical bath is Watt's bath. Direct current, pulsed current and pulsed reversed current techniques can be used for electrodeposition and this choice has a significant impact on the coating properties. Pulsed reversed current and pulsed current techniques have been shown to provide better microhardness and wear resistance of coatings.

Addition of ceramic particles to coatings leads to a smaller grain size, increased microhardness and reduced wear loss of the coatings. Particle incorporation in coating depends on various bath parameters among which bath load, stirring rate, current density and current control have been studied widely. Maximum particle incorporation occurs at an optimal value of these parameters.

At very low stirring rates, kinetic energy of particles is less which slows down their transportation to the cathode surface and may lead to settling down and agglomeration of particles. On increasing stirring rate, agglomerates of particles break and particle movement towards cathode surface becomes easier, leading to increased particle incorporation. At very high stirring rates, particles strike cathode surface and bounce back resulting in very less particle incorporation.

When current density is low, reduction rate is reduced leading to slow deposition of coating. At low current densities, electrostatic attraction between the positive clouds (formed by adsorption of ions on particle surface) and negative cathode surface is reduced that results in less particle incorporation in the coating. On increasing current density, electrostatic force of attraction increases, thereby increasing particle incorporation in growing layers of coating. When current density becomes too high, metal reduction dominates over particle reduction wherein the metal deposition occurs at a much faster rate while the particles are left behind. This reduces amount of particle incorporation.

The selection of current control technique significantly affects the particle incorporation in coating. During constant supply of current, coating layer grows by reduction of ions on metal surface. After some time, formation of a double charged layer at the cathodic surface reduces the diffusion of metal ions leading to slow deposition rates. While using pulsed current and pulsed reversed current techniques, current approaches to zero value and negative value respectively after each charging cycle. This discharges the double layer formed at the cathode surface and replenishes the metal ions in vicinity of cathode, thereby increasing the deposition rates. It is evident that deposition time plays a major role in controlling coating composition and microstructure.

Particle incorporation provides dispersion strengthening of metal matrix resulting in increased microhardness. Particles addition provides heterogeneous sites for nucleation, thereby increasing the nucleation rates which lead to smaller average grain size of the coating. This reduced grain size increases the microhardness values. Additionally, particles hinder grain boundary growth and restrict dislocation movement. This phenomenon reduces the wear loss of coated surface by increasing its resistance against plastic deformation.

Most of the studies available on electrodeposition focus on the effect of current density, current control, bath load, agitation and pH to optimize the coating properties. Some of the key areas that require attention include:

(a) As per the available literature, limited studies have been conducted to assess the effect of deposition time. For achieving the best coating properties, future studies may focus on studying the effect of time on Ni–Co-particles deposition.
(b) Effect of using a combination of different ceramic particles in Ni–Co matrix.
(c) Development of different ternary/quaternary/high-entropy alloys matrix with or without hybrid particles.
(d) Comparison of properties using various electrolytic baths and additives.
(e) Development of machine learning models and materials informatics systems, as done extensively for various materials science applications in recent years [128–134], that can predict properties of novel coatings and act as a guide for selection of process parameters to achieve desired coating properties.
(f) Development of robust interatomic potentials to enable molecular dynamics simulation of electrodeposition [135–140] of Ni–Co based particle reinforced composite coatings.

References

1. Mahdavi S, Allahkaram SR, Heidarzadeh A (2018) Characteristics and properties of Cr coatings electrodeposited from Cr(III) baths. Mater Res Express 6(2):026403. https://doi.org/10.1088/2053-1591/aaeb4f
2. Martin PM (ed) Chapter 1-deposition technologies: an overview. In: Handbook of deposition technologies for films and coatings, 3rd edn. William Andrew Publishing, Boston, pp 1–31. https://doi.org/10.1016/B978-0-8155-2031-3.00001-6
3. Grainger S, Blunt J (2020) Engineering coatings: design and application, 2nd edn. PDF Free Download, epdf.pub. https://epdf.pub/engineering-coatings-design-and-application-2nd-edition.html. Accessed 21 Apr 2020
4. Harper JME (1978) II-5-ion beam deposition. In: Vossen JL, Kern W (eds) Thin film processes. Academic Press, San Diego, pp 175–206. https://doi.org/10.1016/B978-0-12-728250-3.50010-6
5. Lee RE (1998) Microfabrication by ion-beam etching. J Vac Sci Technol 16(2):164–170. https://doi.org/10.1116/1.569897
6. Rehn LE, Picraux ST, Wiedersich H (1986) Surface alloying by ion, electron and laser beams. ASM International, pp 373–388
7. Hartley NEW, Hirvonen J (1980) Ion implantation. Treatise Mater Sci Technol. 18:321
8. Ohno I (1991) Electrochemistry of electroless plating. Mater Sci Eng A 146(1):33–49. https://doi.org/10.1016/0921-5093(91)90266-P
9. Asa Deepthi K et al (2016) Physical and electrical characteristics of NiFe thin films using ultrasonic assisted pulse electrodeposition. Appl Surf Sci 360:519–524. https://doi.org/10.1016/j.apsusc.2015.10.181
10. Koch CC (ed) (2007) Nanostructured materials: processing, properties, and applications, 2nd edn, vol 612. William Andrew Pub, Norwich, NY
11. ASM Handbook (2020) Volume 5: surface engineering-ASM International. https://www.asminternational.org/handbooks/-/journal_content/56/10192/06125G/PUBLICATION. Accessed 22 Apr 2020

12. ASM Handbook (2020) Volume 18: friction, lubrication, and wear technology-ASM International https://www.asminternational.org/bestsellers/-/journal_content/56/10192/275 33578/PUBLICATION Accessed 22 Apr 2020

13. Wojciechowski J, Baraniak M, Pernak J, Lota G (2017) Nickel coatings electrodeposited from watts type baths containing quaternary ammonium sulphate salts. Int J Electrochem Sci 12:3350–3360

14. Karimzadeh A, Aliofkhazraei M, Walsh FC (2019) A review of electrodeposited Ni–Co alloy and composite coatings: microstructure, properties and applications. Surf Coat Technol 372:463–498. https://doi.org/10.1016/j.surfcoat.2019.04.079

15. James J, Heiniger GG (1995) Meeting the pollution prevention challenge at tinker air force base. Fed Facil Environ J 6(3):35–48

16. Shi L, Sun C, Gao P, Zhou F, Liu W (2006) Mechanical properties and wear and corrosion resistance of electrodeposited Ni–Co/SiC nanocomposite coating. Appl Surf Sci 252(10):3591–3599. https://doi.org/10.1016/j.apsusc.2005.05.035

17. Chow GM, Ding J, Zhang J, Lee KY, Surani D, Lawrence SH (1999) Magnetic and hardness properties of nanostructured Ni–Co films deposited by a nonaqueous electroless method. Appl Phys Lett 74(13):1889–1891

18. Amadeh A, Ebadpour R (2013) Effect of cobalt content on wear and corrosion behaviors of electrodeposited Ni–Co/WC nano-composite coatings. J Nanosci Nanotechnol 13(2):1360–1363

19. Dąbrowski A, Hubicki Z, Podkościelny P, Robens E (2004) Selective removal of the heavy metal ions from waters and industrial wastewaters by ion-exchange method. Chemosphere 56(2):91–106

20. Thyssen JP (2011) Nickel and cobalt allergy before and after nickel regulation–evaluation of a public health intervention. Contact Dermatitis 65:1–68

21. Ivanković T, Hrenović J (2010) Surfactants in the environment. Arch Ind Hyg Toxicol 61(1):95–110

22. Jahan K, Balzer S, Mosto P (2008) Toxicity of nonionic surfactants. WIT Trans Ecol Environ 110:281–290

23. Matlosz M (1993) Competitive adsorption effects in the electrodeposition of iron-nickel alloys. J Electrochem Soc 140(8):2272

24. Baker BC, West AC (1997) Electrochemical impedance spectroscopy study of nickel-iron deposition: ii. theoretical interpretation. J Electrochem Soc 144(1):169

25. Zech N, Podlaha EJ, Landolt D (1999) Anomalous codeposition of iron group metals: I experimental results. J Electrochem Soc 146(8):2886–2891

26. Vazquez-Arenas J, Pritzker M (2012) Steady-state model for anomalous Co–Ni electrodeposition in sulfate solutions. Electrochim Acta 66:139–150

27. Chung C-K, Chang WT (2009) Effect of pulse frequency and current density on anomalous composition and nanomechanical property of electrodeposited Ni–Co films. Thin Solid Films 517(17):4800–4804

28. Bai A, Hu C-C (2005) Composition controlling of Co–Ni and Fe–Co alloys using pulse-reverse electroplating through means of experimental strategies. Electrochim Acta 50(6):1335–1345

29. Schario M (2007) Troubleshooting decorative nickel plating solutions (Part I of III installments): any experimentation involving nickel concentration must take into account several variables, namely the temperature, agitation, and the nickel-chloride mix. Met Finish 105(4):34–36

30. Boccaccini AR, Zhitomirsky I (2002) Application of electrophoretic and electrolytic deposition techniques in ceramics processing. Curr Opin Solid State Mater Sci 6(3):251–260

31. Besra L, Liu M (2007) A review on fundamentals and applications of electrophoretic deposition (EPD). Prog Mater Sci 52(1):1–61

32. Hasegawa K, Kunugi S, Tatsumisago M, Minami T (1999) Preparation of thick films by electrophoretic deposition using surface modified silica particles derived from sol-gel method. J Sol-Gel Sci Technol 15(3):243–249. https://doi.org/10.1023/A:1008789025826

33. Sridhar TM, Eliaz N, Kamachi Mudali U, Raj B (2002) Electrophoretic deposition of hydrox-yapatite coatings and corrosion aspects of metallic implants. Corros Rev 20(4–5):255–294. https://doi.org/10.1515/CORRREV.2002.20.4-5.255

34. Guozhong C (2004) Nanostructures and nanomaterials: synthesis, properties and applications. World scientific

35. Guglielmi N (1972) Kinetics of the deposition of inert particles from electrolytic baths. J Electrochem Soc 119(8):1009–1012

36. Bercot P, Pena-Munoz E, Pagetti J (2002) Electrolytic composite Ni–PTFE coatings: an adaptation of Guglielmi's model for the phenomena of incorporation. Surf Coat Technol 157(2–3):282–289

37. Narayan R, Narayana BH (1981) Electrodeposited chromium-graphite composite coatings. J Electrochem Soc 128(8):1704. https://doi.org/10.1149/1.2127714

38. Wang S-C, Wei W-CJ (2003) Kinetics of electroplating process of nano-sized ceramic particle/Ni composite. Mater Chem Phys 78(3):574–580

39. Chang Y-C, Chang Y-Y, Lin C-I (1998) Process aspects of the electrolytic codeposition of molybdenum disulfide with nickel. Electrochim Acta 43(3–4):315–324

40. Liu H, Chen W (2005) Electrodeposited Ni–Al composite coatings with high Al content by sediment co-deposition. Surf Coat Technol 191(2–3):341–350

41. Di Bari GA (2000) Electrodeposition of nickel. Mod Electroplat 5:79–114

42. Gezerman AO, Corbacioglu BD (2010) Analysis of the characteristics of nickel-plating baths. Int J Chem 2(2):124

43. Horkans J (1981) Effect of plating parameters on electrodeposited NiFe. J Electrochem Soc 128(1):45

44. Myung N, Nobe K (2001) Electrodeposited iron group thin-film alloys: structure-property relationships. J Electrochem Soc 148(3):C136–C144

45. Dennis JK, Such TE (1993) Nickel and chromium plating. Elsevier

46. Hansal WE, Tury B, Halmdienst M, Varsanyi ML, Kautek W (2006) Pulse reverse plating of Ni–Co alloys: deposition kinetics of Watts, sulfamate and chloride electrolytes. Electrochim Acta 52(3):1145–1151

47. Baudrand D (1996) Nickel sulfamate plating, its mystique and practicality. Met Finish 94(7):15–18

48. Golodnitsky D, Gudin NV, Volyanuk GA (1998) Cathode process in Nickel–cobalt alloy deposition from sulfamate electrolytes-application to electroforming. Plat Surf Finish 85:65–73

49. Tilak BV, Gendron AS, Mosoiu MA (1977) Borate buffer equilibria in nickel refining electrolytes. J Appl Electrochem 7(6):495–500

50. Tabakovic I, Inturi V, Thurn J, Kief M (2010) Properties of Ni1- xFex (0.1< x< 0.9) and Invar (x= 0.64) alloys obtained by electrodeposition. Electrochim Acta 55(22):6749–6754

51. Marikkannu KR, Kala KS, Kalaignan GP, Vasudevan T (2008) Electroplating of nickel from acetate based bath–Hull Cell studies. Trans. IMF 86(3):172–176

52. Ranjith B, Paruthimal Kalaignan G (2010) Ni–Co–TiO$_2$ nanocomposite coating prepared by pulse and pulse reversal methods using acetate bath. Appl Surf Sci 257(1):42–47. https://doi.org/10.1016/j.apsusc.2010.06.029

53. You YH, Gu CD, Wang XL, Tu JP (2012) Electrodeposition of Ni–Co alloys from a deep eutectic solvent. Surf Coat Technol 206(17):3632–3638

54. Paiva A, Craveiro R, Aroso I, Martins M, Reis RL, Duarte ARC (2014) Natural deep eutectic solvents–solvents for the 21st century. ACS Sustain Chem Eng 2(5):1063–1071

55. Chaudhari AK, Singh VB (2015) Erratum: studies on electrodeposition, microstructure and physical properties of Ni-Fe/In$_2$O3 nanocomposite. J Electrochem Soc 162(10):X21–X21

56. Chaudhari AK, Singh VB (2014) Structure and properties of electro Co-Deposited Ni-Fe/ZrO2 nanocomposites from ethylene glycol bath. Int J Electrochem Sci 9:7021–7037

57. Dolati AG, Ghorbani M, Afshar A (2003) The electrodeposition of quaternary Fe–Cr–Ni–Mo alloys from the chloride-complexing agents electrolyte part I processing. Surf Coat Technol 166(2–3):105–110

58. Saedi A, Ghorbani M (2005) Electrodeposition of Ni–Fe–Co alloy nanowire in modified AAO template. Mater Chem Phys 91(2–3):417–423
59. Tsuru Y, Nomura M, Foulkes FR (2002) Effects of boric acid on hydrogen evolution and internal stress in films deposited from a nickel sulfamate bath. J Appl Electrochem 32(6):629–634
60. Wu Y, Chang D, Kim D, Kwon S-C (2003) Influence of boric acid on the electrodepositing process and structures of Ni–W alloy coating. Surf Coat Technol 173(2–3):259–264
61. Ghorbani M, Dolati AG, Afshar A (2002) Electrodeposition of Ni–Fe alloys in the presence of complexing agents. Russ J Electrochem 38(11):1173–1177
62. Kieling VC (1997) Parameters influencing the electrodeposition of Ni–Fe alloys. Surf Coat Technol 96(2–3):135–139
63. Afshar A, Dolati AG, Ghorbani M (2003) Electrochemical characterization of the Ni–Fe alloy electrodeposition from chloride–citrate–glycolic acid solutions. Mater Chem Phys 77(2):352–358
64. Mockute D, Bernotiene G (2000) The interaction of additives with the cathode in a mixture of saccharin, 2-butyne-1, 4-diol and phthalimide during nickel electrodeposition in a Watts-type electrolyte. Surf Coat Technol 135(1):42–47
65. Bento FR, Mascaro LH (2006) Electrocrystallisation of Fe–Ni alloys from chloride electrolytes. Surf Coat Technol 201(3–4):1752–1756
66. Sharma G, Yadava RK, Sharma VK (2006) Characteristics of electrocodeposited Ni–Co-SiC composite coating. Bull Mater Sci 29(5):491–496. https://doi.org/10.1007/BF02914080
67. Kaisheva M, Fransaer J (2003) Influence of the surface properties of SiC particles on their codeposition with nickel. J Electrochem Soc 151(1):C89. https://doi.org/10.1149/1.1632479
68. Pavlatou EA, Stroumbouli M, Gyftou P, Spyrellis N (2006) Hardening effect induced by incorporation of SiC particles in nickel electrodeposits. J Appl Electrochem 36(4):385–394
69. Stojak JL, Fransaer J, Talbot JB (2002) Review of electrocodeposition. Adv Electrochem Sci Eng 7:193–224
70. Scholl M, Devanathan R, Clayton P (1990) Abrasive and dry sliding wear resistance of Fe-Mo-Ni-Si and Fe-Mo-Ni-Si-C weld hardfacing alloys. Wear 135(2):355–368
71. Guo DZ, Wang LJ, Li JZ (1993) Erosive wear of low chromium white cast iron. Wear 161(1–2):173–178
72. Stojak JL, Talbot JB (2001) Effect of particles on polarization during electrocodeposition using a rotating cylinder electrode. J Appl Electrochem 31(5):559–564. https://doi.org/10.1023/A:1017558430864
73. Shan W et al (2004) Electrophoretic deposition of nanosized zeolites in non-aqueous medium and its application in fabricating thin zeolite membranes. Microporous Mesoporous Mater 69(1):35–42. https://doi.org/10.1016/j.micromeso.2004.01.003
74. Kennedy D, Xue Y, Mihaylova E (2005) Current and future applications of surface engineering. Eng J (Techn) 59:287–292
75. Ghazanlou SI, Farhood AHS, Hosouli S, Ahmadiyeh S, Rasooli A (2018) Pulse and direct electrodeposition of Ni–Co/micro and nanosized SiO2 particles. Mater Manuf Process 33(10):1067–1079. https://doi.org/10.1080/10426914.2017.1364748
76. Chang LM, Guo HF, An MZ (2007) Effect of $Ce_2(SO_4)_3$ on structure and properties of Ni–Co/Al2O3 composite coating deposited by pulse reverse current method. Appl Surf Sci 253(14):6085–6089. https://doi.org/10.1016/j.apsusc.2007.01.010
77. Lozano-Morales A, Podlaha EJ (2008) Electrodeposition of NiCu-matrix nanocomposites using a rotating cylinder Hull cell. J Appl Electrochem 38(12):1707–1714. https://doi.org/10.1007/s10800-008-9620-5
78. Wang YL, Wan YZ, Zhao SM, Tao HM, Dong XH (1998) Electrodeposition and characterization of Al_2O_3–Cu (Sn), CaF_2–Cu (Sn) and talc–Cu (Sn) electrocomposite coatings. Surf Coat Technol 106(2–3):162–166
79. Osborne SJ, Sweet WS, Vecchio KS, Talbot JB (2007) Electroplating of copper-alumina nanocomposite films with an impinging jet electrode. J Electrochem Soc 154(8):D394. https://doi.org/10.1149/1.2744139

80. Balasubramanian A, Srikumar DS, Raja G, Saravanan G, Mohan S (2009) Effect of pulse parameter on pulsed electrodeposition of copper on stainless steel. Surf Eng 25(5):389–392

81. Borkar T, Harimkar SP (2011) Effect of electrodeposition conditions and reinforcement content on microstructure and tribological properties of nickel composite coatings. Surf Coat Technol 205(17–18):4124–4134

82. Sajjadnejad M, Mozafari A, Omidvar H, Javanbakht M (2014) Preparation and corrosion resistance of pulse electrodeposited Zn and Zn–SiC nanocomposite coatings. Appl Surf Sci 300:1–7

83. Mohajeri S, Dolati A, Rezagholibeiki S (2011) Electrodeposition of Ni/WC nano composite in sulfate solution. Mater Chem Phys 129(3):746–750

84. Qu NS, Zhu D, Chan KC, Lei WN (2003) Pulse electrodeposition of nanocrystalline nickel using ultra narrow pulse width and high peak current density. Surf Coat Technol 168(2–3):123–128

85. Chandrasekar MS, Pushpavanam M (2008) Pulse and pulse reverse plating—conceptual, advantages and applications. Electrochim Acta 53(8):3313–3322

86. Dolati A, Sababi M, Nouri E, Ghorbani M (2007) A study on the kinetic of the electrodeposited Co–Ni alloy thin films in sulfate solution. Mater Chem Phys 102(2–3):118–124

87. Hibbard GD, Aust KT, Erb U (2006) Thermal stability of electrodeposited nanocrystalline Ni–Co alloys. Mater Sci Eng A 433(1–2):195–202

88. Chang LM, An MZ, Shi SY (2005) Corrosion behavior of electrodeposited Ni–Co alloy coatings under the presence of NaCl deposit at 800 C. Mater Chem Phys 94(1):125–130

89. Srivastava M, Ezhil Selvi V, William Grips VK, Rajam KS (2006) Corrosion resistance and microstructure of electrodeposited nickel–cobalt alloy coatings. Surf Coat Technol 201(6):3051–3060. https://doi.org/10.1016/j.surfcoat.2006.06.017

90. Marikkannu KR, Kalaignan GP, Vasudevan T (2007) The role of additives in the electrodeposition of nickel–cobalt alloy from acetate electrolyte. J Alloys Compd 438(1–2):332–336

91. Ghahremaninezhad A, Dolati A (2009) A study on electrochemical growth behavior of the Co–Ni alloy nanowires in anodic aluminum oxide template. J Alloys Compd 480(2):275–278

92. Wang L, Gao Y, Xue Q, Liu H, Xu T (2005) Microstructure and tribological properties of electrodeposited Ni–Co alloy deposits. Appl Surf Sci 242(3–4):326–332

93. Imanian Ghazanlou S, Farhood A, Ahmadiyeh S, Ziyaei E, Rasooli A, Saman H (2019) Characterization of pulse and direct current methods for electrodeposition of Ni–Co composite coatings reinforced with nano and micro ZnO particles. Metall Mater Trans A 50:1922–1935. https://doi.org/10.1007/s11661-019-05118-y

94. Ghazanlou SI, Farhood AHS, Hosouli S, Ahmadiyeh S, Rasooli A (2017) Pulse frequency and duty cycle effects on the electrodeposited Ni–Co reinforced with micro and nano-sized ZnO. J Mater Sci Mater Electron 28(20):15537–15551. https://doi.org/10.1007/s10854-017-7442-0

95. Li B, Zhang W (2018) Microstructural, surface and electrochemical properties of pulse electrodeposited Ni–W/Si$_3$N$_4$ nanocomposite coating. Ceram Int 44(16):19907–19918. https://doi.org/10.1016/j.ceramint.2018.07.254

96. Imanian Ghazanlou S, Shokuhfar A, Navazani S, Yavari R (2016) Influence of pulse electrodeposition parameters on microhardness, grain size and surface morphology of Ni–Co/SiO$_2$ nanocomposite coating. Bull Mater Sci 39(5):1185–1195. https://doi.org/10.1007/s12034-016-1256-1

97. Elkhoshkhany N, Hafnway A, Khaled A (2017) Electrodeposition and corrosion behavior of nano-structured Ni-WC and Ni–Co-WC composite coating. J Alloys Compd 695:1505–1514. https://doi.org/10.1016/j.jallcom.2016.10.290

98. Bakhit B, Akbari A (2013) Synthesis and characterization of Ni–Co/SiC nanocomposite coatings using sediment co-deposition technique. J Alloys Compd 560:92–104. https://doi.org/10.1016/j.jallcom.2013.01.122

99. Bahadormanesh B, Dolati A (2010) The kinetics of Ni–Co/SiC composite coatings electrodeposition. J Alloys Compd 504(2):514–518. https://doi.org/10.1016/j.jallcom.2010.05.154

100. Yang Y, Cheng YF (2011) Electrolytic deposition of Ni–Co–SiC nano-coating for erosion-enhanced corrosion of carbon steel pipes in oils and slurry. Surf Coat Technol 205(10):3198–3204. https://doi.org/10.1016/j.surfcoat.2010.11.035

101. Chang LM, An MZ, Guo HF, Shi SY (2006) Microstructure and properties of Ni–Co/nano-Al_2O_3 composite coatings by pulse reversal current electrodeposition. Appl Surf Sci 253(4):2132–2137. https://doi.org/10.1016/j.apsusc.2006.04.018

102. Srivastava M, Grips VKW, Rajam KS (2009) Influence of Co on Si3N4 incorporation in electrodeposited Ni. J Alloys Compd 469(1):362–365. https://doi.org/10.1016/j.jallcom.2008.01.120

103. Srivastava M, William Grips VK, Jain A, Rajam KS (2007) Influence of SiC particle size on the structure and tribological properties of Ni–Co composites. Surf Coat Technol 202(2):310–318. https://doi.org/10.1016/j.surfcoat.2007.05.078

104. Shi L, Sun CF, Zhou F, Liu WM (2005) Electrodeposited nickel–cobalt composite coating containing nano-sized Si3N4. Mater Sci Eng A 397(1):190–194. https://doi.org/10.1016/j.msea.2005.02.009

105. Zamani M, Amadeh A, Lari baghal SM (2016) Effect of Co content on electrodeposition mechanism and mechanical properties of electrodeposited Ni–Co alloy. Trans Nonferrous Met Soc China 26(2):484–491. https://doi.org/10.1016/S1003-6326(16)64136-5

106. Li B, Zhang W, Li D (2020) Synthesis and properties of a novel Ni–Co and Ni–Co/ZrO2 composite coating by DC electrodeposition. J Alloys Compd 821:153258. https://doi.org/10.1016/j.jallcom.2019.153258

107. Medelien V (2002) The influence of B4C and SiC additions on the morphological, physical, chemical and corrosion properties of Ni coatings. Surf Coat Technol 154(1):104–111

108. Ebdon P (1988) The performance of electroless nickel/PTFE composites. https://pascal-francis.inist.fr/vibad/index.php?action=getRecordDetail&idt=7823786. Accessed 25 Apr 2020

109. Krishnaveni K, Sankara Narayanan TSN, Seshadri SK (2008) Electrodeposited Ni–B–Si_3N_4 composite coating: Preparation and evaluation of its characteristic properties. J Alloys Compd 466(1–2).

110. Srivastava M, Grips VKW, Rajam KS (2007) Electrochemical deposition and tribological behaviour of Ni and Ni–Co metal matrix composites with SiC nano-particles. Appl Surf Sci 253(8):3814–3824. https://doi.org/10.1016/j.apsusc.2006.08.022

111. Wu G, Li N, Zhou D, Mitsuo K (2004) Electrodeposited Co–Ni–Al_2O_3 composite coatings. Surf Coat Technol 176(2):157–164. https://doi.org/10.1016/S0257-8972(03)00739-4

112. Vaezi MR, Sadrnezhaad SK, Nikzad L (2008) Electrodeposition of Ni–SiC nano-composite coatings and evaluation of wear and corrosion resistance and electroplating characteristics. Colloids Surf Physicochem Eng Asp 315(1):176–182. https://doi.org/10.1016/j.colsurfa.2007.07.027

113. Yao Y, Yao S, Zhang L, Wang H (2007) Electrodeposition and mechanical and corrosion resistance properties of Ni–W/SiC nanocomposite coatings. Mater Lett 61(1):67–70. https://doi.org/10.1016/j.matlet.2006.04.007

114. Özkan S, Hapçı G, Orhan G, Kazmanlı K (2013) Electrodeposited Ni/SiC nanocomposite coatings and evaluation of wear and corrosion properties. Surf Coat Technol 232:734–741. https://doi.org/10.1016/j.surfcoat.2013.06.089

115. Juneghani MA, Farzam M, Zohdirad H (2013) Wear and corrosion resistance and electroplating characteristics of electrodeposited Cr–SiC nano-composite coatings. Trans Nonferrous Met Soc China 23(7):1993–2001. https://doi.org/10.1016/S1003-6326(13)62688-6

116. Müller C, Sarret M, Benballa M (2003) ZnNi/SiC composites obtained from an alkaline bath. Surf Coat Technol 162(1):49–53. https://doi.org/10.1016/S0257-8972(02)00360-2

117. Shahri Z, Allahkaram SR (2012) Effect of particles concentration and current density on the cobalt/hexagonal boron nitride nano-composite coatings properties. Iran J Mater Sci Eng 9(4):1–7

118. Zarghami V (2014) Alteration of corrosion and nanomechanical properties of pulse electrodeposited Ni/SiC nanocomposite coatings. J Alloys Compd 598:236–242. https://doi.org/10.1016/j.jallcom.2014.01.220

119. Eslami M, Golestani-fard F, Saghafian H, Robin A (2014) Study on tribological behavior of electrodeposited Cu–Si$_3$N$_4$ composite coatings. Mater Des 58:557–569. https://doi.org/10.1016/j.matdes.2014.02.030

120. Lee W-H, Tang S-C, Chung K-C (1999) Effects of direct current and pulse-plating on the co-deposition of nickel and nanometer diamond powder. Surf Coat Technol 120–121:607–611. https://doi.org/10.1016/S0257-8972(99)00445-4

121. Ünal E, Karahan İH (2018) Production and characterization of electrodeposited Ni-B/hBN composite coatings. Surf Coat Technol 333:125–137. https://doi.org/10.1016/j.surfcoat.2017.11.016

122. Abraham M, Holdway P, Thuvander M, Cerezo A, Smith GDW (2002) Thermal stability of electrodeposited nanocrystalline nickel. Surf Eng 18(2):151–156

123. Srivastava M, William Grips VK, Rajam KS (2008) Influence of SiC, Si3N4 and Al2O3 particles on the structure and properties of electrodeposited Ni. Mater Lett 62(20):3487–3489. https://doi.org/10.1016/j.matlet.2008.03.008

124. Yang Y, Cheng YF (2013) Fabrication of Ni–Co–SiC composite coatings by pulse electrode-position—effects of duty cycle and pulse frequency. Surf Coat Technol 216:282–288. https://doi.org/10.1016/j.surfcoat.2012.11.059

125. Sknar YuE, Sknar IV, Savchuk OO, Danilov FI (2020) Electrodeposition of NiCo alloy from methansulfonate electrolyte: the role of the electrolyte pH in the anomalous codeposition of nickel and cobalt. Surf Coat Technol 387:125542. https://doi.org/10.1016/j.surfcoat.2020.125542

126. Allahyarzadeh MH, Aliofkhazraei M, Rouhaghdam ARS, Torabinejad V (2016) Electrode-position of Ni–W–Al2O3 nanocomposite coating with functionally graded microstructure. J Alloys Compd 666:217–226. https://doi.org/10.1016/j.jallcom.2016.01.031

127. Maharana HS, Ashok A, Pal S, Basu A (2016) Surface-mechanical properties of electrode-posited Cu-Al2O3 Composite coating and effects of processing parameters. Metall Mater Trans A 47(1):388–399. https://doi.org/10.1007/s11661-015-3238-0

128. Hart GLW, Mueller T, Toher C, Curtarolo S (2021) Machine learning for alloys. Nat Rev Mater 6(8). https://doi.org/10.1038/s41578-021-00340-w

129. Butler KT, Davies DW, Cartwright H, Isayev O, Walsh A (2018) Machine learning for molecular and materials science. Nature 559(7715). https://doi.org/10.1038/s41586-018-0337-2

130. Ramprasad R, Batra R, Pilania G, Mannodi-Kanakkithodi A, Kim C (2017) Machine learning in materials informatics: recent applications and prospects. Npj Comput Mater 3(1). https://doi.org/10.1038/s41524-017-0056-5

131. Beniwal D, Singh P, Gupta S, Kramer MJ, Johnson DD, Ray PK (2022) Distilling physical origins of hardness in multi-principal element alloys directly from ensemble neural network models. Npj Comput Mater 8(1). https://doi.org/10.1038/s41524-022-00842-3

132. Beniwal D, Ray PK (2021) Learning phase selection and assemblages in high-entropy alloys through a stochastic ensemble-averaging model. Comput Mater Sci 197:110647. https://doi.org/10.1016/j.commatsci.2021.110647

133. Beniwal D, Ray PK (2022) FCC vs. BCC phase selection in high-entropy alloys via simplified and interpretable reduction of machine learning models. Materialia, 101632. https://doi.org/10.1016/j.mtla.2022.101632

134. D. Beniwal, Jhalak, and P. K. Ray, "Data-Driven Phase Selection, Property Prediction and Force-Field Development in Multi-Principal Element Alloys," in *Forcefields for Atomistic-Scale Simulations: Materials and Applications*, A. Verma, S. Mavinkere Rangappa, S. Ogata, and S. Siengchin, Eds. Singapore: Springer Nature, 2022, pp. 315–347. doi: https://doi.org/10.1007/978-981-19-3092-8_16.

135. Brant AM, Sundaram M (2017) A fundamental study of nano electrodeposition using a combined molecular dynamics and quantum mechanical electron force field approach. Proc Manuf 10:253–264. https://doi.org/10.1016/j.promfg.2017.07.054

136. Chaturvedi S, Verma A, Singh SK, Ogata S (2022) EAM inter-atomic potential—its implica-tion on nickel, copper, and aluminum (and their alloys). In: Verma A, Mavinkere Rangappa S,

Ogata S, Siengchin S (eds) Forcefields for atomistic-scale simulations: materials and applications. Springer Nature, Singapore, pp 133–156. https://doi.org/10.1007/978-981-19-3092-8_7

137. Chaturvedi S, Verma A, Sethi SK, Ogata S (2022) Defect energy calculations of Nickel, Copper and Aluminium (and their alloys): molecular dynamics approach. In: Verma A, Mavinkere Rangappa S, Ogata S, Siengchin S (eds) Forcefields for atomistic-scale simulations: materials and applications. Springer Nature, Singapore, pp 157–186. https://doi.org/10.1007/978-981-19-3092-8_8

138. Homer ER, Verma A, Britton D, Johnson OK, Thompson GB (2022) Simulated migration behavior of metastable Σ3 (11 8 5) incoherent twin grain boundaries. IOP Conf Ser Mater Sci Eng 1249(1):012019. https://doi.org/10.1088/1757-899X/1249/1/012019

139. Guymon CG, Harb JN, Rowley RL, Wheeler DR (2008) MPSA effects on copper electrodeposition investigated by molecular dynamics simulations. J Chem Phys 128(4):044717. https://doi.org/10.1063/1.2824928

140. Kaneko Y, Mikami T, Hiwatari Y, Ohara K (2005) Computer simulation of electrodeposition: hybrid of molecular dynamics and Monte Carlo. Mol Simul 31(6–7):429–433. https://doi.org/10.1080/08927020412331332758

Chapter 18
A Mini Review on Active Heat Augmentation Approaches Used for Solar Thermal Collectors

Sahil Tiwari, Sachin Sharma, and Rajesh Maithani

1 Introduction

Solar Air Heater (SAH) is working on free renewable and clean energy. It is a device in which solar energy is used as a primary source, due to which it increases the temperature of the incoming air and provides warm air in agriculture and buildings. The solar air heater consists of a solar collector panel, a duct, and a diffuser, which help to adjust the flow of warm and cool air according to our requirements. It (along with biomaterials) plays an important role in renewable energy and is used in various sectors like industrial, agriculture, and residential, and in energy conservation building codes [1]. To achieve a better performance of the solar heater, we need to choose the best solar collector panel. Performance of the solar collector is dependent on the absorber plate from which air is passed. The convective heat coefficient between the air and the absorber is very poor due to the poor thermal properties of air. So, to enhance the performance of the absorber plate, we need to remove as much heat from the absorber plate as possible. For such cooling, jet impingement is one of the methods which help to reduce the heat of the device or have a high removal rate of heat. For cooling purposes, impingement can be done either in gas or liquid (mostly water and air). Various types of jet impingement have been used for cooling purposes, such as free surface, plunging, submerged, confined, and wall (free surface) (refer to Fig. 1).

Most of the sectors apply three basic configurations of jet impingement, which are the free surface jet, which uses a dense liquid in a less dense medium such

S. Tiwari
Department of Electrical Engineering, School of Engineering, University of Petroleum and Energy Studies, Dehradun, India

S. Sharma · R. Maithani (✉)
Department of Mechanical Engineering, School of Engineering, University of Petroleum and Energy Studies, Dehradun, India
e-mail: rmaithani@ddn.upes.ac.in

© The Author(s), under exclusive license to Springer Nature Singapore Pte Ltd. 2023
A. Verma et al. (eds.), *Coating Materials*, Materials Horizons: From Nature to Nanomaterials, https://doi.org/10.1007/978-981-99-3549-9_18

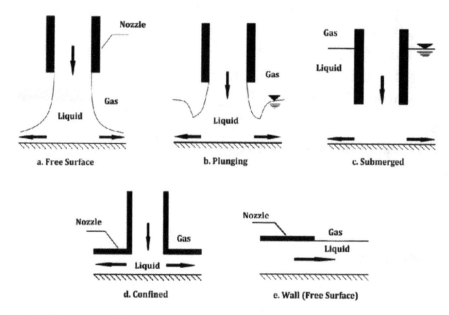

Fig. 1 Different type of jet impingement [1]

as air, the submerged jet configuration, which allows the liquid to impinge on the same liquid medium, and confined jet impingement. The rate of the heat transfer of jet impingement depends on the Reynolds number, jet diameter, surface to nozzle spacing and Prandtl number. One of the pros of jet impingement is that the rate of heat transfer is highest as compared to the other methods due to the stagnation point being in the middle of the jet, which causes a high rate of removal of heat, and easy to locate the location of the jet according to our requirements. The drawback of the jet impingement is that there is significant pressure drop regarding the generation of appropriate jet which places the strict cost, and efficiency.

2   Literature Review

Ranchan Chauhan and Thakur [2] studied the effect of streamwise jet spacing and span due to the in-line arrangement of circular jets in a solar air heater duct, an experimental investigation of heat transfer and friction characteristics. In this, the Reynolds number range covered is 3800–16,000, and the jet spacing and nozzle span have been changed from 0.435 to 1.739 and 0.435 to 0.869, respectively. The aspect ratio of the rectangular duct is 11.6. After comparing the results with a smooth duct under the same circumstances. The findings indicate that as the spanwise and streamwise spacing increase, the heat transfer coefficient and friction factor also increase. There is an increase in heat transfer and friction factor of up to 2.67 and 3.5

times compared to a smooth duct. Abhishek Kumar Goel [3]. Singh focuses on the agricultural industry, where standard drying techniques lead to loss of both quantity and quality. As a result, they introduced the longitudinal fin solar air heater that was studied. By comparing data between smooth and roughened types of solar air heaters and finned solar air heaters, the results indicate an improvement in thermal efficiency for a Reynolds number range (3000–15,000) of up to 3–14.7%. Additionally, heat transfer was shown to improve by up to 28 to 39% with higher Reynolds numbers. Nayak, Singh [4] studied, the performance of a crossflow alternating orifice solar air heater is studied in relation to the effects of flow and channel spacing, Z1 and Z2. This work is carried out with the following parameters: The Reynolds number (Re) is 2700–6900, the depth ratio (Z2/Z1) is 0.75–1.0, the total air depth (Z) is 0.14 m, the nozzle hole diameter (D) is 0.006 m, the number of nozzle holes (N) is 1173, and the tilt angle (h) is 22.5°. Additionally, they compared a typical parallel plate solar air heater with an unconventional jet solar air heater and observed that the performance of the jet plate solar air heater is much better than that of the conventional solar air heater. It was found that at a larger Z2 (0.07 m) than parallel plate solar air heaters and jet plate solar air heaters at lower Z2, outlet air temperature, collector efficiency, and heat transfer are higher (0.06 m). S. N. Singh [5] studied the countercurrent two-pass solar air heater. In this study, a new two-pass solar collector concept was used to achieve greater heat transfer to the flowing air to increase the amount of energy collected and thereby improve the instantaneous efficiency. Furthermore, it was determined what effect Reynolds number and mass flow had on the performance parameters of the solar air heater. After research, it was found that the absorber plate and outlet temperatures of the solar air heater depend on the passage of time. Additionally, remember that the lowest Friction factor is at the highest Reynolds numbered. Rask, Mueller and Pejsa [6] aim was to develop, construct, test, and evaluate a flat solar air heater using the impingement heat transfer method. The primary focus was to compare the efficiency of several arrangements of jet plate collectors with a reference parallel plate collector. At the B HUD Cycle 3 Demonstration House in Minneapolis, Minnesota, 22 prototype jet plate collectors were manufactured and installed. Many nozzle array configurations were investigated and tested, along with cost-effective assembly techniques and collector material selection. A cost and performance analysis were also performed. 211 experimental test runs were completed on four configurations of parallel plate collectors and 15 configurations of nozzle plates. The Y-intercept was shown to be improved by 13% compared to the basic parallel plate collector due to the jet impingement idea. Omar Rafae Alomar, Hareth Maher Abd and Mothana M. Mohamed Salih [1] study and discusses a modified corrugated wave absorber plate to improve the performance of the solar air heater (SAH) collector. To demonstrate the performance of the improved system, the thermal efficiency of the modified and unmodified jet blown corrugated collectors is compared. In the winter of 2021, the experiment was conducted in Mosul, Iraq. According to the results, the thermal efficiency of Model-1 is better than that of Model-2 by 11.5, 14.5, 12.3, and 13.2% for flow rates of 0.009, 0.018, 0.028, and 0.037 kg/s. In addition, although the pressure drop of Model-2 is progressively lower than that of Model-1, the increase in heat transfer rate of Model-1 is greater than that of Model-2. More research papers are discussed in Table 1.

Table 1 Review of various jets investigated

S. No	Author	Parameter	Major finding
1	Molana and Banooni [7]	Liquid impingement jet, Nanofluid, Heat transfer, Nusselt number	This review summarizes research on liquid impingement jet, including their capabilities, limitations, and characteristics. Here are some useful and significant correlations for the Nusselt number. In addition, we show that nanofluids can be used in liquid heat transfer processes in Impingement Jet
2	Amel Raj R and Sabu Kurian [8]	Jet impingement, convective heat transfer, single jet, multiple jets, jet Reynolds number, turbulence models	In the review, many cases of jet impingement with different geometries, crossflow, frequency effects, etc. are studied. Several factors that affect heat transfer at jet impingement have been investigated
3	Sinan Caliskan, Senol Baskaya, Tamer Calisir [9]	The aspect ratios (AR) of elliptic and rectangular jets for 1.0, 2.0 and 0.5, jet Reynolds numbers ranging from 2000 to 10,000, and jet-to-target spacings ranging from 2 to 10	In this study, the effect of beam geometry on flow and heat transfer parameters for elliptical and rectangular arrays of incident beams was investigated experimentally and numerically. Using a thermal infrared camera and laser-Doppler anemometry, heat transfer and flow measurements made by elliptical and rectangular impinging jet fields over a flat surface were investigated

(continued)

Table 1 (continued)

S. No	Author	Parameter	Major finding
4	Colin Glynn, Tadhg O' Donovan, and Darina B. Murray [10]	Jet diameter (d from 0.5 to 1.5 mm), Reynolds number (Re from 1000 to 20,000) and jet-to-target spacing (H from 0.5d to 6d)	Measurement and comparison of heat transfer to closed and submerged air and water jets is the subject of the current study. Surface heat transfer measurements for various test conditions are presented in this paper. For nozzle diameters of 0.5 to 1.5 mm, Reynolds numbers of 1000 to 20,000, and dimensionless nozzle spacings from the target of 1 to 4, heat transfer to simple, axisymmetric, submerged, and closed air–water nozzles was studied
5	Bernhard Weigand and Sebastian Spring [11]	Jet pattern, jet diameter or open area, crossflow effects, separation distance, jet-to-jet spacing, Reynold number and Prandtl number	The aim of this paper is to offer in-depth knowledge for designing such combinations together with a structured description of the variables that affect heat transfer. The mechanics of many nozzle configurations are covered in the first part of the article The characteristics of the heat transfer and the flow field are presented, which are contrasted with the characteristics of the individual incident currents
6	Avijit Bhunia; Sriram Chandrasekaran; Chung-Lung Chen [12]	Jet diameter, variable frequency, variable voltage, 12 power switching devices, (Six insulated gate bipolar transistors and six diodes	This study is the first to directly compare liquid microjet impingement cooling (JAIC) with conventional cooling techniques such as air cooling through finned coolers or Liquid flows in a multi-pass cold plate

(continued)

Table 1 (continued)

S. No	Author	Parameter	Major finding
7	Essam Abo-Zahhad, Shinichi Ookawara, Ali Radwan and ElKady [13]	solar concentration ratio of 1000 Suns, coolant mass flow rate of 50 g/min, e flow rate of 25 g/min, Cell temperature, thermal stresses and Cell efficiency	The development of a thorough three-dimensional model for a high-concentration photovoltaic/thermal (HCPV/T) system was the main objective of the study discussed in this paper. This model combines a thermofluidic model for four different flow-limited heatsink designs and a thermal model for a triple-junction solar cell. And also analysis the thermal and structural analysis of solar concentrator by using the confined jet impingement
8	Kwang-Yong Kim and Sun-Min Kim [14]	Diameter of jet nozzle, D = 2.5 mm, Ratio of dimple diameter to jet. nozzle diameter () Dd/df = 6,and Jet exit to dimpled surface distance = 22.5 mm	In this study, the heat transfer and pressure drop properties of several impinging flow field cooling system configurations were assessed. Steady incompressible laminar flow and heat transfer in a cooling system were investigated using the three-dimensional Navier–Stokes equations
9	Rahul Nadda, Anil Kumar, Rajesh Maithani, [15]	Diameter of jet nozzle, Ratio of dimple diameter to jet. nozzle diameter, and Jet exit to dimpled surface distance	A review of studies using jet impingement to increase the efficiency of solar PV/ solar air collectors is the subject of this article. Several theoretical and experimental studies have been carried out to increase the heat transfer rate in impinging jet solar PV/ solar air collectors

(continued)

Table 1 (continued)

S. No	Author	Parameter	Major finding
10	Dr. Ashwini Kumar, Ravi Kumar, Dr. Arun Kumar Behura [16]	Solar drying, global development, Reynolds number, effective heat transfer coefficient, thermal efficiency	This research investigates a unique jet plate solar air heater that has continuous longitudinal fins under the absorber plate. Performance qualities are represented by important parameters
11	Satyender Singh, Shailendra Kumar Chaurasiya, Bharat Singh Negi [17]	Jet impingement Serpentine wavy channel Solar air heater Thermal efficiency	This research examines two solar air heater designs, both of which include and exclude porous media. Research is carried out numerically and experimentally The extent of thermal performance improvement through beam impingement was revealed by numerical results Compared to Design-II, Design-I achieves excellent thermohydraulic properties without porous media An attractive idea for improving thermal performance is a jet impingement using a perforated metal corrugated sheet
12	Matheswaran et al. [18]	Mass flow rate m = 0.002–0.023 kg/s, stream wise pitch ratio X/Dh = 0.435–1.739, span wise pitch ratio Y/Dh = 0.435–0.869 and jet diameter ratio Dj/Dh = 0.043–0.109	A single-pass two-channel jet plate solar air heater (SPDDJPSAH) is analytically evaluated in terms of its energy efficiency in the current work. Exergy analysis effectively assesses the overall performance of solar air heaters by taking into account both the usable energy gain and the subsequent pumping power requirement

(continued)

Table 1 (continued)

S. No	Author	Parameter	Major finding
13	Choudhury and Garg [19]	'Hole' or 'nozzle' diameter on the jet plate, interspacing, the nozzle height, the distance between the absorber and the jet plate	A thorough theoretical parametric analysis was performed to investigate the effects of various geometrical parameters such as the diameter of the "hole" or "nozzle" on the nozzle plate, their mutual distances, the height of the nozzle, the distance between the absorber and the nozzle plate, and the operating parameter such as velocity of air impinging from the holes/nozzles on the back side of the absorber surface, on the performance parameters of the air heater of the jet impingement concept. To compare the air temperature, rise and performance efficiency of a traditional parallel plate air heater with the temperature rise of a jet plate air heater, a side-by-side study of this device was also conducted
14	Kannan et al. [20]	Jet impingement, pin-fins, solar air collector, solar irradiation, ambient temperature, ambient wind velocity and air mass flow rate	In this study, an experimental investigation of the performance of a JISAC (jet solar air collector) using flat and pin absorbers was carried out
15	Matheswaran et al. [21]	Jet impingement, twisted tape fins, heat transfer, Nusselt number, twist ratio (Y), and pitch ratio (py) are varied from 3500 –13,500, 5.5–9.5, and 0.1– 0.3, respectively	This study used an experimental design to investigate how twisted strip fins integrated into an impinging jet solar air heater could improve heat transfer and friction factors

(continued)

Table 1 (continued)

S. No	Author	Parameter	Major finding
16	Refat Moshery, Tan Yong Chai, Kamaruzzaman Sopian, Ahmad Fudholi, Ali H. A. Al-Waeli [22]	Mass flow (m˙), from 0.007 to 0.039 (kg/s), height of the rib (e) from 0.0017 to 0.0032 (m), relative pitch (P) from 0.01 to 0.04 (m), aspect ratio of the duct was 12, and solar irradiance (I) from 500 to 1000 (W/m2)	Transverse fins on the back side of an absorber plate for a jet-flow solar air heater are used in this research to investigate the theoretical and practical effects of geometrical parameters. The steady-state energy balance equations were developed and solved using the MATLAB software environment
17	Manish Harde, Pratik Samarth, Mohnish Raut1 Ashvin Bhoyar, Krishna Bhandarkar, Pratik Rarokar, Vivek Choudhary [23]	Solar air heater, jet impingement, trapezoidal plate, outlet temperature, and efficiency	In this research, the productivity and test outlet temperature of the solar air heater are controlled. A trial test setup was conducted to evaluate the effects of current impingement on the absorber plate with the effects of a conventional solar air heater on a flat absorber. It used circular nozzles in the ductwork of a solar air heater
18	Alsanossi M. Aboghrara, Mohamed Mansour S. Shukra [24]	Solar Air Heater, Jet Impingement, Corrugated Plat, Outlet Temperature, And Efficiency	The subject of this paper is an experimental investigation of the efficiency of a solar air heater and the outlet temperature with a beam impingement on a corrugated plate absorber. An analysis of the effect of mass air flow and solar radiation on efficiency is carried out
19	Siddhita Yadav, Saini [25]	Finite volume method, Turbulence model, Heat flux, CFD	In this paper, ANSYS FLUENT 18.1 was used in this study to evaluate the thermal behavior of a solar jet impingement (SAHJI) air heater

(continued)

Table 1   (continued)

S. No	Author	Parameter	Major finding
20	Mohammad Salman, Myeong Hyeon Park, Ranchan Chauhan, Sung Chul Kim [26]	Arc angle (ad) from 30 to 75, the relative indented roughness pitch (p/Dh) from 0.269 to 0.810, the relative indented roughness height (e/Dh) from 0.016 to 0.0267	The effect of convective heat transfer and air flow characteristics over an incident current solar heat collector (SHC) with an indentation pit roughness geometry on the absorber plate was experimentally investigated

3   Results and Discussion

The comparison of thermal efficiency with various jets has been discussed in Fig. 2 [27–29]. The results reflected that the system's thermal efficiency with a circular jet is the highest. The apparent reason for the better performance is that the circular jet covers the most heat transfer areas and breaks the sub-viscous layer. The vortices generated by circular jets are powerful, which augment heat transfer more.

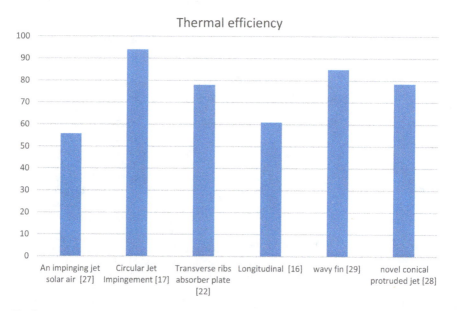

Fig. 2   Comparison of thermal efficiency

4 Conclusions

Following conclusions have been drawn from the comprehensive discussion on active heat transfer augmentation approaches.

1. 1. Heat transfer depends directly upon the shape of roughness.
2. 2. Jet impingement augments heat transfer significantly.
3. 3. Jet impingement increases pumping power requirement; however, the system's thermal efficiency increases significantly.

References

1. Alomar OR, Abd HM, Salih MMM (2022) Efficiency enhancement of solar air heater collector by modifying jet impingement with v-corrugated absorber plate. J Energy Storage 55:105535
2. Chauhan R, Thakur NS (2012) Heat transfer and friction characteristics of impinging jet solar air heater. J Renew Sustain Energy 4(4):043121
3. Goel AK, Singh SN (2019) Performance studies of a jet plate solar air heater with longitudinal fins. Int J Ambient Energy 40(2):119–127
4. Nayak RK, Singh SN (2016) Effect of geometrical aspects on the performance of jet plate solar air heater. Sol Energy 137:434–440
5. Singh SN (2013) Flow and heat transfer studies in a double-pass counter flow solar air heater. Int J Heat Technol 31(2):37–42
6. Rask D, Mueller L, Pejsa J (1977) Jet impingement solar air heater. In: Conference on Solar Energy: Technology Status, p 1760
7. Molana M, Banooni S (2013) Investigation of heat transfer processes involved liquid impingement jets: a review. Braz J Chem Eng 30:413–435
8. Amel Raj R, Kurian S (2018) Jet impingement heat transfer–a review. Int Res J Eng Technol (IRJET) 05:2395–2472
9. Caliskan S, Baskaya S, Calisir T (2014) Experimental and numerical investigation of geometry effects on multiple impinging air jets. Int J Heat Mass Transf 75:685–703
10. Glynn C, O'Donovan T, Murray DB, Feidt M (2005) Jet impingement cooling. In: Proceedings of the 9th UK National Heat Transfer Conference, pp. 5–6
11. Weigand B, Spring S (2009) Multiple jet impingement–a review. In: TURBINE-09. Proceedings of International Symposium on Heat Transfer in Gas Turbine Systems. Begel House Inc.
12. Bhunia A, Chandrasekaran S, Chen CL (2007) Performance improvement of a power conversion module by liquid micro-jet impingement cooling. IEEE Trans Compon Packag Technol 30(2):309–316
13. Abo-Zahhad EM, Ookawara S, Radwan A, El-Shazly AH, ElKady MF (2018) Thermal and structure analyses of high concentrator solar cell under confined jet impingement cooling. Energy Convers Manage 176: 39–54
14. Kim SM, Kim KY (2016) Evaluation of cooling performance of impinging jet array over various dimpled surfaces. Heat Mass Transf 52(4):845–854
15. Nadda R, Kumar A, Maithani R (2018) Efficiency improvement of solar photovoltaic/solar air collectors by using impingement jets: a review. Renew Sustain Energy Rev 93:331–353
16. Kumar A, Kumar R, Behura (2017) Investigation for jet plate solar air heater with longitudinal fins. Int Res J Adv Eng Sci 2(3): 112–119
17. Singh S, Chaurasiya SK, Negi BS, Chander S, Nemś M, Negi S (2020) Utilizing circular jet impingement to enhance thermal performance of solar air heater. Renewable Energy 154:1327–1345

18. Matheswaran MM, Arjunan TV, Somasundaram D (2018) Analytical investigation of solar air heater with jet impingement using energy and exergy analysis. Sol Energy 161:25–37
19. Choudhury C, Garg HP (1991) Evaluation of a jet plate solar air heater. Sol Energy 46(4):199–209
20. Kannan C, Mohanraj M, Sathyabalan P (2021) Experimental investigations on jet impingement solar air collectors using pin-fin absorber. Proc Inst Mech Eng, Part E: J Process Mech Eng 235(1):134–146
21. Matheswaran MM, Arjunan TV, Sahu MK (2021) Influence of twisted tape fins on heat transfer and friction factor characteristics for impinging jet solar air heater. Proc Inst Mech Eng, Part E: J Process Mech Eng 235(4):824–831
22. Moshery R, Chai TY, Sopian K, Fudholi A, Al-Waeli AH (2021) Thermal performance of jet-impingement solar air heater with transverse ribs absorber plate. Sol Energy 214:355–366
23. Harde M, Samarth P, Raut M, Bhoyar A, Bhandarkar K, Rarokar P, Choudhary V (2018) Performance enhancement of solar air heater with jet impingement on trapezoidal absorber plate. Int J Innov Res Stud 8:472–479
24. Aboghrara AM, Shukra MMS (2019) Enhancement efficiency of solar air heater by jet impingement. Int J Sci Eng Technol Res 08: 564−567
25. Yadav S, Saini RP (2020) Comparative study of simple and impinging jet solar air heater using CFD analysis. In: AIP Conference Proceedings, Vol 2273, No. 1. AIP Publishing LLC, p 050043
26. Salman M, Park MH, Chauhan R, Kim SC (2021) Experimental analysis of single loop solar heat collector with jet impingement over indented dimples. Renew Energy 169:618–628
27. Aboghrara AM, Baharudin BTHT, Alghoul MA, Adam NM, Hairuddin AA, Hasan HA (2017) Performance analysis of solar air heater with jet impingement on corrugated absorber plate. Case Stud Therm Eng 10:111–120
28. Das S, Biswas A, Das B (2022) Numerical analysis of a solar air heater with jet impingement—comparison of performance between jet designs. J Sol Energy Eng 144(1)
29. CJ TR, MM M (2022) Analytical investigation on thermo hydraulic performance augmentation of triangular duct solar air heater integrated with wavy fins. Int J Green Energy, 1–11

Chapter 19
Hybrid Feature Selection Techniques to Improve the Accuracy of Rice Yield Prediction: A Machine Learning Approach

C. M. Manasa, Blessed Prince, G. R. Arpitha, and Akarsh Verma

1 Introduction

It has been increasingly evident in recent years that agriculture is essential to the continued success of the economy and the growth of the population. Both developed and developing nations can benefit from utilizing modern GPS and navigational technologies [1], and high-tech sensors to take advantage of the field's inherent variability due to its heterogeneity. Assuring valuable data and eliminating redundant data is a fundamental responsibility in analysis. Data cleansing, data integration, data transformation, and data reduction are all part of the information discovery process that is used to obtain useful information. When a dataset contains inaccurate, incorrectly designed, duplicated, or fragmented information, it must be cleaned. Merging disparate data sets opens a lot of options for accidental data duplication and mislabelling. To create cohesive data arrangements for both practical and intelligent uses, data integration has emerged as the most popular method of combining information from several source frameworks. To create cohesive data arrangements for both practical and intelligent uses, data integration has emerged as the most popular method of bringing together information from several source frameworks. Data transformation is the most widely used method for converting information from one format to

C. M. Manasa · B. Prince
Department of Computer Science and Engineering, Presidency University, Bangalore, India

G. R. Arpitha
Department of Mechanical Engineering, Presidency University, Bangalore, India

A. Verma (✉)
Department of Mechanical Engineering, University of Petroleum and Energy Studies, Dehradun, India
e-mail: akarshverma007@gmail.com

Department of Mechanical Science and Bioengineering, Osaka University, Osaka, Japan

© The Author(s), under exclusive license to Springer Nature Singapore Pte Ltd. 2023
A. Verma et al. (eds.), *Coating Materials*, Materials Horizons: From Nature to Nanomaterials, https://doi.org/10.1007/978-981-99-3549-9_19

another [2]. Transforming a raw data source into a cleaned, reviewed, and usable one is a common example of a change. Limiting the amount of data, you need to retain by eliminating unnecessary details is the most common method. It is possible to increase efficiency while decreasing expenses through information reduction. It's common for businesses to talk about storage capacity in terms of "crude limit" and "compelling limit," both of which refer to data collected after a cut has been made. This is done on a massive database.

Due to the inclusion of unimportant or redundant highlights, a real-world dataset cannot be directly applied to machine learning computation [3–5]. Unimportant features would not contribute anything to the analysis at hand, while redundant features would not add anything to the set of features that have already been decided upon. One method to overcome this is feature selection, which works by reducing the weight of irrelevant features. When talking about farming, the term "precision agriculture" implies that methods for doing it on a small scale have been devised [6]. Data mining is the process of exploring large datasets in search of interesting, previously unknown patterns. Dataset reduction, improved accuracy, reduced processing cost, and enhanced data comprehension are some of the primary goals of feature selection approaches. High-dimensional data sets contain a wealth of information. Superfluous data that has no bearing on the situation, in addition, the data set may contain a relatively large number of variables [7–9]. The following are the disadvantages in prediction taking into consideration the huge number of characteristics in the data set:

1. A longer processing time
2. Excessive use of resources
3. Difficulty during the maintenance phase

Another difficulty is that when the number of features is substantially higher than ideal, many machine learning algorithms perform poorly. For practical reasons, selecting the best feature collection that yields the greatest potential prediction results is desirable. These issues have been thoroughly investigated, and numerous strategies have been devised to decrease the feature set to a tolerable size. Rather than looking for all the crucial characteristics, look for the relevant features. Many different feature selection methods are used by scientists to enhance their data sets. Feature selection is automated along with every other approach. The stepwise regression technique should be used for feature selection. The challenge of feature selection can be characterized as the task of narrowing down a set of candidate features to a manageable subset (M) from a given set (N), where the goal is to minimize the probability of error or some other practical criterion. Feature selection is an active area of study, with numerous publications devoted to filtering, wrapper, and embedding approaches. When working with large datasets, statistical tests help in filtering out irrelevant information by selecting features based on their distribution. Wrapper techniques use greedy algorithms that will try every possible feature combination via a step forward, step backward, or exhaustive search. It employs a cross-validation technique to evaluate the efficacy of each feature combination. Embedded techniques only use one machine learning model when training the features are given priority based on how important they are, while filtering approaches are rapid to compute and relatively inexpensive to implement, they do not produce useful distribution-based features in practice. Wrapper methods fall somewhere in the middle of the computing speeds of filter methods and embedded methods. The execution and selection of the features are challenging. Although decision tree-based algorithms fare poorly in a large feature space, the embedded method incorporates the best of both the filter and embedding approaches. Table 1 compares the three different approaches in terms of how quickly they can process data.

Reviewing the literature allows us to compare the contributions of different efforts in the same area. We can witness the past writers' incredible achievements in this technological sector. These documents would disclose the discovery of new functionalities and their design. Their complicated thoughts and approach to solving the problem would serve as a model for every new employee in the next generation.

To anticipate crops based on soil and environmental characteristics, trait selection methods have been presented by Suruliandi et al. [10] who suggest that the damage to crop output has begun because of the climate change. This means that farmers aren't making informed crop choices based on factors like soil quality and climate, and those manual methods of predicting which plants would thrive in a given area have a high rate of failure. Precision crop forecasting results in increased crop yields. This is where machine learning comes into play in the world of crop forecasting. Soil, regional, and climatic characteristics all enter the harvest predictions. Predicting traits in plants is a vital element of breeding for quality crops, and part of that process includes selecting desirable features. This article provides a high-level summary comparison of the available wrapper functions. Breiman [11] suggests the best response for the system is determined by a vote, which is based on the random forest algorithm's process of building decision trees on several samples of data and making predictions based on those samples. Bagging was employed as a data-gathering technique by Random Forest. The injected randomness, to improve precision, should reduce correlation as much as possible without sacrificing strength. To forecast agricultural yields, Priya et al. [12] constructed a random forest classifier alone. Rainfall, temperature, and the time of year were only a few of the

Table 1 Comparison of feature selection techniques

Existing methods	Computational speed	Cost	Disadvantages
Filter methods	Very fast	Expensive	They do not provide useful characteristics in a practical sense because of their dispersion
Wrapper methods	Fast	Expensive	It is very difficult to carry out and select the features
Embedded methods	Fast	Expensive	If the size of the feature space is large, performance drops off for algorithms based on decision trees
Filter methods	Very fast	Expensive	They do not provide useful characteristics in a practical sense because of their dispersion

factors considered. No other machine learning methods were tried on the data sets. Neither comparison nor quantization could be performed because no other algorithms existed to do so. Zhuo et al. [13] proposed a genetic method using wrapper feature approaches where Overfitting and high processing costs are two potential drawbacks of using high-dimensional feature vectors for the classification of hyperspectral data. Therefore, techniques like feature selection are necessary for dimensionality reduction. These days, developers can choose between filter methods and encapsulation methods when selecting functions. Classification performance can be evaluated in a roundabout way and without eliciting a reaction with this sort of form classifier. The latter evaluates a subgroup of characteristics' "quality" by measuring how well it classifies. Multiple investigations have demonstrated that, despite their high computing cost, wrapper techniques can yield outcomes with improved performance. The Wrapper method uses support vector machines and generalized additive models to classify hyperspectral data. Yoon and Shahabi [14] proposed multivariate time series feature subset selection and feature ranking, the feature subset selection (FSS) approach is widely used to prepare data for subsequent data mining activities, including classification and clustering. Predictions and an improved grasp of the underlying data process are both made more affordable by FSS, As an alternative to traditional supervised methods, also uses unsupervised methods based on principal component analysis to choose feature subsets from MTS. Common FSS methods, like Recursive Feature Elimination (RFE) and Fisher Criterion, have been applied to MTS data sets, including those used for the brain computer interface (BCI). However, the proposed procedures leverage the principal component analysis principles to preserve information regarding the correlation of features, which can be lost when employing the aforementioned methods. Li et al. [15] have developed a hybrid feature extraction and machine learning approach to fruit and vegetable classification. Manual fruit and vegetable detection is easy in small quantities but time-consuming in large ones. Because of this, we employ an automated system to identify them. In the initial stage of processing, the detection was carried out using the fruit and vegetable images. The procedure consisted of three steps: background removal, feature extraction (color and texture), and classification. To get rid of the clutter, the k-means clustering method is applied. Certain hues and tints were isolated using statistical characteristics. Texture features were identified using the oriented gradient histogram (HOG), the local binary model (LBP), and the gray level co-occurrence matrix (GLOM). To train and classify data, support vector machines (SVM) is employed. The study by Shroff and Maheta [16] lays out the standard operating procedure for selecting features and compares the accuracy achieved by various feature selection methods. The effectiveness of constructing a prediction model from a dataset's subset features utilizing wrapper and filter feature selection techniques is explained in detail by Rincy and Gupta [17]. Parmar and Bhatt [18] explored in depth several attribute selection methods and machine learning methods for crop prediction and concluded the best practices for yield prediction. Raja et al. [19] explained the importance of optimal feature selection to ensure that only the most relevant features are accepted as a part of the model. Conglomerating every single feature from raw data without checking for their role in the process of making the model will unnecessarily complicate the model. Maya and Bhargavi [20] used multiple linear regression to assess how well the feature selection methods function. The feature selection algorithms calculate the RMSE, MAE, R, and RRMSE measures. The ideal feature subset was identified using the adjusted R2. The computation took into account the algorithms' temporal complexity as well. The chosen characteristics are used with multilinear regression. Using the characteristics chosen by the SFFS algorithm, MLR offers 85% accuracy. Prameya and Kumar [21] determined the most effective machine learning model to forecast the yield and price of crops and to evaluate the effectiveness of various models. Medar et al. [22] explained the crop selection approach in practice so that it may help farmers and agriculturalists solve a variety of issues. As a result, the rate of crop production is maximized, which benefits our Indian economy. Pavithra and Manasa [23] compared the various big data analytics tools which are useful for the analysis of data.

Most of the reviewed papers used the Filter, Wrapper, and embedded methods for the feature selection, and the selected features are applied to the machine learning algorithms for the prediction of the dependent variable. This paper introduces the hybrid techniques for feature selection which selects relevant features for better accuracy yield prediction.

1.1 Data Source

The area of concentration in this exploration work is Karnataka, the southern territory of the Indian sub-landmass. Karnataka is located in India with the GPS coordinates 20.5937° N and 78.9629° E. The Agricultural Branch in the Administration of Karnataka has responded to the call to accomplish a higher development rate in agribusiness by executing a few advanced plans and progressions in technologies to move forward in production. The Karnataka state is divided into 31 districts and the information was gathered area wise. Seventy years of rice crop related farming measurement information was gathered from the Statistical department and the Horticultural Office of the State Administration of Karnataka and the Meteorological Office in India. The measurable information on farming and climate information were the two sets of information gathered for this review. The individual feature data from each locale are consolidated into a solitary dataset and run through feature selection algorithms and linear regression models, anticipating prediction for rice crop yield. The statistics on agricultural production include information about irrigation systems, planting areas, fertilizer use, and other factors. The climate data includes characteristics like rainfall and maximum and minimum temperatures, as well as solar radiation. The dataset's description is provided in Table 2, and the schematic is given in Fig. 1.

In agreement, an m-layered information collection serves as input data for the highlight feature selection procedure. The data values are referenced using a grid notation of (Data n*m), where n is the total number of information tests and m is the total number of features in the data set. The goal of the feature-choosing technique

Table 2 Description of the dataset

Feature ID	Feature type	Feature category	Description
Tmin	Predictors	Continuous	Average minimum temperature per day recorded for the selected year
Tmax			Average maximum temperature per day recorded for the selected year
MT			Daily average temperature recorded for the selected year
Rainf			Average rainfall of the selected year
N			Total annual nitrogen input utilized in agriculture
K			Total annual potash input utilized in agriculture
Ph			Total annual phosphate input utilized in agriculture
Loamy Alfisol			Soil has relatively high native fertility
Inceptisols			Soil without accumulation of clays, iron oxide, aluminum oxide, or organic matter
Vertisols			Soil has a high content of expansive clay minerals
A			The total area of land under cultivation (hectare)
PRD	Target		Total annual production (ton/hectare)

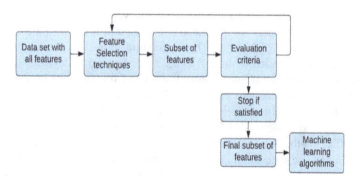

Fig. 1 General procedure for feature selection

is to pick the best possible set of features from among all the options. Consider 'X = X' (i)|'i = 1,2,3,..., m' as a subset of m-aspect capabilities. Once this is done, the ideal subset of the original feature set X can be seen as the new component vector 'Y = y(i)', where 'i' = 1, 2, 3,..., p'. Assuming 'p > m', this means that 'Y X'. After an optimal subset of characteristics has been selected, they are fed into a machine learning classification algorithm, which sorts them into a select few categories. According to this theory, it is possible to predict the number of classes that will be needed. If the feature choice technique is not used, a classifier is viewed as having the capacity 'F: X C.' Once a feature has been selected, it is interpreted as a capability 'F: X C.'

2   Models and Methods

2.1   *Feature Selection and Evaluation*

To increase prediction accuracy in datasets having a lot of features, feature subsets might be used. Feature selection enables machine learning algorithms to simplify and boost the precision of a model built on the appropriate subset. An algorithm for feature selection keeps track of the characteristics that closely connect with crop yield. Different statistical methods can be used in the feature selection such as Feature shuffling and Feature performance. Feature shuffling is an approach that selects the features based on the model's drop when the model's performance is assessed. Feature performance model indicates the importance of the feature on how effectively the target is predicted by a model trained on a particular attribute. Weak or non-predictive features are indicated by poor performance measures. The above-mentioned two algorithms Feature shuffling and Feature performance are used in this research work.

2.2   *Feature Shuffling*

Feature shuffling is the process of assigning priority to a feature, based on the decrease in a model's performance score when a single feature's values are randomly shuffled. A decrease in the model performance score is an indication of how heavily the model depends on that feature because rearranging the feature values (across the dataset's rows) modifies the association between the feature and the target as originally intended. Only one machine learning model will be trained and feature values are randomly given. Predictions are then made using a trained model. If two features are associated, the model will still have access to the data through its correlated variable even if one of the features is shuffled. Due to this, both qualities may have a decreased relevance rating even though they may still be significant. A point of the threshold value is fixed below which, features are deleted. Fewer characteristics will be chosen with higher threshold values. As features are shuffled, there is a chance that distinct subsets of features will be returned with borderline importance. Features with importance values near the threshold considerations in this method are the interpretation of the significance of the feature impacted by correlations. Three-fold cross-validation is used to choose features based on the dip in the R2. Variables will be selected if the performance of the particular feature drops, more than the average drop by all features. Feature shuffling algorithm selects N, K, Ph, Rainf, A, and MT.

2.3 Feature Performance

Feature performance trains a machine learning model only by using that feature to get a clear idea of how important it is. The performance score of the model in this instance determines the "importance" of the feature. Or, how effectively the target is predicted by a model trained on a particular attribute. Weak or non-predictive features are indicated by poor performance measures. For every feature, a machine learning model is implemented. The model predicts the target variable by using a single feature. Then, typically using cross-validation, we assess the model performance and choose features whose performance is above a predetermined threshold. This technique is more computationally expensive since we would train as many models as features in our data set. However, single feature trained models often train quite quickly. Because the performance metric assigns relevance, features are chosen for every necessary model. It is necessary to provide an arbitrary threshold for the feature selection. Pick smaller feature groupings with higher threshold values. Feature performance technique selects N, K, Ph, Rainf, A, and Inceptisols.

2.4 Multiple Linear Regression Model

Multiple Linear Regression is the name of a statistical model that describes how dependent features depend on another set of independent features. Because production (yield) typically depends on several factors, including production area, irrigation, fertilizers, and meteorological parameters, it has been used most frequently for crop yield prediction. The dependent feature (response feature) in this case is production (yield), while the other parameters are independent features.

MLR model is described by.

$Y = B0 + B1 \times 1 + B2X2 + B3X3 + B4X4 + \ldots\ldots BKXK. + E1.$

Where k is the number of features, B0, B1, B2.... is the regression coefficient, and E1 is the error term.

2.5 Multiple Liner Model for Crop Yield Prediction

The regression model is defined with the equation for every pair of the independent and dependent variables. Here, PD is the dependent variable, and the independent variables are Tmin, Tmax, MT, Rainf, N, K, Ph, Loamy Alfisol, Inceptisols, Vertisols, and AR. The slope of a variable shows the expected rise or fall in the variable's predicted value.

$PD = 0.01 + 0.03N + 0.006Ph + 0.15 k + 0.91 inceptisols - 0.008Rainf.$

2.6 Accuracy Metrics

Using accuracy measurements such as the root mean square error (RMSE), mean absolute error (MAE), root relative mean square error (RRMSE), and correlation coefficient, the model's correctness is calculated (R). The discrepancy between the calculated and real value is measured by RMSE. By normalizing the data, RRMSE reduces model error by comparing the mean and predicted value of the model. The linear relationship between the predictions made by the regression model and the actual values is determined by the correlation coefficient (R). The average of estimated discrepancies is known as MAE. Once the features have been fitted into the model, the adjusted R2 value is calculated to examine the correctness of the model and measure the variance.

3 Results and Discussion

The current research uses Python programming to construct feature selection methods for identifying pertinent features. The two techniques used in this research utilize the free, open-source Feature-engine library in Python. The various features were chosen using feature selection algorithms following the selection criteria. The chosen attributes are listed in Table 3. It has been noted that, in comparison to other characteristics, the cultivated area has gained more significance in crop output (yield). Total annual nitrogen input utilized in agriculture, Total annual potash input utilized in agriculture, and Total annual phosphate input utilized in agriculture also play a very important role when compared to other independent variables. Feature performance and Feature shuffling techniques both the techniques selected almost the same set of features when applied to the data set. Both the algorithms rely on threshold value hence resulting in selecting the same set of features.

Feature shuffling trains a machine learning model to evaluate its performance after shuffling the order of values of one feature. The values of one feature are shuffled in sequence. It assesses the performance and makes predictions using the model created. The greatest advantage is only one machine learning model needs to be trained so it is very quick. On the downside, if two features are correlated, when one of the

Table 3 Description of the dataset

Feature selection methods	Tmin	Tmax	MT	Rainf	N	K	Ph	Loamy Alfisol	Inceptisols	Vertisols	A
Feature Shuffling			✓	✓	✓	✓	✓				✓
Feature performance				✓	✓	✓	✓	✓			✓

Table 4 Values with feature shuffling

Tmin	−0.166
Tmax	−0.026
MT	−0.156
Rainf	0.189
N	0.469
K	0.166
Ph	0.189
Loamy Alfisol	−0.129
Inceptisols	0.491
Vertisols	−0.156
A	0.853

features is shuffled, the model will still have access to the information through its correlated variable. In this research work, SVM model is built which helps to select the features.

Feature performance considers every feature for which, a machine learning model is trained. It creates predictions and assesses model performance for each model. It chooses features whose performance metrics are above a threshold level. In this research work, the threshold value of 0.01 is selected. Because we would train as many models as features in our data set, this strategy is more computationally expensive.

The features selected by applying the above two feature selection methods are given in Table 3, and the corresponding values are given in Tables 4 and 5.

To identify the ideal feature subset, the adjusted R^2 was used. The adjusted R^2 value was used to determine the model's fitness. The variance explained by the model is indicated by the adjusted R^2 value. The model is acceptable if the adjusted R^2 is high. As a result, adjusted R^2 was taken into account when choosing the optimum features. When feature selection was employed as compared to using all features as

Table 5 Values with feature performance

Tmin	−0.026
Tmax	−0.129
MT	0.085
Rainf	0.085
N	0.746
K	0.166
Ph	0.018
Loamy Alfisol	−0.109
Inceptisols	−0.235
Vertisols	−0.266
A	0.846

Table 6 Metrics values

	Feature shuffling	Feature performance
R	0.899	0.887
R^2_{adj}	0.86	0.85

input, it has been noted that the adjusted R^2 value is high. The metrics values are given in Table 6. Consequently, feature selection algorithms are crucial for making better predictions. The complexity of time and space will be significantly reduced if the right features have been chosen. It makes a more accurate prediction.

The model validation shows how well a model performs on the data used to train it. In the current research, the model's validation was carried out by the strategy that is by evaluating the model's goodness of fit using adjusted R^2. Here threefold cross-validation was utilized in this method to minimize the loss of critical data. This model performed well and provided an accuracy rating of 86%. According to the literature, the suggested model accuracy is 84%.

4   Conclusions

This research used two feature selection methods to identify the best feature subset for predicting rice crop productivity in the Indian state of Karnataka. The elements that are crucial for predicting crop productivity were found using 30 years' worth of agricultural data. The outcome demonstrates that the Total area under cultivation, the nitrogen, potassium, and Ph of the soil are the crucially required features for the crop yield prediction. Except for minor variations in the RMSE, RRMSE, and MAE values, all feature selection algorithms performed similarly. As a result, the model accuracy was determined using the adjusted R2 value. MLR attained a model accuracy of 86% for both models.

Acknowledgements We extend our sincere thanks to all who contributed to preparing the instructions (especially Dr. Keerthi Kumar N) who supported in editing the paper.

Funding Academic support from the Osaka University (Japan) and monetary support from the Japan Society for the Promotion of Science (JSPS) is highly appreciable (Grant Number: P21355).

Availability of data and materials Not applicable.

Conflicts of Interest There are no conflicts of interest to declare by the authors.

Ethical Approval The authors hereby state that the present work is in compliance with the ethical standards.

Competing Interests The authors declare no competing interests.

References

1. Griggs D, Stafford-Smith M, Gaffney O, Rockström J, Öhman MC, Shyamsundar P, Noble I (2013) Policy: sustainable development goals for people and planet. Nature 495:305–307
2. Wheeler T, Von Braun J (2013) Climate change impacts on global food security. Science 341:508–513
3. Jain N, Verma A, Ogata S, Sanjay MR, Siengchin S (2022) Application of machine learning in determining the mechanical properties of materials. In: Machine Learning Applied to Composite Materials. Springer Nature Singapore, Singapore, pp. 99–113
4. Arpitha GR, Mohit H, Madhu P, Verma A (2023) Effect of sugarcane bagasse and alumina reinforcements on physical, mechanical, and thermal characteristics of epoxy composites using artificial neural networks and response surface methodology. Biomass Convers Biorefinery, 1–19. https://doi.org/10.1007/s13399-023-03886-7
5. Thimmaiah SH, Narayanappa K, Thyavihalli Girijappa Y, Gulihonenahali Rajakumara A, Hemath M, Thiagamani SMK, Verma A (2023) An artificial neural network and Taguchi prediction on wear characteristics of Kenaf-Kevlar fabric reinforced hybrid polyester composites. Polym Compos 44(1):261–273
6. Godfray HCJ, Garnett T (2004) Food security and sustainable intensification. Philos Trans R Soc B 369:20120273
7. Elliott J, Dering D, Müller C, Frieler K, Konzmann M, Gerten D, Glotter M, Flörke M, Wada Y, Best N et al (2014) Constraints and potentials of future irrigation water availability on agricultural production under climate change. Proc Natl Acad Sci USA 111:3239–3244
8. Calzadilla A, Rehdanz K, Betts R, Falloon P, Wiltshire A, Tol RS (2013) Climate change impacts on global agriculture. Clim Chang 120:357–374
9. Haddeland I, Heinke J, Biemans H, Eisner S, Flörke M, Hanasaki N, Stacke T (2014) Global water resources affected by human interventions and climate change. Proc Natl Acad Sci USA 111:3251–3256
10. Suruliandi A, Mariammal G, Raja SP (2021) Crop prediction based on soil and environmental characteristics using feature selection techniques. Math Comput Model Dyn Syst 27(1):117–140
11. Breiman L (2001) Random forests. Mach Learn 45(1):5–32
12. Priya P, Muthaiah U, Balamurugan M (2018) Predicting yield of the crop using machine learning algorithm. Int J Eng Sci & Res Technol 7(1):1–7
13. Zhuo L, Zheng J, Li X, Wang F, Ai B, Qian J (2008) A genetic algorithm based wrapper feature selection method for classification of hyperspectral images using support vector machine. In: Geo informatics 2008 and Joint Conference on GIS and Built Environnent: Classification of Remote Sensing Images, vol 7147, pp 503–511. SPIE
14. Yoon H, Shahabi C (2006) Feature subset selection on multivariate time series with extremely large spatial features. In: Sixth IEEE International Conference on Data Mining-Workshops (ICDMW'06). pp. 337–342. https://doi.org/10.1109/ICDMW.2006.81
15. Li Y, Zhuo L, Hu X, Zhang J (2016) A combined feature representation of deep feature and hand-crafted features for person re-identification. Int Conf Prog Inform Comput (PIC) 2016:224–227. https://doi.org/10.1109/PIC.2016.7949499
16. Shroff KP, Maheta HH (2015) A comparative study of various feature selection techniques in high-dimensional data set to improve classification accuracy. Int Conf Comput Commun Inform (ICCCI) 2015:1–6. https://doi.org/10.1109/ICCCI.2015.7218098
17. Rincy TN, Gupta R (2020) Ensemble learning techniques and its efficiency in machine learning: a survey, In: 2nd International Conference on Data, Engineering and Applications (IDEA), pp 1–6. https://doi.org/10.1109/IDEA49133.2020.9170675
18. Parmar KP, Bhatt T (2022) Crop yield prediction based on feature selection and machine learners: a review. Second Int Conf Artif Intell Smart Energy (ICAIS) 2022:354–358. https://doi.org/10.1109/ICAIS53314.2022.9742891

19. Raja SP, Sawicka B, Stamenkovic Z, Mariammal G (2022) Crop prediction based on character-istics of the agricultural environment using various feature selection techniques and classifiers. IEEE Access 10:23625–23641. https://doi.org/10.1109/ACCESS.2022.3154350
20. Maya Gopal PS, Bhargavi R (2019) Selection of important features for optimizing crop yield prediction. Int J Agric Environ Inf Syst (IJAEIS) 10(3):54–71. https://doi.org/10.4018/IJAEIS.2019070104
21. Hegde PR, Ashok Kumar AR (2022) Crop yield and price prediction system for agriculture application. Int J Eng Res & Technol (IJERT) 11(7)
22. Medar R, Rajpurohit VS, Shweta S (2019) Crop yield prediction using machine learning tech-niques, In: 2019 IEEE 5th International Conference for Convergence in Technology (I2CT), pp. 1–5. https://doi.org/10.1109/I2CT45611.2019.9033611
23. Pavithra N, Manasa CM (2021) Big data analytics tools: a comparative study. In: 2021 IEEE International Conference on Computation System and Information Technology for Sustainable Solutions (CSITSS). pp. 1–6. https://doi.org/10.1109/CSITSS54238.2021.9683711